Lecture Notes in Artificial Intelligence   5040

Edited by R. Goebel, J. Siekmann, and W. Wahlster

Subseries of Lecture Notes in Computer Science

T0135040

Minoru Asada   John C.T. Hallam
Jean-Arcady Meyer   Jun Tani (Eds.)

From Animals
to Animats 10

10th International Conference
on Simulation of Adaptive Behavior, SAB 2008
Osaka, Japan, July 7-12, 2008
Proceedings

 Springer

Series Editors

Randy Goebel, University of Alberta, Edmonton, Canada
Jörg Siekmann, University of Saarland, Saarbrücken, Germany
Wolfgang Wahlster, DFKI and University of Saarland, Saarbrücken, Germany

Volume Editors

Minoru Asada
Osaka University, Graduate School of Engineering, Adaptive Machine Systems
JST ERATO Asada Synergistic Intelligence Project
2-1 Yamadaoka, Suita, Osaka 565-0871, Japan
E-mail: asada@ams.eng.osaka-u.ac.jp

John C.T. Hallam
University of Southern Denmark, The Mærsk Mc-Kinney Møller Institute
Campusvej 55, 5230 Odense M, Denmark
E-mail: john@mmmi.sdu.dk

Jean-Arcady Meyer
Université Pierre et Marie Curie - CNRS, AnimatLab
104 Avenue du Président Kennedy, 75016 Paris, France
E-mail: jean-arcady.meyer@upmc.fr

Jun Tani
RIKEN, Brain Science Institute, Laboratory for Behavior and Dynamic Cognition
2-1, Hirosawa, Wako, Saitama 351-0198, Japan
E-mail: tani@brain.riken.jp

Cover illustration by Jean Solé

Library of Congress Control Number: 2008929600

CR Subject Classification (1998): I.2.11, I.2, I.6, F.1.1-2, K.4, H.5, J.4

LNCS Sublibrary: SL 7 – Artificial Intelligence

ISSN	0302-9743
ISBN-10	3-540-69133-2 Springer Berlin Heidelberg New York
ISBN-13	978-3-540-69133-4 Springer Berlin Heidelberg New York

Springer is a part of Springer Science+Business Media

springer.com

© Springer-Verlag Berlin Heidelberg 2008

Typesetting: Camera-ready by author, data conversion by Scientific Publishing Services, Chennai, India
Printed on acid-free paper   SPIN: 12279081   06/3180   5 4 3 2 1 0

Preface

Welcome to the proceedings of the Tenth International Conference on Simulation of Adaptive Behavior (SAB 2008). A symbolic creature in the SAB 2008 poster is based on GAKUTENSOKU, Japan's first modern robot created in 1928 by Makoto Nishimura. The robot, Gakutensoku (or "learning from natural law"), "was 7' 8" tall, painted gold, could open and close its eyes, could smile, could puff out its cheeks, and at the beginning of each performance would touch its mace to its head and then begin to write (from http://www.robmacdougall.org/index.php/2008/04/gakutensoku/)." Gakutensoku was actuated by pneumatics and seems to have been "a sort of early Japanese animatronics." Designed 80 years ago, it still stimulates researchers' minds.

This year, we received 110 submissions, among which we selected 30 for oral presentations and 21 for posters. In the main conference, we had four very interesting plenary talks: "Modelling Adaptive and Intelligent Behaviour: Some Historical and Epistemological Issues" by Roberto Cordeschi, "Insect-Machine Hybrid System for Understanding an Adaptive Behavior" by Ryohei Kanzaki, "Body Shapes Brain – Emergence and Development of Behavior and Mind from Embodied Interaction Dynamics" by Yasuo Kuniyoshi, and "Thinking and Learning Close to the Sensory-Motor Surface Creates Knowledge That Transcends the Here and Now" by Linda Smith. On the second day, we had a special joint session with the British Council featuring special talks by Giacomo Rizzolatti and Ron Chrisley followed, by a panel discussion. After the main conference, we had a workshop and two tutorials.

Minoru Asada
Jun Tani

Organization

From Animals to Animats 10: The Tenth International Conference on the Simulation of Adaptive Behavior (SAB 2008) was organized by the JST ERATO Asada Project and ISAB (International Society for Adaptive Behavior).

Executive Committee

General Chair   Minoru Asada, Osaka University, Japan
Program Chair   Jun Tani, RIKEN, Japan
General Co-chairs   John Hallam, University of Southern Denmark,
                         Denmark
                     Jean-Arcady Meyer, University of Paris 6 – CNRS,
                         France

Program Committee

Hussein Abbass
Alberto Acerbi
Ronald Arkin
Angelo Arleo
Minoru Asada
Gianluca Baldassarre
Christian Balkenius
Luc Berthouze
Aude Billard
Eleonora Bilotta
Joanna Bryson
Seth Bullock
Angelo Cangelosi
Thomas Collett
Nikolaus Correll
Kerstin Dautenhahn
Marco Dorigo
Michael Dyer
Dario Floreano
Luca Gambardella
Philippe Gaussier
Agnes Guillot
John Hallam
Osam Hanagata
Inman Harvey

Gillian Hayes
Phil Husbands
Fumiya Iida
Hiroyuki Iizuka
Auke Jan Ijspeert
Takashi Ikegami
Akio Ishiguro
Koji Ito
Masato Ito
Naoto Iwahashi
Frederic Kaplan
Kuniaki Kawabata
Toshiyuki Kondo
Robert Kozma
Jeffrey L. Krichmar
Hanspeter Mallot
Davide Marocco
Alcherio Martinoli
Gianluca Massera
Mariagiovanna Mazzapioda
Chris Melhuish
Jean-Arcady Meyer
Marco Mirolli
Francesco Mondada
Kazuyuki Murase

Ryohei Nakano
Chrystopher L. Nehaniv
Stefano Nolfi
Tetsuya Ogata
Pietro Pantano
Frank Pasemann
Jan Peters
Rolf Pfeifer
Eric Postma
Tony Prescott
Miki Sagara
Matthew Schlesinger
Gregor Schoner
Noel Sharkey
Tomohiro Shibata
Olivier Sigaud
Olaf Sporns
Kenji Suzuki
Jun Tani
Charles Taylor
Tim Taylor
Guy Theraulaz
Vadim Tikhanoff
Peter Todd
Vito Trianni

Table of Contents

Perception and Control

Learning and Adaptation

Cognition, Emotion and Behaviour

Collective and Social Behaviours

Adaptive Behaviour in Language and Communication

Applied Adaptive Behaviour

Extended Homeostatic Adaptation: Improving the Link between Internal and Behavioural Stability

Hiroyuki Iizuka[1] and Ezequiel A. Di Paolo[2]

[1] Department of Media Architecture, Future University-Hakodate
116-2 Kamedanakano-cho, Hakodate, Hokkaido, 041-8655, Japan
[2] Centre for Computational Neuroscience and Robotics,
Department of Informatics, University of Sussex
Brighton, BN1 9QH, UK

Abstract. This study presents an extended model of homeostatic adaptation designed to exploit the internal dynamics of a neural network in the absence of sensory input. In order to avoid typical convergence to asymptotic states under these conditions plastic changes in the network are induced in evolved neurocontrollers leading to a renewal of dynamics that may favour sensorimotor adaptation. Other measures are taken to avoid loss of internal variability (as caused, for instance, by synaptic strength saturation). The method allows the generation of reliable adaptation to morphological disruptions in a simple simulated vehicle using a homeostatic neurocontroller that has been selected to behave homeostatically while performing the desired behaviour but non-homeostatically in other circumstances. The performance is compared with simple homeostatic neural controllers that have only been selected for a positive link between internal and behavioural stability. The extended homeostatic networks perform much better and are more adaptive to morphological disruptions that have never been experienced before by the agents.

1 Introduction

The use of homeostatic neurons as the basic building blocks for evolved neurocontrollers proposed in [1] provides a novel approach to modelling adaptivity in artificial systems in a way that resembles the adaptive dynamics of living organisms. The idea behind homeostatic adaptation is based on that of the ultrastable system proposed by Ashby [7]. This is a system – open to interaction with the environment – that will tend to change its own configuration plastically whenever stability is lost and until it finds a new internal dynamics which will make the system stable under the current conditions. Such systems are capable of remarkable adaptation and learning. They have been applied to legged robot locomotion [3], extended to different types of plastic functions [4], applied to the study of the minimal dynamics of behavioural preference [5], and as a model of perseverative reaching in infants (A-not-B error) [10].

In the original model, a neural controller inspired by this system was combined for the first time with the techniques of evolutionary robotics. In this model, local

M. Asada et al. (Eds.): SAB 2008, LNAI 5040, pp. 1–11, 2008.

plastic mechanisms change the incoming synaptic weights only when neural activations move out of a bounded region that is defined in advance by the designer. Plasticity keeps working until the neural activations return to the homeostatic region resulting in a "stable" configuration in the sense that the network weights do not change further as long the firing remains bounded. This mechanism was implemented in a simulated agent evolved with a fitness function simultaneously rewarding phototaxis and the maintenance of neural activations within the homeostatic region. The use of intermittent plasticity in combination with this dual selective pressure allows controllers to evolve where an association is created between internal homeostasis and the desired behaviour. This association is evolved to be positive: *high homeostasis goes together with good performance.* By selective design, once a neurocontroller gives rise to the right sensorimotor coordination within a given environmental situation in a way that results in internal stability, synaptic weight changes no longer happen and the agent behaves as desired. If the situation changes, such as in an inversion of the visual field or some other sensorimotor perturbation, this causes a breakdown of coordination. Under these circumstances some evolved agents also show a breakdown of internal homeostasis demonstrating that some agents evolve at least one negative association: *lack of phototaxis induces lack of homeostasis.* As this happens, the local adaptive mechanism activates until it finds a new synaptic configuration which can sustain the activations within the homeostatic region. In these conditions, some evolved agents are also able to re-form the behavioural coordination (even if they had not been trained to adapt to the induced perturbation). These agents are then able to re-create a positive association: *regaining homeostasis induces a recovery of the original behavioural performances.*

However, the original work has a problem in that these necessary further associations between internal and behavioural stability that allow adaptation to unseen perturbations are *contingent*. They may or they may not evolve in the original setup. This contingency is demonstrated by the high fitness sometimes achieved by solutions for which, under disruption of phototaxis, homeostasis remains unaffected. This problem with the method was first noted in [2] where an alternative model more closely resembling Ashby's homeostat was presented as a proof of concept. That model, however, was limited in that it used Braitenberg-style controllers. Improving the method to avoid contingent solutions using dynamical neural networks remains an important challenge if homeostatic adaptation is to be applied more widely in other areas of autonomous robotics. In this paper we move closer towards this aim. We propose an extended homeostatic neural controller where neurons are biased to have a strong resting membrane potential and an additional fitness condition rewarding not only a positive link between homeostasis and a desired behaviour but also a negative one between the breakdown of homeostasis and undesired behaviour [6].

This paper will present the extended model of homeostatic adaptation and compare it with more basic versions in terms of adaptivity and evolvability. It will be shown that agents evolved in the extended model are more adaptive against unexperienced morphological disruptions and random initial weight connections.

2   Model

Our proposed method is implemented in a simulated mobile agent with a plastic neural controller. The simulated agent is faced with a single light source. The task for the agent is to approach the light source. This task is deliberately simple in order to understand the basic interactions between neural mechanisms, plasticity, homeostasis and selection pressures before moving into more complex applications. Certain modifications are made to the original method for studying homeostatic adaptation [1] with the aim of improving the chances of internal homeostasis being linked with performance both in the positive and negative senses. In all cases, the set of initial weights for a neural controller is given randomly at the beginning of a trial. This modified setting makes the task more difficult than the original one. The agent is expected to adapt to a suitable weight set by the plasticity through interaction with the environment.

Agent. An agent is modelled as a simulated wheeled robot with a circular body of radius 4 and two diametrically opposed motors. The motors can drive the agent forwards in a 2-D unlimited plane. The agent has four light sensors mounted at angles $\pm\pi/4, \pm3\pi/4$ radians to the forward direction. Light from point sources impinges on sensors with a local intensity proportional to the source intensity and the inverse of the distance from sensor to source. The model includes the shadows produced by the agent's body.

Plastic controller. A fully connected continuous-time recurrent neural network (CTRNN) [8] with 8 neurons is used as the agent's controller. The equations are modified from their ordinary form in order to enhance the conditions for homeostasis to be able to activate intermittent plasticity. In particular, a result to avoid is that of evolved agents capable of moving away from lights and remaining homeostatic. The lack on sensory activation (typically leading a CTRNN to converge to some form of asymptotic dynamics) should have as a consequence a breakdown of homeostasis. As before, the local homeostatic condition is fulfilled if a node is firing within a specified range. In the absence of sensory activation, certain parameters controlling resting potentials, synaptic strengths and size of the homeostatic region should be chosen as to enhance the chances for evolution to find solutions with the desired internal/interactive associations. To realize it, two parameters, α and β, are added in the typical CTRNN equations and another parameter γ controls the size of the homeostatic region.

In addition, a common problem in plastic neural networks is that of weight saturation. Typically, Hebbian-like rules, will tend to drive synaptic strengths either to their maximum or minimum values. This can sometimes be avoided with the use of directional damping factors [2]. But the problem is the more general one of loss of variability. In order to expand the range of possible perturbations that the system can adapt to, it is crucial that the ultrastable dynamics be provided with enough variability for its plastic reconfigurations. A possible way of achieving this is to map the weight values specified by the plasticity rules into a continuous but non-monotonic space. Since we are interested in some minimal transformation that will solve this problem while still using simple plastic rules we eschew biological plausibility and make connection strengths a sinusoidal function of the weights. The time evolution of the states of neurons is then expressed as:

$$\tau_i \dot{y}_i = -(y_i - \beta) + \sum_{j=1}^{N} \alpha \sin(w_{ji}) z_j(y_j) + I_i, \tag{1}$$

$$z_i(x) = 1/(1 + e^{-x}), \tag{2}$$

where y_i represents the cell potential of neuron i, z_i is the firing rate, τ_i (range $[0.1, 10]$) is its time constant, I_i represents the sensory input, which is given to only sensory neurons. The sensory input is calculated by multiplying the local light intensity by a gain parameter (range $[1, 100]$), which is genetically encoded. There are two neurons for controlling each motor. The motor output is calculated from the ratio of firing rates of the neurons, which is mapped onto the range $[-5, 5]$ and is then multiplied by a gain parameter (range $[1, 10]$). This is to avoid that the agent cannot move when the dynamics converges to a small value in the absence of sensory stimulations. The *synaptic strength* for the connection from neuron j to i is determined $\alpha \sin(w_{ji})$, where α is a network constant given genetically that regulates how much the pre-synaptic neurons can affect post-synaptic neurons. The parameter β (resting potential) determines the equilibrium point for a node in the total absence of input. The balance between the two parameters, α and β, will become very important, because if α is too small the firing rates converge to their resting potentials (notice that the equation does not include bias terms), and if it is too big, β will not have a very significant effect on the dynamics. Therefore, if β is negative and with an appropriate α, firing rates will tend to converge to a small values unless there is enough stimulus coming from the sensors. The use of such a balanced combination, together with an appropriate choice of homeostatic region, will make it hard for networks to remain within homeostatic bounds in the absence of sensory input.

The connection weights between neurons, w_{ij}, are randomly determined at the beginning of a trial and a plastic mechanism allows for the lifetime modification of the connections. A homeostatic region is described as the finite zero-value set of a plasticity function of the post-synaptic firing rate. This function is modulated by parameter γ (range $[0, 0.5]$), which is genetically determined. Weights from neuron i to j are updated according to :

$$\Delta w_{ij} = \eta_{ij}(1 - z_i) p(z_j), \tag{3}$$

$$p(x) = \begin{cases} 0 & x > \gamma \\ 1 - x/\gamma & \text{else} \end{cases} \tag{4}$$

where z_i and z_j are the firing rates of pre- and post-synaptic neurons, respectively, Δw_{ij} is the change per unit of time to w_{ij}, η_{ij} is a rate of change (range $[-1, 1]$), which is genetically set for each connection, and $p(x)$ is the plastic function that defines the homeostatic region. The reason why this is called homeostatic is that if z_j is more than γ, the weight connection does not change.

Otherwise, the plasticity works and the weight connection keeps changing until z_j is stabilized in the homeostatic region (more than γ). Here we can see that the effect of the balance between α and β can be (if α is small enough and β negative enough) to land the dynamics into the zone of active plasticity. The parameter, γ, is also evolved, however, the trivial solution of evolving $\gamma = 0$ so that neurons are always homeostatic is prevented by the need to use plasticity to organize the weight configuration at the beginning of a trial. The only way this can happen is for firing rates to move out of the homeostatic regions, hence gamma may evolve to be small, but not zero.

3 Evolutionary Setup

A population of agents is evolved using a rank-based genetic algorithm with elitism. All fixed network parameters, τ_i, η_{ij}, α, γ and the gains are represented by a real-valued vector ($[0,1]$) which is decoded linearly to the range corresponding to the parameters (with the exception of gain values which are exponentially scaled). Crossover and vector mutation operators, which adds a small random vector to the real-valued genotype, are used [9].

In the extension of the original method presented here, half of trials during the evaluation process correspond to the presence of light and the other half are carried out in the dark. The light condition consists of the serial presentation of 8 distant light sources that the agent must approach and remain close to. A single source is presented at a time for a relative long time period of 1000. In the dark condition, there is no light in the arena. Consequently, the network receives no sensory stimulus. The agent can move freely in the unlimited arena for the same period as in the light condition. There will be a selective pressure to evolve homeostatic dynamics in the light condition but to *avoid* homeostasis in the dark condition. The scheme is expected to evolve networks for which lack of sensory stimulation leads to non-homeostasis. Therefore, the agents are evaluated by measuring three factors: the proportion of time that the agent spends near the light source (at a distance less than 20 in this paper), f_s, the time-average of the proportion of neurons that have behaved homeostatically in the light condition, f_h, and that have *not* behaved homeostatically in the dark condition, f_{Nh}. The fitness function is given by this, $F = \zeta * f_s * f_h + (1 - \zeta) * avg.(f_s) * f_{Nh}$, where ζ decides the weighted selective pressure between light and dark conditions.

4 Results

The performance of the extended homeostatic neural controller is compared to that of more basic forms of homeostatic neural controllers not using the biased resting potential and that have only been evolved only in the light condition. This basic homeostatic neural controller can be described in our formulations with the two parameters, β and ζ, which are set to 0 and 1.0, respectively. For the extended version, β and ζ are set to -5 and 0.5, respectively. In order to confirm whether the basic homeostatic neural network with the evaluation of light/dark conditions can evolve agents properly to link internal and behavioural stability, a controller with $(\beta, \zeta) = (0, 0.5)$ is also tested. It should be noted that this basic homeostatic neural network has a different form to that of the original homeostatic neural network proposed by Di Paolo [1] but follows the same basic selection scheme.

4.1   Evolvability and Adaptivity

All homeostatic neural controllers are evolved for 4000 generations. Ten independent runs were made for each of the three conditions. Extending the number of generations to 10000 has not produced any observable increase in fitness.

The average of the best fitness values across 10 runs are shown in Fig. 1. It is clear that the extended homeostatic networks evolve better than the other conditions which become saturated at fitness ceilings of about 0.2 and 0.3 in each case. The main observed difference between these conditions and the extended one $(\beta, \zeta) = (-5, 0.5)$ is that agents are not able to sustain a configuration that allows them to perform phototaxis reliably.

In order to study the homeostatic controllers evolved by the three conditions, we select one successful agent at the 4000th generation from each run and investigate them in terms of their adaptivity.

First, we study adaptivity against variation of the initial weights by measuring the performances by the best agents for different initial weight configurations. The performances for phototaxis and neural homeostasis in 100 independent trials are shown in Fig. 2.

The best agents evolved using $\beta = 0$ can in some cases establish phototaxis but fail to adapt against the many initial weight configurations. It can also be

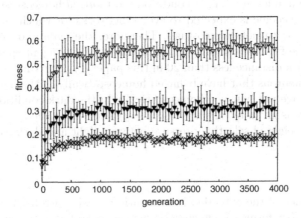

Fig. 1. Average fitness values of the best agents at each GA generation over 10 independent runs evolved with three kinds of the evolutionary regimes. The empty triangles are the results of agents with the extended method, $(\beta, \zeta) = (-5, 0.5)$, and the filled ones are for homeostatic networks without the biased resting potential nor evaluation in the dark condition, $(\beta, \zeta) = (0, 1)$. The crosses are for the one without the biased resting potential and with evaluation of both conditions, $(\beta, \zeta) = (0, 0.5)$.

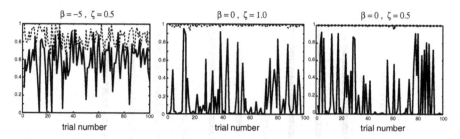

Fig. 2. A comparison of the performances for random initial weights of the best agents evolved with different evolutionary regimes, i.e. our proposed method (left), the basic homeostatic network (center), and the basic one with light/dark evaluation (right). The solid lines show the proportion of time that the agent spends within 30-units distance to light sources (initial distances to the light when it appears are 60). The dashed lines represent the time-averages of the proportion of neurons that have behaved homeostatically in each trial. A single trial consists of the serial presentation of 50 distant light sources. Trial numbers indicate different random seeds for weight connections.

observed that the time-averages of the homeostatic neural firing are always high regardless of the behaviour. This exemplifies the contingency of the link between the two selected evolutionary aims: homeostasis remains unaffected even under disruption of phototaxis since the two are realized independently. On the other hand, the extended condition adapts more widely to the initial random configuration. Depending on the initial weight values, the network has a good chance of establishing an appropriate sensorimotor coordination that results in phototaxis. Notice how the level of homeostasis correlates with phototaxis success. This indicates that agents that do not perform the correct behaviour will also tend to be non-homeostatic.

In the original homeostatic adaptation model agents show adaptation to radical unexperienced perturbations such as swapping sensor positions left and right. Although it is not always guaranteed that such perturbations can be adapted to, the extended model should be able to adapt more reliably than the basic homeostatic controllers because it decreases the chances of internal and behavioural stability being independent of each other. Adaptation to sensory inversion is measured by checking how much the agent can return to phototactic behaviour in the new condition after having adapted to the random initial weights. A recovery of phototaxis is defined as approaching at least 10 lights in sequence at some point after the sensor inversion.

Once having adapted to the initial weights, each pair of the diagonal sensors are swapped left and right ($\pm\pi/4 \leftrightarrow \mp3\pi/4$). If the agent is able to recover phototactic behaviour by the 300th light source presented, it is counted as a success. The success rate is calculated over 100 trials after having adapted successfully to the initial weights. Results are shown in Fig. 3. All controllers evolved with the extended method can return to the phototactic behaviour with high probabilities. Having evolved under a pressure to behave non-homeostatically in the absence of sensor input has induced a condition where plasticity remains active

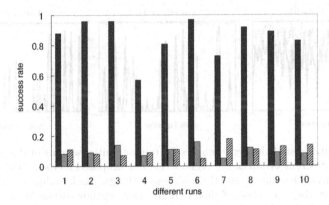

Fig. 3. A comparison of adaptivity to a new sensor configuration after the best agents have established the normal sensorimotor coupling. The new condition is achieved by swapping each pair of diagonal sensors $(\pm\pi/4 \leftrightarrow \mp3\pi/4)$. The vertical axis shows the rate of each best agent successfully showing the phototactic behaviour after the disruption. From the left, the bars means the results of the best agent using the homeostatic neural controller with $(\beta = -5, \zeta = 0.5)$, $(\beta = 0, \zeta = 1.0)$, and $(\beta = 0, \zeta = 0.5)$, respectively. See text for details.

until sensors become active again. The system behaves in the desired ultrastable manner and plastic changes stop only as phototaxis is recovered.

4.2   Homeostatic Adaptation at Work

An example of the lifetime adaptation to the initial weight configuration by the best agent with the extended method is shown in Fig. 4. At the beginning of the trial, weight connections are randomly set so that the resulting behaviour cannot be approaching the light source. If too little light stimulus is provided the neurons cannot maintain the neural dynamics in the homeostatic region. This is the combined effect of using a biased resting potential and the extra fitness constraint for the dark condition. Following the plastic rules, the network starts changing the network structures that can lead to homeostasis and phototaxis at the same time. Regardless of the initial weight values, the best agent can successfully establish both phototactic behaviour and homeostasis. Converged weight values are very different each time the agent is re-initialized since the established sensorimotor coupling is dynamically constructed through the interaction.

The best agents also show adaptivity to inversion of visual field. After having adapted the normal positions of sensors, the left-right swapping is applied. Figure 5(left) shows the distances from the agent to the light sources. The light sources appear at a new place in every 1000 steps and the sensors are swapped when the 13th light source appears. Adaptation to the initial weight set happens before swapping sensors. When the sensors are swapped, the agent moves predictably away from the light source. This causes a breakdown of internal homeostasis and synaptic plasticity is turned on. After a period of adaptation

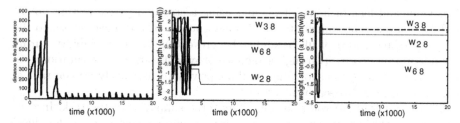

Fig. 4. Left: Distance from robot to sources. Each source lasts for 1000 time steps. Center: Change of synaptic weights corresponding to the same run of the left. For the clarity, only three of weights are shown. Right: Another example of synaptic changes of the same agent starting from different initial configurations.

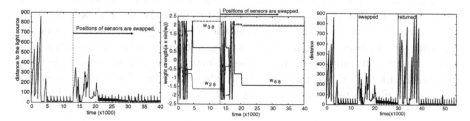

Fig. 5. Left: Distance from robot to sources. The vertical dashed line shows onset of swapping diagonal sensors ($\pm\pi/4 \leftrightarrow \mp 3\pi/4$). All initial configurations are same as Fig. 4. Center: Synaptic changes over time. Right: Re-adaptation to swapped sensor positions. The sensors are swapped at the appearance of 13th light source and are returned to original configuration when the 30th light appears.

the network finds a new homeostatic state that can approach the light. Once an appropriate new sensorimotor coupling has been established, weights becomes stable again (Fig. 5(center)). Re-adaptation is also tested. The measures that enhance dynamical variability in the extended model allow the agent to re-adapt when sensors are returned to the original configuration, something that was not observed in the original model (Fig. 5(right)).

5   Conclusions

This paper presents an extended homeostatic adaptation method capable of improved adaptive performance. The method works by addressing the problem of contingent associations between internal and behavioural stability in the original homeostatic adaptation model. The contingent link between these two requirements could always lead to solutions where behavioural disruption leaves internal stability unaffected. This situation, presumably does not occur so easily in living systems that must behave in a goal-directed manner in most cases *in order to* guarantee continued inner stability. Our proposed method moves closer to this natural situation by explicitly selecting an association between the undesired behaviour and lack of homeostasis and using a strong negative resting potential for neurons that contributes to this association. It also moves closer to the original Ashbyan idea of ultrastability by preventing loss of variability during the adaptation process (through avoidance of synaptic strength saturation). As a result, evolved agents show improved adaptivity to previously unseen radical perturbations such as sensor inversions.

This result leads to wider lessons despite the simplicity of the task. In conventional evolutionary robotics models, sensorimotor connections as low level descriptions are tightly related to the behavioural performances as macro observations. Even in those cases where sensorimotor coordination is allowed to change plastically, the appropriate relation is established by an external process of artificial selection. In the homeostatic adaptation models, the sensorimotor coupling is intermittently reconfigured by the system to maintain the internal requirement of homeostasis, which is associated with the desired behaviour by the external selective process. Therefore, the bottom-up construction from the sensorimotor dynamics and the top-down regulation from the behavioural performances are mutually coupled processes *intrinsic* to the system. It is their 'high-level' link that is constructed by artificial evolution. The careful construction of this link can produce systems that are more adaptive and more lifelike.

Acknowledgements

This research was partially supported by the Japanese Ministry of Education, Science, Sports and Culture, Grant-in-Aid for JSPS Fellows, 17-04443.

References

1. Di Paolo, E.A.: Homeostatic adaptation to inversion in the visual field and other sensorimotor disruptions. In: From Animals to Animats VI: Proceedings of the 6th International Conference on Simulation of Adaptive Behavior, pp. 440–449 (2000)
2. Di Paolo, E.A.: Spike timing dependent plasticity for evolved robots. Adaptive Behavior 10(3/4), 243–263 (2002)
3. Hoinville, T., Henaff, P.: Evolving plastic neural controllers stabilized by homeostatic mechanisms for adaptation to a perturbation. In: Proceedings of the 9th International Conference on the Simulation and Synthesis of Living Systems, pp. 81–87 (2004)
4. Williams, H.: Homeostatic plasticity in recurrent neural networks. In: From Animals to Animats 8: Proceedings of the 8th International Conference on the Simulation of Adaptive Behavior, pp. 344–353 (2004)
5. Iizuka, H., Di Paolo, E.A.: Toward Spinozist robotics: Exploring the minimal dynamics of behavioural preference. Adaptive Behavior 15(4), 359–376 (2007)
6. Iizuka, H.: Evolving homeostatic neural controller without depending on initial weights. In: Proceedings of 13th International Symposium of Artificial Life and Robotics (2008)
7. Ashby, W.R.: Design for a brain: The origin of adaptive behaviour, 2nd edn. Chapman and Hall, London (1960)

8. Beer, R.D.: Intelligence as adaptive behavior: An experiment in computational neuroscience. Academic Press, San Diego (1990)
9. Beer, R.D.: Toward the evolution of dynamical neural networks for minimally cognitive behavior. In: From Animals to Animats 4: Proceedings of the 4th International Conference on Simulation of Adaptive Behavior, pp. 421–429 (1996)
10. Wood, R., Di Paolo, E.A.: New models for old questions: Evolutionary robotics and the 'A Not B' error. In: Proceedings of 9th European Conference on Artificial Life, pp. 1141–1150 (2007)

Evolution of Valence Systems in an Unstable Environment

Matthijs Snel and Gillian M. Hayes

Institute of Perception, Action, and Behaviour
School of Informatics, University of Edinburgh, Edinburgh EH9 3JZ, United Kingdom
m.snel@sms.ed.ac.uk, gmh@inf.ed.ac.uk

Abstract. We compare the performance of drive- versus perception-based motivational systems in an unstable environment. We investigate the hypothesis that valence systems (systems that evaluate positive and negative nature of events) that are based on internal physiology will have an advantage over systems that are based purely on external sensory input. Results show that inclusion of internal drive levels in valence system input significantly improves performance. Furthermore, a valence system based purely on internal drives outperforms a system that is additionally based on perceptual input. We provide arguments for why this is so and relate our architecture to brain areas involved in animal learning.

1 Introduction

Reinforcement learning (RL) is a learning paradigm in which an animal or agent learns to assign valence (positive or negative nature of events or objects) to sensory input, or state, by updating its evaluation of a state based on received reward or punishment following that state. Behaviour can subsequently be adapted by selecting those actions that lead to the highest-valued states, and hence maximise reward and minimise punishment. In animal brains, RL and a sense of pleasure (reward) have been linked to the neurotransmitter dopamine and brain areas involved in the dopaminergic pathway; for example, Schultz et al. [1] showed that dopaminergic neurons are capable of propagating valence information backward in time, thereby learning that a stimulus (e.g. a bell) is predictive of reward (e.g. food). In artificial systems employing RL, reward was traditionally implemented by having a pre-designed reward function provide the agent with a scalar reinforcement signal.

1.1 Towards Truly Autonomous Agents

It is debatable whether an agent that has to rely on a pre-designed reward function to learn can truly be called adaptive and autonomous: the range of environments that the agent will be able to cope with is naturally bounded by the environment types to which the pre-designed reward function applies. In contrast, animals are able to adapt to a wide range of environments by relying on an internal motivational system that is partly innate and partly learned.

M. Asada et al. (Eds.): SAB 2008, LNAI 5040, pp. 12–21, 2008.

It must include some innate component since if it were learned completely, it must have relied on some kind of feedback, and this feedback must have come from somewhere. To avoid an infinite regress, one necessarily ends up with an innate motivational component, shaped by evolutionary processes that selected for valence systems (i.e. systems that evaluate the positive or negative nature of events or objects) that increased a species' chance of survival.

Several approaches for moving RL away from pre-designed reward functions exist; for example, Evolutionary RL [2], that inspired the approach taken in this paper, and, more recently, Intrinsically Motivated RL (IMRL) [3]. IMRL builds a hierarchy of skills by employing behaviour motivated by an intrinsic reward that is based on the novelty of salient events: intrinsic reward decreases with increasing familiarity of an event to the agent. An important limitation of this approach is that event saliency is pre-designed. One attempt to remedy this has been made by [4], who let artificial neural networks (ANNs) determine saliency, and evolved the ANNs by means of genetic algorithm (GA). This approach bears similarities to the one taken in this paper, except that we evolve ANNs that provide reinforcement based on the levels of internal drives.

1.2   Drive-Based Agents

Although IMRL and similar approaches correctly argue that there is more to the motivation of behaviour than the classic Hullian notion of reduction of basic drives such as hunger, thirst and libido, these drives are crucial for a species' survival. It is therefore highly likely that evolutionarily advantageous valence systems are linked to an organism's internal physiology. Hence, an alternative way of increasing an agent's adaptivity might be to evolve its reward function or valence system and to link it to a set of internal drives that the agent should keep within a viable range.

Drive-based architectures for action selection were investigated by e.g. [7], and [8] who took an approach in which an organism is made up of multiple simple, non-intelligent agents that represent basic concepts such as sensors, motivations and emotions. The organism should, by choosing the right actions, keep its internal physiological variables within a viable range. Action selection is mainly driven by the organism's motivations that are in turn influenced by emotions. In [9] agents learn to associate objects with drives by extracting object affordance information, and results showed that agents were able to associate objects with the behaviour that these afforded. However, these approaches pre-designed which action should be taken to satisfy a drive [9] and what kind of objects satisfy which drives [7]. In this article we try to avoid wiring objects to actions or actions to drives by design: the only aspect that is pre-designed (via the fitness function of the GA) is that agents should try to maintain their internal drives within a viable range.

As argued earlier, hard-wiring would not be such an issue if agents only had to deal with one kind of static environment; but since environments are almost never fixed, we aim to create systems that can efficiently deal with changing environments. Recent research with this focus includes e.g. [10], in which adaptivity was increased by genetically encoding learning rules instead of synaptic weights, and [11], in which a locally-recurrent ANN was found to outperform other network types under changing environmental conditions by switching between memory levels and making use of locally recurrent connections.

This article focuses on the inclusion of internal drives in agent architectures to deal with changing environments. We extend the work done in [2] and [5] and investigate the evolution of different innate, non-learning valence systems that should seek to maximise the agent's internal drive levels, by providing a reinforcement signal to a learning ANN that maps sensory input to actions. In particular, to mimic the uncertainty of real-world environments, we investigate the performance of these valence systems in an unstable environment (in which object properties are reversed at regular intervals). We hypothesise that valence systems that are based on internal physiology will have an advantage over systems based purely on sensory input: Essentially, a valence system based solely on internal physiology is decoupled from environmental properties and changes in these properties should therefore not interfere with the reinforcement signal it produces. In contrast, a reinforcement signal that depends (partly) on environmental input is likely to change if environmental properties change, and since evolution has shaped the reinforcement signal to match the old environment, the new signal is likely to be incorrect.

2   Methodology

2.1   Experimental Set-Up

The experimental set-up is largely based on Ackley and Littman's "AL" world [2]: a population of 10 agents with internal drives nutrition, hydration and health lives in a toroidal 20x20 grid world that contains 40 plants (increase nutrition), 40 water "units" (increase hydration), and 40 rocks (decrease health). If an agent enters a square with a plant or water, the relevant drive is updated and the object is randomly relocated in the grid; agents cannot enter squares that contain rocks. Agents have at their disposal the actions North (N), East (E), South (S), and West (W). Moving slightly decrements nutrition and hydration, and slightly increments health. An agent perceives its environment through a state vector that consists of perceptual (the grid world) and physiological (the drives) information. See figure 1 for an explanation of how the state vector is constructed.

Agents' "brains" are made up of two two-layer feed-forward ANNs that work together in an actor-critic reinforcement learning scheme [12]. One ANN, the actor, employs the complementary reinforcement back-propagation (CRBP) algorithm [13] to select actions based on the state vector and to learn from the feedback that it receives from the other ANN, the critic. CRBP adjusts network weights by back-propagation of an error vector that is generated depending on the received reinforcement signal: if the signal is positive, the error is such that the last output of the network is made more likely given the state; if it is negative, the error is such that the network is pushed towards an output that is

	N	E	S	W
Agent	0	0.25	0	0
Plant	0.5	0	0	0
Water	0	0	0	1
Rock	0	0	0.75	0

Nutrition	0.5
Hydration	0.3
Health	0.9

Fig. 1. Construction of the state vector. On the left, a schematic representation of the gridworld, with agent A1's visual field marked in bold. P=plant, W=water, R=rock. Objects within the agent's visual field but behind another object are not visible to the agent. The marked visual field would give rise to the state vector values on the right. Object distance is represented by a number that ranges from 0 (not visible) to 0.25 (edge of the visual field) to 1 (immediately adjacent the agent). The final state vector is formed by placing the numbers on the right into a one-dimensional vector, ordered from top to bottom.

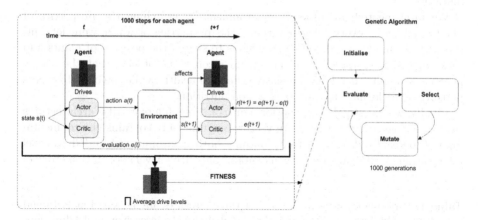

Fig. 2. The complete model. A population of 60 chromosomes is evolved by a GA during 1000 generations. At evaluation, each chromosome is decoded into an actor and critic network and inserted into an agent. Each agent does 1000 steps in the grid world, during which the actor learns by reinforcement signals coming from the fixed critic. Final agent fitness is equal to the product of its average drive levels during the run.

the complement of the last output for this state (see [13] for more details). The critic represents the valence system and computes a scalar temporal-difference (TD) error [14] that is passed to the actor as feedback. The critic does not learn, since we wish to investigate the evolution of an innate valence system that is responsible for primary reinforcement. Additionally, there are no directly perceivable rewards in the environment, nor is there an explicit reward associated with increasing drive levels. Instead, both the critic and the actor weights are evolved by a GA. The complete model is summarised in fig. 2.

The GA evolves a population of 60 real-valued chromosomes for 1000 generations. Fitness is assessed by inserting the chromosome into an agent, setting the agent's ANN weights to the weights encoded by the chromosome, and letting the agent walk around in the grid world for 1000 steps. Since the grid world contains 10 agents, 10 chromosomes are evaluated at a time by inserting one chromosome into each agent. There are 60 chromosomes to assess, so this is repeated 6 times for one complete evaluation of the whole population.

Final agent fitness is calculated by taking the product of the average drive levels (taken over the agents' lives); this ensures that if one of the drives reaches 0, fitness is 0 (since a 0 drive should represent agent death). Fitness is thus assessed continuously, which stimulates the rate of genetic assimilation of behaviour since it implies a cost of learning [15]. Selection is carried out by the tournament selection method with tournament size 2, no crossover takes place, and mutation takes place with probability 0.2 per gene; the high mutation rate might seem surprising, but these parameter values were determined experimentally as optimal for this scenario [6]. Mutation consists of adding a uniformly distributed random variable $\delta \in [-0.2, 0.2]$ to the value of the gene that is being mutated.

We implemented an unstable environment by switching the properties of plants and rocks every 20 generations; the properties of water were left untouched. A simple way of achieving this is to switch the position of plants and rocks in the state vector that the agents receive; the objects that the agents perceive as plants then have a negative effect on health, while the objects agents perceive as rocks have a positive effect on nutrition.

Finally, we investigated the performance of three different critic architectures that each receive a different state vector: one that only contains environmental input ("percept-only"), one that contains both environmental and drive-level information ("full"), and one which contains only drive-level information ("drive-

Table 1. Performance measures. The first three measures are measured at birth and are calculated by averaging the evaluation / probability of approach over all 4 directions in which an agent can face an object.

Measure	Description
Object evaluation	Measure of the critic's evaluation of objects. Positive means object is seen as good. Negative means it is seen as bad.
Drive evaluation	Measure of the critic's evaluation of drives. Positive means high drive level is seen as good. Negative means it is seen as bad.
Object action	Agent tendency to approach an object. 0.25 if agent is equally likely to move in any direction. Higher if agent tends to approach object at birth. Lower if it tends to avoid object at birth.
Delta object action	Measure of how much the actor network has learned w.r.t. objects. Positive if agent object action measure has increased during its lifetime, negative if it has decreased.

only"). Of course, the input dimensionality of the critic changes according to the state vector it receives. The actor network always receives the full state vector and thus always has the same architecture.

2.2 Evaluation

Apart from the usual GA performance measures of average population fitness and fitness of the best individual in the population, we also keep track of the average and final drive levels per generation, averaged over the population. To gain further insight into the underlying causes of agent performance, we defined four additional performance measures, summarised in table 1.

3   Results

Figures 3 and 4 present the results of the experiments using the critic architectures as explained in section 2. All results represent averages over 10 runs of the GA. All population fitness and drive level graphs (fig. 3) exhibit a characteristic

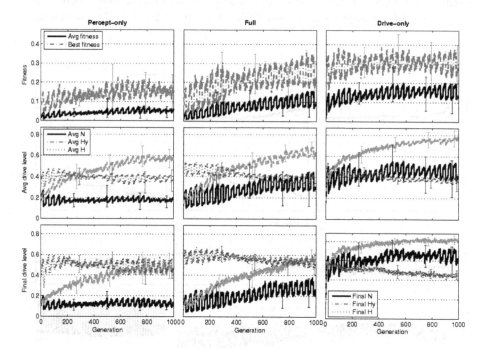

Fig. 3. Population fitness and drive levels. N=nutrition, Hy=hydration, H=health. Columns represent architectures, rows represent performance measures. Trial runs revealed that evolution of the full architecture (middle column) continues slightly beyond 1000 generations but with no significant improvement; we therefore chose to cut off the runs at 1000 generations. Environment switch occurs every 20 generations, as can be seen from the drop in fitness / drive level every 20 generations. Error bars indicate one standard deviation above respectively below the average.

jagged pattern, caused by the switching of object properties (the first environment presented is the "normal" one). As expected, percept-only agents perform worst: their average fitness in both environments fluctuates around 0.05. Drive-only agents do best, reaching a top fitness of around 0.2 and a top fitness in the switched environment of around 0.15. Agents employing the full architecture fall somewhat in between: the fitness starts off low, but increases steadily, reaching a maximum of around 0.15 and a maximum in the switched environment of around 0.08. All differences in performance, although small, are significant (P < 0.05) as determined by a one-tailed Student t-test with 18 degrees of freedom (comparing two groups of 10 samples (i.e., runs), and degrees of freedom determined by $(N_1 - 1) + (N_2 - 1)$, where N_i denotes number of samples in a group) that compares the average fitness and standard deviation taken over all generations.

The graphs of average and final drive levels (fig. 3, middle and last row) show a clear difference between agents whose critic includes perceptual information and those whose critic does not: while percept-only and full agents are on average able to improve their health during their lives (compare the graphs of average versus final drive level), the level of the other two drives generally *decreases* during their lives, in both the normal and switched environment. Drive-only agents are able to improve all drives during their lives, in both the normal and

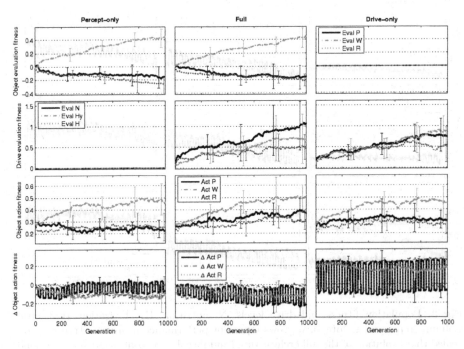

Fig. 4. Performance measures. P=plant, W=water, R=rock. Columns represent architectures, rows represent performance measures. Error bars indicate one standard deviation above respectively below the average.

switched environment; an indication that these agents are able to learn in both environment types.

The performance measures we defined reveal some phenomena shared by all architectures (fig. 4). In terms of objects, water is evaluated as something good (see "object evaluation" graphs, first row), while rocks and plants are evaluated as slightly negative: agents are unable to evolve a positive evaluation of plants because of the insecurity caused by the unstable environment. As expected (e.g. [16]), the tendency to approach water becomes hard-wired into the agents' genes, while plant- and rock-oriented behaviour remain near-random (see "object action" graphs, third row): it is an evolutionary advantage to genetically assimilate traits related to stable properties of the environment, while there is no advantage in hard-wiring traits related to unstable factors.

All architectures (except percept-only, since it does not apply) evolve a positive evaluation of all drives (see "Drive evaluation" graphs, second row). The differentiating factor between the drive-only and other architectures is what is learned with respect to objects ("Δ object action" graphs, last row). The perceptual architectures are only able to improve on rock- and plant-oriented behaviour in one of the environments. Also, learning in these architectures has a negative effect on water-oriented behaviour. In contrast, drive-only agents are able to improve on their plant- and rock-oriented behaviour in both the normal *and* the switched environment, and by a factor that is generally larger than that of the percept-only and full architectures. Furthermore, drive-only agents learn to further increase their tendency to approach water on top of their innate water-approaching behaviour.

4   Discussion

This article is, to the best of our knowledge, the first to investigate the performance of drive- versus perception-based motivational systems in an unstable environment. Results showed that inclusion of internal drive levels in the information that the valence system bases its feedback on significantly improved performance. Furthermore, a valence system based purely on internal drives outperformed a system based on both perceptual input and internal drives, by being able to improve the actor's behaviour-selection policy in both normal and switched environments. This makes sense since, in an environment in which object properties change after a number of generations, a drive-based valence system is not hampered by object evaluations that evolution hard-wired into the network. Instead, it is able to help the agent adapt its behaviour by focusing on its primary goal: surviving by maintaining its drive levels in a viable range.

These results reinforce our belief that the inclusion of drives in agent architectures can be a useful tool to specify agent goals and decouple them from environmental properties. In a real-world example, a drive could be the energy level of a mobile robot's battery levels, and a robot employing a drive-based motivational system could quickly learn to avoid objects that start negatively affecting its battery (e.g., a faulty charge station).

Although extremely simplified, the architecture we implemented bears resemblance to certain brain structures involved in animal learning. The critic is typically associated with the substantia nigra and the part of the basal ganglia called the striasomes, while the actor is associated with the matrisomes, another part of the basal ganglia (e.g. [17]).

In terms of drives, it is known that the perceived reward of gustatory stimuli depends on physiological state (e.g. [19]); an animal would thus e.g. choose food over water when hungry. Results from an earlier, more extensive study that we conducted [6] indicated such drive-influenced decision making might be an emergent property of our architecture. On the neurological level, the hypothalamus is a brain structure that is linked to both drive and reward information (e.g. [18]). The ability of the drive-based critics to teach the actor to re-link object types to drives in the switched environment in our study bears resemblance to the Valenstein effect [20], in which a hypothalamic site that originally elicited feeding behaviour when stimulated became associated with drinking behaviour by repeatedly electrically stimulating the site in the presence of water.

As Barto [21] points out, the critic must include some fixed component, created by evolution, that is responsible for primary reinforcement. Our results and those from other studies support our opinion that, given species evolve in a continuously changing environment, this component must (at least at birth) be linked to an organism's basic drives as opposed to receiving direct input from sensory areas of cortex.

References

1. Schultz, W., Dayan, P., Montague, P.R.: A neural substrate of prediction and reward. Science 275, 1593–1599 (1997)
2. Ackley, D.E., Littman, M.E.: Interactions between learning and evolution. In: Langton, C., Taylor, C., Farmer, D., Rasmussen, S. (eds.) Proc. Second Conf. Artificial Life (1991)
3. Singh, S., Barto, A.G., Chentanez, N.: Intrinsically Motivated Reinforcement Learning. In: Adv. Neural Information Processing Systems, vol. 17. MIT Press, Cambridge (2004)
4. Schembri, M., Mirolli, M., Baldassarre, G.: Evolving internal reinforcers for an intrinsically motivated reinforcement-learning robot. In: Demiris, Y., Mareschal, D., Scassellati, B., Weng, J. (eds.) Proc. 6th Int. Conf. Development and Learning, pp. E1–6. Imperial College, London (2006)
5. Damoulas, T., Cos-Aguilera, I., Hayes, G.M., Taylor, T.: Valency for adaptive homeostatic agents: Relating evolution and learning. In: Capcarrère, M.S., Freitas, A.A., Bentley, P.J., Johnson, C.G., Timmis, J. (eds.) ECAL 2005. LNCS (LNAI), vol. 3630, pp. 936–945. Springer, Berlin (2005)
6. Snel, M.: Evolving a sense of valency. MSc thesis, School of Informatics, University of Edinburgh (2007)
7. Cañamero, D.: Modeling motivations and emotions as a basis for intelligent behavior. In: Proc. Int. Conf. Autonomous Agents, pp. 148–155 (1997)
8. Velásquez, J.: Modeling emotion-based decision making. In: Cañamero, L.D. (ed.) Emotional and intelligent: The tangled of knot of cognition. Papers from the 1998 AAAI Fall Symposium, pp. 164–169. AAAI Press, Menlo Park (1998)

9. Cos-Aguilera, I., Cañamero, L., Hayes, G.M.: Motivation-driven learning of object affordances: first experiments using a simulated Khepera robot. In: Detjer, F., Dörner, D., Schaub, H. (eds.) The logic of cognitive systems: Proc. Fifth Int. Conf. Cognitive Modeling (ICCM 2003), pp. 57–62 (2003)

10. Urzelai, J., Floreano, D.: Evolution of adaptive synapses: robots with fast adaptive behavior in new environments. Evolutionary Computation 9(4), 495–524 (2001)

11. Capi, G., Doya, K.: Evolution of neural architecture fitting environmental dynamics. Adaptive Behavior 13(1), 53–66 (2005)

12. Barto, A.G., Sutton, R.S., Anderson, C.W.: Neuronlike elements that can solve difficult learning control problems. IEEE Transactions on Systems, Man, and Cybernetics 13, 835–846 (1983)

13. Ackley, D.E., Littman, M.E.: Generalization and scaling in reinforcement learning. In: Touretzky, D.S. (ed.) Adv. Neural Information Processing Systems, vol. 2. Morgan Kaufmann, San Mateo (1990)

14. Sutton, R.S.: Learning to predict by the method of temporal differences. Machine Learning 3, 9–44 (1988)

15. Mayley, G.: Landscapes, learning costs and genetic assimilation. Evolutionary Computation 4(3), 213–234 (1996b)

16. Anderson, R.W.: Learning and evolution: a quantitative genetics approach. J. Theor. Biol. 175, 89–101 (1995)

17. Houk, J.C., Adams, J.L., Barto, A.G.: A model of how the basal ganglia generates and uses neural signals that predict reinforcement. In: Houk, J.C., Davis, J.L., Beiser, D.G. (eds.) Models of Information Processing in the Basal Ganglia, pp. 249–274. MIT Press, Cambridge (1995)

18. Olds, J.: Drives and reinforcements: behavioral studies of hypothalamic functions. Raven Press, New York (1977)

19. Conover, K.L., Woodside, B., Shizgal, P.: Effects of sodium depletion on competition and summation between rewarding effects of salt and lateral hypothalamic stimulation in the rat. Behav. Neurosci. 108, 549–558 (1994)

20. Valenstein, E.S., Cox, V.C., Kakolewski, J.W.: Modification of motivated behavior elicited by electrical stimulation of the hypothalamus. Science 157, 552–554 (1968)

21. Barto, A.G.: Adaptive critics and the basal ganglia. In: Houk, J.C., Davis, J.L., Beiser, D.G. (eds.) Models of Information Processing in the Basal Ganglia, pp. 215–232. MIT Press, Cambridge (1995)

Flexible Control Mechanism for Multi-DOF Robotic Arm Based on Biological Fluctuation

Ippei Fukuyori, Yutaka Nakamura, Yoshio Matsumoto, and Hiroshi Ishiguro

Graduate school of Engineering, Osaka University,
2-1 Yamada-oka, Suita, Osaka, Japan
{ippei.fukuyori,nakamura,matsumoto,ishiguro}@ams.eng.osaka-u.ac.jp

Abstract. Controlling a highly dynamics and unknown system by existing control methods would be difficult because of its complexity. Recent biological studies reveal that animals utilize biological fluctuations to achieve adaptability to the environment and high flexibility. In this paper, we propose a simple, but flexible control method inspired by a biological adaptation mechanism. The proposed method is then applied to control robotic arms. The results of simulation indicated that our proposed method can be applied well to the control of a robot with multi-DOF.

1 Introduction

Various robot systems are working in our society, and are indispensable for our lives in these days. However most of them are working at production lines in factories which are designed for robots. In the future, robots are expected to support our daily lives [1,2], however, there are no robots which can work in real, unstructured environments. In order for the robots to work in our daily lives, they are required to have robustness against various disturbances, flexibility in unknown environments, and utility in performing practical tasks. In order to realize such functions, robots should have large degrees of freedom, and achieve complex motions. Humans and animals have complex mechanical structures, and many robot systems inspired by their structures have been developed [3,4].

Untill now various control methods have been studied. The representatives of them are classical control theory which utilizes a transfer function, and optimal control theory such as H_∞ control. However, the higher the complexity of the target system becomes, the harder also the modeling of the system becomes. Learning methods such as reinforcement learning can also be utilized for optimization, however, the number of necessary trials increases drastically when the complexity of the system increases [5,6,7,8]. This paper focuses on the problems in which the system is hard to be modeled due to its complexity and fluctuations of the environment, and we propose a simple and robust control method inspired by biological systems. The biological system is known to have a potential to adapts to new, unknown, and noisy environments. The mechanism of such flexible adaptation is investigated especially in molecular biology, and the importance of the biological fluctuation is made clear [9]. The fluctuation in molecular

M. Asada et al. (Eds.): SAB 2008, LNAI 5040, pp. 22–31, 2008.

science is actually a noise due to the heat fluctuation, which is unavoidable and unpredictable. In conventional control for robot systems, such noise should be removed to the maximum extent. However, it is now believed that biological systems do not necessarily need to remove the noise, but rather make use of it in order to adapt to the environment. In this research, we propose a novel control method inspired by biological fluctuation. This method is expected to handle control problems for complex systems in unknown environments without explicitly modeling them.

Section 2 describes the proposed "adaptive attractor selection" control mechanism and Section 3 describes the formulation and simulation to the feasibility of the proposed algorithm. In Section 5, the proposed method is applied to a multi-DOF robot arm and we show the applicability of the proposed method. Section 6 concludes the paper.

2 Biological Fluctuation

Bacteria can adapt to environmental changes. For example, even if some important nutrients dramatically decrease, bacteria can handle such crisis by alteration of gene expression. Kashiwagi et al. built a model of this adaptation mechanism based on a biological fluctuation, and explained the behavior of the bacteria. In this model, the gene expression is controlled by a dynamical system with some attractors, and this model is called "attractor selection model"[10].

Attractor selection model. The attractor selection model can be represented by Langevin equation as:

$$\tau_x \dot{x} = f(x) \times A + \epsilon, \tag{1}$$

where x and $f(x)$ are the state and the dynamics of the attractor selection model, and τ_x and ϵ are the time constant and the noise, respectively.

A is a variable called "activity" which indicates the fitness of the state x to the environment, and controls the behavior of the attractor selection model. That is, $f(x) \times A$ becomes dominant in Eq. 1 when the activity is large, and the state transition approaches deterministic behavior. On the other hand, the noise ϵ becomes dominant in Eq. 1 when the activity is small, and the state transition becomes more probabilistic. Because $f(x)$ is designed to have some attractors, the state of the system is entrained into one attractor when the activity is large as depicted in Fig. 1(a), and the behavior of the system becomes like a random walk when the activity is small as depicted in Fig. 1(b). The activity is designed to be large (small) when the state x is suited (not suited) to the environment. As a result, the state of the system is entrained into an attractor which is suited to the environment and the activity becomes large. Otherwise the activity remains to be small and a suitable attractor is searched for by a random walk. Our proposed method is an algorithm searching for a sub-optimal attractor where the escape from an unsuited attractor can be achieved by controlling the activity.

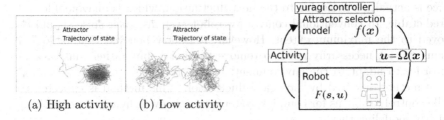

(a) High activity   (b) Low activity

Fig. 1. Behavior of attractor selection model

Fig. 2. Block diagram

Bacteria survive even though their habitat environment is very complex such that many unpredictable disturbances and many dramatic changes occur. This fact suggests that their adaptive mechanism is flexible and would be useful as a control mechanism for a complex system, because the control is a kind of response against the change in state of the target system. In this research, we propose a control method based on the attractor selection model. We apply our method to the control of robot simulators and a human-like robotic arm, and show that our method is well applicable to such types of problems.

3   Control Method Based on the Attractor Selection Model

In this section, we describe the control mechanism using attractor section. We explain our proposed control method, and show the result of the control of a simple robot in simulation. The dynamics of robot is calculated by a simulation software called "Open Dynamics Engine" [11].

The motion of a physical system like a robot can be represented by

$$S' = F(S, u), \tag{2}$$

where S, u, and $F(S, u)$ are the state, the control signal and the dynamics of the system, respectively. In this work, the control signal u is generated from the state of the attractor selection model:

$$u = \Omega(x), \tag{3}$$

where x is the state of the attractor selection model. Fig. 2 shows the block diagram of our control mechanism. The controller based on the attractor selection model receives the activity, which is generated from the state of the robot, and changes its own state according to Eq. 1. The controller outputs the control signal u to the robot based on the state x according to Eq. 3. The robot receives the control signal u, and changes the state S according to its dynamics (2). Ijspeert et al.[12] proposed a control scheme for a robot with large degrees of freedom where a motor primitive is represented by an attractor and its learning mechanism. In our proposed method, each attractor also correponds to a motor primitive, and a motor primitive suited to the situation is searched for by the attractor selection model. In this framework, the attractor selection model is the basis of the controller and we call this a "Yuragi Controller" [1].

3.1 Design of Attractors

In the attractor selection model, $f(x)$ in Eq. 1 has to define the attractors. In this paper, we employ following equations to design attractors:

$$f(x) = \sum_{i=1}^{n_a} \mathcal{N}_i \frac{(X_i - x)}{\|X_i - x\|}, \tag{4}$$

where n_a denotes the number of attractors. X_i $(i = 1, \cdots, n_a)$ is a vector representing the center of the i-th attractor. The function $\mathcal{N}_i = \frac{g_i(x)}{\sum_{j=1}^{n_a} g_j(x)}$ is a normalized Gaussian where $g_i(x) = \exp\{-\beta\|X_i - x\|^2\}$. The behavior of this system is such that the state x approaches to the nearest attractor like a point mass with gravitational pulls from many objects (attractors).

3.2 Design of Activity

In the attractor selection model, the activity controls the behavior of the attractor selection model. The yuragi controller is designed to behave as follows: If the current output is suited to achieve the desired control, the state of the attractor selection model is entrained into an attractor and is only slightly perturbed by noise. Therefore, the desired control signal would be repeated. Otherwise, the state x changes by a random walk and a desired control signal is searched for. This can be achieved if the activity is set up such that the activity becomes large (small) when the current control signal is suited (not suited).

We employ below equations as the activity:

$$A(t) = \alpha(t) - \bar{\alpha}(t - 1), \tag{5}$$

$$\alpha(t) = \{\|\hat{Z} - Z(t)\|\}^{-1}, \tag{6}$$

$$\bar{\alpha}(t - 1) = \frac{\sum_{\tau=1}^{t-1} \gamma^\tau \alpha(t - \tau)}{\sum_{\tau=1}^{t-1} \gamma^\tau}, \tag{7}$$

where $\hat{Z}$ and $Z(t)$ represent the goal of the control task and the current state. α indicates the current attainment level of the task, and $\bar{\alpha}$ is the average of α with forgetting factor γ. Therefore, A represents the instantaneous improvement of the achievement.

In the case of a reaching task for the robotic arm, for example, $\hat{Z}$ and $Z(t)$ are the target position and the current position of the hand, and α is the inverse of the distance between the hand and the target position. The activity becomes larger when the hand approaches the goal while the activity becomes smaller when the hand moves away from the goal.

---

[1] 'Yuragi' means fluctuation in Japanese.

4   Control of 2-Link Robotic Arm

We apply our method to the control of the 2-link robotic arm, shown in Fig. 3(a). This robot consists of 2 rigid links whose length and weight are 0.5[m] and 0.2[kg]. Both links are connected by a hinge joint, and the motion of these links is restricted in the 2-dimensional space. This hinge joint can rotate without restriction except for the friction, and the gravity is ignored. The aim of the control is to achieve a reaching task.

4.1   Conditions of the Simulation

The time constant τ_x in Eq. 1 was set to 0.1[sec]. The calculation of the dynamics was discretized, and was performed by a 4-th order Runge-Kutta method with 0.001[sec] time interval. The goal $\hat{Z}$ in Eq. 7 was the target point shown in Fig. 3(a). Each element of the vector X_i is generated by a uniform random number whose range is $[-1, 1]$, and the number of attractors was $n_a = 10$. The noise term ϵ in Eq. 1 was a Gaussian noise with mean of 0.0 and variance of 1.0. The control signal u is the set of target angles of joints, and is generated from the state x as

$$u = \pi x. \tag{8}$$

The state x basically remains in the region $[-1, 1]$ where attractors exist, and consequently, the control signal u also becomes about $[-\pi, \pi]$.

4.2   Control by the Yuragi Controller

Fig. 3(b) shows the result of the simulation. Vertical axes of these graphs show the distance between the target point and the hand (top), the value of the activity (middle), and the state x (bottom). The horizontal axis is time. The task of this simulation is to make the hand move close to the target point, i.e., the distance (in the top graph) to be close to 0. Fig. 3(b) shows that after a few moments the robot approached the target point, and the activity became large at the same time. As a result, the state x was entrained into an attractor. At about $t = 35$, the robot got away from the target point due to the noise, and the activity became small. As a result, the behavior of the state x became like a random walk. At about $t = 45$, the activity became large again, and the state x was entrained into another attractor, and the robot approached the target point. This result shows that the control for the reaching task can be achieved by our method.

Our proposed method is not an optimal method and the robot can not keep the position and fluctuates around the target position by the noise after the robot achieves to the target position. That is, the robot repeats approaching the target and getting away, but achieves the task robustly. Furthermore, the yuragi controller only utilizes the position information of the hand through the activity. This controller can be applied even if the sensory input is significantly restricted or noisy.

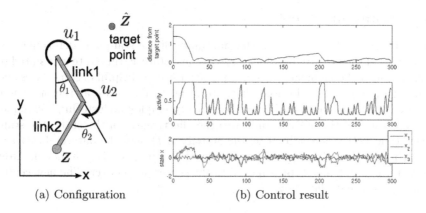

(a) Configuration   (b) Control result

Fig. 3. 2-link robotic arm

4.3   The Number of Attractors

The performance of the control (the average distance from the target) would depend on the arrangement and the number of attractors. In this section, we investigate how the number and the arrangement of attractors affect the performance of the control. Fig. 4(a) shows changes of the performance by the number of attractors. The horizontal axis denotes the number of attractors. The vertical axis denotes the performance of the control. We conducted 30 simulations by altering the position of attractors, and plotted the average distance from the target point to the hand. The average distance was calculated by the arithmetic mean of the distance during the whole period of the simulation. The maximum, 25% point, median, 75% point and minimum value were plotted. This result indicates the performance became the best when the number of the attractor was 36.

In the following, we propose an adaptive allocating method of attractors based on a clustering method called the Gaussian mixture model.

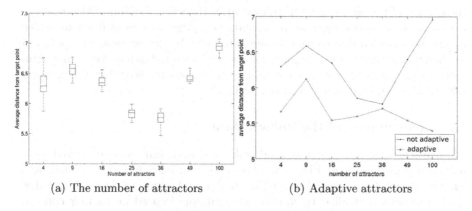

(a) The number of attractors   (b) Adaptive attractors

Fig. 4. Performance of yuragi controller

4.4   Adaptive Attractors

In the attractor selection model, the suitability of the state x is determined as the activity. If the activity becomes large in some state x, this state x is suited to the situation at the moment. This means such a state is helpful in some cases and an attractor should be allocated there. In order to reallocate attractors where the activity became large during the control, we employ a clustering technique.

As the value of the activity indicates the effectiveness of the state, the state of each time $x(t)$ was weighted by the activity $A(t)$, and the dataset $\{x(t); t = 1, \ldots, T\}$ was divided into some clusters by the Gaussian mixture model. After the clustering, each attractor is moved to the center of a cluster. The new center of the cluster is calculated by

$$X_i = \frac{\sum_{t=0}^{T} \mathcal{N}_i A(t) x(t)}{\sum_{t=0}^{T} \mathcal{N}_i A(t)}, \tag{9}$$

$$\mathcal{N}_i = \frac{G_i(x)}{\sum_{j=1}^{n_a} G_j(x)}, \tag{10}$$

$$G_i(x) = \exp\left\{-\frac{\|X_i - x\|^2}{2\sigma^2}\right\}. \tag{11}$$

The update of the position of each attractor was performed once in some constant steps. We compared the performance of this adaptive Yuragi Controller with the vanilla Yuragi Controller described in Section 3. Fig. 4(b) shows median performances of both controllers. The horizontal axis denotes the number of attractors, and the vertical axis denotes the performance. In all cases, the performance of the adaptive yuragi controller was better than that of the vanilla yuragi controller.

5   Control of a Redundant Robot

In the previous section, we showed that our proposed method can be applied to the control of the simple 2-DOF robotic arm. However, the aim of this research is to develop a control method for a robot with large degrees of freedom, which cannot be modeled due to its complex structure. In this section, we apply our method to the control of a complex robotic arm with redundant joints and actuators, as shown in Fig. 5(a). We conduct simulations to show that the complex robotic arm can be controlled by our method.

5.1   Configuration of the Robotic Arm

The robotic arm consists of 4 rigid links and each link was connected by a universal joint shown in Fig. 5(a). The weight of each link was 0.5[kg], and lengths of each link were 0.5[m], 0.4[m], 0.3[m] and 0.2[m] in descending order. Each joint was controlled by four linear actuators located at its four corners. The linear actuator emulates the model of a pneumatic artificial muscle, and

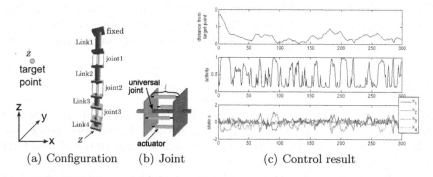

(a) Configuration   (b) Joint   (c) Control result

Fig. 5. Redundant robotic arm

expands or contracts to become the target length given the control signal. Each joint has $[-\pi/3, \pi/3]$-[rad] range of movement, where the origin is the angle of the initial position shown in Fig. 5(a). The shoulder (Link1) was fixed and the gravity acceleration was set to $0.1[\mathrm{m/s^2}]$.

5.2 Conditions of the Simulation

The time constant τ_x in Eq. 1 was set to 0.1[sec]. The coordinate of the target point shown in Fig. 5(a) was used as the target $\hat{Z}$ in Eq. 7. In this simulation, the control signal was generated by a yuragi controller which has 12 dimensional state space. The number of attractors n_a was 30, and the center of each attractor was generated by uniform random numbers. The noise in Eq. 1, ϵ, was generated from a Gaussian whose mean and variance were 0 and 1.0, respectively. The control signal u was a 12 dimensional vector, and each component of the vector was corresponding to the target length of an actuator. The control signal was generated as:

$$u = g \frac{x}{\|x\|} \qquad (12)$$

where g is a constant and was set to 10.0.

5.3 Control Result

Fig. 6 shows the motion of the robot. Fig. 5(c) shows the distance from the target point, the activity and the state of the Yuragi Controller. In the bottom graph

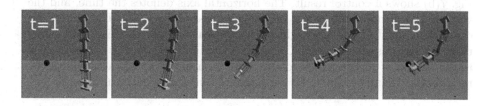

Fig. 6. Snapshots of the 4-link robotic arm at different time instants t

of Fig. 5(c), we only plotted four components out of the state vector, and these components correspond to the control signal for joint1. These graphs indicate that our method was successfully applied to the control of complex robotic arm.

6   Conclusion

In this article, we proposed a simple but flexible control mechanism for a complex system which can not be modeled easily. In this method, the controller is based on the attractor selection model referring to an adaptation mechanism of a bacteria using biological fluctuation, and we call this controller "Yuragi Controller."

We applied our method to the control of robot simulators and confirmed that our method is applicable. Although these robot simulators have a relatively simple structure which might be controlled by an ordinary method, flexible control was able to be achieved without modeling systems by our method. This suggests that our method would be also applicable to the control of a complex system which is hard to be modeled.

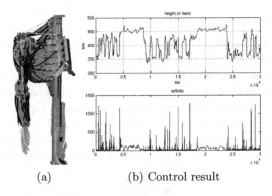

(a)   (b) Control result

Fig. 7. Human-like robotic arm

Currently, we are applying our method to a robotic arm which imitates a human upper limb, shown in Fig. 7(a). This robotic arm has complex bone structure and redundant number of pneumatic actuators, and it seems to be difficult to build a precise kinematic and dynamic model. However, a simple task, i.e. holding its hand at a certain height can be realized by our method. Fig. 7(b) shows a control result. The horizontal axis denotes the time, and the vertical axes denotes the height of the hand and the activity. The aim of the experiment was to keep the height of the hand at 450mm, and Fig. 7(b) shows that this task was achieved well. Our control method is simple, but can be applied to the control of such complex robotic arm. To achieve more practical tasks by the robotic arm and evaluate its performance remains as our future work.

Acknowledgement

This research was supported by "Special Coordination Funds for Promoting Science and Technology: Yuragi Project" of the Ministry of Education, Culture, Sports, Science and Technology, Japan.

References

1. Kanda, T., Ishiguro, H., Ono, T., Imai, M., Nakatsu, R.: An evaluation on interaction between humans and an autonomous robot Robovie. Journal of the Robotics Society of Japan 20(3), 1–9 (2002)
2. Miyake, T.: Wall Climbing Robot for Window Cleaning WallWalker. Journal of the Japan Society of Mechanical Engineers 109(1051), 474–475 (2006)
3. Hirose, S., Takayama, T.: Study on 3D Active Cord Mechanism with Helical Rotational Motion. Journal of the Robotics Society of Japan 22(5), 83–93 (2004)
4. Fukuoka, Y., Kimura, H.: Biologically inspired adaptive dynamic walking of a quadruped on irregular terrain –realization of walking in outdoor environment using a self-contained robot: 'Tekken2'. Journal of the Robotics Society of Japan 25(1), 138–154 (2007)
5. Kaelbling, L.P., Moore, A.W.: Reinforcement Learning:A Survey. Journal of Artificial Intelligence Research 4, 237–285 (1996)
6. Sutton, R.S., Barto, A.G.: Reinforcement Learning: An Introduction. MIT Press, Cambridge (1998)
7. Peters, J., Schaal, S.: Policy Learning for Motor Skills. In: Proceedings of 14th international conference on neural information processing (iconip), p. 10136 (2007)
8. Nakamura, Y., Mori, T., Sato, M., Ishii, S.: Reinforcement learning for a biped robot based on a cpg-actor-critic method. Neural Networks 20(6), 723–735 (2007)
9. Yanagida, T., Ueda, M., Murata, T., Esaki, S., Ishii, Y.: Brownian motion, fluctuation and life. Biosystems 88(3), 228–242 (2006)
10. Kashiwagi, A., Urabe, I., Kaneko, K., Yomo, T.: Adaptive Response of a Gene Network to Environment Changes by Fitness-Induced Attractor Selection. PLoS ONE 1 (2006)
11. Smith, R.L., et al.: Open Dynamics Engine:open source library for simulating rigid body dynamics, http://www.ode.org/
12. Ijspeert, A., Nakanishi, J., Schaal, S.: Learningattractor landscapes for learning motor primitives. Advances in Neural Information Processing Systems 15, 1547–1554 (2003)

Neural Noise Induces the Evolution of Robust Behaviour by Avoiding Non-functional Bifurcations

Jose A. Fernandez-Leon and Ezequiel A. Di Paolo

Centre for Computational Neuroscience and Robotics, University of Sussex, UK
{jf76,ezequiel}@sussex.ac.uk

Abstract. Continuous-time recurrent neural networks affected by random additive noise are evolved to produce phototactic behaviour in simulated mobile agents. The resulting neurocontrollers are evaluated after evolution against perturbations and for different levels of neural noise. Controllers evolved with neural noise are more robust and may still function in the absence of noise. Evidence from behavioural tests indicates that robust controllers do not undergo noise-induced bifurcations or if they do, the transient dynamics remain functional. A general hypothesis is proposed according to which evolution implicitly selects neural systems that operate in noise-resistant landscapes which are hard to bifurcate and/or bifurcate while retaining functionality.

Keywords: Evolutionary robotics; Systemic robustness; Continuous-time neural networks; Neural noise; Bifurcations.

1 Introduction

The role of noise in systems with sensorimotor control has generated a growing interest in bio-inspired robotics – in particular, in its relation to systemic aspects of robust behaviour (e.g., [3, 4]). Neural noise is important to be studied in behavioural systems as it may result in movement inaccuracy (e.g., constant errors) and imprecision (e.g., variable errors and uncertainty) [3]. In the context of adaptive behaviour during goal-oriented tasks, Bays *et al.* (2007) propose that the strategy of the central nervous system for dealing with neural noise, i.e., the spontaneous neural background activity present in most brain tissues, is to 'optimally combine sensorimotor signals'. Despite this broad hypothesis, we have very little idea about how the algorithms underlying the management of the effects of neural noise are realized at the neuronal level because the majority of work in this area neither explains how these mechanisms emerge from sensorimotor interactions, nor analyses how such strategies may have originated during evolution. In the context of artificial evolution, evidence that noise also has some useful properties has been presented several times (e.g., [4, 6]) and this leads us to a second question: whether in natural systems noise should always be considered detrimental. Combining these two ideas, the question of *what sort of control-strategy emerges if neural noise is induced during the evolution of neurocontrollers* becomes one of conceptual and practical interest not only for evolutionary and autonomous robotics but potentially for neuroscience as well.

M. Asada et al. (Eds.): SAB 2008, LNAI 5040, pp. 32–41, 2008.

The use of noise is a widespread practice in evolutionary robotics (ER). Using his *minimal simulations* paradigm, Jakobi (1997) has investigated the uses of noise and parametrical uncertainty in the evolution of neurocontrollers and found that they have a significant rate of success when transferred from simulated solutions into real robots (where direct evolution is impractical or prohibitive). Minimal simulations work by avoiding the accurate but costly replication of the physical complexities of a real-world robot-environment system and instead abstract a base set of factors upon which evolution must rely in order to produce the desired behaviour. All other factors in the robot-environment system are crudely modelled and subject to large amounts of environmental noise and variability between evaluations.

Some of the lessons of the minimal simulations methodology may illuminate questions about natural robustness. Biological systems exhibit phenomena, such as sensorimotor robustness to noise [3] or robustness in functional terms [6], which may relate to the presence of neural noise and therefore warrant investigation. However, in an ER context, it is necessary to address the question of what mechanisms enable robustness of behaviour sustaining functionality in the presence of neural noise. Studying these mechanisms can inform our understanding of what to look for in natural systems and how to build better artificial ones. ER provides a useful, relatively assumption-free paradigm in which to do this (e.g., agent's dynamics maintaining functionally the same during behaviours).

This paper describes an attempt to understand how neural systems can maintain their function while dealing with neural noise. This is an exploratory piece of work aimed largely at generating hypotheses, and the motivations are conceptual as well as practical. We present results from ER simulations exploring the effects of neural noise on neurocontroller dynamics in order to investigate systemic robustness at the behavioural level. In this context, robustness refers to the ability to maintain performance in the face of perturbations (internal or external) and uncertainty [1]. In order to facilitate understanding of the results and comparative analysis, a simple phototaxis task is chosen. In the next section, the methods and experiments are introduced, and in the final section we examine the consequences of the results and discuss questions that remain open.

2 Methods and Experimental Setup

In order to avoid unnecessary complexity at this initial stage, a minimal approach is deliberately used [4, 6]. The aim is to evaluate the consequences of evolving networks under fixed and variable values of neural noise and to test the obtained solutions in terms of behavioural robustness with the purpose of uncovering the mechanisms at play. A population of simulated agents is evolved to perform phototaxis in normal body and environmental conditions while being disrupted by internal neural and external sensorimotor noise. In each test, one light source is presented every time step for an extended period. Limited random noise is applied locally to the dynamics of each neuron. The level of noise in each neuron ($y_0 \in [-A;A]$) is modelled either as a fixed or a constantly changing activation parameter selected every time step, where A is a fixed value for each experiment ($A=0,1,2,3,4$). The range (A) of y_0 is a control parameter in our studies.

Agents are modelled as solid circular bodies of radius 5 (arbitrary units) with two diametrically opposed motors that differentially steer the agent with their output (in range [0;1]) and two frontal light sensors positioned with a separation between sensors of 47.75°. The agents' motors can drive backwards and forwards in an unlimited 2-D arena. Agents have a very small mass, so motor output is directly indicates the tangential velocity at the point of the body where the motor is located. The sensors respond to the closeness of a point light source by linearly scaling the distance from the light to each sensor ((clutteredSensorMiss)*(1-(distanceToLightSource/diagonalArena))). Distance and time units are of an arbitrary scale. The model includes sensor shadowing when an agent body occludes light. When not otherwise specified, each evaluation consists of a serial presentation of 6 light sources for a relatively long fixed time (T_{ls}=50 time steps) during an agent's lifetime (T=300 time steps). An agent's task is to approach light sources as they appear. After T_{ls}, the light source is eliminated and another one appears at a random distance ([10;120]) and angle ([0;2π]). The intensity of each source is fixed and equal among them. Sensory inputs are on the range [0;1].

Agents are controlled by a continuous-time recurrent neural network (CTRNN). The dynamics of the network are governed by the following equations:

$$\tau_i \frac{dy_i}{dt} = -(y_i - y_0) + \sum_j^n w_{ji} z_j + I_i \qquad\qquad z_j = \sigma(y_j + \theta_j) \qquad (1)$$

Using terms derived from an analogy with real neurons, y_i represents the cell potential of the ith neuron (out of N) depending on a decaying time constant τ_i (scaled exponentially in range [1;2+e^2]), θ_i the bias (calculated by center-crossing), z_i the firing rate, w_{ji} the strength of synaptic connection from node j to node i (range [-10;10]), and I_i the degree of sensory perturbation as an incoming current, which is zero for non-input nodes. CTRNN are implemented using center-crossing (see [2, 5]). The center-crossing restriction helps to prevent the incidence of nearly saturated dynamics that would otherwise nullify the effects of neural noise. Time constants τ_i, sensory gain, and synaptic weight w_{ji} are genetically (real-valued) encoded and optimised using a genetic algorithm. The term y_0 represents the level of additive neural noise as described above.

The network topology consists of 2 motor neurons (#0 & #1), 2 input nodes (#2 & #3), and 2 internal neurons (#4 & #5). Full connectivity is used for connecting neurons, but only output neurons include self-connections. Left/right symmetry in synaptic weights is not enforced. We test the role of internal noise in neural systems by randomly biasing the dynamics of the neurons, considering it for each new evaluation in variable configurations of noise only. Even though the addition of the term y_0 could be simply considered as a perturbation on the current input (I_i), this parameter can be also interpreted as influencing the asymptotic behaviour of each neuron including those that receive no sensory input.

A population of 60 individuals is evolved using a steady state, rank-based (selection) genetic algorithm with elitism (50%). Each individual is run for a number of independent evaluations (10), and the fitness of each phenotype is calculated by averaging the fitness obtained in each evaluation. The mutation operator consists of the addition of a small vector displacement selected from a Gaussian distributed value in each gene (with mean 0.0 and standard deviation 1.0). When mutated genes are over or above their range, a non-reflective criterion is applied, generating a new random value for affected genes. Crossover is not used. The network and other simulation variables are integrated with an Euler time step of 0.1. Fitness is calculated in the following manner: $F = 1-D_f/D_i$, (D_f: final distance to source; D_i: initial distance to source), and it is determined for each light source (goal) and then averaged for the whole evaluation. F is taken as 0 if $D_f > D_i$ and F is in the range [0;1]. The search algorithm is run for a fixed number of generations (1000), for a fixed number of iterations per generation (200), generally taking a few hundred generations to achieve a high level of average fitness.

3   Results

3.1   Evolutionary Robustness

We evolve populations with different levels of neural noise using independent random seeds, with either a variable or fixed y_0 value during evolution (A=0,1,2,3,4). The genetic algorithm is run 1000 generations. Each data point in Fig. 1-*left* represents the average fitness of the best 10 neurocontrollers for 20 independent experiments (6 randomly placed light sources) for different A values. The evolutionary algorithm generates relatively better results for phenotypes using variable instead of fixed values of y_0 (around 20% higher with A=4 noise range). Error bars in Fig. 1-*left* indicate standard deviation. Performance in the presence of sensorimotor disruptions (e.g., sensor inversion, sensors removal, etc.) is also higher for neurocontrollers evolved with A=4 than with lower values of A. As described in our previous work [2], neurocontrollers evolved with A=0 obtain 84.8% of robustness against neural noise, while controllers evolved with A=1, A=2, A=3, and A=4 obtain 86.9%, 96.4%, 98.5%, and 99.7% respectively. These percentages indicate the average robustness based on the level of performance under disruptions over the level of fitness in the control case.

Figure 1-*right* shows the performance of specific neurocontrollers evolved with A=0 and A=4 and achieving high fitness. These neurocontrollers, named NC#9, NC#10, NC#3, and NC#7 were selected because they present the highest or the lowest level of fitness during tests after evolution with A=0 or A=4. NC#9 and NC#10 were evolved with A=4, while NC#7 and NC#3 with A=0. All of these neurocontrollers demonstrate robustness against neural noise except NC#3. General better performance despite neural noise during tests after evolution are obtained when using A=4 for neurocontrollers evolved with A=4 (means: 0.89 for NC#9 and 0.87 for NC#10) than with neurocontrollers evolved with A=0 (means: 0.84 for NC#7 and 0.2 for NC#3) (Fig. 1-*right*). For neurocontrollers evolved with A=4 and when neural noise is applied in tests after evolution, agents continue to be able to move coherently toward light sources. Neurocontrollers evolved with A=4 also maintain this high performance with lower values of A (e.g., A=0 during tests after evolution, Fig. 1-*right*).

In summary, neurocontrollers evolved with A=4 neural noise remain robust to disruptions (e.g., sensorimotor and structural disruptions) even when noise is removed during tests, suggesting that the variability of neural noise helps evolution find regions of higher general robustness in parameter space. Nevertheless, while most phenotypes maintain high levels of fitness in spite of induced variability of y_0 (e.g., NC #9 with A=4 during tests in Fig. 1-*right*), other phenotypes show low level of

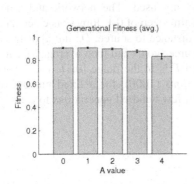 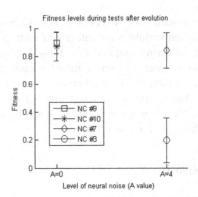

Fig. 1. Evolutionary and phenotypic performance in the presence of variable levels of neural noise (y_0), with randomly positioned 6 lights. *Left*: Fitness values obtained after evolution (avg. of 20 independent experiments), the x-axis is the level of A. *Right*: Comparison between phenotypes evolved with $A=0$ and $A=4$ (x-axis); y-axis is the mean fitness reached by neurocontrollers. Each data point represents the average fitness over 100 independent experiments.

fitness when neural noise is introduced (e.g., NC#3 with $A=4$ during experiments after evolution in Fig. 1-*right*). Understanding why some neurocontrollers perform differently than others under the influence of neural noise can provide clues toward the mechanisms that deal with neural noise.

3.2 Analysis of Results

We analyze here neurocontrollers from Fig. 1-*right* in order to discover the properties which allow robust performance in the presence of neural noise. The analysis focuses both on applying neural noise to single neurons and on the general effects of noise in neurocontrollers. Particularly, neurocontrollers NC#9 and NC#3 are studied in detail because they present robustness (i.e., NC#9) or low performance (i.e., NC#3) during tests after evolution (Fig. 1-*left*). While they represent only particular instances, understanding the difference between these neurocontrollers may shed some light on how evolution works differently in the presence or absence of neural noise.

Most of evolved agents successfully acquired the capacity to perform phototaxis despite neural noise (Fig. 1). As expected, the approaching behaviour of agents is based on maintaining light sensory inputs regardless of neural noise effects, i.e. agents regulate their movements without losing signal from light source. For example, agents depicted in Fig. 2 tend to receive sensory stimulation mainly from one side, which is evidenced in agent's trajectories. The effects of inducing neural noise in neurons (i.e., variable levels noise in range [-4; 4] in neuron #5 and deactivating neuron #4's output), indicate that agents can approach to light or lose the light after turning in the 'wrong direction' and thus sensing it again (Fig. 2). Analysing the asymptotic response of neurocontrollers when inputs are forced to be constantly activated or deactivated for each sensor could show in more detail how different behavioural responses are generated in the presence of neural noise.

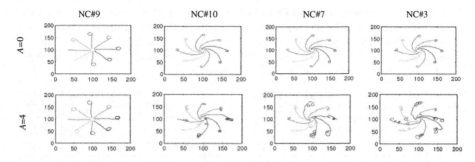

Fig. 2. Examples of behaviours affected by neural noise and neural disruptions. Columns correspond to each neurocontroller indicated at the top of the figure, and each row describes the level of neural noise (A) during tests. Agents start their trajectories to a light source (the small circle at the centre of each figure) from positions separated by 45° on each plot. Top row represents behaviours in normal operation but disabling neuron #4's activity; bottom row represents agent's behaviour after including noise in neuron #5 (neural noise y_0 in [-4; 4]) for each situation in the top row.

Because the noise term y_0 in Eq. 1 is an additive one, this means that nullclines in phase space will tend to be relatively displaced to each other for different values of y_0, but not warped or changed (as would be expected if noise were added to a weight term) [7]. Table 1 shows the asymptotic responses of NC#9 and NC#3 determined by the difference between left and right motor neuron activities (neuron #0's and neuron #1's outputs activations). This difference indicates the action that neurocontrollers generate after inducing different fixed values of y_0 (-4, 0, or 4) in neurons #4 and #5 ($y_0 = 0$ for the remaining neurons). To achieve these results, we forced inputs to be constantly activated (sensor input = 1) or deactivated (sensor input = 0). A positive difference means that agent turns left, while a negative difference that agent turns right in the asymptotic state.

The asymptotic response of neurocontrollers for each sensor and fixed y_0 configuration helps to differentiate the neurocontrollers' strategies for approaching light. Let us first consider NC#9 with no lights (sensor inputs = 0) and without noise in any neuron. In this situation, the agent moves slightly turning right (motor response -0.094) (indicated for (0;0) levels of noise in neurons #4 and #5 in Table 1). In the normal sensing situation, this motion will cause that right sensor to come into contact with the light. Then, the right sensor will start to receive more input, causing the agent to turn left slightly when approaching the light (motor response 0.199)(indicated for (0;1) levels of noise in Table 1). This will produce a decrease of the right sensing input up to a non sensing situation as in the starting condition, because sensor loses contact with the light. After approaching the light, the agent will generate a new movement to right (similar to that described before) generating a new increase of the right input sensing.

This right sensor strategy for approaching light in NC#9 we also observe during normal sensing when the fixed values of y_0 in the interneurons are (0;-4) and (0;4). From an asymptotic perspective, this strategy will cause the right sensor activity to be increased or decreased depending on robot's approach to light and in left and right

Table 1. Asymptotic responses of NC#9's and NC#3's turning behaviours. Table shows the difference between left and right motor activities (motor response), when sensory input nodes #2 and #3 are forced to be constantly activated (sensor input = 1) or deactivated (sensor input = 0) (indicated for each response data); y_0 #4 and y_0 #5 represent fixed levels of neural perturbation (y_0=-4, 0, or 4) in neurons #4 and #5, respectively.

#2	#3	y_0 #4	y_0 #5	NC#9	NC#3
				motor response	
0	0	-4	-4	0.103	0.200
0	0	-4	0	0.170	0.102
0	0	-4	4	0.211	-0.019
0	0	0	-4	-0.114	0.205
0	0	0	0	-0.094	0.108
0	0	0	4	-0.041	-0.012
0	0	4	-4	-0.116	0.210
0	0	4	0	-0.100	0.114
0	0	4	4	-0.075	-0.005
1	1	-4	-4	0.214	0.065
1	1	-4	0	0.233	-0.060
1	1	-4	4	0.242	-0.175
1	1	0	-4	0.214	0.071
1	1	0	0	0.233	-0.053
1	1	0	4	0.242	-0.168
1	1	4	-4	0.214	0.077
1	1	4	0	0.233	-0.046
1	1	4	4	0.242	-0.162

#2	#3	y_0 #4	y_0 #5	NC#9	NC#3
				motor response	
0	1	-4	-4	0.172	0.249
0	1	-4	0	0.202	0.171
0	1	-4	4	0.226	0.064
0	1	0	-4	0.160	0.253
0	1	0	0	0.199	0.176
0	1	0	4	0.225	0.070
0	1	4	-4	-0.099	0.258
0	1	4	0	-0.043	0.182
0	1	4	4	0.128	0.076
1	0	-4	-4	0.200	-0.018
1	0	-4	0	0.225	-0.139
1	0	-4	4	0.238	-0.236
1	0	0	-4	0.200	-0.011
1	0	0	0	0.225	-0.133
1	0	0	4	0.238	-0.230
1	0	4	-4	0.192	-0.004
1	0	4	0	0.223	-0.126
1	0	4	4	0.237	-0.225

responses that depends on right input sensing. However, values of y_0 are changed to (4;0) or (4;4) during normal sensing, the approaching strategy of NC#9 changes toward a left sensor configuration in order to develop phototactic behaviour. In this case, starting from the situation that the agent sees no light, eventually the right sensor will sense first more light producing that the agent moves to right (motor response -0.043) or moves left slightly (motor response 0.128) for (4;0) and (4;4) values for y_0, respectively (Table 1). When the levels of y_0 are (4;0), the agent will move producing a decrease in right sensor input due its approaching angle to the light. The left sensor input will increase generating a left movement (motor response 0.223) instead of a right one as explained before for the right sensor configuration. A similar strategy using the left sensor is observed with (4;4) levels of y_0, but it takes more time for generating such control because the robot turns left more slightly when the right sensor input increases during the approaching behaviour. Therefore, the strategy of NC#9 for maintaining phototaxis behaviour with left of right sensors is based on the combination between motor responses and the approaching behaviour that activate eventually right or left sensors as described before.

As described above, the agent falls into a left or right sensor configuration depending on the value of y_0 in the interneurons. The phototactic behaviour is not demonstrated in other asymptotic configuration of noise during normal sensing approach. For example, adding y_0=-4 to neuron #4, regardless of the level of noise in neuron #5, produces that the agent cannot perform phototaxis because it loses the capacity to turn right and also the ability to maintain right or left sensory inputs (see Table 1). Similarly, inducing y_0=-4 in neuron #5, and y_0=4 in neuron #4 also produces a non-phototactic behaviour, because left movements are not reached correctly during normal sensing approach.

Let us now look at NC#3 defining its input sensing to 0 (meaning that agent sees no lights) and without neural noise. This agent moves turning left (motor response 0.108 indicated for (0;0) levels of noise in Table 1). Eventually, the left sensor will activate first abruptly in normal sensing situations, producing a right movement (motor response -0.133) (see Table 1) that will generate a decrease in left sensor input when agent approaches light. This use of the left sensor is also observed during normal sensing situations when the level of noise in internal neurons are (-4;0) and (4,0). In these cases, phototaxis is performed but not in the noisy configurations. This is the case mainly because the agent loses its capacity for turning left using its left sensor. For example, in (0;-4) y_0 configuration, the agent turns left (motor response 0.205) when it senses no lights, which eventually will produce an increase of left sensing and a slight right movement (motor response -0.011). However, this right movement is not enough to maintain left sensory input while the agent approaches light. This means that eventually the right sensor should become activated, producing a left movement (motor response 0.253). Thus, this also produces a new non-sensing situation with both sensors. According to the asymptotic response of NC#3 in Table 1, the agent sometimes turns in different directions with different levels of noise mainly when both sensors are activated or when no sensing is produced. Thus, the agent will in the end receive inputs from 'the wrong side' causing the agent turns to the wrong direction and cannot perform phototaxis as observed in successful situations.

The asymptotic analysis only provides an indication of how the attractor landscape is affected for different configurations of sensory input and y_0. Observations of the actual transient behaviour for the different configurations indicate that NC#9 performs well in 5 out of 9 cases ((0;0), (0;-4), (0;4), (4;0), and (4;4) values of y_0 in neurons #4 and #5) and NC#3 on 3 out of 9 ((0;0),(-4;0), and (4;0) values of y_0 in neurons #4 and #5). Moreover, NC#9 is able to deploy at least two different behavioural strategies while only one has been observed for NC#3. By taking the values of y_0 investigated as rough representatives of the whole space of variation for y_0, we can conclude that for most levels of noise (but not all) NC#9 will perform phototaxis in a combination of two strategies, but that this is not the case for NC#3. NC#9 is therefore sometimes undergoing bifurcations, but they are most of the time (roughly around two thirds of the time) functional allowing it still to perform phototaxis. Nevertheless, about one third of the time these bifurcations are non-functional as described before. For example, NC#9 is not significatively affected by noise when sensors are simultaneously activated, but it generates wrong long-term responses (e.g., turning left instead of turning right) more frequently than in NC#3 when input sensing are deactivated simultaneously depending on their levels of noise.

The proposed hypothesis is that *those controllers evolved with noise are not undergoing long-term dysfunctional bifurcations because of noise.* In this case, evolution finds networks that operate in regions of phase space for which moderate displacement of the nullclines does not significantly affect the functionality of the system. In fact, because NC#9 has two different strategies for approaching light in the presence of noise, this implies that noise can generate bifurcations but they happen to be also functional, meaning that perturbations in the noise range do not cause qualitative changes to system functionality. Evolution is therefore not only searching for regions of space where bifurcations induced by noise are unlikely to happen, but also for regions where "neighbouring" bifurcations are also functional. As long as the balance of functionality over the possible bifurcations induced by neural noise remains positive, the controller will be able to cope with noise and will moreover be likely to cope with other perturbations (assuming these perturbations induce similar structural changes in the dynamical landscape).

We also observe that noise produces dysfunctional bifurcations in NC#3's dynamics more frequently than in NC#9. NC#9 demonstrates sharp changes in the behaviour of agents, however, indicating that there are still three possibilities to explain how fitness is maintained high despite neural noise. These are: (*I*) those particular bifurcations do not largely affect the transient dynamics of the network; (*II*) all (or most) bifurcations produce different forms of instantaneous phototaxis (they are mostly functional in themselves); (*III*) no bifurcations are produced. These two first possibilities also imply two explanations: (1) negative (non-functional) bifurcation may indeed happen in the range of the noise parameter, but they may be short lived while the agent performs phototaxis; (2) negative (non-functional) bifurcation may occur for significant amounts of time, creating bifurcations that lead asymptotically to non-phototaxis.

In the first case (1), bifurcation during a transient seems to be related with NC#9 because robustness against noise is functionally maintained despite increasing neural noise in most situations. The agent still performs phototaxis because it is held in a transient between attractors that are functional (see [8]). By contrast, NC#3 probably corresponds to case (2) with noise leading to the loss of performance when noise is increased. We have not ruled out transient effects for the situation described for NC#9, however. The pattern of sensor activation and of neural noise may induce bifurcations that are asymptotically non-functional (would not produce phototaxis in the long term) but their change keeps the neural and agent state in a functional transient when $A=0$ during test after evolution.

4   Conclusion

Experiments with neural noise have been presented here from an evolutionary and sensorimotor perspective. The simulation model in itself is minimal but results suggest that, at least in the experimental situations, evolution relies on mechanisms that maintain functional dynamics in transients, as shown for NC#9. Results also indicate that neural systems lose sensitivity to noise when systems are evolved with high levels of neural noise.

From an evolutionary perspective, the interesting lesson is that *neural noise in evolution seems to put pressure for selecting neural systems that are resistant to the effects of bifurcation*, and so their robustness lies in having a dynamic landscape that remains, in the overall balance, functionally the same during behaviour. This is evidenced by the noise robustness of NC#9 and the noise sensitivity of NC#3. The relationship between evolutionary mechanisms selected under noise processes has been minimally investigated in the simulation studies so far. In fact, mechanisms where noise is irrelevant could vary from the simple attractors' view where noise utility is removed because of convergence to stable system dynamics. Our results suggest that the evolutionary process in the presence of neural noise – following the logic of Jakobi's minimal simulations – is finding robust neural dynamics. However, this robustness has a structure. It is a combination of locating the neurocontrollers in regions of parameter space where bifurcations are unlikely to occur and simultaneously where, if and when bifurcations occur, they remain in balance functional. This finding suggests that robustness to other sensorimotor perturbations may be a by-product of locating such regions of parameter space. If this is so, a prediction from this result is that a similar evolutionary process under parametrical uncertainty, but applied to non-additive parameters (such as weight values) may result, if successful, in even higher levels of robustness to sensorimotor perturbations.

In our results, and in accordance to the above explanation, adaptive performance was also observed when noise was removed, indicating that noise is not actively maintaining functionality in the analysed neurocontrollers. Nevertheless, we do not discard the idea that evolution may find solutions for which noise is advantageous, in which case our explanation will need to be appropriately modified. These questions will be further investigated in future work, including comparisons using embodied agents to perform different tasks.

Acknowledgments. Many thanks to Greg Studer and Matthew Egbert for comments of this paper, anonymous reviewers, 'The Peter Carpenter CCNR DPhil Award', and CONICET, Argentina, recognition (Type-I-Res.38-08/01/2004). This research was partially supported by the Programme Alβan, the European Union Programme of High Level Scholarships for Latin America (No. E05D059829AR).

References

1. Kitano, H.: Toward a theory of biological robustness. Mol. Syst. Biol. 3, 137 (2007)
2. Fernandez-Leon, J.A., Di Paolo, E.A.: Neural uncertainty and sensorimotor robustness. In: Almeida e Costa, F., Rocha, L.M., Costa, E., Harvey, I., Coutinho, A. (eds.) ECAL 2007. LNCS (LNAI), vol. 4648, pp. 786–795. Springer, Heidelberg (2007)
3. Bays, P., Wolpert, D.: Computational principles of sensorimotor control that minimize uncertainty and variability. The Journal of Physiology 578(2), 387–396 (2007)
4. Jakobi, N.: Evolutionary robotics and the radical envelope of noise hypothesis. Journal of Adaptive Behaviour 6(2), 325–368 (1997)
5. Mathayomchan, B., Beer, R.: Center-crossing recurrent neural networks for the evolution of rhythmic behavior Source. Neural Comp. 14(9), 2043–2051 (2002)
6. Di Paolo, E.A., Harvey, I.: Decisions and noise: the scope of evolutionary synthesis and dynamical analysis. Adaptive Behavior 11(4), 284–288 (2004)
7. Beer, R.: On the dynamics of small continuous-time recurrent neural networks. Adaptive Behavior 3(4), 469–509 (1995)
8. Iizuka, H., Di Paolo, E.A.: Minimal agency detection of embodied agents. In: Almeida e Costa, F., Rocha, L.M., Costa, E., Harvey, I., Coutinho, A. (eds.) ECAL 2007. LNCS (LNAI), vol. 4648, pp. 485–494. Springer, Heidelberg (2007)

Integration of an Omnidirectional Visual System with the Control Architecture of Psikharpax

Loic Lacheze, Ryad Benosman, and Jean-Arcady Meyer

ISIR, FRE2507, Université Pierre et Marie Curie -Paris 6,
Paris, F-75016, France
{loic.lacheze,jean-arcady.meyer}@lip6.fr
{benosman}@ccr.jussieu.fr

Abstract. This article describes the robotic integration of a robust omnidirectional visual system with a control architecture inspired by neural structures in a rat's brain. The visual system relies on an optimal recursive sampling of images into subimages that remains stable under translation and makes self-localization and object recognition possible. The control architecture affords navigation and action selection capacities. The operationality of both systems is demonstrated through a series of experiments assessing their capacity to maintain the energy level of a robot within the limits of a given viability zone.

Keywords: Sensory-motor coordination, Action selection and behavioral sequencing, Navigation and mapping, Autonomous robotics.

1 Introduction

Previous contributions to the Psikharpax project [13] which aimed at designing a biomimetic artificial rat - were done in simulation [6,7,8] or called upon partial implementations of the robot's control architecture [3,5]. The full sensory-motor implementation of the final Psikharpax platform is still in progress, whereas the various models that will be included in its control architecture are well advanced enough for their complementarity and operationality being worth to be tested in real conditions before final integration.

The work described in this article had two objectives. The first one was to design a robust visual system that can be used in a variety of robotic implementations. The second one was to capitalize on this system to assess the capacity of a rat-inspired control architecture to afford an ad-hoc robotic platform navigation and action selection capacities mandatory for its viability [1].

This article first describes the visual system and control architecture that are used. It then presents the experimental setting in which both were tested, as well as the results that were obtained. It then discusses their applicability to future research efforts.

2 The Visual System

Visual systems allowing navigation and scene recognition can be sorted into two categories: local approaches relying on feature points, like SIFT and Harris [9],

M. Asada et al. (Eds.): SAB 2008, LNAI 5040, pp. 42–51, 2008.

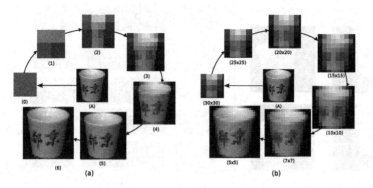

Fig. 1. (a) Optimal sampling of an object A using 7 levels of recursion. (b) Quadtree sampling of the same object. A sharper decomposition of the image is obtained with optimal sampling, as the method aims at preserving information.

and global approaches that consider the whole content of an image, such as histograms [14]. As local approaches are not adapted to the geometry of image transformations occurring with omnidirectional systems, we designed a global approach calling upon an image decomposition that proved to be more robust than those based on traditional quadtrees. Indeed, quadtree algorithms cut an image into four sub images recursively (Fig.1(b)). The *optimal sampling* method that has been used here is based on the same principle, but relies on an entropy measure to divide images which are cut at the location were the difference in the quantity of information characterizing each sub-image is minimal (Fig.1(a)). Information may depend on the application (color, texture, etc) but, in this work, it was measured by the mean grey value of patches. A complete description of the algorithm and its results can be found in [10]. Turns out that the quadtree approach produces results that are not stable under translation. If an object or a location is seen from two slightly translated points of view, the corresponding images will lead to different patches and thus will not be recognized. On the contrary, in the same situation, optimal sampling exhibits a strong stability of decomposition as shown in Fig.2(a). It works even in case of complex backgrounds and with catadioptric omnidirectional images as shown in Fig.2(b). If ng is set to be the mean grey-value of a patch, then the similarity measure between two image locations I_i et I_j is given by:

$$d(I_i, I_j) = \sum_{m=1}^{Nb_{levels}} \sum_{n=1}^{4^m} \|ng_{m,n}^i - ng_{m,n}^j\|$$

with $ng_{a,b}^k$ the mean grey-value of patch I_k, at the level a, placed at the location b. The positions of the different ng in the case of a specific image are given in Fig. 2(b).

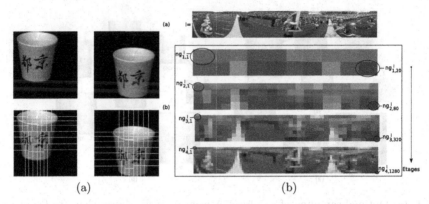

(a)   (b)

Fig. 2. (a) Optimal generation of patches in the case of two images corresponding to a translated object. The patches generated cover the same zones making object recognition possible. (b) Starting from an omnidirectional image, the optimal sampling using 4 steps generates 4 patches sets and most of the initial information is preserved.

3   Control Architecture

The general architecture of Psikharpax (Fig. 3) includes a navigation system that allows a robot to build a *cognitive map* [4,12] making self-localization possible [3], as well as the recording of salient places where resources or potential dangers

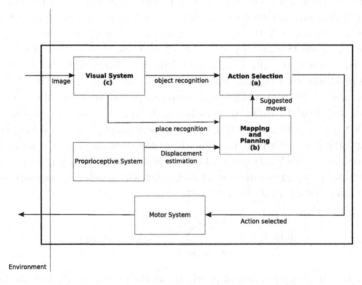

Fig. 3. The control architecture of Psikharpax. The Action Selection module decides which action to execute, according to information provided, on the one side, by the Mapping and Planning module - which may suggest a move towards a given place in the robot's cognitive map - and, on the other side, by the Visual System - which may suggest moving towards a perceived object.

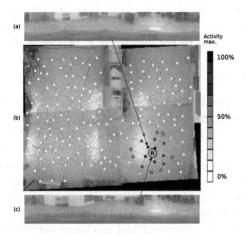

Fig. 4. Example of a cognitive map generated by the robot after exploring its environment (b). A blob of activity in this map indicates the current position of the robot. Two panoramas are shown that respectively correspond to what the robot sees in its current location (c) and to what it previously saw in a nearby location (a).

may be encountered. It also holds an action selection module that selects at every time step the most adapted action ensuring the survival of the artificial rat within its environment [5].

3.1 Navigation System

The navigation system affords two main complementary functionalities: map generation and localization.

Map generation. The system creates and updates a dense topological map relying on a graph where nodes represent locations, with arcs linking each pair of adjacent nodes. Each node stores the optimal decomposition of the omnidirectional image taken at that location, whereas each arc contains the odometric and angular distances between two adjacent nodes.

Localization. The system computes a probabilistic estimate of its current location, using the activities of map nodes combined with information given by the visual system, on the one side, and by odometry and angular data, on the other side. Fig. 4 shows the activity of each node in the map according to the position of the robot.

3.2 Action Selection

To survive, a rat must be able to solve the action-selection problem. Likewise, the robot is innately endowed with an artificial metabolism that imposes it to occasionally find food resources and to return to its nest to be able to digest them [6].

The artificial metabolism. Two essential variables [1] are dealt with in the following experiment, energy (E) and potential energy (E_p). Each action consumes a certain amount of E. When it is close to its nest, the robot can transform part of its E_p in E. To reload E_p, the robot must find in its environment a source of food. The robot is assumed to not survive if E falls to 0. To solve constraints imposed by its metabolism, the action-selection system uses the GPR computational model of basal ganglia described in [5].

Selection action without Navigation. This model (Fig. 5(a)) is implemented as a network of leaky-integrator neurons, and assumes that the numerous segregated channels observed in basal ganglia each correspond to a discrete motor action that is inhibited by default and thus prevented from being executed. Inputs to these channels are so-called *saliences* that take into account both internal and external perceptions to assess the relevance of each action with respect to the robot's needs. Finally, at the output of these circuits, the action that is the least inhibited by others is selected and allowed to be executed by the motor system. In a first series of experiments, the robot had to choose between 5 different behaviors - Digest-in-Nest (E_p becomes E), Eat (increases E_p), Random-Exploration, Go-To-Nest (if visible), Go-To-Food (if visible)- which one to execute at every time step.

Action Selection with Navigation. The action selection model just mentioned has been connected to a navigation model according to recent hypotheses concerning the role of dedicated structures within the basal ganglia, the nucleus accumbens in particular, and the interaction of basal ganglia-thalamus-cortex loops in the rat's brain. The corresponding model is described in [6] and basically involves two such loops (Fig. 5(b)): a ventral loop that selects locomotor actions - like moving north or east - and a dorsal loop that selects non-locomotor

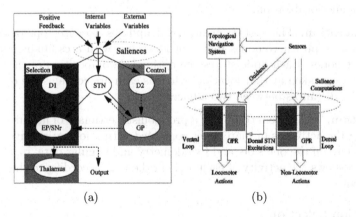

(a)   (b)

Fig. 5. (a) A single channel within the basal ganglia in the GPR model. D1 and D2: striatal neurons with different dopamine receptors; STN: sub-thalamic nucleus; EP/SNr: entopeduncular nucleus and substantia nigra reticula;. GP: globus pallidus. Solid arrows represent excitatory connections, dotted arrows represent inhibitory connections. (b) Interconnection of the ventral and dorsal loops in the basal ganglia. The ventral loop selects locomotor actions, the dorsal loop selects non-locomotor actions. The latter subsumes the former via STN connexions.

actions - like feeding or resting. In a second series of experiments, the robot had a third internal variable that measured its *disorientation* according to the precision of the activity blob that served to determine its position in the map. Depending upon the value of this variable, the robot was urged to return to previously mapped areas when it felt disoriented, or to explore new regions of its environment when it felt both well localized and not lacking energy. Consequently, its ventral loop was endowed with 36 channels - each of them coding for a move in a 10° direction range - , while the dorsal loop was made of 2 channels, one for each type of energy (E or E_p). Hence, the robot had to select the right action among 38.

4   Experiments

The capacity of both the visual system and the control architecture described above to afford the robot survival abilities - according to which it will be able to find food in its environment and digest it in its nest - was tested in a series of experiments including the two mentioned above. The robot was equipped with an omnidirectional catadioptric sensor in an arena observed by four ceiling-mounted cameras. This system allowed an accurate monitoring of the robot's positions and orientations, which were recognized according to colored marks on its roof (Fig 6(b)). Fig 6(a) shows an on-line reconstruction of the scene, when the four camera views are merged in a single one.

Visual system. Visual localization using the optimal sampling procedure was compared with SIFT [11] and with an histogram-based method. In order to check the robustness of each approach, and to study the precision of the corresponding localization, several tests with additive noise were carried out. The acquired images were transformed into cylindrical images of size 1400×140 pixels. The performance of the visual system was assessed in a complete navigation task, where the robot had to explore an unknown environment and to build a cognitive map hopefully making an accurate self-localization possible. Fig. 7(a) shows

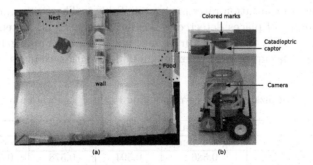

Fig. 6. Upper view of the monitorized arena observed by four camera mounted in the ceiling (a). Close view of the robot (b). The colored marks on the roof serve to monitor the robot's orientation and position.

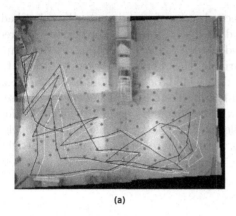
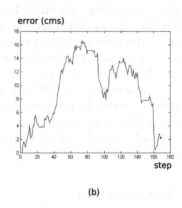

(a)   (b)

Fig. 7. (a) Upper view of the arena showing the estimated trajectory (white) of the robot according to its map (white)) and its actual trajectory (black). (b) The error between the real and estimated position: it is always inferior to the robot's diameter, i.e., 30cm.

the estimated and the real trajectories of the robot, while Fig. 7(b) provides the corresponding localization errors when the optimal sampling approach was used. The nodes were created every 20cm and the robot estimated its position as the barycenter of map nodes' activities. In order to compare the optimal sampling approach with others, the acquired omnidirectional images were modified according to different scenarios:

- Virtual occlusions were added (from 1 to 10 squares) at random positions, their individual size never exceeding 10% of the size of the original image.
- Independent noise was added having a uniform distribution between 0 and 255 and concerning 10% of the maximum number of pixels.
- To simulate illumination variations, a constant noise was added to each pixel in the image. This noise value was chosen randomly in a grey-value interval of 0 to 40.

The evolution of localization rates is given in Table 1. These rates were assessed by the percentage of places detected as similar by the visual system that were actually neighbors in the map. Turns out that optimal sampling leads to the most stable and accurate results, except for illumination changes, which is an

Table 1. Localization rates corresponding to three localization methods

	Localization rate			
	Original images	White noise	Illumination	Occlusions
SIFT	0.586	0.591	0.578	0.558
Histogram	0.395	0.398	0.146	0.326
Optimal (5th level)	0.758	0.781	0.223	0.746

Table 2. Robustness of localization : Effect of orientation accuracy on localization

Robustness of localization					
angles errors	0° (no noise)	5°	10°	15°	20°
Localization rates	0.7261	0.6546	0.533	0.4496	0.4182

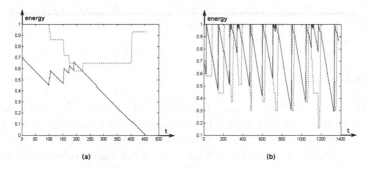

(a)   (b)

Fig. 8. Energy management when map-building is prohibited (a) or allowed (b). Plain lines correspond to energy E, dashed lines correspond to potential energy E_p.

expected result as the method is relying on the mean grey-value of patches. To overcome this limitation, the method could rely on textures to get more stable results, but this approach is likely to be slightly more time-consuming (see [10] for details). Because the optimal sampling procedure uses omnidirectional images that are resampled and transformed into cylindric images, its performance in case of orientation errors was assessed. Several tests were carried out in which various orientation errors were introduced. The corresponding results are given in table 2, and turn out to be very stable, even in case of large errors.

Autonomy. To assess the robot's survival capacities, two experiments have been done that reproduce in reality the simulation settings of [5,6]. In both conditions, the robot had to manage its E and E_p levels to avoid dying from starvation but, in the first case, it relied on mere chance to find food - because it did not use any map - whereas, in the second case, it could build such a map to increase its chances of survival. The corresponding results are given in Fig. 8. It appears that energy is managed in a better way in (b) as it always reaches its maximum each time it is possible, contrary to what is shown in (a). During 6 experiments where map-building was prohibited, the robot survived an average of 848 time steps. During 6 experiments where map-building was allowed, the robot always survived beyond 1800 time steps.

5   Discussion

The optimal sampling procedure that has been used here proved to be applicable to both classical and omnidirectional cameras. It is robust to noise and occlusions, and leads to better results than SIFT and an histogram-based approaches.

Moreover, it is low time consuming compared to SIFT, because its non-optimized program was able to process 5-10 frames/sec on a P4M 2:20GHz/512Mb. As for the autonomy capacities that the control architecture afforded an ad-hoc robot, they may certainly be improved using additional perceptual modalities that are currently implemented on the Psikharpax platform, i.e., 2 moving eyes, 2 moving ears, 2 whisker arrays, a more accurate odometry system, and accelerometers generating vestibular data. In particular, the binocular system relying on wide-angle cameras will provide an omnidirectional coverage with a common front zone that will allow depth retrieval.

6   Conclusions

This article introduced a new visual module that ensures robust perception of scenes based on an optimal sampling of images. Connected with a biomimetic control architecture, it has been used to demonstrate the viability capacities it afforded an ad-hoc robot. Results obtained so far are likely to be improved when new sensors and new controllers will be implemented on the future Psikharpax platform.

Acknowledgement

This research was granted by the EC FP6 integrated Project ICEA (Integrating Emotion, Cognition and Autonomy). The authors would like to thank David Filliat and Benoit Girard for their permission to capitalize on their previous contribution to the Psikharpax project.

References

1. Ashby, R.: Design for a brain, London. Chapman & Hall, Boca Raton (1960)
2. Degris, T., Lacheze, L., Boucheny, C., Arleo, A.: A Spiking Neuron Model of Head-Direction Cells for Robot Orientation. In: Schaal, et al. (eds.) From Animals to Animats: Proceedings of the Eighth International Conference on Simulation of Adaptive Behavior, pp. 255–263. MIT Press-Bradford Books, Cambridge (2004)
3. Filliat, D., Meyer, J.A.: Global localization and topological map learning for robot navigation. In: Hallam, et al. (eds.) From Animals to Animats 7: Proceedings of the Seventh International Conference on Simulation of Adaptive Behavior. MIT Press, Cambridge (2002)
4. Filliat, D., Meyer, J.-A.: Map-based navigation in mobile robots - I. a review of localisation strategies. Journal of Cognitive Systems Research 4(4), 243–282 (2003)
5. Girard, B., Cuzin, V., Guillot, A., Gurney, K., Prescott, T.: A basal ganglia inspired model of action selection evaluated in a robotic survival task. Journal of Integrative Neuroscience 2(3), 179–200 (2003)
6. Girard, B., Filliat, D., Meyer, J.-A., Berthoz, A., Guillot, A.: An integration of two control architectures of action selection and navigation inspired by neural circuits in the vertebrates: The basal ganglia. In: Bowman, H., Labiouse, C. (eds.) Connectionist Models of Cognition and Perception II, Proceedings of the Eighth Neural Computation and Psychology Workshop, pp. 72–81. World Scientific, Singapore (2004)

7. Khamassi, M., Lacheze, L., Girard, B., Berthoz, A., Guillot, A.: Actor-critic models of reinforcement learning in the basal ganglia: From natural to artificial rats. Adaptive Behavior, Special Issue Towards Artificial Rodents 13(2), 131–148 (2005)

8. Khamassi, M., Martinet, L.-E., Guillot, A.: Combining self-organizing maps with mixture of experts: Application to an actor-critic of reinforcement learning in the basal ganglia. In: Nolfi, n Hallam, et al. (eds.) From Animals to Animats 9: Proceedings of the 9th International Conference on the Simulation of Adaptive Behavior (SAB), Rome, Italy, pp. 394–405 (2006)

9. Kosecka, J., Li, F., Yang, W.: Global Localization and Relative Positionning Based on scale Invariant Keypoints. Robotics and Autonomous Systems 52(1), 27–38 (2005)

10. Lacheze, L., Benosman, R.: Visual localization using an optimal sampling of Bags-of-words with entropy. In: International Conference On Intelligent Robots and Systems (IROS), San Diego, pp. 1332–1338 (2007)

11. Lowe, D.: Distinctive image features from scale-invariant keypoints. In: International Journal of Computer Vision (IJCV), pp. 91–110 (2004)

12. Meyer, J.-A., Filliat, D.: Map-based navigation in mobile robots - II. a review of map-learning and path-planing strategies. Journal of Cognitive Systems Research 4(4), 283–317 (2003)

13. Meyer, J.-A., Guillot, A., Girard, B., Khamassi, M., Pirim, P., Berthoz, A.: The Psikharpax Project: Towards Building an Artificial Rat. Robotics and Autonomous Systems 50, 211–223 (2005)

14. Ulrich, I., Nourbakhsh, I.: Appearance Based Place Recognition for topological Localization. In: Proceedings of the IEEE International Conference on Robotics and Automation, pp. 1023–1029 (2000)

Stability of Coordination Requires Mutuality of Interaction in a Model of Embodied Agents

Tom Froese and Ezequiel A. Di Paolo

Centre for Computational Neuroscience and Robotics (CCNR),
Dept. of Informatics, University of Sussex, UK
{t.froese,ezequiel}@sussex.ac.uk

Abstract. We used an evolutionary robotics methodology to generate pairs of simulated agents capable of reliably establishing and maintaining a coordination pattern under noisy conditions. Unlike previous related work, agents were only evolved for this ability and not for their capacity to discriminate social contingency (i.e., a live responsive partner) from non-contingent engagements (i.e., a recording). However, when they were made to interact with a recording of their partner made during a successful previous interaction, the coordination pattern could not be established. An analysis of the system's underlying dynamics revealed (i) that stability of the coordination pattern requires ongoing mutuality of interaction, and (ii) that the interaction process is not only constituted by, but also constitutive of, individual behavior. We suggest that this stability of coordination is a general property of a certain class of interactively coupled dynamical systems, and conclude that psychological explanations of an individual's sensitivity to social contingency need to take into account the role of the interaction process.

Keywords: evolutionary robotics, social cognition, social contingency.

1 Introduction

Evolutionary robotics is typically employed to investigate simulation models of minimally cognitive behavior, namely the simplest behavior that raises issues of genuine cognitive interest [1, 8, 7]. Recently there have been some initial efforts to extend this methodology into the domain of social cognition [4, 9, 10, 11, 3, 14]. While these are important advances within the field of artificial life, they also generate insights of relevance that could form the basis for mutually informing collaborations with the empirical sciences such as psychology.

One promising target for such an endeavor is Murray and Trevarthen's [12] double TV monitor experiment. In this psychological study 2 month old infants were animated by their mothers to engage in coordination via a live double video link. However, when the live video of the mother was replaced with a video playback of her actions recorded previously, the infants became distressed or removed. These results, and those of a more rigorous follow-up study by Nadel and colleagues [13], indicate that 2 month old infants are sensitive to social contingency, i.e. the mutual responsiveness during an ongoing interaction, and that this sensitivity plays a

M. Asada et al. (Eds.): SAB 2008, LNAI 5040, pp. 52–61, 2008.

fundamental role in the unfolding of coordination. Traditional explanations of this sensitivity have focused on innate abilities. For example, Gergely and Watson [6] have postulated the presence of an innate cognitive module which enables the detection of social contingency, and Russell [15] hypothesizes that infants have an innate capacity to understand intentionality and to process agency.

Are these postulations of innate capacities on the part of the infant necessary in order to explain the empirical results? Iizuka and Di Paolo [9] used an evolutionary robotics approach to test whether simpler solutions could also emerge from the dynamics of the interaction process itself. In their simulation model the evolved agents successfully acquired the capacity to discriminate between 'live' (two-way) and 'recorded' (one-way) interaction. Moreover, an analysis of the resulting dynamics suggests that the interaction process itself plays an important role in enabling this behavior. Similar results were also found by the other simulation studies [4, 11, 10].

It could be argued that the result of Iizuka and Di Paolo's [9] simulation study only represents a specific subset of the general solution space, in particular because they used evolutionary robotics to *explicitly* generate agents that terminate interaction when there is a lack of social contingency. We address this issue by testing *whether termination of interaction emerges under more general conditions.* Answering this question is important if the argument is made that these findings might apply more generally and in particular to human interactions. By changing the simulation setup in this manner we thus move closer to the original double TV monitor experiment: the infants presumably did not have the specific goal to detect whether they were dealing with a live video or just a recording. It is more likely that they were simply attempting to establish social coordination with their mothers but were unable to do so.

2 Methods

We implemented a minimal simulation model analogous to Murray and Trevarthen's [12] double TV monitor experiment by building on work by Iizuka and Di Paolo [9]. A schematic of this simulation model is illustrated in Fig. 1.

The goal of the agents is to cross their sensors as far away from their starting positions as possible, a task which requires mutual localization, convergence on a target direction, and movement in that direction while not losing track of each other. This task is non-trivial since sensory stimulation only correlates with the overlapping of position (when the centers of the agents are less than 20 units of space apart); it does not convey the direction or speed of movement of the other agent.

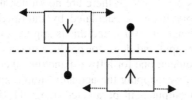

Fig. 1. A schematic view of the model adapted from Iizuka and Di Paolo [9]. The two identical agents are 40 units wide, only able to move in a horizontal direction, and equipped with a single on/off sensor at their centre. They face each other in an unlimited continuous 1-D space.

The agents are controlled by two identical continuous-time recurrent neural networks (CTRNNs), as described by Beer [1]. They were chosen to be clones because work by Iizuka and Ikegami [10] on a related task suggests that genetically similar agents are potentially better at coordination. They face each other in an unlimited continuous 1-D space (i.e. one agent faces 'up' and one agent faces 'down'). Distance and time units are of an arbitrary scale. Each agent can only move horizontally. One on/off sensor is located in the centre of each agent. The sensor is activated (set to 1) when the agents cross each other, otherwise it is set to 0.

Noise is introduced into the simulation for 2 main reasons: (i) since the agents are identical they will need to make use of noise in order to break the symmetry of their movements and converge on a common target direction, and (ii) robustness against noise increases the ability of 'live' agents to cope with playback situations [10]. Accordingly, at each Euler time step there is a 5% probability that the current sensory state is flipped into its opposite state. We add a small perturbation to the motor outputs at each time step drawn from a Gaussian distribution ($\mu = 0$; $\sigma^2 = 0.05$). The noise is applied to the outputs before the application of motor gains.

In order to further increase the robustness of the behavioral strategies, the initial relative displacement between the agents varies (range [-25, 25]). Starting from any of these possible relative positions, the task for the agents is to coordinate their behavior such that they cross each other as far away from position 0 as possible. Since the agents are started in opposite orientation ('up' vs. 'down'), it is not possible for the evolutionary algorithm to hard code any trivial solution (e.g. 'always move left').

2.1 Agents

The agents are 40 units wide, have an on/off sensor at their center, and can only move left or right by controlling the output of their left and right motor nodes (see Fig. 1). Agents are controlled by a CTRNN consisting of 3 fully-connected nodes with self-connections. The time evolution of the node activation follows:

$$\tau_i \dot{y}_i = -y_i + \sum_{j=1}^{N} w_{ji} z_j(y_j) + SI_i \ , \qquad z_i(x) = 1/(1 + e^{-x-b_i}) \tag{1}$$

In this equation y_i represents the cell potential of node i, z_i is the firing rate as calculated by the standard sigmoid function, τ_i (range [1, 100]) is its time constant, b_i (range [-3, 3]) is a bias term, and w_{ji} (range [-8, 8]) is the strength of the connection from the neuron j to i. I_i represents the sensory input to node I and S is the sensor gain. The total number of nodes N is set to 3; there are no hidden nodes (all nodes receive sensory input). The sensory input is calculated by multiplying 1/0 (on/off) by an sensor gain parameter S (range [1, 100]), and this is applied to all nodes. There is one node, which only receives input and does not produce motor output and two actuator nodes for controlling movement; one for leftward and the other for rightward velocity. Each velocity is calculated by mapping the actuator output onto the range [-1, 1] and then multiplying it by an output gain parameter (range [1, 50]). The overall agent velocity is calculated as the difference between the left and right velocities. The time evolution of the simulation environment and each agent's CTRNN controller is calculated by using Euler integration with a time step of 0.1.

Similar settings have already been successfully used by Iizuka and Di Paolo [9]. The main differences are that (i) the agents of the current study only have 3 nodes, (ii) the input is fed to all nodes instead of one dedicated sensory node, (iii) and each actuator node has its own gain parameter. The first difference was chosen to further minimize the conditions of the model and facilitate analysis; differences (ii) and (iii) were implemented to increase the evolvability of the solutions.

2.2 Evolutionary Algorithm

The agents are optimized by using a simple genetic algorithm (GA) which is based on the microbial GA, a steady-state GA with tournament selection [7]. Until some termination criterion is reached, two members of the population are chosen at random, both have their fitness evaluated, and while the 'winner' of the tournament remains unchanged in the population, the 'loser' is replaced by a slightly mutated copy of the 'winner'. Each member is a clonal pair of agents. We define a generation as the number of tournaments required to generate a number of offspring equal to the population size. The population size is 40 and the run terminates at 5000 generations.

All CTRNN parameters and gains are genetically encoded by a real-valued vector which is initialized randomly. The mutation operator changes each gene by a random value drawn from a Gaussian distribution ($\mu = 0$; $\sigma^2 = 0.05$) with reflection at the gene boundaries. Before evaluation, each gene is decoded linearly to the corresponding range (except gains and time constants which are exponentially scaled).

During each fitness evaluation an agent is tested in 15 trials runs; to increase the robustness of the evolving solutions to noise and variations in initial conditions only the lowest score achieved in any of the trials is chosen as the overall score. Each trial run consists of 50 units of time (500 Euler time steps). At the start of each trial agents have their internal node activations set to small random values drawn from a standard Gaussian distribution. The initial distance between the agents varies; agent 'down' always gets placed at position 0, while agent 'up' starts at a different position for each trial (15 different positions evenly distributed across range [-25, 25]).

The fitness score of a trial run is calculated on the basis of a single factor, namely the absolute value of the final crossing position of the two agents divided by a factor of 10. Thus, in contrast to the work done by Iizuka and Di Paolo [9], these agents were not evolved to break off the interaction pattern when detecting a lack of social contingency. Instead, we aimed to generate a simulation model that under normal circumstances results in highly fit coordination behavior. Presumably, such behavior should be more robust when faced with the 'playback' condition.

3   Results

The GA was run 4 times. The fittest agent, with a score of 244.8, was produced during the 4th run in generation 3477. This solution was then tested extensively; agent 'down' was always placed at position 0, while agent 'up' starts at a different position for each trial (101 positions evenly distributed across range [-50, 50]). Each trial is repeated 150 times. The mean score across this range of initial conditions is plotted in Fig. 2 (left). The agents are able to generalize their behavior well beyond the range

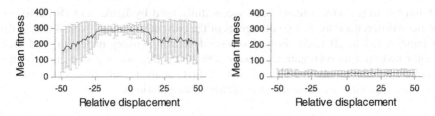

Fig. 2. *Left:* Mean score achieved by the fittest agent starting from various initial positions, with standard deviation. *Right:* Mean score by the fittest agent but this time interacting with non-responsive, recorded movements obtained from the original trials.

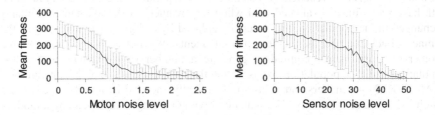

Fig. 3. Robustness to noise: mean fitness score achieved over 150 trials by the fittest evolved agent starting from position 11 for a range of noise levels, with standard deviation. Original noise strength during evolution is 0.05 for motor (left) and 5% for sensor noise (right).

that they were originally evolved to cope with. On average the best initial position for agent 'up' turned out to be at 11 (mean score: 292.9).

In order to demonstrate the general robustness of the evolved agents under this initial condition, we ran another set of trials with agent 'up' starting from position 11 while varying noise levels. The motor noise was varied while the sensor noise remained constant at evolutionary strength (5%), and sensor noise was varied while motor noise remained constant ($\sigma^2 = 0.05$). At each noise level we tested the agents for 150 trials. As shown in Fig. 3, the agents are able to cope with a wide range of perturbations. Indeed, their overall performance degrades gracefully until the sensor and motor signals are completely swamped by noise. In the case of sensor noise, for example, average performance only approaches 0 just before reaching the 50% mark (at which point sensory activation becomes completely arbitrary). This demonstrates that the agents are able to produce highly robust coordination behavior.

Finally, another 150 trials were conducted with agent 'up' at position 11 (under normal noise conditions). The movement of agent 'down' during the best trial (score: 321) was recorded for playback. Another 150 trials were then run under *playback* conditions: the initial conditions reflect those of the recorded best trial run (agent 'up' always starts at position 11 and with the same initial internal activation), and the movement of agent 'down' replicate those which it produced during the recording. While the sensorimotor noise for agent 'up' was different during each of these trials, no additional noise was added to the recorded movement of agent 'down'.

The results are striking: whereas the original 150 trials of mutual (two-way) interaction were highly successful (mean score: 268), the 150 trials of playback (one-way) interaction were a drastic failure (mean score: 19). The severity of this failure is surprising since under normal conditions the active agent is robust against various forms of noise, and able to cope effectively with a wide range of initial conditions. Moreover, during the playback condition its 'partner' performs what had previously been a highly fit behavioral repertoire. Still, the active agent is unable to adapt to the situation of interacting with a non-responsive 'partner'. It could be argued that this result is unique to the chosen situation. However, this is not the case: when testing agent 'up' with each of the original trials we get the same result (see Fig. 2, right).

3.1 Behavioral Analysis

The behavior of the agents under normal conditions can be broken down conceptually into three important aspects: (i) localization, (ii) alignment, and (iii) coordination. We will briefly discuss the first two aspects and then focus on the third. The activity during the first time steps of the best trial run is shown in Fig. 4.

Initially the agents have no knowledge of how their own position relates to that of their partner. Moreover, they have no way of gaining that information except when changing their sensory input by engaging in movement. However, it turns out that one stereotypical behavioral pattern is sufficient to solve the non-trivial problem of reliable localization. First, each agent moves rightwards for a few units of time, and then starts moving leftwards. This sweeping behavior usually takes up to 5 units of time and under evolved conditions always enables the agents to locate each other. In the case of negative initial displacement they will encounter each other during their rightward sweep; otherwise they will cross their positions during their leftward return. Interestingly, the agents always end up with positive *relative* displacement after their initial localization. With this clever maneuver the agents have significantly reduced the complexity of their coordination task: while sensory input is ambiguous (there is

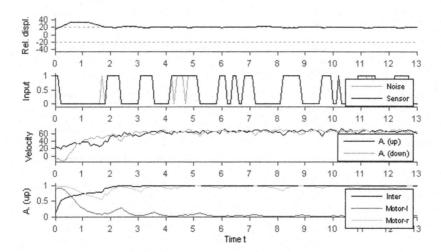

Fig. 4. Initial activity of the two agents during the best trial run. From top to bottom the traces show the evolution over time of (i) their relative displacement, (ii) their noisy input signal and actual sensory contact, (iii) their velocity, and (iv) the CTRNN node outputs of agent 'up'.

no indication about the direction or speed of the other agent's movement), it has now been *co-arranged* as a 'touching on the left' indicator! This change of the sensory meaning is possible because the CTRNN controllers are not symmetric.

How does the final oscillatory coordination pattern emerge out of the relative movements of the agents? Before analyzing the behavior of the agents in more detail it is necessary to briefly describe the evolved CTRNN controller. Most importantly, the sensory input excites all of the nodes with a gain of $S = 10.9$, and the right output gain (44.5) is almost twice as high as the left output gain (24.9). The two motor nodes are inhibited by the non-motor node and they also inhibit each other while hardly affecting the non-motor node. As an example, we can see that the output of the right motor node (z_3) of agent 'up' starts to slightly decrease just before time $t = 8$, due to lack of sensory stimulation. This shift in velocity entails that agent 'down' catches up with agent 'up' and they remain in contact ($I_i = 1$) until just before $t = 9$. During this contact agent 'up' regains its previous rightward velocity due to sensory stimulation. After separating again ($I_i = 0$) the firing of the left motor node goes down followed by the right motor node which eventually leads to the behavioral pattern being reinitiated. Accordingly, agent 'up' should be able to engage with a playback recording. The activity during the *playback* trial run is shown in Fig. 5.

At first the 'live' agent aligns itself with the 'playback' agent as in the original situation (Fig. 4). During mutual (two-way) interaction agent 'down' would always respond to contact by moving away slightly. However, in the playback situation this co-regulation is prevented from occurring. Accordingly, every encounter results in a slight decrease of relative displacement between the two agents, thereby in turn making it more likely that there will be another sensory stimulation. Up to about $t = 3$, agent 'up' is still able to partially regulate this displacement on its own by adjusting the output of its right motor node. However, from that point onwards the right motor node saturates at $z_3 = 1$, and thereafter remains unaffected by further sensory

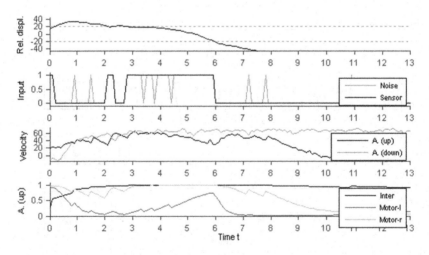

Fig. 5. Initial activity during a *playback* trial run in which the movements of agent 'down' are the same as in Fig. 4. From top to bottom the traces show the evolution over time of (i) the relative displacement between the agents, (ii) their noisy input signal and the actual moments of sensory contact, (iii) their velocities, and (iv) the CTRNN node outputs of agent 'up'.

stimulation. Finally, at around $t = 6$ the positive feedback loop between increasing sensory stimulation and mounting leftward velocity becomes unstable in such a way that recovery from breakdown is impossible. The live agent falls behind the playback agent, and heads in the opposite direction.

Why does this breakdown of coordination not occur when both agents engage in 'live' interaction? The simple answer provided by this model is that the *stability of ongoing coordination requires mutuality of interaction*. After the initial alignment we find that coordinated movement in one direction consists of continuous co-regulated oscillatory behavior. Agents control their respective velocities such that they cross their sensors at relatively regular intervals. This iterative reaction chain constitutes an ongoing pattern of turn-taking; noise perturbations get amplified in a way that requires continuous co-regulated re-establishment of the interaction [11].

3.2 Dynamical Analysis

Can we account for the oscillating pattern in dynamical terms? Since the output of the non-motor node $z_1(y_1) = 1$ during coordination, it can be treated as a fixed parameter. The rest of the system consists only of the two motor nodes. The parameters are $\tau_2 = 1.6$, $b_2 = 2.6$, $w_{12} = -3.7$, $w_{22} = 1.0$, $w_{32} = -7.9$, and $\tau_3 = 1.1$, $b_3 = 2.9$, $w_{13} = -5.8$, $w_{23} = -5.8$, $w_{33} = 2.3$. If agents are not in contact with each other ($I_i = 0$), there is a globally attracting stable equilibrium point at (-3.4, -7.5). Being in this state effectively slows down rightward velocity. Because of this the agents eventually make contact. When $I_i = 1$ the equilibrium point is shifted to (0.3, 1.9). This effectively speeds up the rightward velocity of the agent.

Interestingly, under normal conditions the dynamical system never reaches either of these two equilibrium points, because their existence is made transitory through the ongoing interaction. This is illustrated in Fig. 6 (left) in terms of the motor node firing rates for agent 'up' over a whole run (50 units of time). The trajectory settles down into an oscillatory pattern that traces the corner near point (0, 1), in the middle of the two equilibrium points (located at (0.95, 1) when $I_i = 1$, and at (0.30, 0) when $I_i = 0$). The state trajectory for the playback situation of the same run is displayed in Fig. 6 (right). At first the trajectory moves into the same region of state space but then, during the period of prolonged contact, the left motor node gets saturated while the right motor node remains at 1. This continues until the system almost reaches the equilibrium point at (0.95, 1), but it eventually causes agent 'up' to slow down too much thereby breaking out of the coordination pattern.

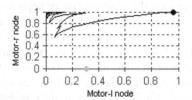   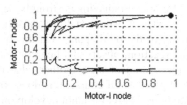

Fig. 6. State trajectory of the outputs for the 2 motor nodes of agent 'up' during 50 units of time. The trace starts at the top right corner of each graph. The gray and black dot represent the globally attracting stable equilibrium point when sensory input $I = 0$ and $I = 1$, respectively. *Left:* mutual (two-way) interaction. *Right:* playback (one-way) interaction.

After agent 'up' slows down enough such that the playback movement of agent 'down' overtakes it, its input I_i stays at 0. This causes its motor system to settle near the equilibrium point at (0.30, 0), from where it is occasionally perturbed by sensor noise. Thus, without the responsive help of the other agent, agent 'up' is unable to regulate its behavior such as to avoid falling into this attractor, an event which limits its further behavior to mere leftward movement.

4 Concluding Remarks

With our simulation study it was found that stable and robust coordination can be reliably established between simulated agents. While the agents were only selected on the basis of this coordination ability (rather than their capacity to detect social contingency), coordination still breaks down when a 'live' agent is forced to interact with a playback of movements from a previous, successful trial. Agents interacting with such a non-responsive 'partner' do not have the capacity to generate and sustain the kind of oscillatory behavior necessary for coordination. Thus, what at first appears to be a *behavioral capacity of the individual agent* emerges out of a combination of the internal dynamics *as well as the interaction process*. This role of the interaction process is also consistently demonstrated in previous models [9, 4, 11, 3].

We are thus faced with a peculiar situation in which the behavior of the individual agents brings forth the interaction process, and that interaction process enables the behavior of the individual agents. This makes a reduction of the coordination breakdown to an individual agent's capacity to detect social contingency impossible. Moreover, it points to the *autonomy* of the interaction process, as postulated by an enactive approach to social cognition [2]. A more detailed analysis of the dynamics of the interaction process in this context is desirable, especially in terms of an artificial life investigation of the systemic basis of "constitutive autonomy" [5].

It is worth emphasizing that we do not claim that our model instantiates the phenomenon which we are investigating or that the baby-mother interaction studied by Murray and Trevarthen [12] is reducible to such a simple system. The model is purely conceptual in that it shows at work a possible explanation that may later be considered and tested in specific empirical cases. Thus, by generating simple models which do not presuppose the methodological individualism which prevails in social cognitive science and psychology, we can re-conceptualize the space of possible explanations [4]. In particular, the model presented in this paper suggests that the capacity for social behavior is strongly dependent on the existence of an appropriate social context, one whose stability is in turn dependent on the active and responsive engagement of the participants. On this basis we propose that an explanation for the distressed reaction, observed when confronting infants with a video recording rather than a live stream of their mother, also needs to take into account the role of the interaction process. Of course, this does not mean that the infants cannot detect social contingency or that they cannot develop this ability, but it does open up the possibility for explanations that do not suppose any necessity for innate behavioral capabilities and/or a complex perceptual strategy on the part of the infant.

References

1. Beer, R.D.: The dynamics of adaptive behavior: A research program. Robotics and Autonomous Systems 20, 257–289 (1997)
2. De Jaegher, H., Di Paolo, E.A.: Participatory sense-making: An enactive approach to social cognition. Phenomenology and the Cognitive Sciences 6, 485–507 (2007)
3. Di Paolo, E.A.: Behavioral coordination, structural congruence and entrainment in a simulation of acoustically coupled agents. Adaptive Behavior 8, 25–46 (2000)
4. Di Paolo, E.A., Rohde, M., Iizuka, H.: Sensitivity to social contingency or stability of interaction? Modelling the dynamics of perceptual crossing. New Ideas in Psychology (in press)
5. Froese, T., Virgo, N., Izquierdo, E.: Autonomy: a review and a reappraisal. In: Almeida e Costa, F., et al. (eds.) ECAL 2007. LNCS (LNAI), vol. 4648, pp. 455–464. Springer, Heidelberg (2007)
6. Gergely, G., Watson, J.: Early social-emotional development: contingency perception and the social biofeedback model. In: Rochat, P. (ed.) Early social cognition: Understanding others in the first months of life, pp. 101–137. Lawrence Erlbaum, Hillsdale (1999)
7. Harvey, I.: Artificial Evolution: A Continuing SAGA. In: Gomi, T. (ed.) ER-EvoRob 2001. LNCS, vol. 2217, pp. 94–109. Springer, Heidelberg (2001)
8. Harvey, I., Di Paolo, E.A., Wood, R., Quinn, M., Tuci, E.A.: Evolutionary Robotics: A new scientific tool for studying cognition. Artificial Life 11, 79–98 (2005)
9. Iizuka, H., Di Paolo, E.A.: Minimal Agency Detection of Embodied Agents. In: Almeida e Costa, F., et al. (eds.) ECAL 2007. LNCS (LNAI), vol. 4648, pp. 485–494. Springer, Heidelberg (2007)
10. Iizuka, H., Ikegami, T.: Adaptability and diversity in simulated turn-taking behavior. Artificial Life 10, 361–378 (2004)
11. Ikegami, T., Iizuka, H.: Turn-taking interaction as a cooperative and co-creative process. Infant Behavior and Development 30, 278–288 (2007)
12. Murray, L., Trevarthen, C.: Emotional regulations of interactions between two-month-olds and their mothers. In: Field, T.M., Fox, N.A. (eds.) Social perception in infants, pp. 177–197. Ablex Publishing, Norwood (1985)
13. Nadel, J., Carchon, I., Kervella, C., Marcelli, D., Réserbat-Plantey, D.: Expectancies for social contingency in 2-month-olds. Developmental Science 2, 164–173 (1999)
14. Quinn, M.: Evolving communication without dedicated communication channels. In: Kelemen, J., Sosík, P. (eds.) ECAL 2001. LNCS (LNAI), vol. 2159, pp. 357–366. Springer, Heidelberg (2001)
15. Russel, J.: Agency: Its role in mental development. Taylor & Francis, Hove (1996)

Internal and External Memory in Neuroevolution for Learning in Non-stationary Problems*

Francisco Bellas, Jose A. Becerra, and Richard J. Duro

Integrated Group for Engineering Research,
Universidade da Coruña, Spain
{fran,ronin,richard}@udc.es
http://www.gii.udc.es

Abstract. This paper deals with the topic of learning through neuroevolutionary algorithms in non-stationary settings. This kind of algorithms that evolve the parameters and/or the topology of a population of Artificial Neural Networks have provided successful results in optimization problems in stationary settings. Their application to non-stationary problems, that is, problems that involve changes in the objective function, still requires more research. In this paper we address the problem through the integration of implicit, internal or genotypic, memory structures and external explicit memories in an algorithm called Promoter Based Genetic Algorithm with External Memory (PBGA-EM). The capabilities introduced in a simple genetic algorithm by these two elements are shown on different tests where the objective function of a problem is changed in an unpredictable manner.

Keywords: evolutionary algorithms, artificial neural networks, learning, non-stationary problems, autonomous robotics.

1 Introduction

Learning in real-world robotics implies that the optimal behavior must be learnt by exploring different actions and observing their consequences, as there are usually no direct targets for a learning process to choose the correct action. The results of this exploration can be stored by obtaining models of the interaction with the world, which are usually complex due to the fact that the real-world is dynamic and the robot state, the environment and the objective may change in time. It is in this context where neuroevolution, that is, to evolve artificial neural networks (ANN) using some type of evolutionary algorithm, becomes a reference learning tool due to its robustness and adaptability to dynamic environments [1] and non-stationary tasks [2].

The episodic nature of these problems, however, must also be considered. This involves that whatever perceptual streams the robot receives could contain sequences

---

* This work was supported by the MEC of Spain through project CIT-370300-2007-18, DPI2006-15346-C03-01 and DEP2006-56158-C03-02.

M. Asada et al. (Eds.): SAB 2008, LNAI 5040, pp. 62–72, 2008.

of information corresponding to different learning processes or models that are intermingled that is, learning samples need not arise in an orderly and appropriate manner. The problem that appears is how to learn all of these different models whose samples are perceived as sequences within the same stream.

The application of evolutionary algorithms to non-stationary tasks or dynamic environments has been extensively studied in the last decade, in most cases focused in optimization problems. According to Trojanowsky [2], in the case of non-stationary optimization problems, three types of time related changes are possible: changes in the variables of the problem and their domains, changes in the objective function to be learnt and changes in the set of constraints that must be satisfied. This work will focus on unpredicted changes in the objective function.

The existing approaches for applying evolutionary algorithms to non-stationary problems can be grouped into two types [3]: memory-based approaches and search-based approaches. In the first group, the algorithm includes some kind of memory structure that stores information that can be used in the future to improve the optimization. This memory may be internal, that is, included in the chromosomes [4][5] and evolved, or external, storing successful individuals that are usually introduced in the population as seeds [6][7]. External memory-based approaches perform better in periodic non-stationary problems, with predictable changes or when the changes are easy to detect, where the individuals can be associated to a given objective function and stored. For problems where the changes of objective functions are not predictable or are hard to detect, most authors have resorted to search-based techniques enhancing their ability for continuously searching, usually by trying to preserve a high level of diversity in the population [3][8]. These techniques have had success to some degree in the case of quasi-stationary changes, but when abrupt changes occur they are basically starting neuroevolution from scratch.

With regards to learning with ANNs, the most relevant neuroevolutionary methods presented in last few years are SANE [9], a cooperative coevolutionary algorithm that evolves a population of neurons instead of complete networks; ESP [10], similar to SANE but which allocates a separate population for each of the units in the network, and where a neuron can only be recombined with members of its own subpopulation; and NEAT [11], nowadays the most used neuroevolutionary algorithm. It can evolve networks of unbounded complexity from a minimal starting point and is based on three fundamental principles: employing a principled method for the crossover of different topologies, protecting structural innovation through speciation, and incrementally growing networks from a minimal structure. Some of these neuroevolutionary techniques have been tested in non-stationary periodic tasks using external memory elements [10] [11].

In order to deal with both predictable and unpredictable non-stationary problems with quasi-stationary or brusque changes in the objective function, in this work we propose an neuroevolutionary algorithm that uses an internal or genotypic memory and that, in addition, tries to maintain diversity using a genotype-phenotype encoding that prevents the loss of relevant information throughout the generations. An external memory is added to improve the adaptive capabilities of the algorithm. The algorithm is called the Promoter Based Genetic Algorithm with External Memory (PBGA-EM).

2 The Promoter Based Genetic Algorithm

There are two basic biologically based approaches to gene expression: diploid representations and promoter based mechanisms. Diploid genotypes are made up of a double chromosome structure where each strand contains information for the same functions. In this work, however, we will concentrate on the use of gene promoters. In prokaryotes (bacteria and other simple cells) the entire DNA coding for a protein is continuous. In more complex, eukaryotic, cells, however, the encoding DNA is generally discontinuous: sequences of encoding DNA (exons) are interspersed with long sequences of non-encoding DNA. This non-encoding DNA sequences, usually about 10-fold longer than the exons, are called introns. The fact that they are so common and have been preserved during evolution leads many researchers to believe that they serve some function.

To control where a protein is encoded, the chromosome contains protein begin and protein end signals called codons. Almost every cell in an organism has a copy of every single gene the whole organism needs. Different genes are expressed in cells corresponding to different organs. Gene promoters are in charge of controlling these effects. Gene promoters are important regulatory structures that control the initiation and level of transcription of a gene.

To include these concepts in an evolutionary algorithm, we have considered a GA that evolves the weights of artificial neural networks. These neural networks, in our case, are encoded into sequences of genes for constructing a basic neural unit. Each of these blocks is preceded by a gene promoter acting as an on/off switch that determines if that particular unit will be expressed or not. In order to simplify the algorithm, we have decided to make use of these gene promoters also as start and end codons due to the position they occupy in the chromosome. For example, in the case of the simple feedforward ANN shown in Fig. 1 left, the genotypic representation used in the PBGA is shown in Fig. 1 top right. All the genes with a value of 1 are promoter genes. Thus, the first two genes represent that the two input neurons are enabled, the third gene represents that neuron unit 3 is enabled (controlling weights W_{13} W_{23}), and so on. Following with the same example, Fig. 1 center right shows the

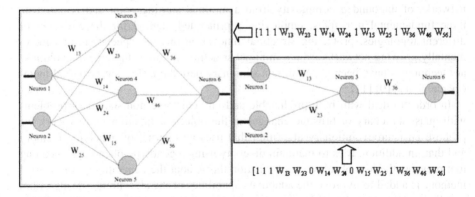

Fig. 1. Phenotypic representations of the ANN genotypes shown

phenotype corresponding to the same genotype with a different promoter setting (the one on the bottom right part), where the promoter genes of neurons 4 and 5 are disabled and consequently these two neurons and their inbound connections are not shown in the phenotype.

The basic functional unit in the PBGA is a neuron with all of its inbound connections. By concatenating units of this type according to the activation genes, the whole network is constructed. This encoding provides a way for the information that is not expressed to still be carried by the genotype during evolution, but shielded from direct selective pressure, preserving this way the diversity in the population. Therefore, a clear difference is established between the search and the solution space, permitting information learnt and encoded into the genotypic representation to be preserved by disabling promoter genes.

Although other approaches are possible, in these first tests, we have chosen to use the same topology for the genotypic representation of all the ANNs in the population to avoid complexities such as a continuous growth in the ANN size (resulting in cpu intensive tasks) or the large number of parameters that are needed to control the combination of the different topologies, associated to other approaches where the topology is completely free like in NEAT [11]. This way, all the ANN genotypes contain the same number of total neurons, in this case, within a two-layer feedforward representation. This does not mean that the ANNs resulting from the genotype-phenotype transformation have the same topology, as the phenotype depends on what functional units are enabled by the promoters. The PBGA usually starts with minimal phenotypic ANNs (just one neuron enabled per hidden layer) and evolution makes different types of ANNs (in terms of enabled neurons) coevolve together.

In the case of reproduction, the main problem that had to be dealt with in the implementation of the algorithm is how to perform crossover and mutation without being extremely disruptive or generating a bias in the evolution of what genes are expressed. Bear in mind that not only are weight values being crossed over, but the whole neuron units that conform the topology of the ANN. It is thus necessary to take care of how disruptive crossover or mutation will be on the information units in the genotype.

Crossover is panmitic in the PBGA, that is, from two parent chromosomes one child chromosome is created. Crossover is performed over the neuron units as blocks. To be statistically neutral regarding the expression of the genes, the crossover for the promoters of functional units is carried out according to the following 2 rules:

1. If both parent units are expressed, the offspring unit is expressed and the weights are obtained applying a BLX-α crossover operator.
2. If both parent units are not expressed, the offspring unit is not expressed and the weights are directly inherited from one of the parents (50% chance)
3. If one unit is expressed and the other isn't, the offspring will inherit with a 50% chance weights and promoters from the neuron unit of one of the parents.

Thus, on average, the number of expressed units is maintained and a bias in this term is avoided. In addition, we follow the strategy of preserving the disabled neuron units and perform information crossover only in cases where both neurons are active, that is, where the crossover effect can be tested when the network is evaluated.

Regarding mutation, things are simpler, and the only consideration that needs to be made is that gene promoters must be mutated at a different rate from that of regular genes. Note that mutating gene promoters may be very disruptive as it seriously affects the composition of the phenotype, whereas mutation of the rest of the genes is, on average, much more gradual on the resulting phenotype. Consequently, we decided to use different mutation rates on the gene promoters (structural mutation) and on the real valued genes (parametric mutation). The structural mutation operator simply inverts the activation value from 0 to 1 or from 1 to 0, and the parametric mutation operator applies a non linear cubed random mutation mechanism ($f(x) = f(x) + rand(0,1)^3$) only to genes belonging to active neuron units. As will be shown in the examples presented later, the values for these mutation probabilities are quite critical for the performance of the algorithm.

Summarizing, the complete PBGA working cycle is:

1. Creation of a population of N individuals using the representation commented above. The values of all the genes are randomly generated.
2. Fitness calculation over the whole population
3. Selection of 2N individuals using a tournament operator
4. Panmitic crossover with a probability of P_c over the 2N population. This operator is applied twice over the same parents and the offspring with highest fitness is selected. This results in an N individual offspring population
5. Mutation with probabilities P_{sm}, P_{pm} over the offspring population
6. Fitness calculation over the offspring population
7. Elitism that substitutes the worst r individuals of the offspring population with the best r individuals of the original population
8. Return to step 3 for n generations

The number of parameters that must be established by the user are 6: maximum number of neurons of the ANNs (size of the genotype), population size, crossover probability, structural mutation probability, parametric mutation probability and the number of generations of evolution. All of them are problem-dependent but, as we will show in the next section, their values are intuitively easy to set up.

An initial approach to the introduction of promoter genes was implemented on the Structured Genetic Algorithm (sGA), developed by Dasgupta and McGregor [4] as a general hierarchical genetic algorithm. They applied a two level interdependent genetic algorithm for solving the knapsack problem and developing application specific neural networks [12]. A two-layer sGA was used to represent the connectivity and weights of a feed-forward neural network. Higher-level genes (connectivity) acted as a switch for sections of the lower level weight representation. Sections of the weight level, whose corresponding connectivity bits were set to one, were expressed in the phenotype. Those whose corresponding bits had the value of zero were retained, but were not expressed. The main difference between the sGA applied to neuroevolution and the PBGA we are presenting here, is that the activation genes in the sGA act at the connection level whereas the PBGA works with neuron units, that is, functional units. This is a very relevant difference as, even though to enable/disable a neuron may seem much more disruptive, it permits preserving complex functional units.

3   Execution Results Using Promoters

To display the capabilities of an algorithm that implements promoters in its structure in the most unfavorable situation, we have resorted to abrupt unpredicted changes in the objective function. To simulate this changing situation, two different 3D functions that the PBGA must learn were used and as time passes they are switched:

$$F_1(x,y) = (x+y)/2 \qquad\qquad x,y \in [-10,10]$$
$$F_2(x,y) = \sin(4x) + y\sin(y) \qquad x,y \in [-10,10]$$

These functions are very different (the second one is much more complex than the first one) in order to test the capability of the PBGA for preserving the learned information in completely new situations. To show the basic features of the PBGA as a consequence of its architecture, in a first experiment we alternatively used objective functions F_1 and F_2 for 100 generations and continued with this cycle. The parameters for this first trial are shown in Table 1. We expect the PBGA to converge faster as the iterations (fitness function switch cycles) progress, because some of the previously learned information has a chance of remaining in the unexpressed part of the genotype. Fig. 2 displays the root mean squared error (RMSE) for the first 1000 generations (10 cycles) of evolution (top graph) and 26000 generations (260 cycles) later (bottom graph), where we can see, firstly, how the F_1 function is learnt with a lower error than the F_2 function, as expected. Furthermore, the RMSE decreases on each 100 generations cycle as in a typical error evolution, but when the change occurs, there is an error peak that decreases rapidly. This peak is larger in the cycle from F_1 to F_2 due to the higher complexity of the second function. In addition, we can observe how the error level at the end of each 100 generation cycle decreases in both functions as more cycles take place until it stabilizes.

In the case of a non-periodic change in the fitness function, this capability of preservation of the learned information with the promoter genes depends on the time each cycle lasts. If the cycle is too short, the algorithm may not be capable of learning the function, on the other hand, if it is too long, there is a higher probability of losing the unexpressed information. For example, in Fig. 3 we have represented the evolution of the RMSE with the same periodical change of fitness function from F_1 to F_2 every 100 generations until cycle 50 (generation 5000) where we maintain function F_1 as

Table 1. Parameter values of the PBGA used in the examples

	Parameter value (Figs. 2, 3)	Parameter value (Fig. 4)
Generations	30000	20000
Population size	2000	2200
Maximum number of neurons in hidden layers	15	12
Crossover probability	70%	70%
Structural mutation probability	2%	2%
Parametric mutation probability	1%	1%
Stability error percentage (S_E), only PBGA-EM	-	20 %
Stability generations (S_G), only PBGA-EM	-	40
Stability similarity percentage (S_S), only PBGA-EM	-	30%

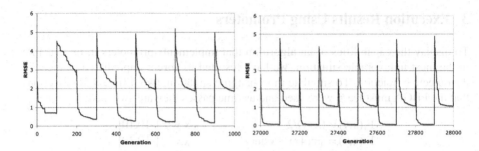

Fig. 2. Evolution of the RMSE for the first 1000 generations (left) and 26000 generations later (right) using the PBGA cycling between functions F_1 and F_2 every 100 generations

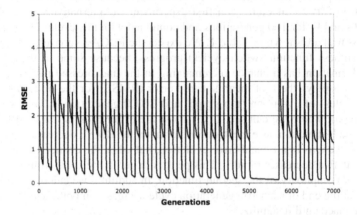

Fig. 3. Evolution of the RMSE with a periodic change of fitness function between F_1 to F_2 every 100 generations until cycle 50 (generation 5000) where the function F_1 is maintained for 7 cycles (700 generations). At the end of this period (generation 5700), the previous periodic change every 100 generations starts again.

objective function for 7 cycles (700 generations). At the end of this period (generation 5700), the previous periodical change every 100 generations starts again. As we can see in Fig. 3, from generation 5000 to 5700 the error level is very low because the PBGA is learning only the simple function F_1. At generation 5700, the PBGA must learn function F_2 again and is able to reach, in this case, an error level of 1.564 RMSE in just 100 generations, where the cycle of function F_1 starts again. We must point out that at the beginning of the run, this error level was achieved for the first time in generation 1186, as observed in Fig. 3, which means that the PBGA required, in this case, 6 cycles of function F_2 to reach this level, and now it only takes one cycle. As a consequence, we can conclude from this experiment that the PBGA is able to maintain information of a fitness function in the chromosomes after a period of learning a completely different function and, thus achieving low error levels in the previous functions earlier.

As a conclusion of this section, we can say that the PBGA is able to work in periodic and non-periodic dynamic tasks or environments. This algorithm uses an internal memory provided by the promoter genes that, together with the preservation of diversity provided by the use of a genotype-phenotype encoding, permits adaptive learning in several dynamic situations without having to explicitly detect these changes of objective function. It is obvious that there is a limit in the information that can be stored in the unexpressed genes, so there is a limit in the number of different changes that the PBGA population can store. In addition, if a learning cycle is too long, the population could converge and the probability of losing the information stored in the unexpressed genes would increase.

4   Adding an External Memory

To solve the kinds of problems derived from the use of internal memories described above, different authors have proposed the addition of external memories to the evolutionary algorithms [6]. These memories directly store individuals that perform successfully in a given situation. In the case of neuroevolution, in [7] the authors apply the real-time version of the NEAT algorithm and study how to increase the probability of obtaining populations that can remember old skills as they learn new ones while in a dynamic video game situation by considering an external memory that stores successful ANNs, which are inserted as seeds in future generations.

However, the main problem of external-memory based approaches to non-stationary problems is that they are limited to predictable or detectable changes as it is necessary to know when an individual must be stored in the memory [7]. That is, it is necessary to identify the different functions the system is learning or, at least when they change, usually by analyzing the evolution of the error. This procedure is very noisy and hard to apply in complex problems [13].

To combine the advantages provided by internal and external memories in non-stationary problems, we have introduced an external memory together with the PBGA (thus creating the PBGA-EM) in order to apply this algorithm to robotic problems in the process of learning world models [14]. We just want to store in the external memory one ANN for each different model of the environment or cycle and, consequently, a management strategy for this external memory was implemented as follows:

1. To decide if an ANN must go into the external memory, a stability criterion was established in terms of the RMSE error in predicting the objective function in a given generation. If this error value is stable (as a percentage S_E) during a number of generations S_G, we assume that the ANN is stable and may be stored in the external memory.
2. Every time a new ANN is selected as stable, we compare it with all those already stored over the current objective function. If both errors are similar within S_S, we assume that the models correspond to the same model and leave the best one in the external memory. If the errors are very different, we assume that we are dealing with a new objective function and the model is automatically stored.
3. All the models in the external memory are introduced in the evolving population as seeds every generation replacing the worst individuals.

A 2D simulator of the Pioneer 2 robot considering a simple task of finding an object with its sonar sensors and reaching it was used to test the PBGA-EM. The world models are represented by ANNs that are learnt with the PBGA and have 5 inputs and 3 outputs. The inputs are: distance given by the nearest sonar, angular position of that sonar, a boolean value that permits distinguishing if the sonar is in the front or back of the robot and the two motor values given as linear and angular speed. The outputs are the predicted distance given by the nearest sonar, the predicted angular position of that sonar and the predicted boolean value.

Fig. 4 displays the evolution of the RMSE when learning the angular position (with respect to the values provided by the simulator) using the PBGA-EM (dotted line) and without the external memory as in the typical PBGA (solid line). In this experiment, the world changed every 2400 generations from a normal operation of the robot to a case where we simulated a hardware failure: the input distance was interchanged with the angle and their ranges were reduced by 80%. The parameters of the PBGA and the PBGA-EM in this case are shown in Table 1.

As shown by the solid line of Fig. 4, this is a hard case for the simple PBGA because the cycles are too long and the functions very complex. Consequently, the information learnt after one cycle will disappear from the population during the next one due to the complexity of the new fitness function, which implies reusing the inactive neuron units to improve the error level. But what is very relevant from this figure is that the PBGA-EM in the second cycle (about generation 4500) achieves an error level below 0.1 RMSE. The best individual in the population at that point is stored in the external memory and the next time the PBGA-EM must learn the same function (cycle 4, generation 7200), the previous error level is achieved immediately. This is because the individual stored in the external memory is introduced in the population every generation so, when the function to be learnt corresponds to the previous model, it performs successfully.

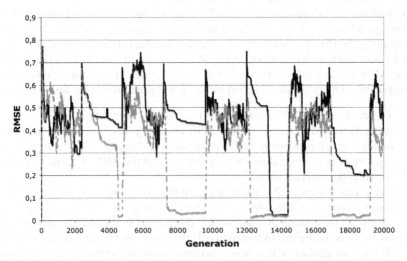

Fig. 4. Evolution of the RMSE in the learning of the angular position using an external memory (dotted line) and without the external memory in the PBGA (solid line). The fitness function is changed every 2400 generations between simulating a hardware failure and normal operation.

In cycle 6 and around generation 13000, the PBGA without external memory (solid line) also reaches this error level below 0.1 RMSE but, as a consequence of the forgetting effect commented before, in cycle 8 (generation 16800) it must relearn the function almost from scratch and does not achieve the same error level as in cycle 6.

5   Conclusions

This paper describes the Promoter Based Genetic Algorithm together with an external memory for learning in non-stationary environments. The PBGA-EM uses promoter genes to create a sort of genetic memory within the genotype. We have shown that the immediate result of this memory is that when a fitness function that has been seen before (or which requires combinations of basic units that were used in previous successful runs) is contemplated again, the GA achieves the desired phenotype much faster than before. To overcome the different problems of internal and external memories operating separately, we have shown the advantage of combining both in order to reduce the time spent on relearning previous situations. This way, the PBGA with external memory is a very powerful neuroevolutionary technique to deal with non-stationary problems and dynamic environments. The algorithm must be improved in the future to deal with non-stationary real tasks, for example, those involved in autonomous robotics, that could contain highly complex non-periodic changes in the environment or in the robot's behavior.

References

1. Yao, X.: Evolving artificial neural networks. Proc. of the IEEE 87(9), 1423–1447 (1999)
2. Trojanowski, K., Michalewicz, Z.: Evolutionary Approach to Non-stationary Optimisation Tasks. LNCS, vol. 1609, pp. 538–546 (1999)
3. Mori, N., Kita, H., Nishikawa, Y.: Adaptation to a Changing Environment by Means of the Feedback Thermodynamical Genetic Algorithm. In: Eiben, A.E., Bäck, T., Schoenauer, M., Schwefel, H.-P. (eds.) PPSN 1998. LNCS, vol. 1498, pp. 149–158. Springer, Heidelberg (1998)
4. Dasgupta, D., MacGregor, D.R.: Nonstationary function optimization using the structured genetic algorithm. Proceedings Parallel Problem Solving from Nature 2, 145–154 (1992)
5. Ryan, C., Collins, J.J., Wallin, D.: Non-stationary Function Optimization Using Polygenic Inheritance. LNCS, vol. 2724, pp. 1320–1331 (2003)
6. Eggermont, J., Lenaerts, T.: Non-stationary Function Optimization using Evolutionary Algorithms with a Case-based Memory, Technical Report TR 2001-11 (2001)
7. D'Silva, T., Janik, R., Chrien, M., Stanley, K., Miikkulainen, R.: Retaining Learned Behavior During Real-Time Neuroevolution. In: Procedings AIIDE (2005)
8. Cobb, H.G.: An investigation into the use of hypermutation as an adaptive operator in genetic algorithms having continuous, time-dependent nonstationary environments, NRL Memorandum Report 6760, 523-529 (1990)
9. Moriarty, D.E.: Symbiotic Evolution of Neural Networks in Sequential Decision Tasks, PhD thesis, University of Texas at Austin, Tech. Rep. UT-AI97-257 (1997)
10. Gomez, F., Burger, D., Miikkulainen, R.: A neuroevolution method for dynamic resource allocation on a chip multiprocessor. In: Proc. Int. Joint Conf. on Neural Networks (2001)

11. Stanley, K.O., Miikkulainen, R.: Evolving neural networks through augmenting topologies. Evolutionary Computation 10, 99–127 (2002)
12. Dasgupta, D., McGregor, D.: Designing application-specific neural networks using the structured genetic algorithm. In: Proceedings COGANN 1992, pp. 87–96 (1992)
13. Branke, J.: Memory enhanced evolutionary algorithms for changing optimization problems. In: Proceedings CEC 1999, vol. 3, pp. 1875–1882 (1999)
14. Bellas, F., Duro, R.J.: Introducing Long Term Memory in an ANN Based Multilevel Darwinist Brain. LNCS, vol. 2686, pp. 590–597 (2003)

Evolving Vision Controllers with a Two-Phase Genetic Programming System Using Imitation

Renaud Barate and Antoine Manzanera

ENSTA - UEI, 32 bd Victor,
75739 Paris Cedex 15, France
renaud.barate@ensta.fr, antoine.manzanera@ensta.fr

Abstract. We present a system that automatically selects and parameterizes a vision based obstacle avoidance method adapted to a given visual context. This system uses genetic programming and a robotic simulation to evaluate the candidate algorithms. As the number of evaluations is restricted, we introduce a novel method using imitation to guide the evolution toward promising solutions. We show that for this problem, our two-phase evolution process performs better than other techniques.

1 Introduction

Our goal is to design vision based obstacle avoidance controllers for mobile robots. The most popular method to perform obstacle avoidance with a single camera uses optical flow and is directly inspired by the flight of insects [1,2]. However, systems based on optical flow don't cope well with thin or lowly textured obstacles. There is also evidence that information based on appearance rather than movement, like texture information, is extracted in the first stages of the vision chain [3]. This kind of information can also be used for obstacle avoidance. For instance, Michels implemented a system to estimate depth from texture information in outdoor scenes [4]. Other systems use this kind of information to discriminate the floor from the rest of the scene and calculate obstacle distances in several directions [5]. Nevertheless those methods suppose that the floor may be clearly discriminated and they neglect potentially useful information from the rest of the scene.

As there is no method that can deal with all contexts, we want our robot to automatically select and adapt an obstacle avoidance method for the current environment. For now, this adaptation is an offline process based on genetic programming which creates original controllers adapted to a given simulation environment. The next step will be either to install this system on a real robot for online evolution or to design a higher level controller able to select in real-time an algorithm adapted to the current context in a database of evolved algorithms. As we only use artificial evolution as an optimization technique, we won't make assumptions about implications on the development of vision processing in the brain of animals by natural selection and evolution. However there is an important issue addressed by this work which is of interest in a bio-inspired framework:

M. Asada et al. (Eds.): SAB 2008, LNAI 5040, pp. 73–82, 2008.
© Springer-Verlag Berlin Heidelberg 2008

how can Imitation increase the learning speed, and what are the consequences on performances and generalization?

Evolutionary techniques have already been widely used for robotic navigation and the design of obstacle avoidance controllers [6] but in general vision is either overly simplified or not used at all. For instance, Marocco used only a 5 × 5 pixels retina as visual input [7]. On the other hand, genetic programming has been proved to achieve human-competitive results in image processing systems, e.g. for the detection of interest points [8]. Parisian evolution has also been shown to produce very good results for obstacle detection and 3D reconstruction but those systems need two calibrated cameras [9].

To our knowledge, only Martin tried evolutionary techniques with monocular images for obstacle avoidance [10]. The structure of his algorithm is based on the floor segmentation technique and the evaluation is done with a database of hand labeled real world image. The advantage of such an approach is that the evolved algorithms are more likely to work well with real images than those evolved with computer rendered images. Nevertheless, it introduces an important bias since the algorithms are only selected on their ability to label images in the database correctly and not on their ability to avoid obstacles.

2   Evolution of the Vision Algorithms

2.1   Structure of the Vision Algorithms

Generally speaking, a vision algorithm can be divided in three main parts: First, the algorithm will process the input image with a number of filters to highlight some features. In our case, this filter chain consists of spatial and temporal filters, optical flow calculation and projection that will produce an image highlighting the desired features. Then these features are extracted, i.e. represented by a small set of scalar values. We implemented this with a mean computation of the pixel values on several windows of the transformed image(s). Finally these values are used for a domain dependent task, here to generate motor commands to avoid obstacles. Fig. 1 shows an example program for a simple obstacle avoidance behavior based on optical flow. Here is the list of all the primitives (transformation steps) that can be used in the programs and the data types they manipulate:

- **Spatial filters** (*input: image, output: image*): Gaussian, Laplacian, threshold, Gabor, difference of Gaussians, Sobel and subsampling filter.
- **Temporal filters** (*input: image, output: image*): pixel-to-pixel min, max, sum and difference of the last two frames, and recursive mean operator.
- **Optical flow** (*input: image, output: vector field*): Horn and Schunck global regularization method, Lucas and Kanade local least squares calculation and simple block matching method. The rotation movement is first eliminated by a transformation of the two images in order to facilitate further use of the optical flow.
- **Projection** (*input: vector field, output: image*): Projection on the horizontal or vertical axis, Euclidean or Manhattan norm computation, and time to contact calculation using the flow divergence.

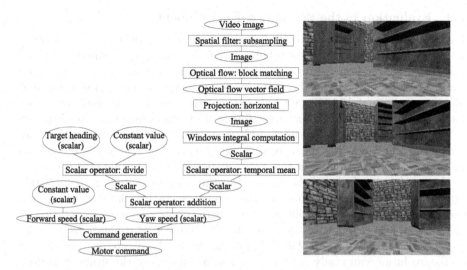

Fig. 1. Left: Algorithmic tree of a program example for obstacle avoidance. Rectangles represent primitives and ellipses represent data. Right: Snapshots of the simulation environment.

- **Windows integral computation** (*input: image, output: scalar*)**:** For this transformation, we define a global coefficient α_0 and several windows on the left half of the image with different positions and sizes. With each window is paired a second window defined by symmetry along the vertical axis. A coefficient α_i and an operator ($+$ or $-$) are defined for each pair. The resulting scalar value R is a simple linear combination calculated with the following formula:

$$R = \alpha_0 + \sum_{i=1}^{n} \alpha_i \mu_i$$
$$\mu_i = \mu_{Li} + \mu_{Ri} \text{ or } \mu_i = \mu_{Li} - \mu_{Ri}$$

where n is the number of windows and μ_{Li} and μ_{Ri} are the means of the pixel values over respectively the left and right window of pair i.
- **Scalar operators** (*input: scalar(s), output: scalar*)**:** Addition, subtraction, multiplication and division operators, temporal mean calculation and simple if-then-else test.
- **Command generation** (*input: two scalars, output: command*)**:** The motor command is represented by two scalar values: the requested linear and angular speeds.

Most of those primitives use parameters along with the input data to do their calculations (for example, the standard deviation value for the Gaussian filter or the number and position of windows for the windows integral computation). Those parameters are specific to each algorithm; they are randomly generated when the corresponding primitive is created by the evolution process.

2.2   Evaluation in the Simulation Environment

For the evaluation of the different obstacle avoidance algorithms, we use a simulation environment in which the robot moves freely during each experiment. The simulation is based on the open-source robot simulator Gazebo. The simulated camera produces 8-bits gray-value images of size 320×160 (representing a field of view of approximately $100° \times 60°$) at a rate of 10 / sec. The simulation environment is a closed room of 36 m^2 area (6 m × 6 m) containing three bookshelves (Fig. 1). All the obstacles are immovable to prevent the robot from just pushing them instead of avoiding them. In each experiment, the goal of the robot is to go from a given starting point to a goal location without hitting obstacles.

Due to the complexity of the vision algorithms, we're limited to about 40,000 evaluations to keep the evolution time acceptable (a few days at most). We therefore face a common problem with genetic programming systems, that is the state space is immense compared to the number of evaluations so large parts of it will never be explored. More, the fitness landscape is very chaotic so the evolution can easily get stuck in a local minima of the fitness function. In previous work we showed that a classical evolution process often produces controllers with a seemingly random trajectory, even if the visual features they use are coherent with the environment [11]. To overcome those problems, we propose a novel approach based on the imitation of a given behavior in a first phase to guide the evolution toward more efficient solutions in the second phase.

2.3   First Phase: Evolution of Algorithms That Imitate a Behavior

In this first phase, the population is initialized with random algorithms and the evolution lasts for 50 generations. The goal will be to match a recorded example behavior. More precisely, we first record the video sequence and parameters of an experiment where we manually guide the robot from the starting point to the target point while avoiding obstacles. We make the trajectory as short and smooth as possible to limit the difficulty of the matching task.

For the evaluation of the algorithms, we replay this sequence and compare the command issued by the evaluated algorithm with the command recorded during the manual control of the robot. The goal is to minimize the difference between these two commands along the recorded sequence. Formally, we try to minimize two variables F and Y defined by the formulas:

$$F = \sqrt{\sum_{i=1}^{n} (f_{Ri} - f_{Ai})^2} \text{ and } Y = \sqrt{\sum_{i=1}^{n} (y_{Ri} - y_{Ai})^2}$$

where f_{Ri} and y_{Ri} are the recorded forward and yaw speed commands for frame i, f_{Ai} and y_{Ai} are the forward and yaw speed commands from the tested algorithm for frame i and n is the number of frames in the video sequence.

2.4   Second Phase: Evolution of Efficient Solutions

In this second phase, the population is initialized with the final population of the first phase and the evolution lasts for 50 more generations. The algorithms will be evaluated on their ability to really avoid obstacles and reach the target location. For that, we place the robot at the fixed starting point and let it move in the environment during 30 s driven by the obstacle avoidance algorithm. Two scores are attributed to the algorithm depending on its performance: a goal-reaching score G rewards algorithms reaching or approaching the goal location, whereas contact score C rewards the individuals that didn't hit obstacles on their way. Those scores are calculated with the following formulas:

$$G = \begin{cases} t_G & \text{if the goal is reached} \\ t_{\max} + d_{\min}/V & \text{otherwise} \end{cases}$$
$$C = t_C$$

where t_G is the time needed to reach the goal in seconds, $t_{\max}$ is the maximum time in seconds (here 30 s), $d_{\min}$ is the minimum distance to the goal achieved in meters, V is a constant of 0.1 m/s and t_C is the time spent near an obstacle (i.e. less than 18 cm, which forces the robot to keep some distance away from obstacles). The goal is hence to minimize those two scores G and C.

For the two phases of evolution, the evaluations consist in fact in two runs with a different starting point and a different goal location. Final scores are the sum of the scores obtained for the two runs. Performing two different runs favors algorithms with a real obstacle avoidance strategy while not increasing evaluation time too much. The starting points are fixed because we want to evaluate all algorithms on the same problem.

2.5   The Genetic Programming System

We use grammar based genetic programming to evolve the vision algorithms [12]. As usual with artificial evolution, the population is initially filled with randomly generated individuals. In the same way that a grammar can be used to generate syntactically correct random sentences, a genetic programming grammar is used to generate valid algorithms. The grammar defines the primitives, data and the rules that describe how to combine them. The generation process consists in successively transforming each non-terminal node of the tree with one of the rules. This grammar is used for the initial generation of the algorithms and for the transformation operators. The crossover consists in swapping two subtrees issuing from identical non-terminal nodes in two different individuals. The mutation consists in replacing a subtree by a newly generated one. Table 1 presents the exhaustive grammar that we used in all our experiments.

The numbers in brackets are the probability of selection for each rule. A major advantage of this system is that we can bias the search toward the usage of more promising primitives by setting a high probability for the rules that generate them. We can also control the size of the tree by setting small probabilities for the rules that are likely to cause an exponential growth (rules like REAL $\rightarrow$ ifThenElse(REAL,REAL,REAL,REAL) for example).

As described previously, we wish to minimize two criteria (F and Y in the first phase, G and C in the second phase). There are different ways to use evolutionary algorithms to perform optimization on several and sometimes conflicting criteria.

Table 1. Grammar used in the genetic programming system for the generation of the algorithms

[1.0] START → COMMAND	[0.14] SPATIAL_FILTER → threshold
[1.0] COMMAND → directMove(REAL,REAL)	[0.14] SPATIAL_FILTER → gabor
[0.1] REAL → targetDistance	[0.14] SPATIAL_FILTER → differenceOfGaussians
[0.1] REAL → targetHeading	[0.14] SPATIAL_FILTER → sobel
[0.1] REAL → scalarConstant	[0.15] SPATIAL_FILTER → subsampling
[0.05] REAL → add(REAL,REAL)	[0.2] TEMPORAL_FILTER → temporalMinimum
[0.05] REAL → subtract(REAL,REAL)	[0.2] TEMPORAL_FILTER → temporalMaximum
[0.05] REAL → multiply(REAL,REAL)	[0.2] TEMPORAL_FILTER → temporalSum
[0.05] REAL → divide(REAL,REAL)	[0.2] TEMPORAL_FILTER → temporalDifference
[0.05] REAL → temporalRegularization(REAL)	[0.2] TEMPORAL_FILTER → recursiveMean
[0.05] REAL → ifThenElse(REAL,REAL,REAL,REAL)	[0.33] OPTICAL_FLOW → hornSchunck(IMAGE)
[0.4] REAL → windowsIntegralComputation(IMAGE)	[0.33] OPTICAL_FLOW → lucasKanade(IMAGE)
[0.3] IMAGE → videoImage	[0.34] OPTICAL_FLOW → blockMatching(IMAGE)
[0.4] IMAGE → SPATIAL_FILTER(IMAGE)	[0.2] PROJECTION → horizontalProjection
[0.15] IMAGE → PROJECTION(OPTICAL_FLOW)	[0.2] PROJECTION → verticalProjection
[0.15] IMAGE → TEMPORAL_FILTER(IMAGE)	[0.2] PROJECTION → euclideanNorm
[0.15] SPATIAL_FILTER → gaussian	[0.2] PROJECTION → manhattanNorm
[0.14] SPATIAL_FILTER → laplacian	[0.2] PROJECTION → timeToContact

For the experiments described in this paper, we chose the widely used multi-objective evolutionary algorithm called NSGA-II. It is an elitist algorithm based on the non-dominance principle. A diversity metric called "crowding distance" is used to promote diversity among the evolved individuals. All the details of the implementation can be found in the paper by K. Deb [13].

In order to prevent problems of premature convergence, we separate the population of algorithms in 4 islands, each containing 100 individuals. Those islands are connected with a ring topology; every 10 generations, 5 individuals selected with binary tournament will migrate to the neighbor island while 5 other individuals are received from the other neighbor island. For the parameters of the evolution, we use a crossover rate of 0.8 and a probability of mutation of 0.01 for each non-terminal node. We use a classical binary tournament selection in all our experiments. Those parameters were determined empirically with a few tests using different values. Because of the length of the experiments, we didn't proceed to a thorough statistical analysis of the influence of those parameters.

3   Experiments and Results

3.1   Comparison with Other Evolution Strategies

We compare here our own system with a classical one-phase evolution process and with two methods commonly used to guide the evolution toward promising solutions, namely incremental evolution and seeding. For these three experiments we only use the evaluation function described in 2.4, to select the controllers on their ability to reach the goal while avoiding obstacles.

Seeding an evolution consists in inserting in the initial population one or several individuals (generally hand written solutions) performing reasonably well on the target problem. This seeding method guides the evolution towards a specific part of the state space where we know that there is a good probability to find interesting solutions. Here we simply inserted our manually designed controller presented before in the initial population of one of the four islands. The evolution then proceeds normally for 100 generations.

Incremental evolution [14] is based on the decomposition of a problem in a set of problems of increasing complexity. Starting with a simple problem, the evolution quickly finds good solutions. The difficulty is then progressively increased and new individuals emerge to adapt to those more and more complex conditions. Here, we proceed for the 20 first generations with a simplified environment with only one bookshelf (the one that is closest to the wall). This way, there is no real obstacle between the starting points and the target points. Then we proceed for 40 more generations with one more bookshelf, and finally for the 40 last generations with the complete environment containing three bookshelves.

In total, all the experiments last for 100 generations, that is 40,000 evaluations. Fig. 2 presents the Pareto fronts at different times of the evolution for the different methods. With the classical one-phase evolution, the first generation contains only individuals with poor performance. After 50 generations we obtain controllers with interesting behaviors, performing better than the hand designed controller. The end of the evolution doesn't improve these results much, we can say here that the evolution process has become stuck in a local minimum of the fitness function. With seeding, the first generations contain better controllers due to the seed individual but the evolution doesn't manage to further improve this behavior. In the incremental evolution experiment, the evolution finds very efficient controllers in the two first environments. But when we introduce the third bookshelf, the evolution doesn't find a way to adapt to those new conditions. Nevertheless the final controllers perform better than in the previous cases. With two-phase evolution, we obtain very good individuals even at the

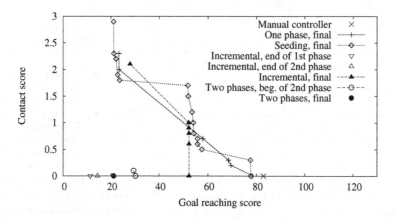

Fig. 2. Pareto fronts at different times of evolution for the different evolution strategies. Those fronts are created by taking the best individuals in the generation in all the four islands and keeping only non-dominated ones. In some cases one individual dominates all others, reducing the front to one point.

end of the first phase. The second phase will then optimize those controllers and the final ones definitely surpass the ones obtained with the other methods.

3.2   Analysis of the Controllers

We represent on Fig. 3 the trajectories of the robot in the environment with controllers issued from the different evolution processes. The hand written controller manages to avoid obstacles but the robot goes too slowly and never reaches the target points. With the one-phase evolution process, the evolved controllers sometimes manage to reach one of the target points without hitting obstacles but generally they move in an almost random way and with a lot of rotations. Turning quickly on themselves prevents the robots from getting stuck against obstacles and thus generally improves their goal reaching score in the first generations. This kind of controllers quickly overcomes the rest of the population and prevents the evolution from finding better controllers. As shown before, the seeded evolution didn't manage to improve the seed individual much. The trajectory obtained with evolved controllers is very close to the trajectory of the hand written controller used as seed.

The incremental evolution produces efficient solutions with one and two bookshelves but when we add the final one, the evolution fails to adapt the controllers. The trajectory clearly shows this problem: one of the bookshelf is quickly avoided but the robot just stops before hitting the second one without trying to avoid it. The first phase here selects very simple controllers going straight on to the target points. To achieve this, they just use the target heading value for the

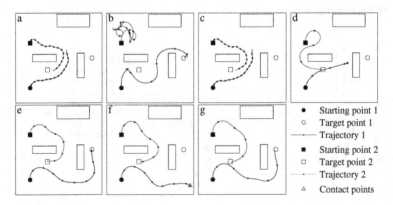

Fig. 3. a) Trajectory of the hand written controller. b) Trajectory of a controller evolved with one-phase evolution. c) Trajectory of a controller evolved with seeded evolution. d) Trajectory of a controller evolved with incremental evolution. e) Trajectory of the recorded sequence. f) Trajectory of a controller evolved with only example matching evaluation during 50 generations (end of the first phase). g) Trajectory of the controller evolved with two-phases evolution process (end of the second phase). The small dots on the trajectories are placed every 2 seconds and thus indicate the speed of the robot.

yaw speed command. Although it is a good strategy without obstacles, these controllers are difficult to adapt afterward for more complex problems.

With two-phase evolution, the first phase evolves controllers with smooth trajectories very similar to the recorded sequence. Some of those controllers even reach the two target points although most of them get stuck on an obstacle as shown on Fig. 3. With this initialization, the second phase quickly evolves very efficient controllers reaching the two target points without hitting obstacles and even faster than when the robot is manually controlled. The structure of the controllers evolved in this environment is generally quite simple (mainly based on contrast information obtained with simple threshold filters) and there's almost no structure difference between the end of the first phase and the end of the second phase. This means that the imitation phase is crucial for the evolution of the structure to produce efficient controllers. The second phase mainly optimizes the parameters of the algorithms to obtain a faster and more robust behavior.

The main limitation of all those controllers is that they don't generalize well to different conditions. If we move around the obstacles, the starting point and the target point, the performance of the controllers decreases quickly. The controller issued from two-phase evolution still shows some obstacle avoidance abilities but it doesn't move toward the target point. Some other controllers just seem to move randomly in the environment. There are two reasons for this mediocre generalization behavior: First, it seems that the choice of only two trajectories in the evaluation environment was too optimistic. The controllers overlearn these trajectories and this decreases their generalization performance. Second, the evolved controllers often use the target heading information in quite complicated ways or even not at all. In this case when we move the target point they are not able to find it unless by chance. To overcome these problems we have designed experiments with more trajectories during the evolution and with a different grammar which facilitates an efficient use of the target heading for the controllers. The generalization performance of the controllers evolved this way is presented in [15].

4   Conclusion

In this paper, we present a system that selects and adapts automatically an appropriate obstacle avoidance method to a given visual context. This system is based on genetic programming and we introduced a novel evolution process to guide the evolution toward promising solutions in the state space. We have shown that for our target application, it improves the performance of the final controllers greatly compared to classical evolution, seeding or incremental evolution. This method could be adapted for many evolutionary robotic systems and it would be interesting to validate it on other applications.

Our goal is now to test this method on a wide range of environments (including real-world environments) to check if some primitives are selected more often than others and how they are combined by the evolution. This will lead to a better understanding on how to combine different vision information in a robotic navigation system in order to obtain more adaptive and robust behaviors.

References

1. Muratet, L., Doncieux, S., Brière, Y., Meyer, J.A.: A contribution to vision-based autonomous helicopter flight in urban environments. Robotics and Autonomous Systems 50(4), 195–209 (2005)
2. Zufferey, J., Floreano, D.: Fly-Inspired Visual Steering of an Ultralight Indoor Aircraft. IEEE Transactions on Robotics 22(1), 137–146 (2006)
3. Malik, J., Perona, P.: Preattentive texture discrimination with early vision mechanisms. Journal of the Optical Society of America A: Optics and Image Science, and Vision 7(5), 923–932 (1990)
4. Michels, J., Saxena, A., Ng, A.: High speed obstacle avoidance using monocular vision and reinforcement learning. In: Proceedings of the 22nd international conference on Machine learning, pp. 593–600 (2005)
5. Ulrich, I., Nourbakhsh, I.: Appearance-based obstacle detection with monocular color vision. In: Proceedings of AAAI Conference, pp. 866–871 (2000)
6. Walker, J., Garrett, S., Wilson, M.: Evolving controllers for real robots: A survey of the literature. Adaptive Behavior 11(3), 179–203 (2003)
7. Marocco, D., Floreano, D.: Active vision and feature selection in evolutionary behavioral systems. From Animals to Animats 7, 247–255 (2002)
8. Trujillo, L., Olague, G.: Synthesis of interest point detectors through genetic programming. In: Proceedings of the 8th annual conference on Genetic and evolutionary computation, pp. 887–894 (2006)
9. Pauplin, O., Louchet, J., Lutton, E., De La Fortelle, A.: Evolutionary Optimisation for Obstacle Detection and Avoidance in Mobile Robotics. Journal of Advanced Computational Intelligence and Intelligent Informatics 9(6), 622–629 (2005)
10. Martin, M.: Evolving visual sonar: Depth from monocular images. Pattern Recognition Letters 27(11), 1174–1180 (2006)
11. Barate, R., Manzanera, A.: Automatic Design of Vision-based Obstacle Avoidance Controllers using Genetic Programming. In: Proceedings of the 8th International Conference on Artificial Evolution, EA 2007. LNCS, vol. 4926, pp. 25–36 (2008)
12. Whigham, P.: Grammatically-based genetic programming. In: Proceedings of the Workshop on Genetic Programming: From Theory to Real-World Applications, pp. 33–41 (1995)
13. Deb, K., Pratap, A., Agarwal, S., Meyarivan, T.: A fast and elitist multiobjective genetic algorithm: NSGA-II. IEEE Transactions on Evolutionary Computation 6(2), 182–197 (2002)
14. Gomez, F., Miikkulainen, R.: Incremental Evolution of Complex General Behavior. Adaptive Behavior 5(3-4), 317–342 (1997)
15. Barate, R., Manzanera, A.: Generalization Performance of Vision Based Controllers for Mobile Robots Evolved with Genetic Programming. In: Proceedings of the Genetic and Evolutionary Computation Conference (to appear, 2008)

Embodiment and Perceptual Crossing in 2D
A Comparative Evolutionary Robotics Study

Marieke Rohde and Ezequiel Di Paolo

Centre for Computational Neuroscience and Robotics (CCNR)
University of Sussex, Brighton, BN1 9QH, UK
{m.rohde,ezequiel}@sussex.ac.uk

Abstract. We present the results from an evolutionary robotics simulation model of a recent unpublished experiment on human perceptual crossing in a minimal virtual two-dimensional environment. These experiments demonstrate that the participants reliably engage in rhythmic interaction with each other, moving along a line. Comparing three types of evolved agents with radically different embodiment (a simulated arm, a two-wheeled robot and an agent generating a velocity vector in Euclidean space), we identify differences in evolved behaviours and structural invariants of the task across embodiments. The simulation results open an interesting perspective on the experimental study and generate hypotheses about the role of arm morphology for the behaviour observed.

1   Introduction

In a recent unpublished study, Lenay et al. (personal communication) from the perceptual supplementation group (GSP) at the UT Compiègne tested human subjects on their capacity to discriminate static and mobile objects from other intentional sensing entities in a minimal two-dimensional virtual environment. This paper presents the results from an evolutionary robotics simulation model of this experiment that aims at exploring the space of possible behavioural strategies afforded by the experimental design, in order to help explain their results.

In the experiment, the blindfolded participants' task was to indicate via mouse-clicks whether an object encountered by moving the mouse (tactile stimulus to the finger) was another sensing entity (i.e. another participant) or not, where the only two other objects in the toroidal simulated two-dimensional environment were a mobile lure that shadows the other participants' movements and a static lure (see Sect. 2 for details of the task). This paradigm was a direct extension of a previous experiment on perceptual crossing in a one-dimensional simulated environment by the same group [1], to test whether the experimental results transfer qualitatively or quantitatively from a one-dimensional to a two-dimensional scenario. Some preliminary results from their study are that the results transfer qualitatively (i.e. 65% correct clicks), and that the behavioural strategies are strikingly and unexpectedly similar to those observed in the 1D version of the experiment. In particular, even though participants search for interaction exploring both dimensions, they move back and forth on a line once they encounter an interaction partner or object.

M. Asada et al. (Eds.): SAB 2008, LNAI 5040, pp. 83–92, 2008.
© Springer-Verlag Berlin Heidelberg 2008

Having shown in our previous work [2,3] that evolutionary robotics simulations of this kind of minimal sensorimotor experiments (in particular of the one-dimensional version of the same task [2]) can be a rich source of clarification, inspiration and illustration, we conducted the present simulation in order to understand the dynamical principles of the task and to generate hypotheses for the analysis and evaluation of the experimental data based on the analogy with the simpler and more controllable simulated agents.

Our intuition was that the rhythmic one-dimensional interaction is related to the morphology of the human arm. To explore this intuition, we compared a simple model of a human arm moving a mouse on a desk surface with two other types of agents that are controlled by the same control network, but have radically different bodies and sensorimotor dynamics: a two-wheeled robotic agent and an agent generating a velocity vector in Euclidean space with a vertical and a horizontal component (similar to a joystick; model described in Sect. 2).

The results (Sect. 3) allow to identify interesting differences and commonalities between the solutions evolved for these different types of agents and the results from the original experiment. Firstly, the *same behavioural patterns* (e.g., independent realisation of search and interaction, rhythmic oscillations) evolve for all agents. Secondly, the *realisation* of these behaviours varies a lot with morphology, which includes the production of near-linear rhythmic trajectories in arm agents, as hypothesised. Thirdly, the evolution of a *viable yet unintuitive strategy* (avoiding interaction with each other) in some of the agents. This refutation of our hypothesis that the results would qualitatively resemble those from the experiment is not a shortcoming: the existence of a counter-intuitive efficient strategy opens a different perspective on the experimental data (see Sect. 4).

Analysing sensorimotor data from experimental studies like the one we modelled is a difficult task because of the multitude of factors influencing human behaviour. Exploring the space of possible solutions in simulation (the task and the simulated environment used here are, apart from parametric details, identical with those used in the experiment), we hope that our results will directly facilitate the data analysis and interpretation for the researchers at the GSP.

2   The Model

The artificial agents evolved, just as the experimental participants, move around a virtual plane (200×200) that wraps around in both dimensions (i.e., a torus; see Fig. 1 (A)). In this plane, there are six different objects. Two circular simulated agents (circular objects in Fig. 1 (A)), two mobile lures that are attached to the agents and two fixed lures that are statically installed at $(50, 50)$ and $(150, 150)$ respectively. All objects are circular of diameter 20, even if the lures are represented as boxes in Fig. 1 (A) and other figures. The attached lures shadow the trajectories of each of the agents at a distance of 93 units, being attached in perpendicular directions (see Fig. 1 (A)).

The only sensory signal S that the agents receive is a touch signal, i.e. if the distance d between the agent and something else is $d < 20$, an input S_G

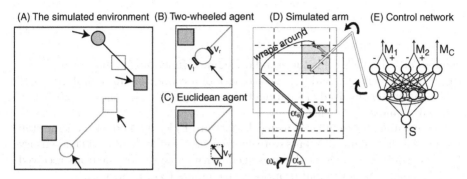

Fig. 1. Schematic diagram of the simulation model and control network. (A) The simulated environment with the two agents (circles), the attached lures (boxes attached with a line) and the fixed lures (boxes). Diagrams of the two-wheeled agent (B), the agent moving in Euclidean space (C) and the simulated arm agent, with the space in which they can act (D). The control network (E).

(sensory gain, evolved) is fed into the control network. Each agent can only perceive the other and one of each kind of lure, i.e., the dark agent can perceive all light objects in Fig. 1 (A), but not the dark ones, and vice versa, in order for interaction between the agents not to be mediated through another object.

Three different types of agents were evolved: 1.) A two-wheeled agent that generates the velocity $v_{l,r} \in [-20M_G, 20M_G]$ for each wheel (Fig. 1 (B); M_G is the evolved motor gain; velocities are specified in units/s). 2.) An agent that generates a horizontal and a vertical velocity vector $v_{h,v} \in [-30M_G, 30M_G]$ that are summed up ('Euclidean agent', Fig. 1 (C)). 3.) A simple simulated arm with two segments of length 400 units that is steered through angular velocity signals $\omega_{e,s} \in [-0.05M_G, 0.05M_G]$ to the elbow and the shoulder joint (see Fig. 1 (D)).

In order to approximate the dynamics of human mouse motion, the arm agent is restricted in its movements in two ways: through joint stops $\alpha_s \in [0.1\pi, 0.6\pi]$ and $\alpha_e \in [0.2\pi, \pi]$ and through the delimitation of movement to an area of 600×600 units that represents the 'desk' surface (i.e., the area within which a human participant would move the mouse), whose bottom left corner is fixed at $(-200, 200)$ taking the shoulder joint as the origin. The desk area is translated randomly with respect to both the desk area of the other agent and the simulated virtual environment (see Fig. 1 (D)) to avoid that agents evolve to meet in the middle of the desk.

For purpose of comparison, all three kinds of agent are controlled by structurally identical neural network controllers (Fig. 1 (E)), i.e., continuous-time recurrent neural network (CTRNN, see e.g., [4]) with one input neuron, four fully connected interneurons and 5 output neurons. Four output neurons regulate the two motor outputs ($M_1 = a_{M1} - a_{M2}, M_2 = a_{M3} - a_{M4}$) that are interpreted as $v_{l,r}$, $v_{h,v}$ or $\omega_{e,s}$ respectively. The fifth output neuron generates the categorisation signal M_C. CTRNN dynamics are governed by

$$\tau_i \frac{da_i(t)}{dt} = -a_i(t) + \sum_{j=1}^{N} w_{ij}\sigma(a_j(t) + \theta_j) + I_i \tag{1}$$

where $\sigma(x) = 1/(1+e^{-x})$ is the standard sigmoid function, θ_i a bias term, τ_i the activity decay constant and w_{ij} the strength of a connection from unit j to unit i. The parameters evolved (74 parameters) are: $S_G, M_G \in [1, 50]$. $\tau_i \in [20, 3000]$, $\theta_i \in [-3, 3]$ and $w_{i,j} \in [-6, 6]$.

The dynamics are simulated using the forward Euler method with a time step of 1 ms. All three kinds of agents were evolved with and without a 100 ms sensory delay. A modified version of the arm agent with three sensory neurons received the joint angles as additional proprioceptive inputs ($S_{2,3} = S_G \alpha_{e,s}$).

Each trial lasts $T \in [6000, 9000]$ ms. The task is to interact with something and correctly classify the object encountered as either of the lures ($M_C \leq 0.5$) or the other agent ($M_C > 0.5$). Agents are matched with clones of themselves in the task. The starting positions are random for the wheeled and the Euclidean agent and random within the centre area for the arm agent. The starting angle for the wheeled agents is random. For the arm agent and the Euclidean agent, the relative orientation of the agents to each other is random $\in \{\frac{-\pi}{2}, 0, \frac{\pi}{2}, \pi\}$. The fitness $F(i)$ of an individual i in each trial is given by the following function

$$F(i) = \begin{cases} 1 & \text{if } (d_s \leq D) \wedge (d_o > D) \wedge (M_C > 0.5) \text{ (true positive)} \\ 1 & \text{if } (d_s > D) \wedge (d_o \leq D) \wedge (M_C \leq 0.5) \text{ (true negative)} \\ 0.25 & \text{if } (d_o < D) \wedge (d_s < D) \qquad\qquad \text{(ambiguity)} \\ 0.1 & \text{if false classification and } S > 0 \qquad \text{(touch)} \\ 0 & \text{else} \end{cases} \tag{2}$$

where $D = 30$, d_o the distance to the closest of the two lures and d_s the distance to the other agent. Agents are tested on eight trials and fitness is averaged.

The search algorithm used to evolve the parameters of control networks (1000 generations) is a generational genetic algorithm (GA) using a population of 30 individuals with a real-valued genotype $\in [0, 1]$, truncation selection (1/3), vector mutation of magnitude $r = 0.6$ and reflection at the gene boundaries. The weights w_{ij} and the bias θ_i are mapped linearly to the target range, the sensor gain S_G, the motor gain M_G and the time constants τ_i are mapped exponentially.

3   Results

3.1   Evolvability

All agents evolved to a higher level of performance with delays than without (see Fig. 2 (A)). A similar counter-intuitive benefit of sensory delays had already been observed in the one-dimensional scenario [2]. As explained in the following section 3.2, this is because overshooting of the target bootstraps the evolution of active perceptual strategies.

The wheeled agent and the Euclidean agent evolve to a much higher level of performance (see Fig. 2 (A)), with the best individual from the best evolutionary run achieving nearly perfect performance (Fig. 2 (B)), whilst even the best

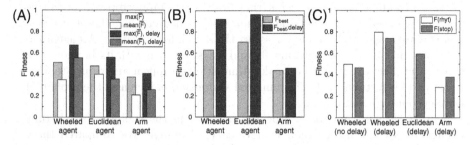

Fig. 2. (A): Population fitness average $\bar{F}$ (mean and maximum of 10 evolutionary runs). (B): Performance average across 100 evaluations for the best individual from the best evolution. Light: no delay, dark: 100 ms sensory delay. (C) Fitness for rhythmic solutions (white) is on average much higher than that for non-rhythmic solutions (grey). (No rhythmic activity was evolved for Euclidean or arm agents without delay; note that the measure for rhythmicity is an approximation as explained in section 3.2.)

evolved arm agent stays well below the 50 % performance level. As explained in Sect. 3.2, this is because evolution of search strategies is much more difficult for the arm agent than for the other two.

3.2   Evolved Behaviours

Two large classes of behaviour dominate the fitness landscape for the evolved task, irrespective of embodiment. The more successful strategy (1) is to always output 'no' ($M_C \leq 0.5$) and seek contact with the fixed lure, avoiding any interaction with any mobile entity (up to perfect fitness). This perfectly viable yet slightly 'autistic' behaviour differs a lot from the participants' behaviour, who avoid the fixed lure and seek interaction with each other. This discrepancy is interesting, not lamentable, because it illustrates alternative solutions afforded by the paradigm. Replicating human behaviour was not the main objective. The other predominating strategy (2) is to interact with anything indiscriminately and constantly output 'yes' ($M_C > 0.5$) and yields a fitness of up to ca. 40%. The arm agents nearly exclusively evolve strategy (2), whereas the Euclidean and wheeled agents evolve strategy (1), frequently passing during evolution through a phase of strategy (2). What evolved, therefore, were preferences rather than discriminatory capacity: only four agents (one arm, one wheeled, two Euclidean) evolved additionally a classification output contingent on stimulation.

Both strategy (1) and strategy (2) can in principle be realised by rhythmical interaction with the target or simply by halting. It appears that rhythmic behaviour is more effective: as an approximation, let rhythmic behaviour be activity confined to a radius of $d = 50$ around an entity with least five inversions of sensory state during the last second of a trial. Within each condition for which both rhythmic and non-rhythmic solutions evolved, the rhythmic ones were on average 9% more successful (see also Fig. 2 (C); some rhythmically interacting arm agents that were not captured by the approximate measure). The reason for this advantage is that an agent that evolves to simply stop will be clueless upon unexpected cessation of the stimulus, e.g. when crossing the object at an unfortunate angle, and therefore re-start the search for sensation. An agent that interacts with an object rhythmically and stays in touch with its boundary can reverse the effect of actions that make stimulation disappear and return to where it last had been stimulated. This minimal *spatialisation* increases the probability to re-encounter a lost object.

Sensory delays seem to be crucially involved in the evolution of rhythmic behaviour, accounting for their evolutionary advantage: We compared the 60 best individuals from all evolutionary runs (across bodies). Only two of the 30 individuals evolved without delays behaved rhythmically at least once in 10 trials (opposed to 16 out of 30 evolved with delay). Sensory delays lead the agents to overshoot an object, which triggers the evolution of return trajectories. This, in turn, facilitates the evolution of rhythmic interaction as effective and active perceptual strategy, and thus helps to overcome the local fitness maximum to stop upon any stimulation and start the search anew if stimulation ceases.

Despite these abstract commonalities, the exact realisation and behavioural dynamics vary quite significantly between the different agents. The following subsections analyse in detail the strategies evolved (with delays) to explore the strategy space and how it is constrained by different embodiments.

The Two-Wheeled Agent. Two behavioural phases that can be variably realised independent of each other can be identified: search and interaction. Wheeled agents evolved a variety of strategies to establish contact: some shoot off straight, others drive around in large circles, yet others in arches or spirals. Interaction is initiated immediately on contact, or, otherwise, the agent backs off and comes back to see if the stimulating object is still there. All wheeled agents evolved to drive in circles (of variable size) around the encountered entity, most of them aiming at a distance from the object that makes stimulation rhythmically appear and disappear.

Figure 3 depicts a sample behaviour of the best agent evolved ($F(i) = 0.92$). Whilst agent 1 is in stable interaction with the fixed lure, agent 2 is momentarily trapped in an interaction with agent 1's attached lure ($t = [500, 1500]$), eventually abandons it, passes the other agent twice and then finds the fixed lure (Fig 3 (A)). Stimulation received by either of the mobile stimuli is not long and rhythmic enough for interaction to stabilise. Even if interaction with the attached lure is maintained over a number of crossings, the irregularity and intermittence of the sensation, which becomes amplified through gradual modification of return trajectories, eventually allows the agent to move on. This strategy only fails in very exceptional cases in which interaction with a mobile entity is phase-locked in a way that resembles interaction with a fixed lure.

The Euclidean Agent. The Euclidean agents evolved mostly scan the space by infinitely going straight around the torus in a direction that produces slightly inclined thin stripes (see best agent with $F(i) = 0.96$ in Fig. 4). This extremely effective search strategy is made possible by the fact that their motor outputs

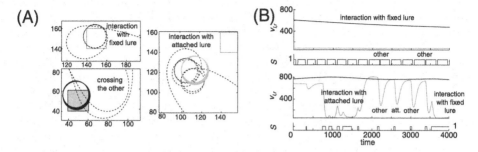

Fig. 3. Best evolved agent. (A) Trajectories (dotted line) of agent 2 during interaction with the fixed lure $t > 3500$, crossing the other $t = [2000, 3000]$ and interaction with attached lure (grey) $t = [750, 1900]$. (B) Corresponding sensorimotor values: $v_{l,r}$ and S (rectangular) for agent 1 (top) and agent 2 (bottom).

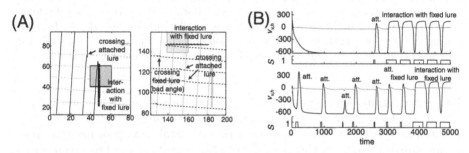

Fig. 4. Best evolved agent. (A) Trajectories of agent 1 (solid) and agent 2 (dotted) leading up to stabilised interaction with the fixed lure. Crossing the attached lure (grey) stimulates for a short time only and thus negligibly slows down the agents (kinks in trajectories). (B) Corresponding sensorimotor values $v_{h,v}$ and S (rectangular) for agent 1 (top) and agent 2 (bottom).

define absolute direction in space. Only two agents evolved to start search in a large curve. About half of the agents evolve to simply stop upon stimulation (hence the slight across population disadvantage compared to the wheeled agents), while the other half engages in rhythmic interaction along one dimension, making stimulation continually appear and disappear. Some of the agents evolved to slowly grind past objects encountered, or to move a bit further away with each oscillation. With such strategies, interaction with the fixed lure is not permanently stable, even if it lasts much longer than interaction with a mobile object (thus avoiding mobile objects). Due to the efficient search behaviour, the chances that to re-encounter the fixed object are still very high. This strategy, again, is very effective and fails only in exceptional cases. Fig. 4 shows how the best agent is hardly perturbed by encountering the fixed lure of the stabilised other because both of them move so fast that stimulation is too short to induce an actual return to the locus of stimulation.

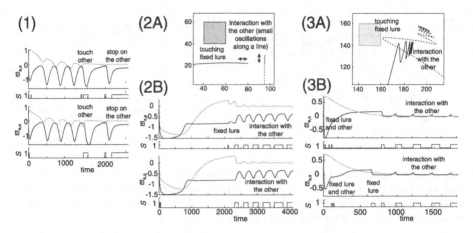

Fig. 5. (1) Agent with neural oscillator as CPG. (2) Best evolved agent. (3) Example agent with proprioception. Trajectories (2A, 3A): Both diagrams depict the rhythmic linear interaction of both agents. In (2A) these oscillations are not visible, because moving just one joint leads to the *exact* inversion of the path just made. Sensorimotor diagrams (1, 2B, 3B): $\omega_{e,s}$ and S (rectangular) for agent 1 (top) and agent 2 (bottom).

The Arm Agent. For the arm agent, scanning the environment is much more difficult than for either of the other agents. Without proprioceptive feedback it has no way of telling where it is and whether it is still moving or has run up to a joint stop or the edge of the 'desk'. Most of the agents evolved to sweep across the surface just once (either by running up to the desk edge in a large arch and then grinding down or by running up to a joint stop and then back in a large arch, using slow and fast τs). If during their sweep no object is touched, they will indefinitely remain immobile and receive $F = 0$, as there is no environmental feedback to guide or inform further action. This enters an element of chance into fitness evaluation, which makes arm evolution noisy and probably accounts for the performance disadvantage. The only way to reliably sustain motion is to evolve a neural oscillator as central pattern generator (CPG), as it was found in one of agents (Fig. 5 (A)), which, by virtue of this CPG, is the second best agent, despite just stopping when being stimulated.

There are, in principle many possibilities to mitigate this problem. A modified evolution with proprioception (joint angles), for instance, immediately produced much higher results (population average/best after 1000 generations in 10 runs: 0.33/0.70; see Fig. 5 (3) for example behaviour). However, as stated earlier, it is not the primary objective of the paper to reproduce human behaviour but to explore dynamical principles given the simulation set-up. Even in the sub-optimal solutions evolved, nearly all arm agents evolve to rhythmically interact with any entity encountered (which is not always recognised by the criterion specified in Sect. 3.2), making the sensory stimulation constantly appear and disappear (see Fig. 5 (2) and (3)). This rhythmic activity is realised by varying one of joint angles that control the arm, which leads to the generation of one-dimensional near straight trajectories, similar to those observed in the human participants. Our hypothesis that this kind of behaviour can result from arm morphology is thus supported by the present results.

4   Discussion

The behavioural strategies that the different kinds of evolved agents employ, even though they differ a lot in immediate appearance and quantitative aspects, follow the same dynamical principles. The behaviour evolved for all three agents is dominated by the two local maxima in the fitness landscape (i.e., 'say yes and indiscriminately interact' and 'say no and seek out the fixed lure'). Also, for all embodiments, a scanning strategy and an interaction strategy can be distinguished and independently realised. Furthermore, all three types of agents have a tendency to evolve rhythmic interactions that make sensations appear and disappear, which is more likely to evolve with delayed sensory feedback and implies an advantage in fitness, irrespective of preference for the other or for the fixed lure. This is because agents that stay in touch with the boundary of their stimulant are more likely to adjust interaction patterns to a certain degree and maintain interaction against perturbations.

Comparing the different sensorimotor diagrams, there are some remarkable similarities as to how such rhythmic interaction is realised. Irrespective of embodiment, once a stimulation is encountered, one of the motor signals is frozen (statistically significant $p \ll 0.01$ difference between variance in M_1 and M_2 during the last second of a trial across conditions), in order to keep interaction in one place, and only becomes active again if stimulation ceases for too long a period of time. The motor neurons that generate this 'frozen' output also tend to have slower time constants (difference not statistically significant). The other motor signal is used to implement local motion and quickly reacts to changes in the rhythmic inputs, actively maintaining interaction. In the wheeled agent, implementing this principle results in small circular trajectories, whereas in the Euclidean and the arm agents, it results in oscillation along a line, just like those observed in human participants. The simulation results thus support our hypothesis that arm morphology plays a role in constituting the one dimensional rhythmic interaction, as the arm-specific implementation of a more general dynamical principle of dimensionality reduction during rhythmic interaction. These simulation results generate the hypotheses that the direction in which the participants oscillate should be orthogonal to the orientation of the arm, and that dimensionality reduction should serve rhythmic interaction.

Two phenomena already observed in the model for the one-dimensional version of the experiment have been found to occur again: the role of delays in the evolution of oscillatory scanning behaviour and the distinction between the fixed lure and the other agent on the basis of integrated stimulation time, which propose the investigation of dependencies between a) latencies in sensorimotor action and frequency of oscillations and b) the variation in integrated stimulation time due to anti-phase co-ordination and its role in behavioural preference.

In our earlier simulation model [2], we had evolved agents to seek interaction with one another, presuming a preference for live interaction. Leaving artificial evolution to determine an agent's preference, however, favours 'autistic' agents that avoid any interaction with moving entities, because stable interaction with a static object is easier established and more reliably maintained. This opens an interesting perspective on the previous experiment and simulation, for which we had concluded that perceptual crossing is a nearly inevitable result from the mutual search for each other. In the light of the present simulation results, it becomes clear that the dynamics of the task alone (i.e., leaving aside motivational factors such as boredom) do not favour perceptual crossing, but much rather interaction with the static lure, and that perceptual crossing is established *despite* this strong basin of attraction. Interestingly, a recent modified replication of the one-dimensional experiment with humans (Di Paolo, personal communication) appears to produce results that resemble more the 'autistic' behaviour reported here than the social behaviour observed in the original study in some subjects.

The results presented identify dynamical principles in all evolved solutions and variations in how these principles are realised across different embodiments. Our simulation experiments have generated a number of hypotheses for analysis of the empirical results and thus reaffirm our conviction that evolutionary robotics simulation models are a rich source of illustration and proofs of concept to aid minimalist experimental research on human sensorimotor dynamics.

References

1. Auvray, M., Lenay, C., Stewart, J.: Perceptual interactions in a minimalist virtual environment. New Ideas in Psychology (forthcoming)
2. Di Paolo, E., Rohde, M., Iizuka, H.: Sensitivity to social contingency or stability of interaction? modelling the dynamics of perceptual crossing. New Ideas in Psychology (forthcoming); Special Issue on Dynamical Systems approaches
3. Rohde, M., Di Paolo, E.: Adaptation to sensory delays: An evolutionary robotics model of an empirical study. In: Almeida e Costa, F., Rocha, L., Costa, E., Harvey, I., Coutinho, A. (eds.) Proceedings of the 9th European Conference on Artificial Life. LNCS (LNAI), pp. 193–202. Springer, Heidelberg (2007)
4. Beer, R.: Toward the evolution of dynamical neural networks for minimally cognitive behavior. In: Maes, P., Mataric, M., Meyer, J., Pollack, J., Wilson, S. (eds.) From Animals to Animats 4, pp. 421–429. MIT press, Cambridge (1996)

Adaptive Optimal Control for Redundantly Actuated Arms

Djordje Mitrovic, Stefan Klanke, and Sethu Vijayakumar

Institute of Perception, Action & Behavior, University of Edinburgh,
The King's Buildings, Edinburgh EH9 3JZ, United Kingdom
{d.mitrovic,s.klanke,sethu.vijayakumar}@ed.ac.uk

Abstract. Optimal feedback control has been proposed as an attractive movement generation strategy in goal reaching tasks for anthropomorphic manipulator systems. Recent developments, such as the iterative Linear Quadratic Gaussian (iLQG) algorithm, have focused on the case of non-linear, but still analytically available, dynamics. For realistic control systems, however, the dynamics may often be unknown, difficult to estimate, or subject to frequent systematic changes. In this paper, we combine the iLQG framework with learning the forward dynamics for a simulated arm with two limbs and six antagonistic muscles, and we demonstrate how our approach can compensate for complex dynamic perturbations in an online fashion.

Keywords: Adaptive optimal control, learning dynamics, redundant actuation.

1 Introduction

In this work, we focus on the issues related to planning and control of reaching movements for anthropomorphic manipulators with redundant actuation based on antagonistic muscles. While such systems are becoming more and more popular especially where compliance and interaction with humans is required, controlling these systems remains a big challenge: Apart from the problem of often highly non-linear and hard to model system dynamics, the controller has to make a choice between many different possible trajectories (kinematics) and a multitude of applicable motor commands (dynamics) for achieving a particular task. How do we resolve this redundancy?

Optimal control theory [1] answers this question by establishing a certain cost function, and selecting the solution with minimal cost (e.g., minimum jerk [2]). Quite often these control schemes are only concerned with trajectory *planning* and an "open loop" optimisation of the control commands, while the correction of errors during *execution* is left to simple PID controllers.

As an alternative, closed loop optimisation models are aimed at providing a control law which is explicitly based on feedback from the system. In the ideal case, the system state is directly mapped to control signals during execution, and the form of this mapping is again governed by a cost function. A key property

M. Asada et al. (Eds.): SAB 2008, LNAI 5040, pp. 93–102, 2008.
© Springer-Verlag Berlin Heidelberg 2008

of such optimal feedback controllers (OFC) is that errors are only corrected if they adversely affect the task performance (minimum intervention principle [3]). This is important especially in systems that suffer from control dependent noise, since task-irrelevant correction could destabilise the system beside expending additional control effort. Empirically, OFC also accounts for many motion patterns that have been observed in natural, redundant systems and human experiments [4] including the confounding trial-to-trial variability in individual degrees of freedom that, remarkably, manages to not compromise task optimality [5,6]. Therefore, this paradigm is potentially a very attractive control strategy for artificial anthropomorphic systems (i.e., many degrees of freedom, redundant actuation, flexible lightweight construction, variable stiffness).

Unfortunately, finding a globally valid optimal control law is a very hard problem especially for non-linear and high-dimensional systems. We therefore resort to hybrid algorithms that present a compromise between open loop and closed loop optimisation, that is, algorithms which iteratively compute an optimal trajectory together with a locally valid feedback law. Examples of these are differential dynamic programming (DDP) [7,8], iterative linear-quadratic regulator designs [9], or the recent iterative Linear Quadratic Gaussian (iLQG) framework [10], which will form the basis of our work.

A major shortcoming of iLQG (and DDP) is the dependence on an analytic form of the system dynamics, which often may be unknown or subject to change. We overcome this limitation by learning an adaptive internal model of the system dynamics using an online, supervised learning method. We consequently use the learned model to derive an iLQG formulation that is computationally efficient, reacts optimally to transient perturbations, and most notably adapts to systematic changes in the plant dynamics.

The idea of learning the system dynamics in combination with iterative optimisations of trajectory or policy has been explored previously in the literature, e.g., for learning to swing up a pendulum [11] using some prior knowledge about the form of the dynamics. Similarly, Abeel et al. [12] proposed a hybrid reinforcement learning algorithm, where a policy and an internal model get subsequently updated from "real life" trials. In contrast to their method, however, we employ a second-order optimisation method, and we refine the control law solely from the internal model. To our knowledge, learning dynamics in conjunction with control optimisation has not been studied in the light of adaptability to changing plant dynamics. In this paper, we successfully apply our adaptive control formalism to a movement system with six antagonistic muscles, which exhibits large redundancies and complex non-linearities of the dynamics.

2   A Simulation Model of Redundant Actuation

We wish to study a two degrees of freedom (DoF) planar human arm model, which is actuated by four single-joint and two double-joint antagonistic muscles (Fig. 1, left). The arm model described in this section is based on [13]. Although kinematically simple, the system is over-actuated and therefore an interesting testbed for our control scheme, because large redundancies in the dynamics

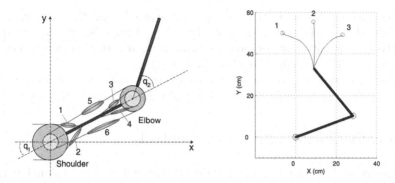

Fig. 1. Left: Human arm model with 6 muscles (adapted from [13]). Right: Same arm model with three selected targets (circles) and iLQG generated trajectories as benchmark data. The physics of the model is simulated using the Matlab Robotics Toolbox [14].

have to be resolved. The dimensionality of the control signals makes adaptation processes (e.g., to external force fields) quite demanding.

The dynamics of the arm is in part based on standard equations of motion. For our planar 2-DoF manipulator the joint torques τ are given by

$$\tau = \mathbf{M}(\mathbf{q})\ddot{\mathbf{q}} + \mathbf{C}(\mathbf{q}, \dot{\mathbf{q}})\dot{\mathbf{q}}, \tag{1}$$

where $\mathbf{q}$ and $\dot{\mathbf{q}}$ are the joint angles and velocities, respectively; $\mathbf{M}(\mathbf{q})$ is the two-dimensional symmetric joint space inertia matrix and $\mathbf{C}(\mathbf{q}, \dot{\mathbf{q}})$ accounts for Coriolis and centripetal forces.

Given the antagonistic muscle-based actuation, we can not command joint torques directly, but rather we have to calculate effective torques from the muscle activations $\mathbf{u}$. For the present model the corresponding transfer function is given by

$$\tau(\mathbf{q}, \dot{\mathbf{q}}, \mathbf{u}) = -\mathbf{A}(\mathbf{q})^T \mathbf{T}(\mathbf{l}, \dot{\mathbf{l}}, \mathbf{u}), \tag{2}$$

where $\mathbf{A}$ represents the moment arm. For simplicity, we assume $\mathbf{A}$ to be constant and independent of the joint angles $\mathbf{q}$:

$$\mathbf{A}(\mathbf{q}) = \mathbf{A} = \begin{pmatrix} a_1 & a_2 & 0 & 0 & a_5 & a_6 \\ 0 & 0 & a_3 & a_4 & a_7 & a_8 \end{pmatrix}^T. \tag{3}$$

The muscle lengths $\mathbf{l}$ depend on the joint angles $\mathbf{q}$ through the affine relationship $\mathbf{l} = \mathbf{l}_m - \mathbf{A}\mathbf{q}$, which also implies $\dot{\mathbf{l}} = -\mathbf{A}\dot{\mathbf{q}}$. The term $\mathbf{T}(\mathbf{l}, \dot{\mathbf{l}}, \mathbf{u})$ in (2) denotes the muscle tension, for which we follow the Kelvin-Voight model [15] and define:

$$\mathbf{T}(\mathbf{l}, \dot{\mathbf{l}}, \mathbf{u}) = \mathbf{K}(\mathbf{u})(\mathbf{l}_r(\mathbf{u}) - \mathbf{l}) - \mathbf{B}(\mathbf{u})\dot{\mathbf{l}}. \tag{4}$$

Here, $\mathbf{K}(\mathbf{u})$, $\mathbf{B}(\mathbf{u})$, and $\mathbf{l}_r(\mathbf{u})$ denote the muscle stiffness, the muscle viscosity and the muscle rest length, respectively. Each of these terms depends linearly on the motor commands $\mathbf{u}$, as given by

$$\mathbf{K}(\mathbf{u}) = diag(\mathbf{k}_0 + k\mathbf{u}), \qquad \mathbf{B}(\mathbf{u}) = diag(\mathbf{b}_0 + b\mathbf{u}), \qquad \mathbf{l}_r(\mathbf{u}) = \mathbf{l}_0 + r\mathbf{u}. \tag{5}$$

The elasticity coefficient k, the viscosity coefficient b, and the constant r are given from the muscle model. The same holds true for k_0, b_0, and l_0, which are the intrinsic elasticity, viscosity and rest length for $u = 0$, respectively. For the exact values of these coefficients please refer to [13].

Please note that in contrast to standard torque-controlled robots, here the dynamics (1) is *not* linear in the control signals, since u enters (4) quadratically.

3   Locally-Optimal Feedback Control

Let $x(t)$ denote the state of a plant and $u(t)$ the applied control signal at time t. In this paper, the state consists of the joint angles q and velocities $\dot{q}$ of the arm, and the control signals u are the muscle activations. If the system would be deterministic, we could express its dynamics as $\dot{x} = f(x, u)$, whereas in the ·
presence of noise we write the dynamics as a stochastic differential equation

$$dx = f(x, u)dt + F(x, u)d\omega. \tag{6}$$

Here, $d\omega$ is assumed to be Brownian motion noise, which is transformed by a possibly state- and control-dependent matrix $F(x, u)$. We state our problem as follows: Given an initial state x_0 at time $t = 0$, we seek a control sequence $u(t)$ such that the system's state is x^* at time $t = T$. Stochastic optimal control theory approaches the problem by first specifying a cost function which is composed of (i) some evaluation $h(x(T))$ of the final state, usually penalising deviations from the desired state x^*, and (ii) the accumulated cost $c(t, x, u)$ of sending a control signal u at time t in state x, typically penalising large motor commands. Introducing a policy $\pi(t, x)$ for selecting $u(t)$, we can write the *expected cost* of following that policy from time t as [10]

$$v^\pi(t, x(t)) = \left\langle h(x(T)) + \int_t^T c(s, x(s), \pi(s, x(s)))ds \right\rangle. \tag{7}$$

One then aims to find the policy π that minimises the total expected cost $v^\pi(0, x_0)$. Thus, in contrast to classical control, calculation of the trajectory (planning) and the control signal (execution) is not separated anymore, and for example, redundancy can actually be exploited in order to decrease the cost. If the dynamics f is linear in x and u, the cost is quadratic, and the noise is Gaussian, the resulting so-called LQG problem is convex and can be solved analytically [1].

In our case of non-linear dynamics, global solutions can in theory still be found by applying dynamic programming methods [16] based on the Hamilton-Jacobi-Bellman equations. However, in their basic form these methods rely on a discretisation of the state and action space, an approach that is not viable for large DoF systems. Some research has been carried out on random sampling in a continuous state and action space [17], and it has been suggested that sampling can avoid the curse of dimensionality if the underlying problem is simple enough [18], as is the case if the dynamics and cost functions are very smooth.

As an alternative, one can compute linear and quadratic approximations to the dynamics and the cost, respectively, and iteratively solve a "local" LQG problem to improve the control solution, until at least a local minimum of the cost function is found. The resulting iLQG algorithm has only recently been introduced [10], so we give a brief summary in the following[1].

One starts with an initial time-discretised control sequence $\bar{\mathbf{u}}_k \equiv \bar{\mathbf{u}}(k\Delta t)$ and applies the deterministic forward dynamics to retrieve an initial trajectory $\bar{\mathbf{x}}_k$, where

$$\bar{\mathbf{x}}_{k+1} = \bar{\mathbf{x}}_k + \Delta t\, \mathbf{f}(\bar{\mathbf{x}}_k, \bar{\mathbf{u}}_k). \tag{8}$$

Linearising the discretised dynamics (6) around $\bar{\mathbf{x}}_k$ and $\bar{\mathbf{u}}_k$ and subtracting (8), one gets a dynamics equation for the deviations $\delta\mathbf{x}_k = \mathbf{x}_k - \bar{\mathbf{x}}_k$ and $\delta\mathbf{u}_k = \mathbf{u}_k - \bar{\mathbf{u}}_k$:

$$\delta\mathbf{x}_{k+1} = \left(\mathbf{I} + \Delta t \frac{\partial \mathbf{f}}{\partial \mathbf{x}}\Big|_{\bar{\mathbf{x}}_k}\right)\delta\mathbf{x}_k + \Delta t \frac{\partial \mathbf{f}}{\partial \mathbf{u}}\Big|_{\bar{\mathbf{u}}_k}\delta\mathbf{u}_k + \sqrt{\Delta t}\left(\mathbf{F}(\mathbf{u}_k) + \frac{\partial \mathbf{F}}{\partial \mathbf{u}}\Big|_{\bar{\mathbf{u}}_k}\delta\mathbf{u}_k\right)\boldsymbol{\xi}_k. \tag{9}$$

Similarly, one can derive an approximate cost function which is quadratic in $\delta\mathbf{u}$ and $\delta\mathbf{x}$. Thus, in the vicinity of the current trajectory $\bar{\mathbf{x}}$, the two approximations form a "local" LQG problem, which can be solved analytically and yields an affine control law $\delta\mathbf{u}_k = \mathbf{l}_k + \mathbf{L}_k \delta\mathbf{x}_k$ (for details please see [10]). This control law is fed into the linearised dynamics (eq. 9 without the noise term) and the resulting $\delta\mathbf{x}$ are used to update the trajectory $\bar{\mathbf{x}}$. In the same way, the control sequence $\bar{\mathbf{u}}$ is updated from $\delta\mathbf{u}$. This process is repeated until the total cost cannot be reduced anymore. The resultant control sequence $\bar{\mathbf{u}}$ can then be applied to the system, whereas the matrices $\mathbf{L}_k$ from the final iteration may serve as feedback gains.

In our current implementation we do not utilise an explicit noise model $\mathbf{F}$ for the sake of clarity of results; in any case, a matching feedback control law is only marginally superior to one that is optimised for a deterministic system [10].

4 iLQG with Learned Dynamics (iLQG–LD)

In order to eliminate the need for an analytic dynamics model and to make iLQG adaptive, we wish to learn an approximation $\tilde{\mathbf{f}}$ of the real plant forward dynamics $\dot{\mathbf{x}} = \mathbf{f}(\mathbf{x}, \mathbf{u})$. Assuming our model $\tilde{\mathbf{f}}$ has been coarsely pre-trained, for example by motor babbling, we can refine that model in an online fashion as shown in Fig. 2.

For optimising and carrying out a movement, we have to define a cost function (where also the desired final state is encoded), the start state, and the number of discrete time steps. Given an initial control sequence $\bar{\mathbf{u}}^0$, the iLQG iterations can be carried out as described in the previous section, but utilising the learned model $\tilde{\mathbf{f}}$. This yields a locally optimal control sequence $\bar{\mathbf{u}}_k$, a corresponding desired state sequence $\bar{\mathbf{x}}_k$, and feedback correction gain matrices $\mathbf{L}_k$. Denoting the plant's true state by $\mathbf{x}$, at each time step k, the feedback controller calculates the required correction to the control signal as $\delta\mathbf{u}_k = \mathbf{L}_k(\mathbf{x}_k - \bar{\mathbf{x}}_k)$. We then use

---

[1] DDP works similarly, but requires quadratic approximations of both the dynamics and the cost function.

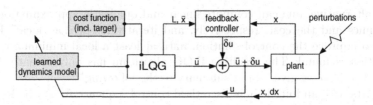

Fig. 2. Illustration of our iLQG–LD learning and control scheme

the final control signal $\mathbf{u}_k = \bar{\mathbf{u}}_k + \delta\mathbf{u}_k$, the plant's state $\mathbf{x}_k$ and its change $d\mathbf{x}_k$ to update our internal forward model $\tilde{\mathbf{f}}$. As we show in Section 5, we can thus account for (systematic) perturbations and also bootstrap a dynamics model from scratch.

The domain of real-time control demands certain properties of a learning algorithm, namely fast learning rates, high prediction speeds at run-time, and robustness towards negative interference if the model is trained incrementally. Locally Weighted Projection Regression (LWPR) has been shown to exhibit these properties, and to be very efficient for incremental learning of non-linear models in high dimensions [19]. In LWPR, the regression function is constructed by blending local linear models, each of which is endowed with a locality kernel that defines the area of its validity (also termed its receptive field). During training, the parameters of the local models (locality and fit) are updated using incremental Partial Least Squares, and models can be pruned or added on an as-need basis, for example, when training data is generated in previously unexplored regions.

LWPR learning has the desirable property that it can be carried out online, and moreover, the learned model can be adapted to changes in the dynamics in real-time. A forgetting factor λ [19], which balances the trade-off between preserving what has been learned and quickly adapting to the non-stationarity, can be tuned to the expected rate of external changes.

5   Experiments

We study movements of our arm model (Section 2) for a fixed motion duration of one second, which we discretise into $K = 50$ steps ($\Delta t = 0.02$s). The manipulator starts at an initial position $\mathbf{q}_0$ and reaches towards a target $\mathbf{q}_{tar}$. During movement we wish to minimise the amount of muscle activation ($\approx$ energy consumption) of the system. We therefore use the cost function

$$v = w_p \,|\, \mathbf{q}_K - \mathbf{q}_{tar} \,|^2 + w_v \,|\, \dot{\mathbf{q}}_K \,|^2 + w_e \sum_{k=0}^{K} |\, \mathbf{u}_k \,|^2 \Delta t, \tag{10}$$

where the factors for the target position accuracy (w_p), the final target velocity accuracy (w_v), and for the energy term (w_e) weight the importance of each component.

5.1 Stationary Dynamics

In order to make iLQG–LD work for our three reference targets (see Fig. 1, right) we coarsely pre-trained our LWPR model with a focus on a wide coverage of the workspace. For the arm model we use in this paper, the training data are given as tuples consisting of $(\mathbf{q}, \dot{\mathbf{q}}, \mathbf{u})$ as inputs (10 dimensions in total), and the observed joint accelerations $\ddot{\mathbf{q}}$ as the desired two-dimensional output. We stopped training once the normalised mean squared error (nMSE) in the predictions reached ≤ 0.005. At this point LWPR had seen $1.2 \cdot 10^6$ training data points and had acquired 852 receptive fields, which is in accordance with the previously discussed high non-linearity of the plant dynamics.

We carried out a reaching task to the three reference targets using the feedback controller (feedback gain matrix $\mathbf{L}$) that falls out of iLQG(-LD). To compare the stability of the control solution, we simulated control dependent noise by contaminating the muscle commands $\mathbf{u}$ just before feeding them into the plant. We applied Gaussian noise with 50% of the variance of the signal $\mathbf{u}$.

Figure 3 depicts the generated control signals and the resulting performance of iLQG–LD and iLQG over 20 reaching trials per target. Both methods show similar endpoint variances and trajectories which are in close match. As can be seen from the visualisation of the control sequences, antagonistic muscles (i.e., muscle pairs 1/2, 3/4, and 5/6) are never activated at the same time. This is a direct consequence of the cost function, which penalises co-contraction as a waste of energy. Table 1 quantifies the control results of iLQG–LD and iLQG for each target with respect to the number of iterations, the generated running costs and the end point accuracy.

5.2 Adaptation Results

A major advantage of iLQG–LD is that it does not rely on an accurate analytic dynamics model; consequently, it can adapt 'on-the-fly' to external perturbations

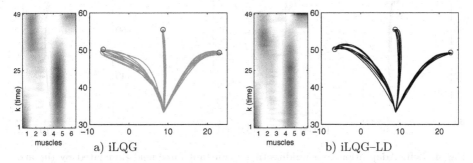

Fig. 3. Illustration of an optimised control sequence (left) and resulting trajectories (right) when using a) the known analytic dynamics model and b) the LWPR model learned from data. The control sequences (left target only) for each muscle (1–6) are drawn from bottom to top, with darker grey levels indicating stronger muscle activation.

Table 1. Comparison of the performance of iLQG–LD and iLQG with respect to the number of iterations required to compute the control law, the average running cost, and the average Euclidean distance to the three reference targets (left, center, right).

		iLQG			iLQG–LD	
Targets	Iter.	Run. cost	d (cm)	Iter.	Run. cost	d (cm)
Center	19	**0.0345** ± 0.0060	**0.11** ± 0.07	14	**0.0427** ± 0.0069	**0.38** ± 0.22
Left	40	**0.1873** ± 0.0204	**0.10** ± 0.06	36	**0.1670** ± 0.0136	**0.21** ± 0.16
Right	41	**0.1858** ± 0.0202	**0.57** ± 0.49	36	**0.1534** ± 0.0273	**0.19** ± 0.12

and to changes in the plant dynamics that may result from altered morphology or wear and tear. We carried out adaptive reaching experiments (towards the center target) in our simulation similar to the human manipulandum experiments in [20]. We generated a constant unidirectional force field (FF) acting perpendicular to the reaching movement (see Fig. 4). Using the iLQG–LD model from the previous experiment, the manipulator gets strongly deflected when reaching for the target because the learned dynamics model cannot yet account for the "spurious" forces. However, using the resultant deflected trajectory as training data, updating the dynamics model online brings the manipulator nearer to the target with each new trial. In order to produce enough training data, as is required for a successful adaptation, we generated 20 slightly jittered versions of the optimised control sequences, each with length $K = 50$. We then ran those 20 trajectories on the plant, and trained the LWPR model with a total of $K \times 20 = 1000$ samples. We repeated this procedure until the iLQG–LD performance converged successfully, which was the case after 27000 training samples. At that point, the internal model successfully accounted for the change in dynamics caused by the FF. Then, we switched off the FF while continuing to use the adapted LWPR model. This resulted in an overshooting of the manipulator to the other side,

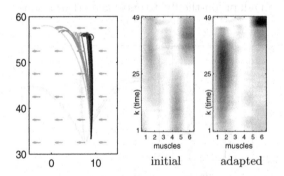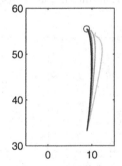

Fig. 4. Left: Adaptation to a unidirectional constant force field (indicated by the arrows). Darker lines indicate better trained models. In particular, the left-most trajectory corresponds to the "initial" control sequence, which was calculated using the LWPR model *before* the adaptation process. The fully "adapted" control sequence results in a nearly straight line reaching movement. Right: Resulting trajectories during re-adaptation after the force field has been switched off.

trying to compensate for non-existing forces. Just as before, we re-adapted the dynamics online over repeated trials. The arm reached the target again after 7000 training points.

For accelerating the adaptation process, we set LWPR's forgetting factor to $\lambda = 0.95$ (instead of the default 0.999), which allows the learner to weight the importance of new data more strongly [19]. It is interesting to note that since the iLQG–LD control scheme always tries to correct the system towards the target, it produces relevant dynamics training data in a way that could be termed "active learning".

Figure 4 summarises the results of the sequential adaptation process just described. Please note how the optimised "adapted" control sequence contains considerably stronger activations of the extensor muscles responsible for pulling the arm to the right (denoted by "2" and "6" in Fig. 1, left), while still exhibiting practically no co-contraction.

6 Conclusion

In this work we introduced iLQG–LD, a method that realises adaptive optimal feedback control by incorporating a learned dynamics model into the iLQG framework. Most importantly, we carried over the favourable properties of iLQG to more realistic control problems where the analytic dynamics model is often unknown, difficult to estimate accurately or subject to changes. As with iLQG control, redundancies are implicitly resolved by the OFC framework through a cost function, eliminating the need for a separate trajectory planner and inverse kinematics/dynamics computation.

Using a non-linear arm model actuated by six antagonistic muscles, we empirically showed that iLQG–LD performs reliably in the presence of noise and that it is adaptive with respect to systematic changes in the dynamics; hence, the framework has the potential to provide a unifying tool for modelling (and informing) non-linear sensorimotor adaptation experiments even under complex dynamic perturbations.

Acknowledgements. This work has been carried out within the SENSOPAC project which is supported by the European Commission through the Sixth Framework Programme for Research and Development.

References

1. Stengel, R.F.: Optimal control and estimation. Dover Publications, New York (1994)
2. Flash, T., Hogan, N.: The coordination of arm movements: an experimentally confirmed mathematical model. Journal of Neuroscience 5, 1688–1703 (1985)
3. Todorov, E., Jordan, M.: A minimal intervention principle for coordinated movement. In: Advances in Neural Information Processing Systems, vol. 15, pp. 27–34. MIT Press, Cambridge (2003)

4. Shadmehr, R., Wise, S.P.: The Computational Neurobiology of Reaching and Ponting. MIT Press, Cambridge (2005)
5. Li, W.: Optimal Control for Biological Movement Systems. PhD dissertation, University of California, San Diego (2006)
6. Scott, S.H.: Optimal feedback control and the neural basis of volitional motor control. Nature Reviews Neuroscience 5, 532–546 (2004)
7. Dyer, P., McReynolds, S.: The Computational Theory of Optimal Control. Academic Press, New York (1970)
8. Jacobson, D.H., Mayne, D.Q.: Differential Dynamic Programming. Elsevier, New York (1970)
9. Li, W., Todorov, E.: Iterative linear-quadratic regulator design for nonlinear biological movement systems. In: Proc. 1st Int. Conf. Informatics in Control, Automation and Robotics (2004)
10. Todorov, E., Li, W.: A generalized iterative LQG method for locally-optimal feedback control of constrained nonlinear stochastic systems. In: Proc. of the American Control Conference (2005)
11. Atkeson, C.G., Schaal, S.: Learning tasks from a single demonstration. In: Proc. Int. Conf. on Robotics and Automation (ICRA), Albuquerque, New Mexico, vol. 2, pp. 1706–1712 (1997)
12. Abbeel, P., Quigley, M., Ng, A.Y.: Using inaccurate models in reinforcement learning. In: Proc. Int. Conf. on Machine Learning, pp. 1–8 (2006)
13. Katayama, M., Kawato, M.: Virtual trajectory and stiffness ellipse during multi-joint arm movement predicted by neural inverse model. Biol. Cybern. 69, 353–362 (1993)
14. Corke, P.I.: A robotics toolbox for MATLAB. IEEE Robotics and Automation Magazine 3(1), 24–32 (1996)
15. Özkaya, N., Nordin, M.: Fundamentals of biomechanics: equilibrium, motion, and deformation. Van Nostrand Reinhold, New York (1991)
16. Bertsekas, D.P.: Dynamic programming and optimal control. Athena Scientific, Belmont, Mass (1995)
17. Thrun, S.: Monte carlo POMDPs. In: Advances in Neural Information Processing Systems 12, pp. 1064–1070. MIT Press, Cambridge (2000)
18. Atkeson, C.G.: Randomly sampling actions in dynamic programming. In: Proc. Int. Symp. on Approximate Dynamic Programming and Reinforcement Learning, pp. 185–192 (2007)
19. Vijayakumar, S., D'Souza, A., Schaal, S.: Incremental online learning in high dimensions. Neural Computation 17, 2602–2634 (2005)
20. Shadmehr, R., Mussa-Ivaldi, F.A.: Adaptive representation of dynamics during learning of a motor task. The Journal of Neurosciene 14(5), 3208–3224 (1994)

Monostable Controllers for Adaptive Behaviour

Christopher L. Buckley[1], Peter Fine[2], Seth Bullock[1], and Ezequiel Di Paolo[2]

[1] School of Electronics and Computer Science, University of Southampton, UK
[2] School of Informatics, University of Sussex, UK
clb05r@ecs.soton.ac.uk, p.a.fine@sussex.ac.uk,
sgb@ecs.soton.ac.uk, ezequiel@sussex.ac.uk

Abstract. Recent artificial neural networks for machine learning have exploited transient dynamics around globally stable attractors, inspired by the properties of cortical microcolumns. Here we explore whether similarly constrained neural network controllers can be exploited for embodied, situated adaptive behaviour. We demonstrate that it is possible to evolve globally stable neurocontrollers containing a single basin of attraction, which nevertheless sustain multiple modes of behaviour. This is achieved by exploiting interaction between environmental input and transient dynamics. We present results that suggest that this globally stable regime may constitute an evolvable and dynamically rich subset of recurrent neural network configurations, especially in larger networks. We discuss the issue of scalability and the possibility that there may be alternative adaptive behaviour tasks that are more 'attractor hungry'.

Keywords: Global stability, echo state networks, evolvability.

1 Introduction

Certain regions of the cortex are organised into neural microcolumns. It has been suggested that the computational power of these cortical microcolumns stems from their transient dynamics rather than their attractor structure [1]. This could be the result of weak coupling between the neurons, both in terms of their connectivity and weight strength, which can confer global stability on their dynamics [2]; consider that ripples on the surface of a liquid are only a temporary echo of a stone dropped into a it. Any "computation" undertaken by such a system may only be achieved by exploiting interaction between its environmental input and its transient dynamics within a single basin of attraction. Recently, the echo state and liquid state approaches have employed artificial neural networks that share this constrained dynamics [1,3]. They have been shown to perform well on a range of machine learning tasks [4]. Despite this, and notwithstanding the biological heritage of these artificial neural networks (ANNs), it is not immediately obvious that they can serve as effective control systems for adaptive behaviour since, in general, such controllers must cope with an agent that is embodied and situated in a changing environment that demands different modes of behaviour in different circumstances.

Here we explore whether continuous-time recurrent neural networks (CTRNNs) constrained to exhibit transient dynamics around a globally asymptotically stable fixed

M. Asada et al. (Eds.): SAB 2008, LNAI 5040, pp. 103–112, 2008.
© Springer-Verlag Berlin Heidelberg 2008

point attractor are capable of successfully completing a task that demands different behavioural modes. First, we briefly recap the role of attractor dynamics in the use and understanding of ANNs. Subsequently, we introduce a state-hungry task and evolve a solution that employs bistability. Section 3 presents results from evolutionary studies in which systems constrained to exhibit a single attractor are demonstrated to solve the task. After making a preliminary investigation of the difference between the two kinds of solution, we go on to consider the scalability of this type of constrained neural network. We conclude by discussing the implications of these results.

1.1   Mechanisms of State Retention in Neural Networks

Much of the early work on artificial neural networks focused on feedforward architectures [5]. Here, information is processed along a unidirectional pipeline mapping sensory input onto motor output. This process is necessarily atemporal and does not involve internal state. Stateful networks only really arrived in the 1980s, when Hopfield popularised recurrent neural networks (RNNs). Hopfield's (1982) RNNs were constrained such that each was guaranteed to exhibit a number of fixed point attractors. In such networks, an input is a static or slowly time-varying signal and the system is allowed to converge to an equilibrium. The particular attractor that is achieved is interpreted as a recognition or recall event. The success of Hopfield style systems resulted in the attractor becoming the dominating neurological metaphor for memory and state within biological organisms.

More recent work in neuroscience and adaptive behaviour research have reconsidered organisms as dynamical systems, and foregrounded the role of time [7]. This has led to the frequent observation that, far from settling into stable attractor states, neural systems often spend most of the time exhibiting *transient* dynamics, often far from equilibrium. Systems that tend to settle into static states when examined in isolation (either in a petri dish or as computational models) may in fact be far from equilibrium when coupled to bodies and environments. However, while one need not subscribe to the idea that attractors within the agent's control system dynamics are 'representational', intuitively, it may still be attractive to interpret the presence of distinct attractors as enabling for tasks that demand distinct behavioural modes.

Here we probe this intuitive correspondence between behavioural modes and distinct attractors by comparing control systems that are constrained to exhibit a single attractor with those that are capable of exhibiting multiple fixed-point and/or cyclic attractors. Before defining the task, agent architecture and evolutionary scheme that will be employed for the remainder of the paper, we introduce and formalise some constraints on global asymptotic stability.

1.2   Single Fixed-Point Attractors and the Echo State Property

An echo state machine, comprising a large neural network "reservoir" and a simple feedforward network "readout", must satisfy the echo state property, typically achieved by enforcing global asymptotic fixed-point stability (GAS). Informally, the echo state property demands that the dynamics of the recurrent neural network (the reservoir) will "wash out" all information from the initial conditions after some sufficient period of time [3]. One formulation of this condition demands that the distance between any two trajectories in the system is a monotonically decreasing function of time, i.e., the system exhibits a single limit set [3]. Jaeger provides a *sufficient* condition for GAS, however, it is difficult to check efficiently. Furthermore, it is believed to be overly restrictive since many systems which do not meet this strict condition may exhibit GAS. In practice, a simpler but only *necessary* condition can be derived for systems with an equilibrium coincident with the centre of each node's transfer function (i.e. centre-crossing networks [8]). While not sufficient for GAS, this condition has been shown to confer the echo state property in the majority of echo state applications [3].

In this paper we modify this condition so that it can be used with a CTRNN equation given by:

$$\tau_i \dot{y}_i = -y_i + \sum_j \omega_{ji} \tanh(y_j + \theta_j) + I_i \tag{1}$$

Here y_i represents the activation at the ith neuron, ω_{ji} is the weight of the connection from neuron j to neuron i, θ_i is the bias value at the ith neuron, τ_i defines the rate of leakage or decay of the activation, and tanh is the transfer function. While this equation is not identical to either Jaeger or Beer's formulations, all such networks fall in the larger class of Cohen and Grossberg networks [9] to which the following stability results will apply.

As stated above, we first require that the network is in the centre-crossing configuration. While this is not trivial to impose in general [8], for Equation (1) it can be achieved by setting all of the system's biases to zero ensuring that there is an equilibrium at the zero state, i.e., $\mathbf{y}^* = \bar{\mathbf{0}}$, where $\mathbf{y}^*$ is a vector describing the position of an equilibrium point. We then construct a criterion for local stability by linearising the system around this point. Given that the slope of the tanh function is equal to unity at its centre, the linear dynamics is completely described by a Jacobian at the equilibrium point given by:

$$J = \begin{pmatrix} \frac{\omega_{11}-1}{\tau_1} & \cdots & \frac{\omega_{1N}}{\tau_1} \\ \vdots & & \vdots \\ \frac{\omega_{N1}}{\tau_N} & \cdots & \frac{\omega_{NN}-1}{\tau_N} \end{pmatrix} \tag{2}$$

This system will be locally stable around the zero state if all real parts of the eigenvalues of the Jacobian are negative, otherwise it will be unstable [10]. Local stability is a necessary but not sufficient precondition for global stability of the full, nonlinear centre-crossing system [11]. For a full discussion of the condtions for local and global stability in CTRNNs see [10].

2 Methods

In order to experimentally investigate the role of attractor structure in an embodied, behaving agent, we needed to construct a task that demands state and encourages multiple modes of behaviour. We carried out experiments using a simulated agent required to perform phototaxis towards a sequence of lights using a single light sensor. The task was made more challenging by alternating the location of the light sensor intermittently between the front and back of the agent's body throughout its lifetime. Since the agent cannot reverse, the nature of the agent's phototactic task is fundamentally altered by this unsignalled change to its sensor. In previous work on this task evolved solutions employed bistable controllers [12].

The experiment consists of a circular agent with radius equal to unity located in an infinite 2-D plane. (Distance and time are measured in arbitrary units.) At the start of each trial, a single light source is presented, located at a random distance, d from the agent, drawn from a uniform random distribution over the range [10, 15], in any random direction. After a variable time period drawn uniformly at random from the range [40, 60], the trial finishes, and a new trial commences with a new randomly re-positioned light. Less frequently, the sensor is switched from the front to the back of the agent (or vice-versa) at the start of a trial.

The agent is controlled by a neural network receiving input from the light sensor and driving two motors, which differentially steer the robot with their output (in range [0, 1]). The light sensor accepts incoming light so long as it is not occluded by the agent's body, and provides a value, I, in [0, 1] varying inversely with the distance between the sensor and the light source up to a maximum of 150% of the largest possible initial distance between the agent and the light.

The agent is controlled by a network of continuous time recurrent neurons governed by Equation (1). Time constants (τ_i) were scaled as e^{x_i} with x_i drawn from the uniform distribution [0, 5], and weights w_{ij} and biases θ_i drawn from the uniform distribution [-10, 10]. The sensor value is scaled by a sensor weight s uniform in [-10, 10], and is made available to the first neuron, only. The outputs of each of the last two neurons were used to generate left and right motor speeds, respectively.

Motor neuron outputs are first scaled by a motor weight, ω_r, and translated by a motor bias, θ_r, before being squashed and rescaled in the range [0, 1] to prevent reversing. Thus, the speed of the right motor, r, was derived from the output of the right motor neuron, y_r, as $r = \frac{1}{2}[1 + \tanh(\omega_r y_r + \theta_r)]$. The network (and other simulation variables) are integrated with an Euler time-step of 0.1 during optimisation of the agent's controller, and 0.01 during analysis (to ensure stability).

Network parameters were optimised using a genetic algorithm employing pairwise tournament selection and a population of 50, for up to 6000 generations. A losing genotype was replaced with a copy of the winner subject to parameter mutation via Gaussian perturbation (zero mean, variance scaled between 0.02 and 0.05 as the previous generation's elite genotype fitness varies between 0.4 and 0.8). Fitness was calculated as the normalised average distance of the agent from the light during the last 25 time units of each trial. The first trial after the sensor was switched does not contribute towards fitness, allowing for a possible adaptation phase to occur without punishing the agent.

Evolution progressed according to a shaping scheme. When the best agent of each of the 15 prior generations had attained a fitness greater than 0.8, the phase was advanced. During phase one, the sensor was solely located on the front of the agent, so standard phototaxis was all that was required. During its lifetime, the agent was subject to eight sets of six sequential light presentation trials, with its internal state reset between each set. The average of these eight sets was taken as the overall fitness score. Phase two consisted of the phase one presentations followed by an additional eight sets of six trials with the sensor located on the back of the agent. The overall fitness score was the mean of all 16 sets. A third phase proceeded as per phase two, but with double the number of trials per set (i.e., 12). During each set of trials the sensor was switched to the opposite side of the agent at the start of a random trial number drawn uniformly from the range [4, 8]. In the final phase, three changes of sensor location took place at random intervals within each of 16 sets, each comprising 16 trials. As such, the successful completion of phase four demands that an agent must be able to cope with repeated alternation between front-mounted and rear-mounted sensors.

3   Results

Before exploring the behaviour of different recurrent networks on the task described above, we ran experiments with feedforward neural networks that are unable to exploit internal state. These networks comprised nodes governed by Equation (1). However, all recurrent connections were removed, all time constants were set to unity ($\tau_i = 1$), and each network's Euler integration step was also set to unity. Of 70 runs, not a single evolved controller was able to progress beyond phototaxis (the first phase of the shaping scheme). Based on these results, we conclude that it is difficult, or perhaps impossible for a reactive control system to solve the full behavioural task for the agent/environment combination explored here.

Unconstrained CTRNNs. We wish to determine whether the two behavioural modes that were exploited in previously reported work [12] result from bistability in the autonomous dynamics of such controllers, and whether this is a necessary property of evolved CTRNN solutions. To this end, we examined successfully evolved four-node CTRNNs and compared them with four-node CTRNNs that are biased towards having global stability and hence are less likely to express bistable dynamics.

It proved somewhat difficult to evolve controllers in this scenario. Of the 50 evolutionary runs evolving standard CTRNNs, only three agents were produced that were able to successfully complete the final phase of the task. In order to investigate whether the intrinsic dynamics of the successful control systems were bistable, we first considered their autonomous dynamics, i.e., in the absence of input. We tested for *autonomous global stability* by allowing each network to relax from 50 different random initial conditions ($y_i \in [-10, 10]$). This analysis revealed that the networks were not globally stable, revealing the presence of two fixed point attractors within the intrinsic dynamics of all three of the successful control systems.

The fact that more than one fixed-point attractor is present in a network's dynamics does not necessarily imply that the agent's internal state spends time in more than one basin of attraction during its behaving lifetime, i.e., a network might exhibit *lifetime global stability* in the absence of autonomous global stability. We tested for this possibility by running trials of each successful agent, and removing any sensory input at various stages during these trials. For all of the successful solutions, the agent's dynamics settled to one attractor when its light sensor was front-mounted, and the other attractor when it was rear-mounted.

Stable CTRNNs. We have described how, when permitted to explore the full CTRNN parameter space, successful solutions exhibited bistable lifetime dynamics. In order to determine whether it was possible to satisfy this task in a network that possessed only a single attractor in its autonomous dynamics, the stability constraint developed above (Section 1.2) was applied during a further 50 evolutionary runs. Recall: this is only a *necessary* condition for stability, and therefore can only bias the population towards globally stable fixed point (monostable) solutions.

The stability constraint was implemented by initialising all weights $\omega_{ij} \in [-0.3, 0.3]$, and rejecting any of these random networks that did not satisfy the constraint. While evolution was permitted to explore a full CTRNN weight range ($\omega_{ij} \in [-10, 10]$), any mutations that generated a network which failed to satisfy the stability constraint were rejected, and a new offspring was attempted.

From 50 runs, two evolved working solutions. In order to confirm that these solutions exhibited autonomous and/or lifetime global stability, we repeated the tests described above for unconstrained networks. For both solutions, the network's autonomous internal state always settled to same fixed point from every initial condition tested. Furthermore, we ensured that they exhibited autonomous global stability in the presence of constant input (over the range experienced by the agent during its lifetime). Straightforwardly, this carried over to stability in their lifetime dynamics, also.

3.1   Contrasting Solutions

How did the bistable and globally stable solutions differ from one another? Figure 1 depicts a projection of the internal dynamics of two successful controllers undergoing a series of trials during which the sensor's position is switched. The bistable solution (on the left) transitions between two basins of attraction, exhibiting two modes corresponding to the two types of phototactic behaviour demanded by the task. By contrast, while the dynamics of the stable system also exhibit two distinct quasi-oscillatory modes,

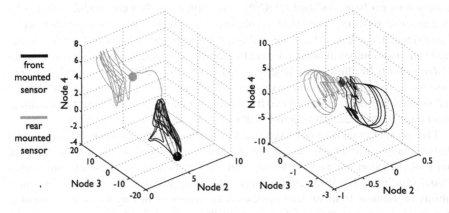

Fig. 1. Phase space plots of a typical trajectory during several consecutive trials of the best evolved unconstrained (left) and constrained (right) networks. Half way through, the sensor switches from the front to back. The discs indicate the attractor locations (in the absence of input); one (the origin) in the constrained case, and two in the bistable case.

these trajectories all occur within the same basin of attraction. In both cases, the transition from one basin to another is prompted by "pathological" sensory input, which occurs soon after the sensor location is changed. In the former case, this is responsible for moving the controller's internal state across a separatrix in the system's dynamics. In the latter case, rather than using environmental input to transition across a separatrix, the monostable controller uses it to shuttle between different regions of the same basin of attraction. Whereas a monostable system must rely solely upon this environmental coupling to maintain at least one behavioural mode, a bistable solution may exploit its *autonomous* dynamics to maintain two distinct modes of behaviour. However, accounting for the importance of the intrinsic dynamics in the complete brain/body/environment system is problematic. We will return to this issue in Section 4.

Figure 2 presents time series for two runs of the same monostable controller, one of which undergoes sensor relocation at the beginning of the second trial. In this trial the trajectories proceed identically until the light is first detected, at which point they begin to diverge. The unchanged agent's behaviour proceeds as before, with the positive slope of the bursts of sensory input indicating that the agent is approaching the light. However the neural trajectories diverge as the sensory input of the agent whose sensor has been switched at the start of this trial diminishes, due to the now maladaptive behaviour. By the third trial a different and adaptive pattern of behaviour is achieved and maintained.

3.2   Scalability and Stability

The four-node networks that we have considered so far are clearly much smaller than typical liquid state machines and echo state networks, and even smaller than the cortical

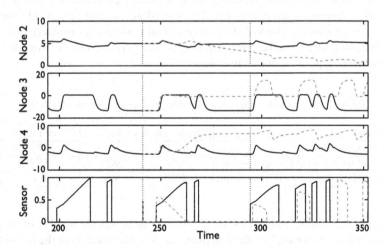

Fig. 2. Time series for three of the four nodes plus sensory input, during three consecutive trials (separated by dotted lines). The solid line depicts front-mounted sensor behaviour in three consecutive trials. The dashed line depicts rear-mounted sensor behaviour in trials two and three. At the outset of trial two, in both cases the light is positioned such that it is initially occluded from the sensor by the agent's body.

networks that inspired them. Like other kernel methods, the high dimensionality of these networks is core to their ability to use transient dynamics for pattern recognition and machine learning tasks. By contrast, high-dimensional CTRNNs are not typically employed for adaptive behaviour tasks. This stems partly from the computational demands of simulating large networks and partly from the analytic challenge that must be overcome in understanding their behaviour.

In addition, as network size increases, significant portions of the search space exhibit saturating dynamics that may be unhelpful for the production of interesting behaviour [13]. One possible method for encouraging interesting generic dynamics in large networks is to bias evolutionary search toward interesting regions of parameter space. Specifically, recall that in order to obtain the echo state property in the networks considered here, we have placed them in the centre-crossing configuration. In general, such networks have been shown to oscillate readily, making them an appropriate substrate for the evolution of, e.g., pattern generators [8]. Furthermore, it has been suggested that networks in the centre-crossing configuration will exhibit rich dynamics because their nodes interact at the most sensitive parts of their transfer functions. However, this oscillatory behaviour, which becomes more pronounced in large CTRNNs, can be disruptive, interfering with effective signal transduction. One possible solution is to further constrain them to exhibit global asymptotically stable fixed point behaviour, as we have done here. Such a constraint mitigates against destructive reverberative oscillation while retaining sensitivity to input. This has been shown to encourage effective signal propagation in large networks and has been conjectured to constitute a computationally rich subregion of CTRNN space [14].

To explore this we perform a preliminary study of the scalability of constrained versus unconstrained networks by repeating the evolutionary experiments reported above for networks comprising ten nodes. We also evolved unconstrained networks with all bias values set to zero in order to distinguish the contribution to performance of the stability constraint from that of the centre-crossing property. The results of 50 evolutionary runs per network type are reported in table 1.

Table 1. Success and stability rates for three classes of 10-node CTRNN

(50 runs)	Unconstrained	Constrained	Centre crossing
(i) % Successful Runs	28%	60%	44%
(ii) % of (i) that were Lifetime Stable	50%	90%	55%
(iii) % of (ii) that were Globally Stable	43%	89%	8%

Increasing the number internodes in the networks increases the number of successful evolutionary runs for each of three network classes. However, both the constrained and unconstrained centre-crossing systems produced more successful solutions than the unconstrained CTRNNs, with the constrained networks producing the greatest number overall. Furthermore, the majority of solutions even for unconstrained CTRNNS employed a single basin of attraction.

4   Discussion

One of the initial demands that we placed on our agent is the requirement for a control system that maintains some kind of internal state. It is important to be clear that we are not talking about any form of representation within the agents, but merely the ability to retain information over temporally extended periods. The construction of a task that *demands* state to be held specifically *within* a control system is problematic. For example, doubt has been cast on how much can be understood about the cognitive limitations of an agent's behaviour from the limitations on its internal dynamics. It has been shown that even purely reactive systems with no internal state are capable of behaviour that can be interpreted as non-reactive [15]. However, while many tasks could theoretically be satisfied with a reactive controller in interaction with a rich and dynamic environment, in practice, both biological and artificial control systems frequently exploit internal state. Consequently, while it may be hard to specify tasks that can *never* be satisfied with a purely reactive controller, it is likely that many tasks could be described as being at least 'state hungry', possibly because the agents involved do not have the necessary privileged access to their environment.

Adaptive behaviour research typically discusses the dynamics of internal state in terms of transients around the attractors of a system (even if the equilibirum associated with an attractor is never reached). In this paper we have made an attempt to understand the dynamics of transients in the absence of complex attractor structure. To this end we have examined a task that demands multiple behavioural modes. Given the bistability exhibited by evolved CTRNN solutions, it might be tempting to equate distinct attractors with distinct behavioural modes. However, we were subsequently able to evolve networks able to satisfy the same task with only a single fixed point attractor. Closer inspection reveals that both kinds of solution exhibit two distinct bundles of transients corresponding to the two behavioural modes. Given that the agent has recourse to some kind of environmental interaction with which to separate these two bundles, it need not rely on a seperatrix to differentiate these behaviours in its *autonomous* dynamics.

While monostable solutions to the task were evolved, for small CTRNNs they did not arise readily under a conventional evolutionary robotics methodology. Indeed, in order to obtain such solutions we had to explicitly encourage stable controllers. However, in larger networks (see Section 3.2) solutions that utilised a single basin of attraction evolved readily, even when the networks were unconstrained. One possible implication is that the utility of monostable versus multi-stable controllers may be sensitive to network size.

Lastly, might there be classes of behavioural task that *cannot* be satisfied without the presence of multiple attractors? In the same way that one can conceive of a task as being 'state hungry', might a particular subset of tasks be 'attractor hungry'? Consider the challenge posed by multiple time-scales of adaptive behaviour. While a bistable controller can retain certain state information indefinitely, by virtue of relaxing to one of many stable states, a monostable controller must rely on environmental stimuli on an appropriate timescale. Conversely, opting to solve a task by employing a minimal number of basins of attraction may also afford certain advantages in terms of evolvability, tunability and generalisability. This is supported by the fact that, for the task considered here, unconstrained CTRNNs often evolve to exploit a single basin of attraction, and that constraining CTRNNs in this way improves their chances of evolving a successful controller. However, further studies of the interaction between network scaling and evolvability are necessary before we can confirm that in order for larger networks to achieve their full computational potential, they will benefit from mechanisms that constrain them into a stable centre-crossing configuration.

Acknowledgements. We would like to acknowledge Eduardo Izquierdo for valuable discussion. This work was funded by EPSRC grants EP/C51632X/1 and EP/C513711/1.

References

1. Maass, W., Natschläger, T., Markram, H.: Real-time computing without stable states: A new framework for neural computation based on perturbations. Computation 14, 2531–2560 (2002)
2. Hirsch, M.W.: Convergent activation dynamics in continous time networks. Neural Networks 2, 331–349 (1989)
3. Jaeger, H.: The echo state approach to analysing and training recurrent neural networks. GMD-Report 148, German National Research Institute for Computer Science (2001)
4. Jaeger, H., Maas, W.: Special issue on echo state networks and liquid state machines. Neural Networks 20, 287–289 (2007)
5. Rumelhart, D.E., McClelland, J.L.: Parallel Distributed Processing: Explorations in the Microstructure of Cognition. MIT Press, Cambridge (1986)
6. Hopfield, J.J.: Neural networks and physical systems with emergent collective computational abilities. In: Proceedings of the National Academy of Sciences, USA, vol. 79, pp. 2554–2558 (1982)
7. Beer, R.D.: The dynamics of active categorical perception in an evolved model agent (with commentary and response). Adaptive Behavior 4(11), 209–243 (2003)
8. Mathayomchan, B., Beer, R.D.: Center-crossing recurrent neural networks for the evolution of rhythmic behavior. Neural Computation 14, 2043–2051 (2002)
9. Cohen, M., Grossberg, S.: Absolute stability of global pattern formation and parrallel memory storage by competetive neural networks. IEEE Trans on Syst, Man and Cyber 13, 815–826 (1983)
10. Strogatz, S.H.: Nonlinear Dynamics & Chaos. Addison-Wesley, Reading (1994)
11. Jaeger, H., Lukosevicius, M., Popovici, D., Siewart, U.: Optimization and application of echo state networks with leaky-integrator neurons. Neural Networks 20, 335–352 (2007)
12. Fine, P., Di Paolo, E., Izquierdo, E.: Adapting to your body. In: Costa, F.A.e., Rocha, L.M., Costa, E., Harvey, I., Coutinho, A. (eds.) Ninth European Conference on Artificial Life, pp. 203–212. Springer, Heidelberg (2007)
13. Williams, H., Noble, J.: Homeostatic plasticity improves signal propagation in continuous time recurrent neural networks. Biosystems 87(2-3), 252–259 (2007)
14. Buckley, C.L., Bullock, S.: Sensitivity and stability:a signal propagation sweet spot in a sheet of recurrent centre crossing neurons. Submitted to Biosystems (2007)
15. Izquierdo-Torres, E., Di Paolo, E.: Is an embodied system ever purely reactive? In: Capcarrere, M., Freitas, A., Bentley, P.J., Johnson, C.G., Timmis, J. (eds.) Eighth European Conference on Artificial Life, pp. 252–261. Springer, Heidelberg (2005)

Bifurcation Angles in Ant Foraging Networks: A Trade-Off between Exploration and Exploitation?

Luc Berthouze and Alexander Lorenzi

Centre for Computational Neuroscience and Robotics,
University of Sussex,
Brighton BN1 9QH, UK
L.Berthouze@sussex.ac.uk
http://www.informatics.sussex.ac.uk/users/lb203/

Abstract. The distribution of bifurcation angles found in ant foraging networks has been shown to give polarity to the networks so that nest-bound ants reaching a bifurcation can choose the appropriate direction. In this paper, we use an individual-based model to test the hypothesis that this distribution is an emergent property of a population of foraging ants optimising the trade-off between exploitation of the existing network to maximise food intake and exploration of the environment to maximise the population's ability to rapidly adapt to novel or changing environments. We identify a parameter regulating an ant's drives to forage existing trails and explore uncovered areas of the environment as a collective variable controlling the distribution of bifurcation angles in the foraging network and we show that when the exploration-exploitation trade-off is realised, the resulting distribution exhibits the same informational characteristics as that found in the original study.

Keywords: Exploration-exploitation trade-off, ant foraging, network polarisation, bifurcation angle, stigmergy, self-organisation.

1 Introduction

Whilst many ant species use a wide variety of cues to orientate themselves (see [1,2,3,4], for some examples), some ants rely primarily on chemical trails. For those species, reorientation when displaced may be harder unless trails are polarised. In a recent study on Pharaoh's ants, Jackson et al. [6] showed the distribution of bifurcation angles in their foraging networks to follow a normal distribution with mean 53° and standard deviation 15°. They argued that this value (which is shared by other trail-laying species such as the leaf-cutters and seed harvester ants studied by Acosta et al. [5]) is particularly information rich since a returning forager would have a choice of two paths, one only of which deviating greatly from its current heading. In a series of neat experiments, they validated their geometry hypothesis by showing that whilst in straight trails reorientation was as likely to be incorrect as correct, with a bifurcation angle

M. Asada et al. (Eds.): SAB 2008, LNAI 5040, pp. 113–122, 2008.

of 55°, the ratio of correct reorientations to incorrect reorientations was a significant 5.63. Forty three percent of fed ants heading away from the nest (the 'wrong' direction) made a U-turn on meeting a bifurcation, and only 8% when heading in the correct direction. Similarly, 47% of unfed ants made a course change when heading in the wrong direction, and again, only 8% made incorrect changes when heading away from the nest. Testing their hypothesis further, the authors showed that the ability of ants to make correct course changes degraded as the bifurcation angle was increased to 120 degrees, at which point the trail lost its polarisation. These results clearly suggest that ants can use the trail geometry to orientate themselves. As Collett and Waxman [7] pointed out, however, the study does not answer the question of how these angles are formed. In particular, it does not explain whether they occur as a result of directional decisions made by individuals when they first branch from a trail, or whether they are formed when the trails become well-travelled.

Central to this question are the concepts of stigmergy [8] and self-organisation, that is, the processes by which large numbers of agents interacting on a local level can coordinate these interactions to produce the complex global behaviours that are apparent in social insects colonies. Ants leave chemical trails in their environment so as to recruit more ants to forage newly discovered food sources, this in turn recruits more ants to the trail in a positive feedback loop. This method of communication through the environment was first described by Grassé [8] and is known as stigmergy. Importantly it does not need any form of centralised control, and individuals need only to react to local changes left by others within their environment. Holland and Melhuish [9] described two minimal qualities needed by an agent and its environment to support stigmergic interactions: (i) the agent must be able to move within the environment, and modify parts of it; (ii) the environment must be able modifiable, and the modifications should persist long enough for other agents acting within it to be affected by them. With such a small set of key features, it is no surprise that we should see many cases of stigmergic interactions with nature. Many of the amazing displays by social insects can in part be explained by stigmergy, however, on its own, it is not enough. Stigmergy only describes how individual colony members are able to communicate indirectly, but it does not explain how these interactions are coordinated to produce the complex colony level behaviours. For this, we need to include the ideas of self-organisation, that is, the "set of dynamical mechanisms whereby structures appear at the global level of a system from interaction among its lower-level components" [10]. Bonabeau et al. [10] described four basic ingredients needed for self-organising patterns to emerge (positive and negative feedback, amplification of fluctuations, and the presence of multiple interactions), the signatures of which are the creation of spatiotemporal structures in an initially homogeneous medium, the possibility of multistability and the existence of parametrically determined bifurcations. Applied to ant foraging, the positive feedback relates to trail recruitment through the leaving of chemical markers. Negative feedback can occur when food sources become depleted, or through saturation when ants are physically unable to access a food source or to enter the nest owing to the number of other ants. Amplification of fluctuations occurs in situations where ants lose the trail they are following and stumble upon a new food source, the resulting trails will be amplified as more ants are recruited to the new source. The presence of multiple interactions naturally results from the sheer size of ant colonies. Bifurcation angles then become the signatures of the resulting self-organised foraging network. These ideas are central to the concept of quantitative stigmergy [11] which gives us a process by which we can coordinate and regulate stigmergic interactions.

In this paper, we put forth the hypothesis that the distribution of bifurcation angle observed by Jackson et al. is an emerging property of a population of foraging ants optimising the trade-off between exploitation of the existing network to maximise food intake and exploration of the environment to maximise the population's ability to rapidly adapt to novel or changing environments. To test this hypothesis, we constructed a simple individual-based model that incorporates the qualities necessary for self-organisation and stigmergy to take place.

2   Model

The methods used to test our hypothesis are based on common techniques used to examine ant behaviour and pattern formation in natural sciences. The simulation used is a cellular automata (CA), of which the two major components of our model are the ants and the pheromone concentration map. In the context of CA, the ants are modeled as a Lattice Gas class of CA, in which particles operate in a bi-directional grid, and their movement contains a random component. The pheromone concentration map on the other hand falls in the deterministic class of CA, in which states are associated with sites on the grid, and a state's change is a deterministic function of its surrounding sites [12,18]. The environment in which the ants operate is a 500x400 lattice (the unit of distance used throughout this study is the length of one ant). The single nest is located at the centre of the grid, and has a radius of 7.5 units. Ants are released from the nest at a rate of 1 ant every 5 updates. The food sources are randomly placed in the environment, and have a finite amount of food. Once a food source is depleted, it is removed from the environment, and replaced by a new food source in a different location.

Each ant is described by its position on the lattice, its heading and velocity and two coefficients that represent how fatigued the ant is (c_f) and the food load it is carrying (c_{fl}). At each update, the ant's fatigue coefficient increases by 0.0005 until its maximum value of 1 when the ant is considered dead from exhaustion and is removed from the environment. Only when the ant is either at a food source, or in the nest, does the fatigue coefficient decrease (multiplied by a factor 0.9 at each update). The food load coefficient increases when an ant is at a food site (+0.1 unit per update), and decreases when the ant is in the nest (−0.1 unit per update).

Environmental stimuli and internal variables modulate five drives that control the ant's heading and velocity. For simplicity, each of these drives will be described in a polar coordinate system where the pole is the position of the ant on the lattice, and the polar axis is the ant's current heading. Radial coordinates will denote the velocity component of each drive.

Exploratory drive: This drive implements an ant's 'urge' to change to a new heading deviating randomly from its current heading by $\pm 15°$. The trajectory is smoothed by spreading the change over α_r updates.

$$d_1 = (0, \frac{-15 + rnd * 30}{\alpha_r}) \tag{1}$$

where rnd is a random value between 0 and 1 drawn from a uniform distribution.

Exploitation drive: The movement of the ant is affected by the pheromone concentrations ϕ of its eight Moore neighbours

$$d_2 = (\sum_{i=1}^{8} \phi_i(1 - c_{fl})(1 - \frac{|\theta_i|}{180}), \theta_i) \tag{2}$$

where θ_i is the angular deviation to the ith Moore neighbour. Note that this expression favours pheromone sites ahead of the ant. This is to prevent the ant from turning around to follow its own trail.

'Move to food' drive: Each food source has a draw area (of diameter $f_d = 3$) in which ants become 'aware' of the food source, either visually or, in the case of the blind army ants, through olfactory perception. This drive is only activated when the ant gets within this area.

$$d_3 = ((1 - c_{fl})(1 - \frac{d_f}{f_d}), \theta_f) \tag{3}$$

where θ_f is the angular deviation to the food source and d_f is the distance to the food source. The attraction to the food source is inversely proportional to the distance to the food source.

'Return to nest' drive: An ant will be drawn back to the nest either when it has collected food, or when its fatigue coefficient increases above a threshold. For simplicity, ants maintain a memory of the exact location of the nest relative to their current position. This isn't biologically plausible, of course, but since ants can use path integration and other cues to locate their nest (see [19], for example), it is an acceptable compromise.

$$d_4 = (\sum_{i=1}^{8} \phi_i c_{fl}(1 - \frac{|\theta_i|}{180})^r, \theta_i) + (e^{-k(1-c_f)}, \theta_n) \tag{4}$$

where θ_n is the angular deviation to the nest, θ_i is the angular deviation to the ith Moore neighbour, r regulates the ant's directional selectivity, and k determines at which fatigue level the drive will kick-in. The bias toward stronger pheromone concentrations in front of the ant corresponds to observations that ants move quicker on trails than on unmarked areas [11]. Since the fatigue coefficient is updated at each step, this is a plausible mechanism from an energetic viewpoint. A value of $r = 4$ was used that implements a good compromise between accuracy and energy saving by way of trail following. Parameter k was set to 10 throughout the study, which corresponds to a fatigue coefficient of 0.5. The drive to return to the nest increases in a non-linear fashion with the fatigue coefficient.

'Move forward' drive: This drive simply 'motivates' the ant to go forward.

$$d_5 = (min_v, 0) \tag{5}$$

where min_v is a constant ant velocity (set to 0.3 units per update in all runs). It guarantees that all ants will move at a minimum speed, and also puts a limit to the ant's angular velocity (a similar constraint was used in [20]).

The ant's actual movement is given by the weighted sum

$$d = \gamma d_1 + (1 - \gamma)d_2 + d_3 + d_4 + d_5 \tag{6}$$

where γ is in the range $[0, 1]$ and is used to regulate how exploratory/exploitative the ant is. At $\gamma = 1.0$, the ant is fully exploratory and ignores pheromone concentrations. Conversely, at $\gamma = 0.0$, the ant will fully exploit existing chemical trails. Consistent with animal observations [14,15] and other models [16,17,18], ants lay pheromones both when leaving the nest and returning with food, although returning ants do so at a higher concentration ($0.1c_{fl}$ unit of food per location visited for ants returning to the nest compared with a fixed 0.01 for ants leaving the nest). In addition, the level at which pheromones are no longer deposited differs whether ants are returning to the nest (maximum concentration of 1) or leaving the nest (maximum concentration of 0.2). A similar constraint was used in [16,17].

Finally, two constraints complete the model: (i) as in [16,17], an ant's maximum velocity is limited to 1 unit per update; (ii) a single lattice location can hold a maximum of 15 ants. If an ant's move takes it into a location that contains 15 ants, it attempts to move to the next grid location 1 unit away from its current position while retaining its desired heading. This minimises the occurrence of unwanted branching that might confound our results. It no site is available, the ant remains in place.

3   Results

To examine the relationship between our putative controlling variable γ and the distribution of bifurcation angle in the resulting foraging trail network, we run simulations varying γ in the interval $[0, 1]$ in steps of 0.1. The environments in each simulation had two food sources. To avoid the random bias of differing food source locations, the locations of food sources stayed constant between simulations. Simulations were run for 10000 iterations, with a colony size of 1500 and food source size of 2000.

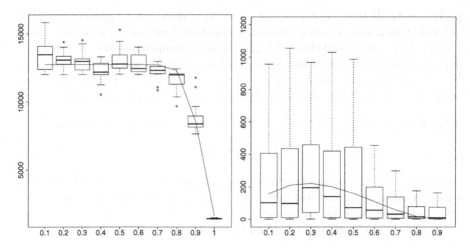

Fig. 1. Left: Cumulated food intake (vertical axis, per run of 10,000 iterations) as a function of γ (horizontal axis) over 16 runs. Food intake is given in unit of food where 1 is the maximum food load of an ant entering the nest at a given time. Solid line: logistic regression (y=1274.00/(1+exp(-(0.93-x)/0.038), d.f.=157, $p < 0.0001$). Right: Recovery times (per run of 10,000 iterations) as a function of γ (horizontal axis) over 16 runs. No data is available at $\gamma = 1.0$ since no food sources got depleted. Times are given in number of iterations. Solid line: cubic fit by nonlinear regression (y=1788.08x^3-3190.93x^2+1367.34x+49.88, d.f.=6, $p < 0.01$).

The trade-off between exploitation of the network and exploration of the environment was assessed through two 'fitness' measures based on the amount of food entering the nest at each time step. Cumulated food intake over the 10000 iterations of each run provided a measure of the ability of the network to exploit the existing network. Robustness of the population to environmental changes was assessed by the average time (recovery time) it took for the population to find a new food source once a food source had been depleted. As shown by Figure 1, simulations confirmed the role of γ in controlling the overall behaviour of the population. Lower γ settings (highly exploitative) resulted in higher cumulated food intake, whilst higher γ settings (highly explorative) resulted in faster recovery times. The fact that recovery times were faster at $\gamma = (0.1, 0.2)$ than at $\gamma = 0.3$ was unexpected but finds its origin in a flaw of the model owing to an absence of published data on the behaviour of the first ant off the nest in a novel environment. The model being an individual-based model, such data is critical to determine what the behaviour of a highly exploitative ant should be when no established trail network exists. To circumvent this problem, ants were sent off the nest at an angle drawn from a uniform distribution, with a minimal amount of exploratory drive. In a small environment, when food sources are not too far from the nest, this actually amounts to a fairly effective systematic exploration of the environment, thus the fast recovery times. For a given colony size, increasing the size of the environment or placing the food source further from the nest should suppress this artefact.

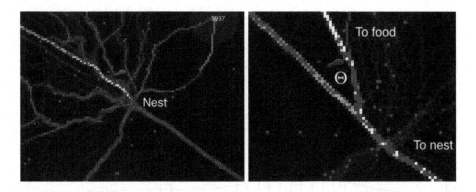

Fig. 2. Left: Snapshot of a foraging network. The nest is found at the centre of the figure. In the upper right quadrant, a new food source has just been discovered. The brightness of a trail is proportional to its concentration in pheromone. The purple lines denote bifurcations as used to construct Figure 3. Right: Close-up on a bifurcation. Only Y junctions were considered, and a bifurcation angle θ was determined as the angle between branches away from the nest.

To determine the optimal trade-off between exploration and exploitation, we derived a 'fitness' function y=I(1-1/RT) maximising food intake I for low recovery time RT with I and RT approximated by the nonlinear regressions given in Figure 1, after normalisation to $[0, 1]$. Theoretically, the maximum of this function is obtained when $I = 1/RT$. Experimentally, the maximum was obtained at $\gamma = 0.8$, a highly explorative setting.

As in Jackson et al. [6], bifurcation angles were measured from well-structured foraging networks over three of the 16 simulation runs collected[1]. Only networks that showed well-established Y junctions were considered (see Figure 2, left). A bifurcation angle was defined as the angle between the two branches away from the nest (see Figure 2, right). The distribution of bifurcation angle for each γ was determined using normal mixture modelling [21]. At $\gamma = 0.8$, the putative optimal trade-off between exploration and exploitation, the density function showed a mean angle of 40.84° and a standard deviation of 12.69° ($n = 67$). In comparison, Jackson et al. [6] found a mean of 53.48°, and a standard deviation of 14.88°. The model used in this study being an abstract model with parameters having little or no biological relevance, any outright comparison of the numerical values of the means and standard deviations would have little significance. Nevertheless, the data shown in Figure 3 (top) display three important characteristics. First, the probability of bifurcation angles of more than 80° is very small, which means that the resulting network is polarised. It is not the case in more exploitative settings. Second, the number of clusters (or components) as identified by Bayesian Information Criterion (BIC) decreases as γ increases. As in Jackson et al.'s study, the distribution of the bifurcation angles at high γ (highly explorative population) is a normal distribution. Third, as shown by

---

[1] This is a very laborious activity which we are trying to automatise.

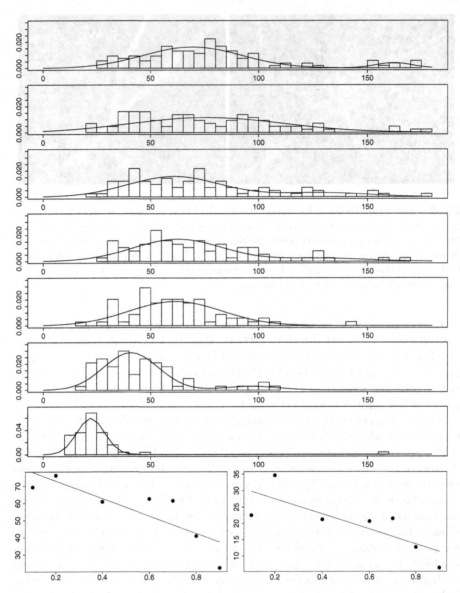

Fig. 3. Top: Densities of bifurcation angles for $\gamma = (0.1, 0.2, 0.4, 0.6, 0.7, 0.8, 0.9)$ (from top to bottom) from three separate runs ($n > 69$ for all γ setting except for $\gamma = 0.9$ when $n = 50$. Note the different scale in vertical axis for $\gamma = 0.9$.). Horizontal axis denotes angular values in degrees. Solid line, fit from normal mixture modelling [21]. Bottom left: Mean (in degrees) of the largest cluster as a function of γ. Fit by linear regression ($R^2 = 0.700; d.f. = 5, F = 11.69, p = 0.019$). Bottom right: Standard deviation (in degrees) of the largest cluster as a function γ. Fit by linear regression ($R^2 = 0.637; d.f. = 5, F = 8.76, p = 0.032$).

Figure 3 (bottom), both mean and standard deviation of the main cluster (component) show a statistically significant correlation with γ. This suggests that γ, an individual-based control parameter, is indeed a control variable for the distribution of bifurcation angle in the foraging network.

4 Conclusion

In providing evidence that an individual-based parameter can turn into a collective variable controlling the geometry of the foraging network, this paper aims to answer an important question raised by Jackson et al.'s study, namely, whether bifurcation angles occur though directional decisions made by individuals ants when they first branch from a trail, or whether they form when travelled by many ants [7]. Our results support the latter view, that is, bifurcation angles are an emergent property of the stigmergic system. Our results also support our hypothesis that the distribution of bifurcation angles observed by Jackson et al. results from an optimisation of the trade-off between exploitation of the existing network to maximise food intake and exploration of the environment to maximise the population's ability to rapidly adapt to novel or changing environments. Whilst such an hypothesis is not unreasonable from an evolutionary perspective, the limitations of using an abstract model with parameters have little or no biological relevance are obvious. Yet, and as often when using the animat approach, constructing a model raises interesting questions that warrant further study: (i) what is the behaviour of the first ant off the nest? could individual ants' trajectories be modelled by correlated random walk? (ii) how could the γ parameter be implemented: switching individual behaviour or changing ratio of scouts to foragers?

Acknowledgments. We are grateful to Pierre Nouvellet, David Waxman, and Paul Graham (all from the Department of Biology and Environmental Science at University of Sussex) for useful discussions and advices.

References

1. Carthy, J.D.: The orientation of two allied species of british ant. I. Visual direction finding in Acanthomyops (Lasius) niger. Behaviour 3, 275–303 (1951)
2. Banks, A.N., Syrgley, R.B.: Orientation by magnetic field in leaf-cutter ants Atta colombica (hymenopetra foricidae). Ethology 109, 835–846 (2003)
3. Carthy, J.D.: The orientation of two allied species of british ant. II. Odour trail laying and following in Acanthomyops (Lasius) niger. Behaviour 3, 304–318 (1951)
4. Hölldobler, B.: Recruitment behaviour, home range orientation and territoriality in harvester ants, Pogonomyrmex. Behavioral Ecology and Sociobiology 1, 3–44 (1976)
5. Acosta, F.J., López, F., Serrano, J.M.: Branching angles of ant trunk trails as an optimization cue. Journal of Theoretical Biology 160, 297–310 (1993)
6. Jackson, D.E., Holcombe, M., Ratnieks, F.L.W.: Trail geometry gives polarity to ant foraging networks. Nature 432, 907–909 (2004)

7. Collett, T.S., Waxman, D.: Ant navigation: Reading geometrical signposts. Current Biology 15(5), 171–173 (2005)
8. Grassé, P.P.: La reconstruction du nid et les coodinations interindividuelles chez bellicositermes natalensis et cubitermes sp. La théorie de la stigmergie: Essai d'interprétation des termites constructeurs. Insectes Sociaux 6(1), 41–80 (1959)
9. Holland, O., Melhuish, C.: Stigmergy, self-organization, and sorting in collective robotics. Artificial Life 5(2), 173–202 (1999)
10. Bonabeau, E., Theraulaz, G., Fourcassié, V., Deneubourg, J.: The phase-ordering kinetics of cemetery organization in ants. Physical Review E 57(4), 4568–4571 (1998)
11. Bonabeau, E., Dorigo, M., Theraulaz, G.: Swarm Intelligence: From Natural to Artificial Systems. Oxford University Press, New York (1999)
12. Ermentrout, B., Edelstein-Keshet, L.: Cellular automata approaches to biological modeling. Journal of Theoretical Biology 160, 97–133 (1993)
13. Resnick, M.: Turtles, Termites, and Traffic Jams. Explorations in massively parallel mircoworlds. MIT Press, Cambridge (1997)
14. Aron, S., Pastells, J.M., Deneubourg, J.L.: Trail-laying behaviour during exploratory recruitment in the argentine ant, Iridomyrmex humilis. Biology of Behavior 14(3), 207–217 (1989)
15. Fourcassié, V., Deneubourg, J.L.: The dynamics of collective exploration and trail-formation in Monomorium pharaonis: experiments and model. Physiological Entomology 19, 291–300 (1994)
16. Deneubourg, J.L., Gross, S., Franks, N., Pasteels, J.M.: The blind leading the blind: Modeling chemically mediated army ant raid patterns. Journal of Insect Behavior 2, 719–725 (1989)
17. Solé, R.V., Bonabeau, E., Delgado, J., Fernández, P., Marn, J.: Pattern formation and optimization in army ant raids. Artificial Life 6(3), 219–227 (2000)
18. Watmough, J., Edelstein-Keshet, L.: Modelling the formation of trail networks by foraging ants. Journal of Theoretical Biology 176(3), 357–371 (1995)
19. Santschi, F.: Observations et remarques critiques sur le mécanisme de l'orientation chez les fourmis. Revue Suisse de Zoologie 19, 305–338 (1911)
20. Couzin, I.D., Frank, N.R.: Self-organized lane formation and optimized traffic flow in army ants. Proceedings of the Royal Society B: Biological Sciences 270(1511), 139–146 (2003)
21. Fraley, C., Raftery, A.E.: MCLUST Version 3 for R: Normal Mixture Modeling and Model-Based Clustering. Technical Report, No. 504, Department of Statistics, University of Washington (2006)

Episodes in Space: A Modeling Study of Hippocampal Place Representation*

Balázs Ujfalussy, Péter Erős, Zoltán Somogyvári, and Tamás Kiss**

KFKI Research Institute for Particle and Nuclear Physics of the Hungarian Academy
of Sciences,
Konkoly-Thege Miklós út 29 – 33. Budapest, H-1121, Hungary
bognor@mail.kfki.hu
http://cneuro.rmki.kfki.hu

Abstract. A computer model of learning and representing spatial locations is studied. The model builds on biological constraints and assumptions drawn from the anatomy and physiology of the hippocampal
formation of the rat. The emphasis of the presented research is on the
usability of a computer model originally proposed to describe episodic
memory capabilities of the hippocampus in a spatial task. In the present
model two modalities – vision and path integration – are contributing to
the recognition of a given place. We study how place cell activity emerges
due to Hebbian learning in the model hippocampus as a result of random
exploration of the environment. The model is implemented in the Webots
mobile robotics simulation software. Our results show that the location
of the robot is well predictable from the activity of a population of model
place cells, thus the model is suitable to be used as a basic building block
of location-based navigation strategies. However, some properties of the
stored memories strongly resembles that of episodic memories, which do
not match special spatial requirements.

Keywords: space representation, spatial memory, episodic memory, grid
cell, place cell, hippocampus, computational neuroscience.

1 Introduction

Animals spend considerable amount of time moving from one place to an other
in their habitat looking for food and avoiding possible dangers. Their survival
depends upon how well they are able to find or identify certain places and return
to them later. Although, they often make mistakes, routes they choose may not
be optimal from a mathematical point of view still they are able to learn rapidly
and adapt their behavior to changing environments in a highly flexible way.
Thus, the study of brain structures responsible for spatial behavior might offer
a unique opportunity to ameliorate artificial navigational algorithms and mobile
robotic applications.

---

* This research was supported by the EU Framework 6 ICEA project (IST 027819).
** Corresponding author.

M. Asada et al. (Eds.): SAB 2008, LNAI 5040, pp. 123–136, 2008.
© Springer-Verlag Berlin Heidelberg 2008

Early laboratory experiments have already drawn the attention to the good navigation capabilities of rodents, which are able to quickly find the location of food sources even in complicated mazes. The idea of the formation of a cognitive map was first proposed by Tolman [45] in the late 40s and was later supported by the discovery of place cells by O'Keefe and Dostrovsky [35]. According to their definition an animal possesses a cognitive map if it is able to make novel short-cuts. For interpreting our present results, however, a weaker definition of a cognitive map is sufficient: any neural representation of the space is called a cognitive map [16] (for a comparison of different cognitive map definitions see eg. Bennett [6]).

One of the most widely accepted theory places the cognitive map of rodents into their *hippocampal* regions [36]. Electrophysiological experiments with rats in test environments led to the discovery that the firing rate of hippocampal neurons was correlated with the location of the animal in a test environment (these cells are referred to as place cells) [35, 32, 34]; and that hippocampal damage causes spatial learning deficits [31] (for a review see [17]).

In humans and rodents the hippocampus plays a key role encoding and storing episodic memories [41]. The question whether the primary function of the hippocampus of rodents is the generation and storage of a cognitive map or it is only a side-effect of a more general memory have arisen in a number or of researchers. For example the experiments by Eichenbaum and co-workers [48] supported the results that hippocampal place cells can code the location of the animal, however, pointed out that similarly, non-spatial properties (desired actions to specific stimuli) might be coded by the hippocampal cells as well. They trained rats to perform a non-matching-to-sample odor recognition task where cups with scented sand in them were placed at different locations in an open platform. They found cells that showed correlated activity with spatial variables but also cells that changed their activity in relation with non-spatial variables like odor, match/non-match, approach of the cup. In a review paper Eichenbaum [11] concludes that "the defining features of hippocampal representation in rodents, as in humans, lie not in the modality of the information processed, but in the organization of the information that supports a capacity for flexible memory expression."

In this article we study the performance of a hippocampus model of the rat [37] – originally created to describe the episodic memory capabilities of the hippocampus – in spatial learning. The basic theory building on detailed anatomical and physiological experiments originates in the early work by Marr [26]. According to Marr's theory the hippocampal formation acts as a temporary memory store that rapidly acquires new memory traces and encodes them by a sparse code. During encoding Hebbian learning takes place in its synaptic pathways, while during recall the recurrent collaterals of the Cornu Ammonis 3 (CA3) region of the hippocampus mediate associative retrieval of the memories initiated by noisy or partial recall cues. This system would enable a gradual consolidation of memories by transferring them into the neocortex for long-term storage.

In our experiments the above theory was implemented in a mobile robot simulator, which moved in a test arena. The simulated robot was equipped with cameras providing visual inputs to the hippocampus model, distance sensors, representing whisker inputs to detect obstacles and wheel rotation sensors used to guide grid cell-based path integration.

2 Premises of Our Computer Model

2.1 Hippocampal Memory Models

Our model of the hippocampal formation benefits mostly from the theory by Treves and Rolls [47] and its numerical realization [37], which is closely related to the highly influential theory proposed by Marr [26]. Models [27, 37, 20, 22, 5], which build on Marr's theory follow some key assumptions. First, the dentate gyrus (DG) region of the hippocampal formation serves the function of a pre-processing stage for the CA3 acting as a competitive network that creates a sparse and clustered code of the pre-synaptic entorhinal cortical (EC) input, which – similarly to other neocortical regions – realizes a denser representation. This sparse, orthogonal code is in turn used as a teaching signal for the subsequent CA3 region. Second, the CA3 region acts as an autoassociation memory, which stores memory traces in its extensive recurrent collateral (RC) network for later retrieval. Third, many previous hippocampal models assume that the hippocampus operates in two distinctly different modes during learning and retrieval, which we also incorporated into our model. In most of the models (but see the suggestions by Hasselmo [19] or Lisman and Otmakhova [25]) switching between the two modes is done manually, as in our model.

2.2 Models of Hippocampal Place Cell Generation

Since the discovery of the place cells in 1971 by O'Keefe and Dostrovsky many different models have emerged to explain the generation of place cell activity in the hippocampus.

The first models used distance from and bearing to identified landmarks as input to the hippocampus [49, 42, 46, 10]. Hippocampal place cells were activated when the currently perceived scene matched the stored landmark configuration. However, these models requires exact object recognition, and representation of distances and angles between objects in the EC.

Other models [1, 14] trained their multi-layered neural networks with filtered visual information and showed that place correlated cell activity emerges in these simple feed-forward networks. Computations performed by these cell layers, however, are hardly analogous to that of neuron populations in the hippocampus. Furthermore, in the Arleo and Gerstner [1] model resetting of the idiothetic place representation by the allothetic information incorporates heuristic mechanisms.

An other class of the models uses continuous attractors in the CA3 region of the hippocampus According to these models [40, 22] path integration takes place in the CA3 region of the hippocampus, where the pre-configured recurrent connections provide continuous attractor representation of the location. Associations of the attractors with sensory input can be used to reset the path integration system. In contrast to the grid cell (see Sect. 2.4 below) system in the EC, the hippocampus realizes a multichart attractor map, i.e. there are different relations among place fields in different environments. However, the development of specific internal connections required for this multichart architecture is still an open question.

To introduce the continuous properties of space into our model two steps are made. First, the simulated robot moves continuously in its environment and as a result the camera image changes continuously. Second, we implemented the entorhinal grid cell system as a part of the input to the hippocampal model. In the next section we give a brief summary of the properties of grid cells and in Sect. 3.4 its mathematical implementation is described.

2.3 Hippocampally-Inspired Robotic Models

Bio-inspired spatial mapping and navigation algorithms have a long history in robotics. Milford and co-workers studied an effective mapping and navigation system (RatSLAM) [29] on real robots in indoor and outdoor environment. They used place cells modelled as a two dimensional competitive attractor network [30] motivated by hippocampal anatomy. Brain-based devices were designed and tested by Krichmar and colleagues using computational neuroscience techniques (eg. [12]). The Darwin series of robots utilize intricate cortico-hippocampal interactions, while the physical device moves in a real environment [23].

Our motivation is primarily routed in the neurosciences besides building a functional algorithm to be used in navigating robots. Similarly to the study of Banquet et al. [2], our question is how the hippocampal formation can fulfill its role in the formation and possible storage of episodic memories, and, at the same time place representation.

2.4 Entorhinal Cortical Grid Cell Models

Since the discovery of the existence of grid cells in 2005 [15, 18] several theories and models of the generation and function of this cell type were proposed [33, 8, 44, 39, 13, 7]. Grid cells are located in the medial EC and are characterized by special firing properties: they fire selectively when the rat visits points of a hexagonal lattice. So far this is the only neural system we know about that represents a metric space in the brain [21].

Grid cells are characterized by their spatial frequency or scale, the orientation and the phase of the grid [18]. From our perspective, grid cells are good candidates to be used in the path integration mechanism of animals as suggested by Fuhs and Touretzky [13] or to form synapses with hippocampal principal cells to yield place cells as in the model of Rolls et al. [39] or help in location determination in an economical way as in the theory of [9].

3   Methods

Our model – used in computer simulations – aims at understanding how place cell activity is generated in the hippocampus using pre-processed sensory information coming from the EC. In our work the model proposed by Rolls [37] was used and modified to accommodate it with spatial inputs. We simulated topographically mapped, pre-processed sensory information in the EC and two regions of the hippocampus the DG and the CA3 regions (see Fig. 1 for an overview).

Below we demonstrate that using only local learning rules, a realistic connection scheme and realistic sparsity of cell activities, associating visual, tactile and proprioceptive information together in an autoassociative memory network, cell activities show spatial correlates.

3.1   Biological Constraints and Assumptions

Similarly to previous modelling efforts, we incorporated the following biological constraints: 1. Representation of memory traces (coding) is sparse in the DG and in the CA3, while denser in the EC. 2. All hippocampal regions receive pre-processed sensory information from the neocortex via its perforant path (PP) input. 3. The EC innervates both the DG and the CA3, the DG innervates the CA3 and the CA3 innervates itself. Furthermore, to set up the model the following assumptions are made: 1. Only local, Hebbian learning rules are used throughout the model. 2. The DG translate the dense code of the EC to a sparser code used throughout the hippocampus. 3. The mossy fiber (MF) synapses are an order of magnitude stronger than the PP or the RC synapses. 4. The hippocampus operates in two distinct modes: **learning** (also used as encoding) and **recall** (also used as retrieval). During learning modification of synaptic strengths is allowed, during recall synaptic strengths are fixed (see Fig. 1). 5. During encoding activity of CA3 cells is determined by their MF input from the DG cells. 6. Synaptic strengths of the PP – both in the DG and in the CA3 – and of the RC are modifiable, while the strength of MF synapses do not change in time. 7. During retrieval activity of CA3 cells is cued by its PP input. This subsystem

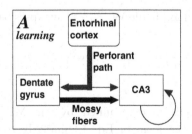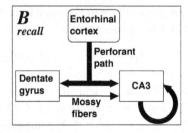

Fig. 1. Modelled areas and pathways. Thick arrows represent active pathways, ie. pathways that add to the activation of the target region. Red arrows represent pathways being modified.

operates as a hetero-associative network. 8. Memory recall is refined by the RC synapses of the CA3 region.

In general, the above outlined scheme serves as a framework to store and recall memories. In particular, the information composing the memories in our model have spatial aspects, which results in the generation of place cell-like activity.

3.2   Representation of the Hippocampal Input

In the simulations we accounted for the following modalities: dead reckoning via the grid cell activities, whisker activity and vision coded as firing rate vectors of real values in the [0..1] interval, respectively. In the EC firing rate of cells is immediately calculated from the simulated sensors.

3.3   Hippocampal Computation

In this section we separate the two working modes of the hippocampus (see Sect. 3.1) to clearly indicate what processes take place during learning and during recall, respectively.

Learning: Treves and Rolls [47] suggested that any new event to be memorized is represented in the CA3 as a firing rate pattern (vector) of pyramidal cells. To create this pattern, sensory information of the EC is first processed by the DG and its sparse, orthogonalized version is generated. First, the activation of DG cells is calculated:

$$h_j^{EC \rightarrow DG} = \sum_i W_{i,j}^{(EC, DG)} EC_i, \tag{1}$$

using the synaptic weight matrix and the activity of EC cells. Then a nonlinear activation function $(f_j(\cdot))$ is applied on the activation vector to calculate the firing rate of each cells:

$$DG_j = f_j(\mathbf{h}^{EC \rightarrow DG}, s^{DG}), \tag{2}$$

where s^{DG} is the desired sparseness of the code in the DG area. The sparseness of a given pattern defined by [38] is based on averaging the r firing rate distributions of cells over the stored patterns p ($< r >_p$): $a = < r >_p^2 / < r^2 >_p$.

Learning takes place in the associatively modifiable $\mathbf{W}^{(EC, DG)}$ synapses based on the following learning rule:

$$\Delta W_{i,j}^{(EC, DG)} = \alpha^{(EC, DG)} DG_j \left(EC_i - W_{i,j}^{(EC, DG)} \right), \tag{3}$$

where $\alpha^{(EC, DG)}$ is the learning rate.

During learning activation of CA3 cells is determined by the MF (ie. $\mathbf{W}^{(DG, CA3)}$) input ($h_j^{DG \rightarrow CA3}$) following the form of Eq. 1. Similarly to the case of the DG, activity of CA3 cells is computed by an equation of the form of Eq. 2.

Learning takes place in all synaptic pathways of the CA3 region as it is hypothesized that due to the activation by the DG afferents the membrane potential of CA3 pyramidal cells is depolarized and enables plastic changes in synapses originating from the PP and the RC:

$$\Delta W_{i,j}^{(\text{EC, CA3})} = \alpha^{(\text{EC, CA3})} CA3_j \left(EC_i - W_{i,j}^{(\text{EC, CA3})} \right), \tag{4}$$

$$\Delta W_{i,j}^{(\text{CA3, CA3})} = \alpha^{(\text{CA3, CA3})} CA3_j CA3_i \left(1 - W_{i,j}^{(\text{CA3, CA3})} \right) - \beta^{(\text{CA3, CA3})} W_{i,j}^{(\text{CA3, CA3})}, \tag{5}$$

where αs are the learning rates, β is the "forgetting" rate, respectively.

Recall: During recall the EC input is used to initiate retrieval of a stored memory pattern. First, the activation of CA3 cells resulting from the PP ($h_j^{\text{EC} \rightarrow \text{CA3}}$) is calculated based on Eq. 1, and used to compute the **CA3** activity vector by an equation of the form Eq. 2. Second, this initial activity vector was used as the cue to retrieve the memory trace in the autoassociative network. In this process activation vectors $\mathbf{h}^{\text{EC} \rightarrow \text{CA3}}$ and $\mathbf{h}^{\text{CA3} \rightarrow \text{CA3}}$ resulting from the PP and the RC input, respectively, were calculated and normalized to unitary length. Finally, activation of CA3 cells were computed:

$$CA3_j = f_j(\mathbf{h}^{\text{EC} \rightarrow \text{CA3}} + \chi \mathbf{h}^{\text{CA3} \rightarrow \text{CA3}}, s^{\text{CA3}}), \tag{6}$$

where χ is a scaling factor, s^{CA3} is the sparseness of coding in area CA3. Calculation of $\mathbf{h}^{\text{EC} \rightarrow \text{CA3}}$, $\mathbf{h}^{\text{CA3} \rightarrow \text{CA3}}$ and **CA3** were iterated a number of times (τ_R) to allow the network to find a stable attractor corresponding to the memory being recalled.

3.4   The Grid Cell System

To represent grid cell activity in our model we used a formula motivated by the cosine grating model of Blair et al. [7]. Firing rate of a grid cell can take a value from the $G(i) = \{0.0, 0.0, 0.1, 0.2, 0.4, 0.6, 0.8, 0.9, 1.0, 1.0\}$ set where the index i is given by:

$$i(\mathbf{r}) = 20 \left(\sqrt{\frac{1}{3} \sum_k \cos^2(\omega_k \cdot (\frac{\mathbf{r}}{\zeta} - \mathbf{c}))} + 0.5 \right), \tag{7}$$

where $\mathbf{r}$ is the position of the robot, ζ sets the scale, the direction of the ω_k vectors which were 60° apart from each other sets the orientation and $\mathbf{c}$ sets the phase of the grid.

We used five scales ($\{36, 41, 47, 53, 61\}$ centimeters), five uniformly distributed orientations and five phases to cover the space.

3.5   Parameters and Initial Values of the Model

In the simulations all-to-all connections were applied in all synaptic matrices except for the $\mathbf{W}^{(\text{DG, CA3})}$ connections. The learning rules we used in the simulations ensure that while activities are in the $[0..1]$ interval, synaptic weights are also constrained to the [0..1] interval. Initial matrix values are generated randomly from a normal distribution in all four cases (the DG → CA3 matrix was constant in time):

$$P(W_{i,j}^{(X,\ Y)} = w) = \frac{1}{\sqrt{2\pi}\sigma} e^{-\frac{(w-\mu)}{2\sigma^2}}, \quad \text{if } 0 < w < 1 \tag{8}$$

with parameter values listed in Table 1.

Parameters of our model are summarized in Table 2.

3.6   Numerical Simulation

The model was simulated in the Webots [28] mobile robotics simulation software. We constructed a simple, wheel driven model robot equipped with a camera of 360° field of vision, 10 distance sensors, representing whiskers used partly to detect obstacles. The robot moved in an open arena bounded by vertically striped walls . The robot was restricted to move in the middle of the arena and not allowed to approach the wall closely to assure high variability in the visual signal. The motion of the robot was a random walk with basic obstacle avoidance

Table 1. Parameter values used to generate the initial synaptic weight matrix elements according to Eq. 8

synapse (X → Y)	μ	σ
EC → DG	0.6	0.1
EC → CA3	0.1	0.05
CA3 → CA3	0.1	0.05
DG → CA3	0.2	0.05

Table 2. Default parameter values used in simulations of our model

Parameter	Value	Description
N^{EC}	265	Number of cells in the EC
N^{DG}	1000	Number of cells in the DG
N^{CA3}	1000	Number of cells in the CA3
$C^{DG \to CA3}$	30	Convergence of DG cells to CA3 cells
$\alpha^{(EC,\ DG)}$	0.05	Learning rate of PP synapses from EC to DG
$\alpha^{(EC,\ CA3)}$	0.05	Learning rate of PP synapses from EC to CA3
$\alpha^{(CA3,\ CA3)}$	0.05	Learning rate in the RC of the CA3
$\beta^{(CA3,\ CA3)}$	$2 \cdot 10^{-5}$	"Forgetting" rate of the CA3 RC
s^{EC}	0.2	Sparseness of coding in the EC
s^{DG}	0.01	Sparseness of coding in the DG
s^{CA3}	0.02	Sparseness of coding in the CA3
χ	3	Scaling factor of RC activation relative to PP activation in CA3
τ_R	5	Number of iterations in the CA3 RC

and was in no connection with the hippocampal model. The EC input vector was composed of a row of the camera image matrix, the value of the distance sensors and the activity vector of the grid-cell system. In the first phase of the simulation the robot moved around the arena for 2000 simulation steps (equaling 4 minutes), which usually allowed the robot to acquire a sufficient sampling of the continuous space. During the first phase the robot was in the learning mode. In the second phase we switched to recall mode and let the robot move around for an other 2000 time steps on a different route. Although routes were different, the EC activity pattern at near-by places were similar due to similar visual and grid inputs.

4   Results

Simulation results show that although learning occurs simultaneously in the DG and the CA3 stable place representation evolves in both the DG and the CA3 under 4 minutes of exploration (Fig. 2, left panel). We found that in the DG majority of the cells exhibit one place field, out of the 1000 cells simulated only one had two and one had three place fields. In the CA3 50% of the cells had one single place field, about 32% had two, 15% had three and the rest had more than three place fields (Fig. 2, left panel). Place fields were identified as continuous places not smaller than $64\,cm^2$, where a given cell had high activity. Activities below 25% of the maximal firing rate were discarded from the calculation.

The population of the place cells redundantly covered the space and could reliably be used to predict the location of the robot. For this calculation first, firing rate map of each CA3 cells were calculated on a $4 \times 4\,cm$ lattice (see Fig. 2, left panel, column 2 and 4). In every step of the recall process, active cells were selected, and their rate maps were multiplied in every spatial point. The maximum value in the resulting matrix was considered to give the most probable location of the robot on Fig. 2, right panel.

The input to the hippocampus model arises from sensory readings obtained during continuous motion. As a result, temporally close EC representations are similar. In his paper Rolls [37] showed that recall of different memories in the CA3 region is sufficiently good in certain circumstances. However, input patterns were decorrelated in his model, while highly correlated in ours. To clarify the significance of input correlations in spatial memories we analyzed the similarity of the hippocampal representation of positions.

Our simulations show that the initially temporally uncorrelated CA3 activity vectors (AVs) that follow each other in time become correlated as shown in Fig. 3. First, due to competitive learning in the DG similar EC activity patterns are grouped and evoke the same DG activity, which in turn determines the CA3 activity during learning. Second, modification of the $\mathbf{W}^{(EC,\ CA3)}$ synapses temporally close AVs become correlated. Third, this correlation is further increased by iterating the CA3 activity via the $\mathbf{W}^{(CA3,\ CA3)}$ matrix and results in a partitioning of the space into areas within which AVs are highly correlated (Fig. 3B, C).

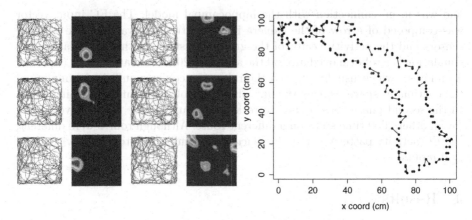

Fig. 2. *Left panel*: Cell activity in the CA3 region. Images in the first and third column show the path of the robot in the 1 m x 1 m square arena. Lines represent the path of the robot, green dots show where a certain cell had non-zero activity. In the second and fourth columns color coded images show the firing rate. On the left columns place cells with one single place field, on the right place cells with multiple places fields are shown. *Right panel*: Estimation of the robot's position based on the population of place cells. On a short section of the path we predicted the position based on the current activity of place cells. Real position is denoted by black dots, estimated position by red dots connected to the real positions they estimate. Mean and std of the difference between real and estimated positions was 2.5 ± 1.4 cm.

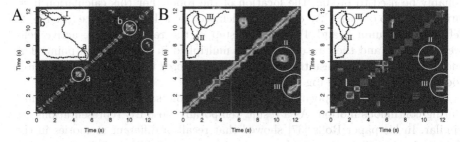

Fig. 3. Temporal correlation of activity vectors in the CA3 area. Correlation among AVs composed of the firing rates of CA3 cells were calculated along the random path of the robot (inset). *A*: during learning the AVs are decorrelated (note that increased correlation is seen at sharp turns (a, b) and at points I., II. and III. where the path crosses itself. The robot started from the bottom left corner.) *B*: during recall the EC input initiates memory retrieval. AVs in nearby time points show high correlations. *C*: (same time points are analyzed as in *B*) after the EC input promoted some of the CA3 cells the CA3 RC refine the cell activities and drive the associative memory system into an attractor, which classifies space into distinct categories within which correlations are high. Note the increased correlation at points I and II, where the robot crosses its path, or moves on parallel to it.

5   Discussion

We have set up and analyzed a biologically constrained computational model of associative memory in the hippocampal formation based on previous works by Treves and Rolls [47] and Rolls [37], which was used in a spatial context. Extero- and interoceptive sensory signals were used as inputs to the hippocampal model. Specifically, our model is composed of the entorhinal cortex, which represented the source of inputs segregated by their type of modality, the dentate gyrus serving as a pre-processing stage, where learning keys are generated from the entorhinal input and the CA3 region of the hippocampus, which served as an attractor network. We showed that combining pre-processed visual and tactile sensory information with grid cell activity, cells in the DG and the CA3 showed place cell behavior.

We studied the properties of the system of place cells and concluded that – even though we have not implemented a biologically based neural system to decode the place cell code – using their activity vector a good prediction of the current position of the animat can be gained.

The model presented and the behaviour it produces agree with the classical view of hippocampal processing and the previous physiological measurements. In the light of the most recent electrophysiological studies, however, the assumptions made in Sect. 3.1 need to be reviewed. Experimental results [24] show that while principal cells in the CA3 region, which exhibit place correlated activity mostly have a single place field, DG granule cells usually have multiple place fields. This phenomena is hard to explain in the framework presented previously as DG cell activities are used as the teaching signal for the CA3 cells in the learning mode of the system. It might, however, be possible to solve the problem of missing multiple place fields in the DG. Solstad et al. [44] showed that summing the activity of a few tens of grid cells with completely random phase would produce a rate maps with multiple poorly defined and scattered place fields even in the absence of any learning. An other solution is that places corresponding to place fields of a given DG cell are indistinguishable for that cell. This means that the grid pattern of afferent grid cells is the same, which would be the case if there were too few grid scales (for an indication see the Supplementary material of Barry et al. [3]). Finally, the morphology of dentate granule cells might enable that synapses on separate dendritic branches of this cell type would be modified independently. Thus, a given DG cell would be active if any of its dendritic branches were activated.

The goodness of the position reconstruction also depends on the quality or the number of place fields of CA3 cells. If co-activation of multiple DG cells is required to activate a post-synaptic CA3 cell the firing probability of CA3 cells would decrease relative to the case when a single DG cell can activate a CA3 cell. Indeed, cells with a single place field at well defined positions would better encode location that cells with several poor place fields.

The analysis of Fig. 3 reveals a fundamental difference between models designed to describe place cell formation (eg. [22, 40, 43]) and our approach. To introduce the continuous nature of space into computer models of place cells, a usual approach is to use continuous attractor models to describe the CA3 region. The model by Rolls [37] and similar auto-associative memory models generate discrete attractors. In our model the continuity of space is introduced via the grid cell system, which exhibits a space-periodic activity, the hippocampus itself does not reflect any property of space. As a result, instead of a smoothly changing representation of space nearby positions are clustered as shown in Fig. 3, C. Indeed, compared with measurements (eg. see Fig. 6 in [4]) a CA3 without attractor dynamics (Fig. 3, B) is closer to experimental results.

In summary, the presented model is capable of generating place cell-like behavior but misses some of the fine-scale properties of the real hippocampal place representation. Using the resulting space code reliable prediction of the robot's position can be achieved, thus the model might be used as a basic building block.

References

[1] Arleo, A., Gerstner, W.: Spatial cognition and neuro-mimetic navigation: a model of hippocampal place cell activity. Biological Cybernetics 83, 287–299 (2000)
[2] Banquet, J.P., Gaussier, P., Quoy, M., Revel, A., Burnod, Y.: A hierarchy of associations in hippocampo-cortical systems: cognitive maps and navigation strategies. Neural Computation 17, 1339–1384 (2005)
[3] Barry, C., Hayman, R., Burgess, N., Jeffery, K.J.: Experience-dependent rescaling of entorhinal grids. Nature Neuroscience 10, 682–684 (2007)
[4] Battaglia, F.P., Sutherland, G.R., McNaughton, B.L.: Local sensory cues and place cell directionality: Additional evidence of prospective coding in the hippocampus. Journal of Neuroscience 24, 4541–4550 (2004)
[5] Becker, S.: A computational principle for hippocampal learning and neurogenesis. Hippocampus 15, 722–738 (2005)
[6] Bennett, A.T.D.: Do animals have cognitive maps? The Journal of Experimental Biology 199, 219–224 (1996)
[7] Blair, H.T., Welday, A.C., Zhang, K.: Scale-invariant memory representations emerge from moiré interference between grid fields that produce theta oscillations: a computational model. Journal of Neuroscience 27, 3211–3229 (2007)
[8] Burak, Y., Fiete, I.: Do we understand the emergent dynamics of grid cell activity? Journal of Neuroscience 26, 9352–9354 (2006)
[9] Burak, Y., Brookings, T., Fiete, I.: Triangular lattice neurons may implement an advanced numeral system to precisely encode rat position over large ranges. arXiv, arXiv:q-bio/0606005 (2006)
[10] Burgess, N., Recce, M., O'Keefe, J.: A model of hippocampal function. Neural Networks 7, 1065–1081 (1994)
[11] Eichenbaum, H.: Is the rodent hippocampus just for 'place'? Current Opinion in Neurobiology 6, 187–195 (1996)
[12] Fleischer, J.G., Gally, J.A., Edelman, G.M., Krichmar, J.L.: Retrospective and prospective responses arising in a modeled hippocampus during maze navigation by a brain-based device. Proceedings of the National Academy of Sciences 104, 3556–3561 (2007)
[13] Fuhs, M.C., Touretzky, D.S.: A spin glass model of path integration in rat medial entorhinal cortex. Journal of Neuroscience 26, 4266–4276 (2006)

[14] Fuhs, M.C., Redish, A.D., Touretzky, D.S.: A visually driven hippocampal place cell model. In: Bower, J. (ed.) Computational Neuroscience: Trends in Research, pp. 379–384. Plenum Publishing, New York (1998)

[15] Fyhn, M., Molden, S., Witter, M.P., Moser, E.I., Moser, M.B.: Spatial representation in the entorhinal cortex. Science 305, 1258–1264 (2004)

[16] Gallistel, C.R.: Animal cognition, the representation of space, time and number. Annual Review of Psychology 40 (1989)

[17] Good, M.: Spatial memory and hippocampal function: Where are we now? Psicológica 23, 109–138 (2002)

[18] Hafting, T., Fyhn, M., Molden, S., Moser, M.-B., Moser, E.I.: Microstructure of a spatial map in the entorhinal cortex. Nature 436, 801–806 (2005)

[19] Hasselmo, M.E.: Neuromodulation: acetylcholine and memory consolidation. Trends in Cognitive Sciences 3, 351–359 (1999)

[20] Hasselmo, M.E., Wyble, B.P., Wallenstein, G.V.: Encoding and retrieval of episodic memories: role of cholinergic and GABAergic modulation in the hippocampus. Hippocampus 6, 693–708 (1996)

[21] Jeffery, K.J., Burgess, N.: A metric for the cognitive map: found at last? Trends in Cognitive Sciences 10 (2006)

[22] Káli, S., Dayan, P.: The involvement of recurrent connections in area CA3 in establishing the properties of place fields: a model. Journal of Neuroscience 20, 7463–7477 (2000)

[23] Krichmar, J.L., Seth, A.K., Nitz, D.A., Fleischer, J.G., Edelman, G.M.: Spatial navigation and causal analysis in a brain-based device modeling cortical-hippocampal interactions. Neuroinformatics 3, 197–221 (2005)

[24] Leutgeb, J.K., Leutgeb, S., Moser, M.B., Moser, E.I.: Pattern separation in the dentate gyrus and CA3 of the hippocampus. Science 315(5814), 961–966 (2007)

[25] Lisman, J.E., Otmakhova, N.A.: Storage, recall, and novelty detection of sequences by the hippocampus: elaborating on the socratic model to account for normal and aberrant effects of dopamine. Hippocampus 11, 551–568 (2001)

[26] Marr, A.: Simple memory: a theory for archicortex. Philosophical transactions of the Royal Society of London. Series B Biological sciences 262, 23–81 (1971)

[27] McNaughton, B.L., Morris, R.G.M.: Hippocampal synaptic enhancement and information storage within a distributed memory system. Trends in Neurosciences 10, 408–415 (1987)

[28] Michel, O.: Webots: Professional mobile robot simulation. Journal of Advanced Robotics Systems 1(1), 39–42 (2004), http://www.ars-journal.com/ars/SubscriberArea/Volume1/39-42.pdf

[29] Milford, M.J.: Robot Navigation from Nature. Springer Tracts in Advanced Robotics. Springer, Heidelberg (2008)

[30] Milford, M.J., Wyeth, G.: Spatial mapping and map exploitation: a bio-inspired engineering perspective. In: Spatial Information Theory. Springer Lecture Notes in Computer Science, pp. 203–221. Springer, Heidelberg (2007)

[31] Morris, R.G., Garrud, P., Rawlins, J.N., O'Keefe, J.: Place navigation impaired in rats with hippocampal lesions. Nature 24, 681–683 (1982)

[32] O'Keefe, J.: Place units in the hippocampus of freely moving rat. Experimental Neurology 51, 78–109 (1976)

[33] O'Keefe, J., Burgess, N.: Dual phase and rate coding in hippocampal place cells: theoretical significance and relationship to entorhinal grid cells. Hippocampus 15, 853–866 (2005)

[34] O'Keefe, J., Conway, D.H.: Hippocampal place units in the freely moving rat: Why they fire when they fire. Experimental Brain Research 31, 573–590 (1978)

[35] O'Keefe, J., Dostrovsky, J.: The hippocampus as a spatial map. Preliminary evidence from unit activity in the freely moving rat. Brain Research 34, 171–175 (1971)

[36] O'Keefe, J., Nadel, L.: The Hippocampus as a Cognitive Map. Oxford University Press, Oxford (1978), http://www.cognitivemap.net/

[37] Rolls, E.T.: A model of the operation of the hippocampus and entorhinal cortex in memory. International Journal of Neural Systems 6, 51–71 (1995)

[38] Rolls, E.T., Treves, A.: The relative advantages of sparse versus distributed encoding for associative neuronal networks in the brain. Network 1, 407–421 (1990)

[39] Rolls, E.T., Stringer, S.M., Elliot, T.: Entorhinal cortex grid cells can map to hippocampal place cells by competitive learning. Network 17, 447–465 (2006)

[40] Samsonovich, A., McNaughton, B.L.: Path integration and cognitive mapping in a continuous attractor neural network model. Journal of Neuroscience 17(15), 5900–5920 (1997)

[41] Scoville, W.B., Milner, B.: Loss of recent memory after bilateral hippocampal lesions. Journal of Neurology, Neurosurgery and Psychiatry 20, 11–21 (1957)

[42] Sharp, P.E.: Computer simulation of hippocampal place cells. Psychobiology 19(2), 103–115 (1991)

[43] Sharp, P.E., Blair, H.T., Brown, M.: Neural network modeling of the hippocampal formation spatial signals and their possible role in navigation: a modular approach. Hippocampus 6, 720–734 (1996)

[44] Solstad, T., Moser, E.I., Einevoll, G.T.: From grid cells to place cells: a mathematical model. Hippocampus 16, 1026–1031 (2006)

[45] Tolman, E.C.: Cognitive maps in rats and man. Psychological Review 55, 189–208 (1948)

[46] Touretzky, D.S., Redish, A.D.: Theory of rodent navigation based on interacting representations of space. Hippocampus 6(3), 247–270 (1996)

[47] Treves, A., Rolls, E.T.: Computational analysis of the role of the hippocampus in memory. Hippocampus 4, 374–391 (1994)

[48] Wood, E.R., Dudchenko, P.A., Eichenbaum, H.: The global record of memory in hippocampal activity. Nature 397, 613–616 (1998)

[49] Zipser, D.: A computational model of hippocampal place fields. Behavioral Neuroscience 99(5), 1006–1018 (1985)

Modelling the Cortical Columnar Organisation for Topological State-Space Representation, and Action Planning

Louis-Emmanuel Martinet[1,2,3], Benjamin Fouque[1,2,3], Jean-Baptiste Passot[2,3], Jean-Arcady Meyer[1], and Angelo Arleo[2,3]

[1] Université Pierre et Marie Curie - Paris 6, FRE2507, ISIR, Paris, F-75016, France
[2] UPMC Univ Paris 6, UMR 7102, F-75005, Paris, France
[3] CNRS, UMR 7102, F-75005, Paris, France
louis-emmanuel.martinet@isir.fr

Abstract. We present a neuromimetic navigation system modelling the columnar structure of the cortex to mediate spatial learning and action planning. The model has been validated on a spatial behavioural task, namely the Tolman & Honzik's *detour* protocol, which allowed us to test the ability of the system to build a topological representation of the environment, and to use it to exhibit flexible goal-directed behaviour (i.e., to predict the outcome of alternative trajectories to avoid blocked pathways). First, it is shown that the model successfully reproduces the navigation performance of rodents in terms of goal-directed path selection. Second, we report on the neural response patterns characterising the learnt columnar space representation.

1 Introduction

This paper presents a biomimetic model of action planning inspired by the columnar organisation of the mammalian neocortex. Planning is defined here as the ability, given a state space S and an action space A, to "mentally" explore the $S \times A$ space to infer an appropriate sequence of actions leading to a goal state $s_g \in S$. This definition calls upon the capability of (i) predicting the consequences of actions, i.e. the most likely state $s' \in S$ to be reached when an action $a \in A$ is executed from a state $s \in S$, (ii) evaluating the effectiveness of the selected plan on-line. The model generates a topological representation of the environment, and it employs an activation-diffusion mechanism to plan goal-directed trajectories. The activation-diffusion process is based on the propagation of a reward-dependent activity signal from the goal state s_g through the entire topological network. This propagation process enables the system to generate sequences of actions (i.e., trajectories) from the current state s towards s_g.

Topological map learning and path planning have extensively been studied in biomimetic robotics [1]. We focus on models inspired by the anatomical organisation of the cortex, and implementing an activation-diffusion planning principle. The existence of cortical columns was first reported by Mountcastle [2], who observed vertical groups of neurones responding to the same external stimuli simultaneously. Neuroanatomical findings suggest that these "functional columns" can be further divided into several

M. Asada et al. (Eds.): SAB 2008, LNAI 5040, pp. 137–147, 2008.

"minicolumns", i.e. vertical bundles of neurones across the layers of the cortex separated from each other by a cell-poor area [3].

Burnod [4] proposed one of the first models of the cortical column architecture, called "cortical automaton". He also described a "call tree" process that can be seen as a neuromimetic implementation of the activation-diffusion principle. Several action selection models were inspired by Burnod's hypothesis. Some of these works employed the cortical automaton concept explicitly [5,6,7]. Others used either more classical connectionist architectures [8,9,10] or Markov decision processes [11]. Yet, none of these works took into account the multilevel coding property offered by the possibility to refine the cortical organisation by adding a sublevel to the column, i.e. the minicolumn. The topological representation presented here exploits this idea by associating the columnar level to a compact representation of the environment, and by employing the minicolumn level to characterise the agent's behaviour.

In order to validate the preliminary version of the model, we have implemented it on a simulated robot, and tested it on the classical navigation task designed by Tolman & Honzik [12]. This protocol allowed us to assess the ability of the system to learn topological representations, and to exploit them to perform flexible goal-directed behaviour (e.g., planning optimal *detour* trajectories). The Tolman & Honzik's task is a purely spatial navigation protocol. Our middle-term goal is to extend the cortical model presented here to elaborate more abstract contextual representations. For example, besides learning the spatial properties of the environment, the system shall be able to encode multidimensional information, such as motivation-dependent memories, multiscale spatio-temporal correlates, and action cost/risk constraints.

2   Methods

2.1   Single Neurone Model

The elementary computational units of the model are artificial firing-rate neurones i, whose mean discharge $r_i \in [0, 1]$ is given by:

$$r_i(t) = f\Big(V_i(t) \cdot (1 \pm \epsilon)\Big) \tag{1}$$

where $V_i(t)$ is the membrane potential at time t, f is the transfer function, and ϵ is a random noise uniformly drawn from $[0, 0.01]$. The potential V_i varies according to:

$$\tau_i \cdot \frac{dV_i(t)}{dt} = -V_i(t) + I_i(t) \tag{2}$$

where $\tau_i = 10$ ms is the membrane time constant, and $I_i(t)$ is the synaptic drive generated by all the afferent inputs at time t. Eq. 2 is integrated by using a time step $\Delta_t = 1$ ms. Both the synaptic drive $I_i(t)$ and the transfer function f are characteristic of the different types of model units, and they will be defined thereafter.

2.2   Encoding Space and Actions: The Minicolumn and Column Model

The main inputs to the cortical model are the location- and orientation-selective activities of hippocampal place (HP) and head-direction cells, respectively [13,14]. The HP

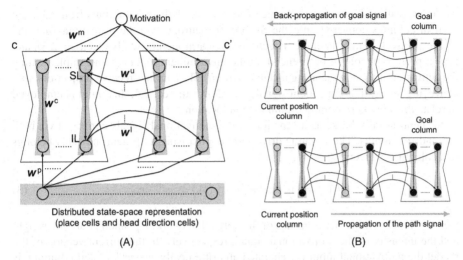

Fig. 1. The cortical model and the activation-diffusion process. (**A**) The architecture of two column units c and c'. Columns are composed of sets of minicolumns (vertical grey regions), each of which consists of a supragranular layer unit (SL) and an infragranular layer unit (IL). (**B**) Top: back-propagation of the motivational signal through the network of SL neurones. Bottom: forward-propagation of the goal-directed action signal through the IL neurones.

field representation is built incrementally as the animat explores the environment, and it provides the system with a continuous distributed and redundant state representation S [15,16]. A major objective of the cortical model is to build a more compact state-action representation $S \times A$ suitable for topological map learning and action planning.

In the model, the basic component of the columnar organisation is the minicolumn (vertical grey regions in Fig. 1). An unsupervised learning scheme (see Sec. 2.3) is employed to make the activity of each minicolumn selective to a specific state-action pair $(s, a) \in S \times A$. Notice that a given action $a \in A$ represents the allocentric motion direction of the animat when it performs the transition between two locations $s, s' \in S$. According to the learning algorithm, all the minicolumns selective for the same spatial location $s \in S$ are grouped to form a higher-level computational unit, i.e. the column (see c and c' in Fig. 1A). This architecture is inspired by biological data showing that minicolumns inside a column have similar selectivity properties [17]. Thus, columns consist of a set of minicolumns that are incrementally recruited to encode all the state-action pairs $(s, a_{1...N}) \in S \times A$ experienced by the animat at a location s. During planning (see Sec. 2.4), all the minicolumns of a column compete with each other to locally infer the most appropriate goal-directed action.

Every minicolumn of the model consists of two computational units, representing supragranular layer (SL) and infragranular layer (IL) neurones (Fig. 1A). The discharge of SL and IL units simulates the mean firing activity of a population of cortical neurones in layers II-III, and V-VI, respectively. Each minicolumn receives three different sets of afferent projections (Fig. 1A): (*i*) Hippocampal inputs conveying space coding activity converge onto IL neurones; these connections are plastic, and their synaptic efficacy is determined by the weight distribution w^p (all the synaptic weights of the model are within the maximum range of $[0, 1]$). *(ii)* Collateral afferents from adjacent cortical columns converge onto the SL and IL neurones via the projections $\boldsymbol{w}^u$ and $\boldsymbol{w}^l$, respectively. These lateral connections are learnt incrementally (see Sec. 2.3), and play a prominent role in both encoding the environment topology and implementing the activation-diffusion planning mechanism. *(iii)* SL neurones receive projections $\boldsymbol{w}^m$ conveying motivation-dependent signals. As shown in Sec. 2.4, this input is employed to relate the activity of a minicolumn to goal locations.

SL neurones discharge as a function of the motivational signals conveyed via both $\boldsymbol{w}^u$ and $\boldsymbol{w}^m$ inputs. The synaptic drive $I_i(t)$ depolarising a SL neurone i that belongs to a column c is given by:

$$I_i(t) = \max_{i' \in c' \neq c} \left\{ w_{ii'}^u \cdot r_{i'}(t) \right\} + w_i^m \cdot r_m \tag{3}$$

where i' indexes other SL neurones of the cortical network; w_i^m and r_m are the weight and the intensity of the motivational signal, respectively. In the current version of the model the motivational input is generated algorithmically, i.e. $w_i^m = 1$ if column c is associated to the goal location, $w_i^m = 0$ otherwise, and the motivational signal $r_m = 1$. The membrane potential of unit i is then computed according to Eq. 2, and its firing rate $r_i(t)$ is obtained by means of an identity transfer function f.

Within each minicolumn, SL neurones project onto IL units via non-plastic projections $\boldsymbol{w}^c$ (Fig. 1A). Thus, IL neurones are driven by HP cells p (via the projections $\boldsymbol{w}^p$), by IL neurones belonging to adjacent columns (via the collaterals $\boldsymbol{w}^l$), and by SL units i (via $\boldsymbol{w}^c$). The synaptic drive of a IL neurone $j \in c$ is:

$$I_j(t) = \max\left\{ \sum_{p \in HP} w_{jp}^p \cdot r_p(t) \, , \, \max_{j' \in c' \neq c} \left\{ w_{jj'}^l \cdot r_{j'}(t) \right\} \right\} + w_{ji}^c \cdot r_i(t) \tag{4}$$

where j' indicates other IL neurones of the network; $w_{ji}^c = 1$ if the SL neurone i and the IL neurone j belong to the same minicolumn, $w_{ji}^c = 0$ otherwise. Then, the membrane potential $V_j(t)$ is computed by Eq. 2, and a sigmoidal transfer function f is employed to calculate $r_j(t)$. The parameters of the transfer function change online to adapt the electroresponsiveness properties of IL neurones j to the strength of their inputs [18].

2.3   Unsupervised Growing Network Scheme for Topological Map Learning

The topological representation is built incrementally as the animat explores the environment. At each location visited by the agent at time t the cortical network is updated if-and-only-if the infragranular layers of all existing minicolumns remain silent, i.e. $\sum_j \mathcal{H}(r_j(t) - \rho) = 0$, where j indexes all the IL neurones, $\mathcal{H}$ is the Heaviside function (i.e., $\mathcal{H}(x) = 1$ if $x \geq 0$, $\mathcal{H}(x) = 0$ otherwise), and $\rho = 0.1$. If at time t the novelty condition holds, a new group of minicolumns (i.e., a new column c) is recruited to become selective to the new place. Then, all the simultaneously active place cells $p \in HP$ are connected to the new IL units $j \in c$. Weights w_{jp}^p are initialised according to: $w_{jp}^p = \mathcal{H}(r_p - \rho) \cdot r_p$. For $t' > t$, the synaptic strength of these connections is changed by unsupervised Hebbian learning combined to a winner-take-all scheme. Let

c be the column selective for the position visited by the animat at time t', i.e. let all the $j \in c$ be the most active IL units of the network at time t'. Then:

$$\Delta w_{jp}^p = \eta \cdot r_p \cdot (r_j - w_{jp}^p) \tag{5}$$

with $\eta = 0.005$. Whenever a state transition occurs, the collateral projections $\boldsymbol{w}^l$ and $\boldsymbol{w}^u$ are updated to relate the minicolumn activity to the state-action space $S \times A$. For instance, let columns c and c' denote the animat position before and after a state transition, respectively (Fig. 1A). A minicolumn $\theta \in c$ becomes selective for the locomotion orientation taken by the animat to perform the transition. A new set of projections $w_{j'j}^l$ are then established from the IL unit $j \in \theta$ of column c to all the IL units j' of the column c'. In addition, at the supragranular level, a new set of connections $w_{ii'}^u$ is learnt to connect all the SL units of column c', i.e. $i' \in c'$, to the SL unit i of the minicolumn $\theta \in c$. The strengths of the lateral projections are initialised as: $w_{j'j}^l = w_{ii'}^u = \beta_{LTP}$, $\forall i', j' \in c'$, with $\beta_{LTP} = 0.9$. Finally, in order to adapt the topological representation online, a synaptic potentiation-depression mechanism can modify the lateral projections $\boldsymbol{w}^l$ and $\boldsymbol{w}^u$. For example, if a new obstacle prevents the animat from achieving a previously learnt transition from column c to c' (i.e., if the activation of the IL unit $j \in \theta \in c$ is not followed in the time by the activation of all IL units $j' \in c'$), then a depression of the $w_{j'j}^l$ synaptic efficacy occurs: $\Delta w_{j'j}^l = -\beta_{LTD} \cdot w_{j'j}^l$, $\forall j' \in c'$, where $\beta_{LTD} = 0.5$. The projections $w_{ii'}^u$ are updated similarly. A compensatory potentiation mechanism reinforces both $\boldsymbol{w}^l$ and $\boldsymbol{w}^u$ connections whenever a previously experienced transition is performed successfully: $\Delta w_{j'j}^l = \beta_{LTP} - w_{j'j}^l$, $\forall j' \in c'$. The weights $w_{ii'}^u$ are updated similarly. Notice that $\boldsymbol{w}^l, \boldsymbol{w}^u \in [0, \beta_{LTP}]$.

2.4 Action Planning

This model aims at developing a high-level controller determining the agent's behaviour based on action planning. Yet, a low-level reactive module enables the animat to avoid obstacles. Whenever the proximity sensors detect an obstacle, the reactive module takes control and prevents collisions. Also, the simulated animal behaves in order to either follow planned pathways (i.e., exploitation) or improve the topological map (i.e., exploration). This exploitation-exploration tradeoff is governed by an ϵ-greedy selection mechanism, with $\epsilon \in [0, 1]$ decreasing exponentially over time [16].

Fig. 1B shows an example of activation-diffusion process mediated by the columnar network. During trajectory planning, the SL neurones of the column corresponding to the goal location s_g are activated via a motivational signal r_m (see Eq. 3). Then, the SL activity is back-propagated through the network by means of the lateral projections $\boldsymbol{w}^u$ (Fig. 1B, top). During planning, the responsiveness of IL neurones (Eq. 4) is decreased to detect coincident inputs. In particular, the occurrence of the SL input r_i is a necessary condition for a IL neurone j to fire. In the presence of the SL input r_i, either the hippocampal signal r_p or the intercolumn signal r_j' are sufficient to activate the IL unit j. When the back-propagated goal signal reaches the minicolumns selective for the current position s this coincidence event occurs, which triggers the forward propagation of a goal-directed path signal through the projections $\boldsymbol{w}^l$ (Fig. 1B, bottom).

Goal-directed trajectories are generated by reading out the successive activations of IL neurones. Action selection calls upon a competition between the minicolumns

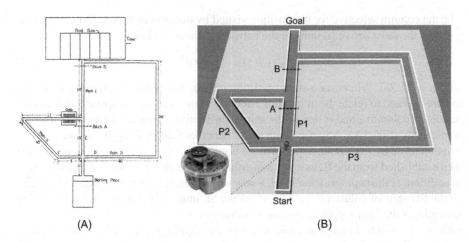

(A)   (B)

Fig. 2. (A) Tolman & Honzik's maze [12]. The gate near the second intersection allowed the rats to go from left to right only. **(B)** The simulated maze and robot. The dimensions of the simulated maze were taken so as to maintain the proportions of the real Tolman & Honzik's experimental setup. Bottom-left inset: the real *e-puck* mobile robot has a diameter of 70 mm and is 55 mm tall.

encoding the $(s, a_{1...N}) \in S \times A$ pairs, where s is the current location, and $a_{1...N}$ are the transitions from s to adjacent positions s'. For sake of robustness, competition occurs over a 10-timestep cycle. It is worth stressing that each SL synaptic relay attenuates the goal signal by a factor $w_{ii'}^u$ (Eq. 3). That is, the smaller the number of synaptic relays, the stronger the goal signal received by the SL neurone corresponding to the current location s. Because the model column receptive fields are distributed rather uniformly over the environment, the intensity of the goal signal at a given location s is correlated to the distance between s and the target position s_g.

2.5 The Behavioural Task and the Animat

In order to validate our navigation planning system, we chose the classical experimental task proposed by Tolman & Honzik [12]. The main objective of this behavioural protocol was to demonstrate that rodents undergoing a navigation test are able to show some "insights", e.g. to predict the outcome of alternative trajectories leading to a goal location in the presence of blocked pathways. The original Tolman & Honzik's maze is shown in Fig. 2A. It consisted of three narrow alleys of different lengths (Paths 1, 2, and 3) guiding the animals from a starting position (bottom) to a feeder location (top).

We implemented our model by means of the Webots© robotics simulation software. Fig. 2B shows a simulated version of the Tolman & Honzik's apparatus, and the simulated robot. We emulated the experimental protocol designed by Tolman & Honzik to assess the subjects' navigation performance. The overall protocol consisted of a training period followed by a probe test. Both training and probe trials were stopped when the subject had found the goal.

Training period: it lasted 14 days with 12 trials per day. The subjects could explore the maze and learn a navigation policy by developing their preferences for P1, P2, and P3.

– During Day 1, a series of 3 *forced runs* was carried out, in which additional doors were used to force the subjects to go successively through P1, P2, and P3. Then, during the remaining 9 runs, all additional doors were removed, and the subjects could explore the maze freely. At the end of the first training day, a preference for P1 was expected to be already developed [12].

– From Day 2 to 14, a block was introduced at place A (Fig. 2B) to require a choice between P2 and P3. In fact, additional doors were used to close the entrances to P2 and P3 to force subjects to go first to the Block A. Then, doors were removed, and subjects were forced to decide between P2 and P3 on their way back to the first intersection. Each day, there were 10 "Block at A" runs that were mixed with 2 non-successive free runs to maintain the preference for P1.

Probe test period: It lasted 1 day (Day 15), and it consisted of 7 runs during which a block was placed at position B to interrupt the common section (Fig. 2B). The subjects were forced to decide between P2 and P3 when returning to the first intersection point.

For these experiments, Tolman & Honzik used 10 male rats of mixed breed, from 5 to 8 months old, with no previous training. In our simulations, we used a population of 100 animats, and we assessed the statistical significance of the results by means of an ANOVA analysis (the significant threshold was set at 10^{-2}, i.e. $p < 0.01$ was considered significant).

3   Results

3.1   Behavioural Analysis

Day 1. During the first 12 training trials, the animats learnt the maze topology, and planned their trajectory in the absence of both block A and B (Fig. 2B). Similar to Tolman & Honzik's findings, our results show that the model learnt to select the shortest pathway P1 significantly more frequently than the alternative paths P2, P3 (ANOVA, $F_{2,297} = 168.249$, $p < 0.0001$). The quantitative and qualitative analyses reported on Fig. 3 (left) describe the path selection performance averaged over 100 experiments.

Days 2-14. During this training phase (consisting of 156 trials), a block was introduced at location A (Fig. 2B), which forced the animats to update their topological maps dynamically, and to plan a *detour* to the goal. The results reported by Tolman & Honzik provided strong evidence for a preference for the shortest *detour* path P2. Consistently, in our simulations (Fig. 3, centre) we observed a significantly larger number of transits through P2 compared to P3 (ANOVA, $F_{1,198} = 383.068$ $p < 0.0001$), P1 being ignored in this analysis (similar to Tolman & Honzik's analysis) because blocked.

Day 15. In agreement with Tolman & Honzik's protocol, seven probe trials were performed during the 15th day of the simulated protocol, by removing the block A and adding a new block B (Fig. 2B). This manipulation aimed at testing the "insight" working hypothesis: after a first run through the shortest path P1 and after having encountered the unexpected block B, will rats try P2 or will they go directly through P3? According to Tolman & Honzik's results, the rats behaved as predicted by the insight hypothesis, i.e. they tended to select the longer but effective P3. The authors concluded

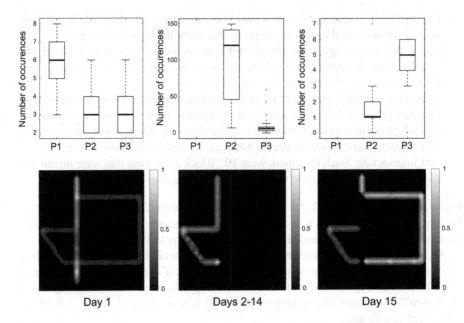

Fig. 3. Behavioural results. **Top row:** mean number of transits through P1, P2, and P3 (averaged over 100 experiments). **Bottom row:** occupancy grid maps.

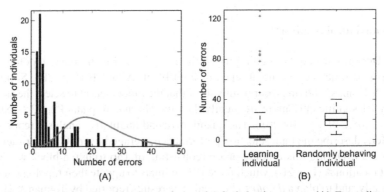

Fig. 4. Comparison between a learning and a randomly behaving agent. **(A)** Error distribution of learning (black histogram) versus random (grey line) animats. **(B)** Boxplots of the number of errors done on average by our model and by a randomly behaving agent.

that rats were able to inhibit the previously learnt policy (i.e., the "habit behaviour" consisting of selecting P2 after a failure of P1 during the 156 previous trials). Our probe test results are shown in Fig. 3 (right). Similar to rats, the animats exhibited a significant preference for P3 compared to P2 (ANOVA, $F_{1,198} = 130.15$, $p < 0.0001$). Finally, in order to further assess the mean performance of the system during the probe trials, we compared the action selection policy of learning animats with that of randomly behaving (theoretical) animats. Fig. 4A provides the results of this comparison by showing

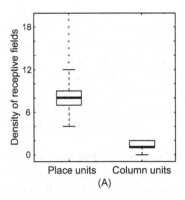 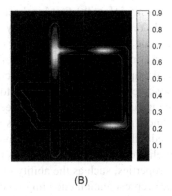

(A)   (B)

Fig. 5. (A) Spatial density (i.e., mean number of units selective for a place) of the receptive fields of HP cells and cortical column units. **(B)** Samples of receptive fields of three column units.

the error distribution over the population of learning agents (black histogram) and randomly behaving agents (grey curve). The number of errors per individual are displayed in the boxplot of Fig. 4B. These findings indicate a significantly better performance of learning animats compared to random agents (ANOVA, $F_{1,196} = 7.4432, p < 0.01$).

3.2 Analysis of Neural Activities

A series of additional analyses were done to begin to characterise the underlying processes (e.g., neural activities) subserving the action selection behaviour of the model. We measured the mean spatial density of the receptive fields of HP cells and cortical column units of the model. We recall that one of the aim of our cortical column model was to build a less redundant state-space representation, compared to the HP field representation. Fig. 5A shows that the cortical network permitted to reduce the redundancy of the learnt spatial map significantly, compared to the upstream hippocampal space code (ANOVA, $F_{1,316} = 739.2, p < 0.0001$). Finally, Fig. 5B displays some samples of cortical column receptive fields of the model.

4   Discussion

We presented a navigation model based on the columnar organisation of the mammalian cortex. It builds a topological map of the environment incrementally, and it uses it to plan an efficient course of actions leading to a goal location. The model was successfully employed to solve the classical Tolman & Honzik's behavioural task [12]. As aforementioned, other models have been proposed to solve goal-directed navigation tasks. They are mainly based on the properties of hippocampal (e.g., Samsonovich and Ascoli 2005, [19]), and prefrontal cortex (e.g., Hasselmo 2005, [7]) neural assemblies. However, most of these models do not perform action *planning* as defined in this paper (see Sec. 1). Samsonovich and Ascoli [19] rather implement a local path finding mechanism to select the most suitable orientation leading to the goal. Similarly, Hasselmo's model [7] does not plan a sequence of actions from the current location s to the goal s_g but only infers the first local action to be taken, based upon a back-propagated goal signal. Yet, these two models rely on discretized state spaces (where predefined grid units code for places), whereas our model uses a distributed population of HP cells providing a continuous representation of the environment [16]. Also, our model learns topological maps coding for the state-action space $S \times A$ simultaneously. In the model by Samsonovich and Ascoli (2005) no topological information is represented, but only a distance measure between each visited place and a set of potential goals. Likewise, in Hasselmo's model states and actions are not jointly represented, which generates a route-based rather than a map-based navigation system [20].

The preliminary version of the model enabled us to investigate some basic computational properties, such as the ability of the columnar organisation to learn a compact state-space representation encoding topological information, and the efficiency of the activation-diffusion planning mechanism. Further efforts will be put to extend the current model to integrate multiple sources of information. For example, the animat should be able to learn maps that encode all the reward (subjective) values, and action-cost constraints. Also, these maps should be suitable to represent multiple spatio-temporal scales to overcome the intrinsic limitation of the activation-diffusion mechanism in large scale environments. Additionally, these multiscale maps should allow the model to infer high-level shortcuts to bypass the low-level constraints of the environment.

To conclude, although the model has been based upon biological knowledge, some of our working hypotheses are still under debate. First, the existence of cortical columns has been questioned recently [21]. Second, the hippocampus has also been proposed as a likely brain structure encoding topological maps [22]. Yet, the HP cell representation seems too redundant and distributed to constitute a suitable substrate for compact topological map learning [23]. Also, the evidence for high-level spatial representations mediated by neocortical areas (such as the prefrontal cortex, PFC [24]) corroborates the hypothesis of an action planning processing shared among multiple cortical regions [25]. In particular, several experimental observations [24,26] point towards a role of the PFC in abstract map building and action selection.

Acknowledgments. Funded by the EC Integrated Project ICEA, IST-027819-IP.

References

1. Meyer, J.A., Filliat, D.: Cogn. Syst. Res. 4(4), 283–317 (2003)
2. Mountcastle, V.B.: J Neurophysiol. 20(4), 408–434 (1957)
3. Buxhoeveden, D.P., Casanova, M.F.: Brain. 125(5), 935–951 (2002)
4. Burnod, Y.: An adaptative neural network: the cerebral cortex. Masson (1989)
5. Bieszczad, A.: Proc. World Congr. Comput Intell WCCI 1994, vol. 3, pp. 1313–1318 (1994)
6. Frezza-Buet, H., Alexandre, F.: In: IEEE Int Jt Conf Neural Netw., vol. 1, pp. 252–257 (1999)
7. Hasselmo, M.E.: J Cogn. Neurosci. 17(7), 1115–1129 (2005)
8. Schmajuk, N.A., Thieme, A.D.: Biol. Cybern. 67(2), 165–174 (1992)
9. Dehaene, S., Changeux, J.P.: Proc. Natl. Acad. Sci. USA 94(24), 13293–13298 (1997)
10. Banquet, J.P., et al.: Neural Comput. 17, 1339–1384 (2005)
11. Fleuret, F., Brunet, E.: Neural Computation 12(9), 1987–2008 (2000)
12. Tolman, E.C., Honzik, C.H.: Univ. Calif Publ. Psychol. 4(14), 215–232 (1930)

13. O'Keefe, J., Nadel, L.: The Hippocampus as a Cognitive Map. Oxford Univ Press, Oxford (1978)
14. Wiener, S.I., Taube, J.S.: Head Direction Cells and the Neural Mechansims of Spatial Orientation. MIT Press, Cambridge (2005)
15. Arleo, A., Gerstner, W.: Neurocomputing. 38(40), 1059–1065 (2001)
16. Arleo, A., Smeraldi, F., Gerstner, W.: IEEE Trans. Neural Netw. 15(3), 639–651 (2004)
17. Rao, S.G., Williams, G.V., Goldman-Rakic, P.S.: J Neurophysiol. 81(4), 1903–1916 (1999)
18. Triesch, J.: Neural Comput. 19(4), 885–909 (2007)
19. Samsonovich, A., Ascoli, G.: Learn Mem. 12, 193–208 (2005)
20. Arleo, A., Rondi-Reig, L.: J. Integr. Neurosci. 6(3), 327–366 (2007)
21. Horton, J.C., Adams, D.L.: Philos. Trans R Soc. Lond B Biol. Sci. 360(1456), 837–862 (2005)
22. Poucet, B., et al.: Rev. Neurosci. 15(2), 89–107 (2004)
23. Wilson, M.A., McNaughton, B.L.: Science, vol. 261, pp. 1055–1058 (1993)
24. Hok, V., et al.: Proc. Natl. Acad. Sci. U S A. 102(12), 4602–4607 (2005)
25. Knierim, J.J.: Learn Mem. 13(4), 405–415 (2006)
26. Granon, S., Poucet, B.: Behav Neurosci. 109(3), 474–484 (1995)

Adaptive Olfactory Encoding in Agents Controlled by Spiking Neural Networks

Nicolas Oros, Volker Steuber, Neil Davey, Lola Cañamero, and Rod Adams

Science and Technology Research Institute
University of Hertfordshire
AL10 9AB
United Kingdom
{N.Oros,V.Steuber,N.Davey,L.Canamero,R.G.Adams}@herts.ac.uk

Abstract. We created a neural architecture that can use two different types of information encoding strategies depending on the environment. The goal of this research was to create a simulated agent that could react to two different overlapping chemicals having varying concentrations. The neural network controls the agent by encoding its sensory information as temporal coincidences in a low concentration environment, and as firing rates at high concentration. With such an architecture, we could study synchronization of firing in a simple manner and see its effect on the agent's behaviour.

Keywords: spiking neural network, neural encoding, firing rate, temporal coincidence.

1 Introduction

Animals are able to react to chemicals (odours, pheromones...) present in the environment. The key sense to detect these chemical cues is smell [18]. Almost all animals have a similar olfactory system including olfactory sensory neurons (OSN) that are exposed to the outside world and linked directly to the brain. Pheromones and other odour molecules present in the environment are converted into signals in the brain by first binding to the olfactory receptor protein situated in the cell membrane of the OSN. Spikes are then sent down the axon of the OSN [10]. A chemical blend is composed of many molecules that can be detected with tuned odour receptors and therefore, activates a large range of olfactory sensory neurons. Odours are coded by which neurons emit spikes and also by the firing patterns of those neurons sending spikes to others during and after the stimulus. In many vertebrates and insects, oscillations of the neural activity have been recorded in the olfactory systems [18]. Therefore, the synchronization of firing between different sensory neurons seems to be very important for odour perception and interpretation. The firing rate and the number of sensory neurons are also important in odour recognition when stronger stimuli increase the frequency of firing of individual sensory neurons but also stimulate a larger number of them.

M. Asada et al. (Eds.): SAB 2008, LNAI 5040, pp. 148–158, 2008.

Different studies have been done on the perception of simulated chemicals using artificial neural networks where neural synchronization occurs [2, 6, 7] and also using robots [11, 13, 15-17]. We were interested in studying the behaviour of an agent in response to changes of its environment. The primary research question is how two encoding strategies can be used to integrate sensory information in order to control a simulated agent. To the best of our knowledge, no neural architecture, controlling a simulated agent, has been created that encodes the sensory information onto both the firing rate and the synchronization of firing (temporal coincidence) depending on the environment. As the interaction between the two encoding strategies is complex, we decided to create a simple architecture using a spiking neural network. This model could encode the sensory information onto both the firing rate and the synchronization of firing depending on the environment. The neural network controlled the agent by encoding the sensory information onto temporal coincidences in a low concentration environment, and firing rates at high concentration.

2 Environment

We created a simulation of a continuous world including an agent and a maximum of two chemicals. We decided to use a simple model of chemicals that are not diffused and evaporated but with concentrations that can be calculated directly at any given point. In this experiment, each chemical source had a circular shape and the same fixed value all over its surface. Our agent was equipped with two antennae and a differential steering system using two wheels. The two antennae were separated widely enough to detect the presence of the chemical concentration (Fig. 1 & 2). The left and right wheels were situated in the appropriate position (middle of each side) so the gravity centre was in the middle of the agent's body (excluding the antennae).

To control the agent, we had to decide which neurons' model to use in order to study firing synchronization of the sensors.

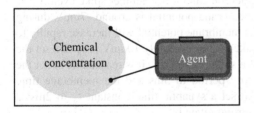

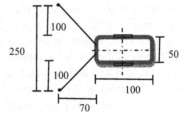

Fig. 1. An agent equipped with two wheels and two antennae used to detect chemicals

Fig. 2. Properties of an agent. Units are arbitrary.

3 Neural Network

There are three main ways to encode the intensity of sensory information into spiking neurons based on biological evidences [3-5, 8, 9, 12] . The most commonly used method consists of mapping the stimulus intensity to the firing rate of the neuron (firing rate encoding). Another method encodes the intensity of the stimulation into the number of spikes sent by different neurons arriving at a pre-synaptic neuron at the same time (firing synchronization or temporal coincidence encoding). The last main encoding scheme maps the strength of the stimulation in the firing delay of the neuron (delay encoding). As we saw earlier, spatial configuration is an important feature in odour recognition of neurons as is the synchronization of firing between neurons [10, 14, 18]. J. Hopfield and C. Brody [2, 6] created simple neural networks using spiking neurons to simulate an olfactory process. In their system, the recognition of an odour was signalled by spike synchronization in artificial glomeruli. In our system, the neural network was supposed to detect the blend of two different chemicals and modify the agent's behaviour. We used a model of neural network that allowed us to study synchronization of firing in a simple manner. The neural network could control the agent by encoding the sensory information onto temporal coincidences in a low concentration environment, and firing rates at high concentration.

a) Models of Spiking Neurons

It is well known that compared to the complex and computationally slow Hodgkin and Huxley model, simple spiking models like integrate-and-fire neurons can run quickly enough and have a more realistic behaviour than firing rate ones [4, 5, 8, 9, 12]. This is why more and more researchers are implementing spiking neurons in robots and simulated agents. Therefore, we decided to use a simple model of a spiking neuron. Our model is based on a leaky-integrator model which includes synaptic integration and conduction delays. The idea is that a spike sent by a neuron will take some time to arrive at another neuron. This time delay depends on the distance between the sender and the receiver. All the spikes arriving at a neuron are summed to calculate the neuron's input current (in Amperes per Farad) and membrane potential (in Volts) after every time step ($\Delta t = 0.1 ms$). Once the membrane potential reaches a certain threshold θ, the neuron will fire and then will be set to 0 for a certain time (refractory period). During this time, the neuron cannot fire another spike even if it is highly stimulated. Many real neurons' membrane potential is around -70mV during resting state [10]. When a neuron fires, its membrane potential will increase rapidly to about 30mV, so the height of a typical spike is approximately 100mV [10]. We set the resting potential to 0 and the potential of a spike to 100mV. It is reasonable to set the neuron's threshold at 20mV, the refractory period to 3ms and the membrane time constant τ_m to 50ms. We also decided to set a synaptic time constant τ_s to 2ms: a spike that arrives at a synapse triggers a current given by:

$$I_j(t) = \left(\frac{t-(t_{spike}+delay)}{\tau_s}\right) \exp\left(1 - \frac{t-(t_{spike}+delay)}{\tau_s}\right) \qquad (1)$$

where $I_j(t)$ is the synaptic input current, t_{spike} corresponds to the time a spike has been sent to the neuron, *delay* is the time delay in seconds before the spike arrives to the neuron (*delay = coeff_delay * distance*) with *coeff_delay* = 5.10^{-5}.

The change of membrane potential is given by:

$$\frac{dV}{dT} = -\left(\frac{V}{\tau_m}\right) + \Sigma(I_j W_j) \tag{2}$$

where V is the membrane potential, τ_m is the membrane time constant and W_j the synaptic weight.

b) Sensory Neurons

We created a model of a spiking sensory neuron in which the chemical concentration is processed so that a quasi-linear relationship between the concentration and the firing rate of the sensor is produced. Such relationships exist in biological systems. For example in humans, the relationship between the frequency of firing and pressure on the skin is linear [10]. We used a two step process where two biologically realistic non-linear mappings between sensory information and input current and between input current and firing rate results in a linear relationship. The sensory neurons used in our model are able to encode the stimulus intensity, measured at the tip of the antenna, into sensory input current using a biologically plausible sigmoid function. In this paper, we are using a very simple model chemical concentration that has only one value. Therefore, the sensory neurons encode this value onto the appropriate firing rate. The sensors were configured in order to distinguish a large range of concentrations between 1 and 300. Over this 300, they were saturating.

c) Motor Neurons

We decided that, in order to move, the agent should be driven by two wheels each controlled by two motor neurons: one to go forward, one to go backward. We created sensors able to detect a chemical gradient. But an agent equipped with such sensors will not move without any stimulus. So we decided for simplicity that an agent should always move forward in the absence of any external input. We performed this by adding a small baseline input current (0.5 A/F) in the motor neurons responsible to go forward. The final velocity of the wheels was calculated by subtracting the firing rate of the motor neurons, responsible for moving the agent forward and backward, running over a certain period of time.

4 Experiments

We used the agent and world described in Section 2. The world contained either one or two chemicals denoted by A or B. One agent, placed in the world, was controlled by a simple spiking neural network implementing the neurons described in Section 3. The neural controller was based on a Braitenberg vehicle (anger behaviour) [1] where an agent moves faster toward a stimulus when it detects it (Fig. 3).

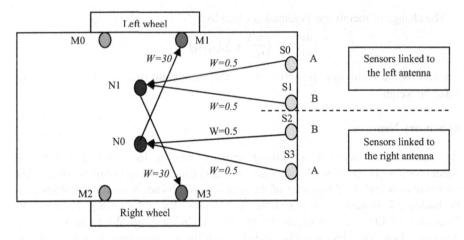

Fig. 3. Agent's neural controller. The sensors S0 and S3 detect the chemical A and the sensors S1 and S2 detect the chemical B. The sensory axons' lengths are all similar (delays = 2.5ms). The motor neurons M1 and M3 are responsible to move the agent forward. The threshold of the neurons (N0 and N1) was set to 4.6 mV. W is the synaptic weight.

Our hypothesis was that by using this architecture, the sensory neurons needed to encode the sensory information onto the firing rates, and also onto temporal coincidences between spikes sent by sensors. To verify this hypothesis, we performed three series of tests to study the effect of the starting positions, the sensory delays and the value of the concentrations on the agent's behaviour.

a) Experiment I

The first test was to study the effect of the agent's starting position on its behaviour. Both concentration values for the chemicals A and B were set to be low. In all the experiments described in this paper, the concentration range was from 1 to 300. In this instance, A and B concentrations were set to 1 or 2. We tried ten different starting positions and five different settings for the environment: with one chemical A, one chemical B, and finally one concentration of the chemical A overlapping with one concentration of the chemical B. Each run lasted 600 seconds and the neural network was updated every 0.1ms. Every 10ms, the agent was moved and the sensory inputs updated. In these experiments, the agent could detect double concentrations of one chemical (A or B) but did not react to it. However, the agent was able to react only to the blend of both chemicals A and B, where it stayed inside the overlapping concentrations. We recorded the agent's neural activity during each run. Figure 4 shows an example where an agent starts from the position P2. In this case, the agent was able to stay in the overlapping area.

By looking at Figure 5, we can see that the agent begins by moving horizontally left to right until its right antenna detects the chemicals A and B (T1, Fig. 4 and 5). At this point, the sensors S2 and S3 fire and the temporal coincident arrival of their

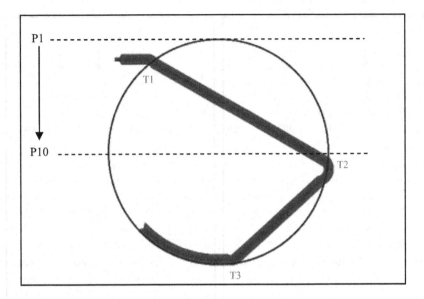

Fig. 4. Path of the agent starting from the left at P2. The circle in the centre represents the two overlapping concentrations of chemicals A and B.

spikes causes N0 to fire. M1 is then stimulated and increases its firing rate turning the left wheel faster than the right one. Soon after this, both antennae detect the chemicals causing also the neuron N1 to fire so the agent moves straight forward again. At T2, the left antenna of the agent goes outside the overlapping area so the sensors S0 and S1 stop to fire and therefore, do not stimulate the neuron N1. The motor neuron M3 then fires at a lower rate than M1 resulting in a left turn of the agent to stay inside the area. Finally, from T3, the interaction between the left antenna and the concentration causes the edge-following behaviour.

We also recorded the current density and membrane potential of the neuron N0 during a small interval of time when the agent was inside the blend of chemicals A and B. The input current of the neuron N0 was increasing when spikes coming from both S2 and S3 arrived at the same time. Then, the membrane potential also increased and reached the threshold θ (0.0046 Volts) making the neuron N0 fire. The potential was then set to 0 during the refractory period. As the sensors were synchronized and the delay between them and the neurons were the same, the spikes arrived at the same time to the neuron allowing it to detect them and fire (Fig. 6).

b) Experiment II

The second experiment was to test our hypothesis by modifying the sensory response delays to verify that our architecture necessarily needed to encode the sensory

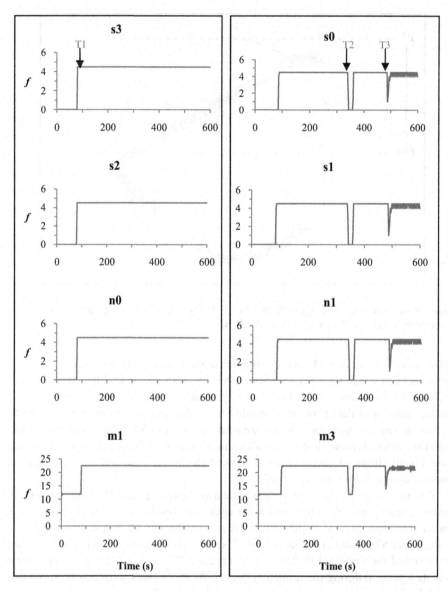

Fig. 5. Firing rates *f* (in spikes/s) of the neural network cells recorded every 2s during one run (Experiment shown in Fig. 4). The motor neurons M0 and M1 are not shown here as they do not fire. On the left panel, the sensors detecting the chemicals A (S3) and B (S2) from the right antenna activates the neuron N0 that stimulates the motor neuron M1 controlling the left wheel to move forward. On the right panel, the sensors detecting the chemicals A (S0) and B (S1) from the left antenna activates the neuron N1 that stimulates the motor neuron M3 controlling the right wheel to move forward.

information onto temporal coincidence. We changed the delays by modifying the position of the sensors therefore modifying the length of their axons linked to the neurons. We only changed the delays of the sensors detecting the chemical B (S1 and S2). We used the same set up as for the experiment shown in Figure 4.

We tried different values of delays (from 1ms to 50ms) and we noticed that a small change (up to 7.5ms) did not modify the agent's behaviour. But a further change in the delays (from 7.5ms) made the agent unable to react to the blend of chemicals A and B so it could not stay inside the concentrations.

As in the Experiment I, we recorded the current density and membrane potential of the neuron N0 during 0.5s when the agent was inside the chemical blend. In Figure 6, we can see that the current of the neuron N0 increases when a spike coming from both S2 and S3 arrive but as the delay has been changed, the spikes do not arrive at the same time so the current is lower than in Experiment I. Therefore, the neuron's potential increases but never reaches the threshold so the neuron does not fire (Fig. 6).

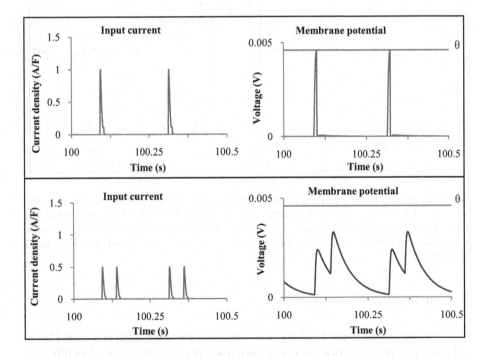

Fig. 6. Current density (in Amperes per Farad) and membrane potential (in Volts) of the neuron N0 recorded between 100s and 100.5s. On the top panel (Experiment I), the spikes sent by the sensors arrived at the same time increasing the current density to 1 A/F. The membrane potential was then increased and reached the threshold making the neuron N0 fire. On the bottom panel (Experiment II), the spikes sent by the sensors were not coincident as the delays were changed to 50ms in this case, so the current was never above 0.5 A/F and therefore, the membrane potential could not reach the threshold to make the neuron N0 fire.

c) Experiment III

In order to investigate the use of firing rate encoding, we used only one concentration of either A or B and increased it. When the concentration was augmented from 1 to above 50, the agent was then able to react to it. Therefore, the neural network showed much more sensitivity to two chemicals than to one. We also realized when using two overlapping chemicals A and B, as the concentration value increased, modifying the delays had a minor effect and the agent was still able to react to the chemicals. The firing rates were increasing too so the agent was moving faster. In these experiments, the temporal coincidence encoding was not necessary. The sensory information was encoded onto the firing rates of the sensors.

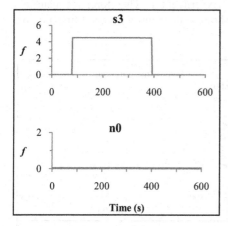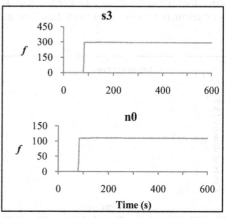

Fig. 7. Firing rates of the sensor S3 and neuron N0 recorded every 2s. On the left panel, as the environment contained a low concentration (=1) of chemical A only, the neuron could not detect it and therefore, the agent did not stay within the chemical source area. On the right panel, the concentration was high (=300) so the neuron could detect it and the agent stayed inside the area.

5 Conclusion

We presented in this paper a simple neural architecture where temporal coincidence and firing rate encoding strategies were both important mechanisms used in different environmental settings. In a low concentration setting, synchronization of spikes sent by the sensors was essential to allow the agent to detect the blend of two chemicals. We changed the sensory delays and noticed that the agent was then not able to react to the chemicals anymore. In a high concentration setting, the temporal coincidence between sensors firing was not a necessary condition and the agent was able to stay inside the chemical concentration using just the firing rate encoding strategy.

Interestingly, the model showed much more sensitivity to the presence of two chemicals than a single chemical. In principle, more than two chemicals can be detected and processed. The architecture presented here also works when the chemical concentration has a linear gradient.

Future work will investigate evolving such architectures using a developmental model (evolving the number of neurons and their connections, the synaptic weights, and delays of the neural network). Moreover, we will add noise to the neural network and use a more complex environment.

References

1. Braitenberg, V.: Vehicles: Experiments in Synthetic Psychology. MIT Press, Cambridge (1984)
2. Brody, C.D., Hopfield, J.J.: Simple Networks for Spike-Timing-Based Computation, with Application to Olfactory Processing. Neuron 37, 843–852 (2003)
3. Floreano, D., Mattiussi, C.: Evolution of Spiking Neural Controllers for Autonomous Vision-Based Robots. In: Gomi, T. (ed.) ER-EvoRob 2001. LNCS, vol. 2217, pp. 38–61. Springer, Heidelberg (2001)
4. Florian, R.V.: Biologically inspired neural networks for the control of embodied agents. Technical report Coneural-03-03 Version 1.0 (2003)
5. Gerstner, W., Kistler, W.M.: Spiking Neuron Models. Single Neurons, Populations, Plasticity. Cambridge University Press, Cambridge (2002)
6. Hopfield, J.J.: Odor space and olfactory processing: Collective algorithms and neural implementation. PNAS 96, 12506–12511 (1999)
7. Hoshino, O., Kashimori, Y., Kambara, T.: An olfactory recognition model based on spatio-temporal encoding of odor quality in the olfactory bulb. Biological Cybernetics 79, 109–120 (1998)
8. Izhikevich, E.M.: Simple model of spiking neurons. IEEE Transactions on Neural Networks (2003)
9. Izhikevich, E.M.: Which model to use for cortical spiking neurons. IEEE Transactions on Neural Networks (2004)
10. Kandel, E.R., Schwartz, J.H., Jessell, T.M.: Principles of Neural Science. McGraw-Hill, New York (2000)
11. Kanzaki, R., Nagasawa, S., Shimoyama, I.: Neural Basis of Odor-source Searching Behavior in Insect Brain Systems Evaluated with a Mobile Robot. Chemical Senses 30, 285–286 (2005)
12. Koch, C.: Biophysics of Computation: Information Processing in Single Neurons. Oxford University Press, New York (1999)
13. Kuwana, Y., Shimoyama, I.: A Pheromone-Guided Mobile Robot that Behaves like a Silkworm Moth with Living Antennae as Pheromone Sensors. The International Journal of Robotics Research 17, 924–933 (1998)
14. Laurent, G., Wehr, M., Davidowitz, H.: Temporal Representations of Odors in an Olfactory Network. Journal of Neuroscience. 16, 3837–3847 (1996)
15. Payton, D., Daily, M., Estowski, R., Howard, M., Lee, C.: Pheromone Robotics. Auton. Robots. 11, 319–324 (2001)

16. Pyk, P., Bermúdez i Badia, S., Bernardet, U., Knüsel, P., Carlsson, M., Gu, J., Chanie, E., Hansson, B., Pearce, T.J., Verschure, P.: An artificial moth: Chemical source localization using a robot based neuronal model of moth optomotor anemotactic search. Autonomous Robots 20, 197–213 (2006)
17. Webb, B.: Robots crickets and ants: models of neural control of chemotaxis and phonotaxis. Neural Networks 11, 1479–1496 (1998)
18. Wyatt, T.D.: Pheromones and Animal Behaviour, Communication by Smell and Taste. Cambridge University Press, Cambridge (2003)

Theta Phase Coding and Acetylcholine Modulation in a Spiking Neural Network

Daniel Bush, Andrew Philippides, Phil Husbands, and Michael O'Shea

Centre for Computational Neuroscience and Robotics, University of Sussex, Brighton, England
{d.bush,philh,andrewop,m.o-shea}@sussex.ac.uk

Abstract. Theta frequency oscillations are a prominent feature of the hippocampal EEG during active locomotion and learning. It has also been observed that the relative timing of place cell firing recedes as its place field is traversed – a phenomena known as phase precession. This has led to the development of a theory of theta phase coding, whereby spatial sequences being encountered on a behavioural timescale are compressed into a firing sequence of place cells which is repeated in each theta cycle and stored in an auto-associative network using spike-timing dependent plasticity. This paper provides an abstract, descriptive model of theta phase coding in a spiking neural network, and aims to investigate how learning and recall functions may be separated by the neuromodulatory action of Acetylcholine (ACh). It is demonstrated that ACh is not essential for concurrent learning and recall without interference in this case, thanks to the robust nature of the theta phase coding implementation. However, the neuromodulation of synaptic plasticity offers other advantages, and may be essential to avoid continually consolidating false predictions when learning new routes.

Keywords: Acetylcholine, attractor network, cognitive map, Hippocampus, neuromodulation, place cells, spatial memory, STDP, theta phase coding.

1 Introduction

The hippocampus has long been identified with spatial and episodic learning and memory. This theory has been bolstered by the discovery of several distinct groups of cells throughout the region whose activity corresponds directly to an animal's location (place, grid and spatial view cells) or idiothetic inputs (head direction and vestibular information cells). This in turn has led to the notion that the hippocampus may function as a cognitive map, which integrates environmental cues, past experience and self-motion input in order to aid efficient navigation [1, 2]. The mechanisms by which the cognitive map might operate have often been modelled using auto-associative networks – recurrent neural architectures with synaptic plasticity which can store input patterns and recall them from incomplete or noisy cues [3, 4, 5, 6]. These models are inspired by the presence of a large number of recurrent collaterals within the hippocampal formation, and the ease with which synaptic plasticity can be induced and observed in the region. The corresponding neural networks have been very successful in replicating the update of head direction cell activity from idiothetic

M. Asada et al. (Eds.): SAB 2008, LNAI 5040, pp. 159–168, 2008.

cues and path integration over a learned environment in the absence of sensory input, as well as some more abstract functions of episodic memory. Elsewhere, the anatomy and postulated function of the hippocampus has also inspired several successful simultaneous localisation and mapping (SLAM) implementations on robotics platforms [7, 18].

However, these models have often utilised rate-coded neural and synaptic dynamics, while it has become clear from neurobiology that changes in the strength of synapses in the hippocampus are primarily mediated by the temporal sequence, rather than frequency, of neuronal firing. According to this spike-timing dependent plasticity (STDP), only those pre-synaptic inputs which have been active in a short time window (~50ms) before post-synaptic spiking are potentiated, while those which are active within a similar time window after post-synaptic spiking are depressed [8, 9]. In order to implement STDP within an associative network, a spiking model which can replicate the dynamics of real neurons as accurately as possible is required. Neurons in the hippocampus demonstrate one of the most well known dynamic firing patterns in the mammalian EEG. Pyramidal cells throughout the region exhibit theta (~8Hz) frequency oscillations in their local field potential whenever an animal is actively locomoting, attending to external stimuli or during REM sleep [10, 11]. Furthermore, it has been established that the firing of place cells is not simply modulated by this oscillation, but advances in phase relative to theta as their place field is traversed [10, 12, 13]. This produces a compressed temporal firing sequence within each theta cycle which corresponds directly to the current sequence of locations being navigated on a behavioural timescale – a firing sequence which is ideally suited for storage in an associative network using STDP. It has been suggested that this theta-phase coding may be the mechanism by which the hippocampus processes continuous spatial information [10, 11, 14, 15].

It has also been noted that the release of the neuromodulator Acetylcholine (ACh) is closely linked to the theta oscillation [16]. ACh acts on muscarinic and nicotinic receptors within the hippocampus, and is known to be involved in learning and recall processes. This is demonstrated by experiments in which the infusion of ACh antagonists impairs performance on spatial tasks [17]. Neurobiological research has revealed several effects of ACh on neurons and synapses in this brain region, among them the enhancement of afferent input relative to excitatory feedback, and the enhancement of synaptic plasticity. These properties have led to the theory that the role of ACh is to separate phases of learning and recall within each theta cycle [10, 11]. Associative memory models can encounter significant problems if learning and recall processes are concurrently active. During learning, for example, the activity in an auto-associative memory model must approximate external input, or the patterns which are stored will be a combination of novel experience and the recall of earlier, similar experience [4, 10, 11]. If recurrent connections are made too weak to provoke neural activity, this interference will disappear, but recall of an activity pattern from a partial cue is rendered impossible. The changes in neural and synaptic dynamics which are incurred by ACh suggest that it may act as a trigger to switch between functions of learning and recall. When the neuromodulator is present, afferent input (from the entorhinal cortex or dentate gyrus) dominates the dynamics of the auto-associative CA3 network, and this activity is maintained while information is stored via enhanced synaptic plasticity. When ACh is absent, feedback from recurrent collaterals dominates and plasticity is vastly reduced, allowing the network to make predictive recall from the patterns stored in the synaptic weights of the CA3 network [10, 14, 16]. This posits an elegant and biologically plausible solution to the problem of interference between the processes of learning externally applied input patterns and recall via recurrent feedback which has been encountered in previous auto-associative memory models, and which may occur in the hippocampus.

This aim of this research is to construct an abstract, descriptive model of theta phase coding, in order to investigate the possible advantages of ACh modulation in storing and recalling temporal sequences on a behavioural time scale. To our knowledge, this is the first attempt to investigate the phenomena of both phase precession and Acetycholine modulation in a spiking neural network which implements STDP. Previous research has examined theta phase coding in a similar network, but with more simplified - and therefore, less biologically realistic - models of neural or synaptic dynamics [10, 15]. To our knowledge, neither theta phase coding nor STDP have yet been examined on a robotics platform with a similar focus [7, 18].

2 Methods

2.1 Network Properties and Neural Dynamics

The neural network consisted of N=20 neurons, whose activity corresponds to that of place cells in the CA3 region of the hippocampus. Each had a randomly assigned axonal delay in the range 1 : 5ms. The network was fully recurrently interconnected by excitatory synapses except for self-connections. The neurons operated according to the Izhikevich (2004) spiking model, which dynamically calculates the membrane potential (v) and a membrane recovery variable (u) based on the values of four dimensionless constants (a,b,c and d) and a dimensionless current input (I), according to Eqn. 1. This model can exhibit firing patterns of all known types of cortical neurons by variation of the magnitude of applied current and the parameters a – d [19]. The values used for tonic spiking in a standard excitatory neuron are a=0.02, b=0.2, c=-65 and d=6.

$$v' = 0.04v^2 + 5v + 140 - u + I$$
$$u' = a(bv - u)$$
$$\text{if } v \geq +30\,\text{mV then} \begin{cases} v \leftarrow c \\ u \leftarrow u + d \end{cases}$$

(1)

2.2 Synaptic Dynamics

At the beginning of each simulation, all of the recurrent synaptic weights were assigned a value of w=0.1. Mathematically, with $s = t_{post} - t_{pre}$ being the time difference between pre- and post- synaptic spiking, the change in the weight of a synapse (Δw) due to STDP can be calculated using equation 2.

$$\Delta w = F(s) = \begin{cases} P_+ = A_+ \exp(-s/\tau_+) & \text{for } s > 0 \\ P_- = A_- \exp(-s/\tau_-) & \text{for } s < 0 \end{cases}$$

(2)

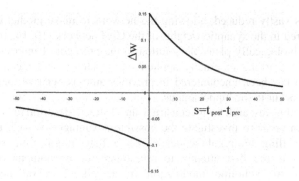

Fig. 1. The asymmetric time window of STDP

The parameters A_+ and A_- effectively correspond to the maximum possible change in the weight of a synapse per spike pair, while τ_+ and τ_- denote the decay constants of potentiation and depression increments respectively (see Fig. 1). Previous research suggests that the window of depression should be set larger than that of potentiation – in order to ensure that the STDP model depresses chance spike pairings and thus operates stably [20]. This constraint was observed throughout our simulations, by setting τ_+ =20ms and τ_-=40ms and constraining the relative size of A_+ and A_-.

2.3 Network Input

In order to replicate the phenomena of phase precession, input to our network was formulated as a combination of theta frequency inhibition and gradually increasing excitation [21]. Every neuron in the network was fed with inhibitory input which oscillated sinusoidally between a value of I=-1nA and I=-3nA at a frequency of 8Hz. A route which consisted of a series of N=20 overlapping place fields was then traversed. Each place field was divided into seven equal segments, and each place field overlapped with five segments of those on either side. The level of excitation in the corresponding place cell would increase from a value of I=2nA as the place field was entered, by increments of I=0.25nA as each segment was traversed. This method is not intended to replicate the possible generation mechanisms of phase precession *in vivo*, but to replicate all the key features of those activity patterns – including that the phase of firing corresponds with distance travelled through the place field, rather than time spent within it [2]. Once a place field is exited, excitation for that place cell was reset to zero. The weights of the recurrent synapses were adjusted according to the temporal sequence of the firing of these place cells, as the route was traversed.

2.4 Acteylcholine Modulation

When neuromodulation was employed, the concentration of ACh oscillated uniformly across the network in the range 0 : 1, in synchrony with inhibitory input. Excitatory synaptic currents from the recurrent collaterals were inversely modulated by this concentration in the range 0 : f. The plasticity of recurrent synapses was also dynamically adjusted, by directly scaling each weight change by the instantaneous concentration of neuromodulator. Hence, in the presence of ACh, the strength of recurrent excitatory feedback tended towards zero, while synaptic plasticity was active in order to store incoming activity patterns without interference. When ACh was absent, the excitatory feedback was enhanced and synaptic plasticity was suppressed, in order to allow predictive recall activity which would not be stored.

3 Results

3.1 Theta-Phase Coding without Acetylcholine Modulation

Initial tests of the model aimed to establish how well activity in the network approximated what is known of phase precession *in vivo*. Figure 2 illustrates a typical spike raster from our network, whereby a section of the place field sequence is translated into a compressed sequence of place cell firing. Because one behavioural sequence is repeated many times on the theta temporal scale, it is possible to store a spatial route in a single trial. The learning rate in the network is effectively determined by the parameters A_+ and A_- in the STDP model. The higher these values are, the more quickly synaptic weights re-arrange to a stable distribution which reflects the input sequence. One of the weaknesses of our model is that it does not replicate the magnitude of phase precession seen *in vivo*, which can closely approach 360 degrees [2]. However, in the absence of ACh modulation, this provokes associations between the initial and final place cells firing at each stage of the route, creating artificial, circular associations and thus corrupting the ideal final weight matrix (data not presented).

External input corresponding to five laps of a circular route consisting of twenty overlapping place fields was then applied to the network, and the resultant weight matrix and spike raster is shown in Fig. 3. In the absence of a mechanism to differentiate

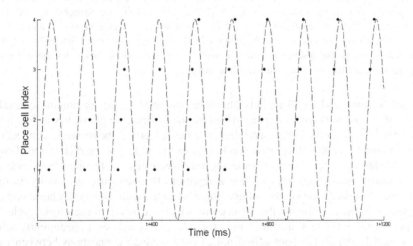

Fig. 2. The theta phase coding mechanism. As the place fields of place cells 1-4 are sequentially traversed, the phase at which the neurons fire precesses, and thus a compressed, representative firing pattern (i.e. sequential firing in place cells 1-4) is generated in each cycle.

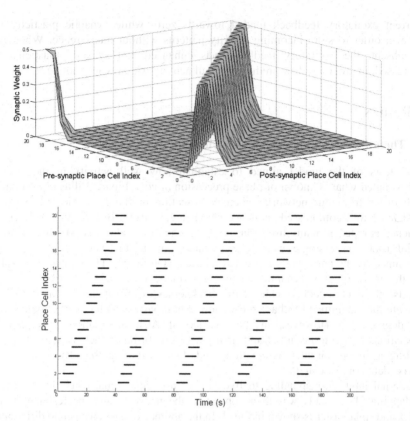

Fig. 3. Final synaptic weight matrix and overall spike raster for the network when traversing a circular route five times, with $w_{max}=0.5$; $A_+=0.012$ and $A_-=0.01$. (a) Synaptic weights from each neuron to those which follow it on the learned route have been potentiated, and to those which immediately precede it have been depressed. The majority are unchanged from their initial value, due to the absence of neural noise. (b) The spike raster illustrates the sequential firing activity, and the absence of any recall activity.

between learning and recall periods, the value assigned to the maximum achievable weight of a synapse becomes critical. Although the synaptic weight matrix in Fig. 3 has re-arranged to reflect the behavioural sequence being learned, the recurrent connections are not powerful enough to provoke spiking activity, and so no predictive recall from these weights can occur. Hence, some separate mechanism is required to decode the weight matrix and effectively produce a representative sequence of activity from it when cued. Similarly, the number of upcoming locations with which each place cell can associate is limited by the number of neurons which are active in each theta cycle. As the spike raster in figure 3 illustrates, only four place cells were ever concurrently active in our model, and so (without neural noise) only synaptic connections between each neuron and the three which follow or precede it can be modified.

If w_{max} is increased, however, then predictive recall becomes possible, and, surprisingly, the remarkably robust nature of this theta phase coding implementation means that this process does not interfere with the ideal structure of the synaptic

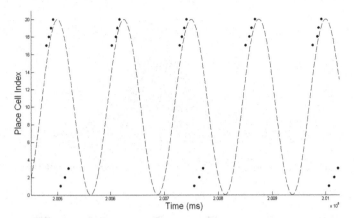

Fig. 4. Recall activity in the network with w_{max}=5, A_+=0.12 and A_-=-0.1 following a single traversal of the entire route. Place cells 17 – 20 were being stimulated sequentially by external input, and activity in place cells 1-3 being recalled immediately after this activity had terminated. The sequence of firing is maintained during recall by the increasing level of recurrent synaptic input from activity in the preceding neurons and the axonal delays.

weight matrix. As Fig. 4 illustrates, the spiking dynamics, gradually increasing level of recurrent excitation from preceding place cells and axonal delays conspire to concentrate recalled activity (in place cells 1-3) after that generated by external input (in place cells 17-20) within each theta cycle. This recall activity is further illustrated by the spike raster for this simulation (Fig. 5), which also shows that our model replicates a well known property of phase precession *in vivo* – the experience-dependent expansion of place fields against the direction of motion, so that place cell firing starts earlier in each successive traversal [2]. The increased number of concurrently active place cells which results from this recall activity means that each neuron can alter the strength of its connections with a greater frequency of those adjacent to it on the behavioural sequence – and this is clearly illustrated by the greater spread of potentiated synapses in the weight matrix (see Fig. 5).

3.2 Phase Precession with Acetylcholine Modulation

When ACh modulation is introduced into the network, the value of the maximum synaptic weight becomes less important (unless it is set trivially low), and predictive recall and the experience-dependent expansion of place fields are present in all incarnations of the network. The final weight matrix for a typical simulation (using an identical route to that examined above) is shown in Fig. 6, and is remarkably similar to that produced by the model in the absence of neuromodulation, but with a large maximum weight limit (see Fig. 5). The one key difference lies in the spread of the peak of the weight matrix. Because plasticity is absent during recall when ACh is present in the network, no associations are generated between those neurons which are active due to external input, and those which are concurrently active due to recurrent (recall) input. The spike rasters observed with ACh modulation are also very similar to those obtained from the network in the absence of neuromodulation, but

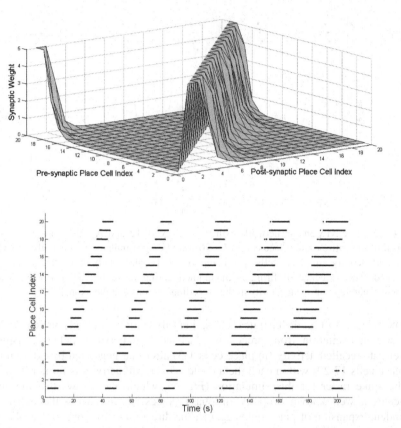

Fig. 5. Final synaptic weight matrix and overall spike raster for the network when traversing a circular route five times, with $w_{max}=5$; $A_+=0.12$ and $A_-=0.1$. (a) A comparison with the weight matrix in Fig.3 illustrates how each place cell has become associated with a greater number ahead of it on the route, due to the concurrent learning and recall activity. (b) The experience dependent expansion of place fields against the direction of travel, and predictive recall are also clearly visible in the spike raster (the activity corresponding to the beginning of the sixth run was not externally applied). These features can also be seen with neuromodulation.

with a higher value of w_{max} (see Fig. 5), as recall activity and the experience-dependent expansion of place fields are clearly visible (data not presented).

4 Conclusions

We have presented an abstract, descriptive model of theta-phase coding in place cells within the auto-associative CA3 network of the hippocampus. This implementation of theta-phase coding is remarkably robust, and can function effectively even in the presence of significant recall activity without interference. Concurrent learning and recall activity is possible both with and without Acetylcholine modulation, provided that the maximum weight limit is set sufficiently high. This result leaves us with the question of what significance this form of neuromodulation may have *in vivo*. While

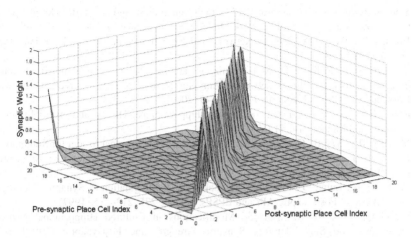

Fig. 6. Final synaptic weight matrix for theta phase coding simulation with ACh modulation. The weight matrix retains its ideal structure, but has a narrower spread, as plasticity is absent during the recall phase and hence external input is not associated with recurrent activity.

this research is still in its early stages, it suggests that one advantage of ACh modulation may be to prevent circular associations between the first and last active place cells in sections of the behavioural sequence. Also, since our model does not incorporate the suspected mechanisms which generate phase precession *in vivo*, it seems likely that the neuromodulation also offers control over the timing and sequence of recall activity – conspiring to place it after that stimulated by the current external input, and in the correct order. Finally, although it has not been examined here, previous research has suggested that the main role of ACh may be to allow the elimination of redundant learned spatial sequences, and their subsequent replacement with new navigational routes which make use of the same place fields [10]. This makes intuitive sense, as synaptic plasticity is almost absent during the recall phase when ACh is present, and hence the predictions of future location which are made are not stored in the recurrent weights. This is illustrated by the narrower peak of potentiation in the synaptic weight matrix of Fig. 6. Without this scaling of synaptic plasticity, redundant sequences will be continually recalled and simultaneously consolidated. Although the ACh modulated network may still predict future locations based on past experience, new associations (and thus new predictions) will be rapidly acquired. The next step in this research, therefore, is to assess how incarnations of the network with and without neuromodulation can learn, recall, un-learn and re-learn a wider variety of complex behavioural sequences.

References

1. O'Keefe, J., Nadel, L.: The Hippocampus as a Cognitive Map. Oxford University Press, Oxford (1978)
2. McNaughton, B.L., Battaglia, F.P., Jensen, O., Moser, E.I., Moser, M.-B.: Path integration and the neural basis of the cognitive map. Nat. Rev. Neuroscience 7, 663–678 (2006)

3. Samsonovich, A., McNaughton, B.L.: Path Integration and Cognitive Mapping in a Continuous Attractor Neural Network Model. J. Neuroscience 17, 5900–5920 (1997)
4. Rolls, E.T., Treves, A.: Neural Network and Brain Function. Oxford University Press, Oxford (1998)
5. Stringer, S.M., Rolls, E.T.: Self-organizing path integration using a linked continuous attractor and competitive network: Path integration of head direction. Network: Computation in Neural Systems 17, 419–445 (2006)
6. Rolls, E.T.: An attractor network in the hippocampus: Theory and neurophysiology. Learning and Memory 14, 714–731 (2007)
7. Milford, M.J., Wyeth, G., Prasser, D.: RatSLAM: A Hippocampal Model for Simultaneous Localization and Mapping. In: International Conference on Robotics and Automation, New Orleans, United States (2004)
8. Markram, H., Lubke, J., Frotscher, M., Sakmann, B.: Regulation of Synaptic Efficacy by Coincidence of Postsynaptic APs and EPSPs. Science 275, 213–215 (1997)
9. Bi, G.-Q., Poo, M.-M.: Synaptic Modifications in Cultured Hippocampal Neurons: Dependence on Spike Timing, Synaptic Strength, and Postsynaptic Cell Type. J. Neuroscience 18, 10464–10472 (1998)
10. Hasselmo, M.E., Bodelon, C., Wyble, B.P.: A proposed function for hippocampal theta rhythm: separate phases of encoding and retrieval enhance reversal of prior learning. Neural Computation 14, 793–817 (2002)
11. Hasselmo, M.E.: What is the Function of Hippocampal Theta Rhythm? Linking Behavioral Data to Phasic Properties of Field Potential and Unit Recording Data. Hippocampus 15, 936–949 (2005)
12. O'Keefe, J., Recce, M.L.: Phase relationship between hippocampal place units and the EEG theta rhythm. Hippocampus 3, 317–330 (1993)
13. Skaggs, W.E., McNaughton, B.L., Wilson, M.A., Barnes, C.A.: Theta phase precession in hippocampal neuronal populations and the compression of temporal sequences. Hippocampus 6, 149–172 (1996)
14. Wagatsuma, H., Yamaguchi, Y.: Neural dynamics of the cognitive map in the hippocampus. Cognitive Neurodynamics 1, 119–141 (2007)
15. Yamaguchi, Y., Sato, N., Wagatsuma, H., Wu, Z., Molter, C., Aota, Y.: A unified view of theta-phase coding in the entorhinal–hippocampal system. Current Opinion in Neurobiology 17, 197–204 (2007)
16. Hasselmo, M.E.: The role of acetylcholine in learning and memory. Current Opinion in Neurobiology 16, 710–715 (2006)
17. Bunce, J.G., Sabolek, H.R., Chrobak, J.J.: Intraseptal infusion of the cholinergic agonist carbachol impairs delayed-non-match-to sample radial arm maze performance in the rat. Hippocampus 14, 450–459 (2004)
18. Fleischer, J.G., Krichmar, J.L.: Sensory Integration And Remapping In A Model Of The Medial Temporal Lobe During Maze Navigation By A Brain-Based Device. Journal of Integrative Neuroscience 6, 403–431 (2007)
19. Izhikevich, E.M.: Which Model to Use for Cortical Spiking Neurons? IEEE Transactions on Neural Networks 15, 1063–1070 (2004)
20. Song, S., Miller, K.D., Abbott, L.F.: Competitive Hebbian learning through spike-timing-dependent synaptic plasticity. Nature Neuroscience 3, 919–926 (2000)
21. Mehta, M.R., Lee, A.K., Wilson, M.A.: Role of experience and oscillations in transforming a rate code into a temporal code. Nature 417, 741–746 (2002)

Interest of Spatial Context for a Place Cell Based Navigation Model

Nicolas Cuperlier[2], Philippe Gaussier[1,2], and Mathias Quoy[2]

[1] Member of the Institut Universitaire de France
[2] Neuro-cybernetic team, Image and Signal processing Lab.,
Cergy Pontoise University / CNRS UMR 8051 / ENSEA

Abstract. After a short review of properties of biological place cells, mainly found in the hippocampal region of rodents, and a brief presentation of a biologically inspired navigation architecture relying on these cells, we will show how contextual information could facilitate scale changes to large environments. We thus present a simple model of spatial context allowing to both reduce noise effects on place cells (in biological model) and increase its computational performance.

1 Introduction

The startling discovery by O'Keefe & Dostrovsky [15] of the spatial correlates of neural activity in the hippocampal system of rodents was a first step to understand the mechanisms by which the brain processes spatial information. This pioneer work has stimulated a substantial body of computational models to understand the role of these place sensitive cells in spatial representation and navigation. These cells exhibit the property to fire selectively in different regions of an environment and were named place cells (PC). PC regions of high activity are termed place fields. Later, PC like property neurons have been detected in other structures near the rodent hippocampus. Indeed, neurons of these regions also show significant activity in localized regions of spatial environments while performing spatial tasks: the superficial [16,17] and deep [7] entorhinal cortex (EC), the dentrate gyrus (DG) [11] and the subiculum (SUB) [18] where also grid cells are found [10,8].

Thus, ensembles of PC are thought to form spatial representations that can be used in navigation and numerous models have been proposed, see [4] for a short rewiew of navigational model based on PC.

In previous papers [3,5,4], we had proposed a control architecture based on transition cells and a cognitive map for planned navigation tasks (see figure 1). Whereas our model relies on transition cells to navigate, place cells are still a key element of the system, since PC are constitutive elements of transition cells and successful spatial task performance is associated with stable place fields (like in rodents). Without describing here the whole model in detail, we just mention that this architecture has been already successfully tested on a robot with a panoramic camera in several indoor environments (one room, two rooms and a corridors).

M. Asada et al. (Eds.): SAB 2008, LNAI 5040, pp. 169–178, 2008.

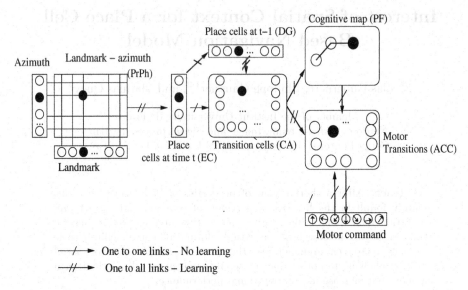

Fig. 1. Control architecture for planned navigation involving a simplified view of the loop between the hippocampus, the prefrontal cortex and the basal ganglia (here limited to ACC)

Instead of describing the global architecture, we will focus in this paper on its two main scaling problems impairing navigation in large environments which have been shown by our last robotics experiments:

1. The computational cost of the PC layer is linear in the number of learned place cells, which itself increases with the number of environments explored. Moreover this growing number of place cells involved in the competitions increases the noise level.
2. In our architecture, planning relies on a single cognitive map. Problems appear when one wants to code several different maps coding for different environments or subsets of a big environment.

A contextual information could overcome these difficulties for large scale navigation:

- First, contextual information could increase the reliability of PC response by selecting only a subset of the PC population. Appropriate context activity and connectivity reduce the number of place cells in competition and thus avoid false detection. Furthermore, from a computational point of view, we can also use this contextual information to increase the simulation speed (reducing the number of place cells that have to be computed).
- Second, each map could be linked with a kind of context signal (linked for instance to the global recognition of a room) that should be able to "reload" the previous learned map (or a part of it) into the different neural structures used here.

Place fields of rodent hippocampal place cells are strongly determined by the local geometry of the environment. Thus spatial representations in the hippocampus proper are highly context-specific. It is well established that modifications in environment change the spatial ring properties of PC, a phenomenon also known as remapping. The dependencies of the place code on geometric information (visual cues) have been investigated extensively [13,14]. However, place representations also exhibit strong dependence on nongeometric information such as sounds, odors, somatosensory stimuli and behavioral context [1,6]. This suggests the existence of a context-representation system which biases the hypocampal representations. We have chosen to begin our study by focusing on the geometric information since they seem to be of importance for this context code [12]. Nethertheless, we still have in mind that context doesn't rely exclusively on this sole information [1]. Future works will try to integrate nongeometic information in the model (see section 4).

In the next sections, we will show how a simple mechanism, based on a PC system with several resolution scales, could be a fisrt step toward the constitution of this contextual information. We will show some experiments enlightening the interest of this model. Finally, we will discuss the results and the future extentions of the model.

2   Model

We present a model of modulation between two levels of PC (context and EC_s). We show how such a modulation could be modified by experience in a Hebbian manner, thus explaining the context specificity of PC.

In order to capture and learn geometric contextual information, context neurons are modelled by a specific PC layer. We name context neurons ($X^{Context}$), see figure 2. This new layer takes its inputs from a network (PrPh) merging recognized landmarks and azimuths in a product space. One can refer to [4] for more details on this process.

Activity of the the jth context neuron is expressed with the same equation than in our previous work, as follows:

$$X_j^{Context}(t) = \frac{1}{W_j} \left(\sum_{kl}^{N_{PrPh}} W_{j,kl}^{PrPh-Context} . X_{kl}^{PrPh}(t) \right) \tag{1}$$

with $W_j = \sum_{kl}^{N_{PrPh}} W_{j,kl}^{PrPh-Context}$, $X_{kl}^{PrPh}(t)$ the activity of the kl^{th} neuron in the product space (PrPh) merging landmark recognition and their azimuth. The learning of context neurons follows a Hebbian like rule:

$$\frac{dW_{j,kl}^{PrPh-Context}(t)}{dt} = -\lambda_1 . W_{j,kl}^{PrPh-Context}(t) . X_{kl}^{PrPh}(t) + \lambda_2 . \left(1 - \sum_{kl \in N_{PrPH}} W_{j,kl}^{PrPh-Context}(t)\right) . X_{kl}^{PrPh}(t) . X_j^{Context}(t) \tag{2}$$

With λ_1 a decay term and λ_2 a learning constant.

The recruitment of a new neuron for encoding a new location occurs during the exploration of an unknown environment. This mechanism is performed

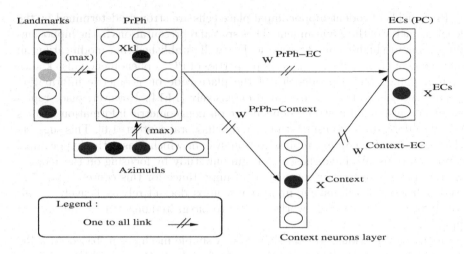

Fig. 2. Our model with a PC layer linked to the context layer. Context neurons also exhibit PC properties but with a larger firing field.

autonomously, without any external signal, relying only on the PC population activity. If the activity of any previously learned place cell is below a *fixed* recognition threshold (RT), then a new neuron is recruited for coding this new location. We need PCs of the Context layer with bigger place field than PCs on EC_s. Place field size of the winning PC (or generalization) can be set by adjusting this RT parameter. We thus choose a RT for the context population lower than the one chosen for the EC_s population (see section 3). If at a given place, several PC respond with an activity greater than the recognition threshold, a competition takes place so that the most activated cell wins and codes the current location. However, place fields are overlapping.

The PC population previously identified in our model as cells in the enthorinal cortex (EC_s) still also receive inputs from PrPh. But EC_s has now another input from the context layer (possibly located in a cortical area). PCs are linked to contextual PC with a weigth equal to one. Hence, at the beginning, each PC can receive an activity from all the context neurons. This choice has been ruled by the following modification of the activity equation on the EC_s layer: activity of a neuron X_i^{EC} is modulated by active context neurons. Thus in the beginning, since all weigths $W_{i,j}^{Context-EC_s}$ are equal to 1, any PC on EC_s can be activated and recruited. The context modulation operates under a multiplicative form, according to the following equation:

$$X_j^{EC_s}(t) = \left(\sum_i^{N_{Context}} W_{j,i}^{Context-EC_s}.X_i^{Context}(t) \right) . \frac{1}{W_j} \left(\sum_{kl}^{N_{PrPh}} W_{j,kl}^{PrPh-EC_s}.X_{kl}^{PrPh}(t) \right)$$

$$(3)$$

At this state, since there is always at least one context neuron activated (a large PC), the activity of the PCs on EC_s remains the same as without the context layer. But we want that each PC belongs only to coactivated contexts and not to the others. This specialization process occurs online via the learning of links $W_{i,j}^{Context-EC_s}$ following a hebbian learning rule. Coactivation of a context neuron and a PC is detected and memorized (*coact*). This information modulates the modification applied to the corresponding synaptic weight:

$$W_{j,i}^{Context-EC_s}(t) = (W_{j,i}^{Context-EC_s}(t-1) + (X_j^{Context}) * X_i^{EC} * c)) - \alpha.(1 - coact)$$
(4)

with α a penality term. If a coactivation has been already detected ($c = 1$) and both input (context) and PC are activated, the weight is increased otherwise the weight remains inchanged. Otherwise, if no coactivation has been yet detected ($c = 0$), the weight is decreased.

After some time exploring the environment, the PC only keep connections ($W^{Context-EC_s} > 0$) with very few context neurons. A majority of PC are linked with one context neurons. We thus has constructed a two resolutions PC system. Context neurons code for a broad and large zone of the environment, whereas EC_s can have a much better resolution. Only two levels of hierarchy have been used, but this could be generelized to other level (for instance the landmark recognition,or a more generalcontext for PCs)...

3   Experiments

As our context neurons are themselve place cells, they are sensible to the same parameters: proximity of landmarks (closer landmarks have greater angular displacement when moving) and the complexity of the environment (number of room, gates etc...). Actually, more locations are learned near walls or doors due to the fast changes in the angular position that can occur near landmarks, or in the (dis)appearance of landmarks caused by these obstacles.

We have thus performed several experiments on differents large simulated environments (one room, three rooms and nine rooms) with serveral RT for the context neurons see fig 3.

For each one, we measure the number of PC linked to each context neuron and the number of context neuron linked to each PC.

This preliminary statistical study implies very long experiment that can't be easely performed on a real robot. We thus use simulation performed over more than 400000 simulation step (around 56h on bi-core bi-processor PC with 4Gb of ram). Simulation is alway different from real robot experiment; netherthe-less simulations are based on the results of real robotics experiments and some previous simulations were confirmed by real experiments. Moreover real data experiment are currently performed on an image data base > 100 giga octets corresponding to a loop of more than 30km and robotic tests will be performed to definitively validate this approach. We use in our simulation a fixed RT_{PC} since we already have studied the impact of this parameter over a population of PC [4].

Table 1 show the mean number of cells recruited over several tries with ($\alpha = 1$, $RT_{PC} = 0.966$):

Fig. 3. Example of two enviroments. Blue crosses are landmarks. Each colored region represents the place field of a particular place cell. After a complete exploration of the environment, the entire environment is covered by the place cell population. Context place field are delimited in black.

Table 1. Mean results for different values of context R.T. and several environments. (See text for an explanation.)

Context R.T=0.65	1 room	3 rooms	9 rooms
PC recruited on ECs	243	525	647
Context neurons recruited	8	26	61
PC over context neurons	43.87	26.38	23.87
Context R.T=0.75	1 room	3 rooms	9 rooms
PC recruited on ECs	238	531	760
Context neurons recruited	13	51	101
PC over context neurons	43.87	26.38	23.8689
Context R.T=0.85	1 room	3 rooms	9 rooms
PC recruited on ECs	212	501	720
Context neurons recruited	29	106	244
PC over context neurons	17.51	12.38	10.36

These results confirm several expectations:

- the number of PC associated to a given context increases when the context RT decreases (Context PCs have larger place field)
- this number of PC associated to a given context decreases with the number of rooms. This was expected since PC are sensible to the complexity induced by obstacle (hiding landmarks). Moreover the global size of the whole environment remain inchanged for all the environments. Hence, combining the fact that rooms in the three (and nine) rooms environment are much smaller

and that landmarks are placed along the wall leads to an increase of context and place cell neurons.

This last point can balance the apparently medium result found in the three and nine rooms environments, since we know that, for a given RT, our PC place fields are homothetic: bigger in a large environment than in a smaller one [9].

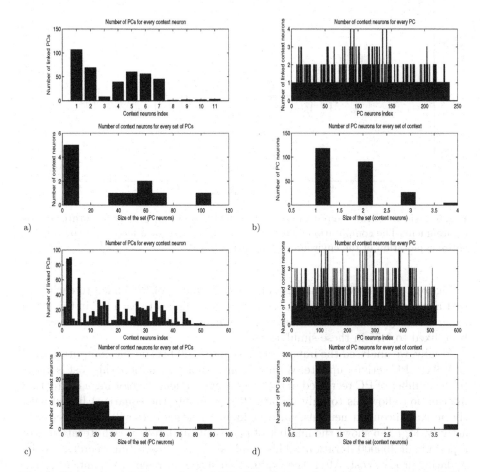

Fig. 4. Histograms of the number of PCs linked to each context neuron and of the number of context neurons linked to each PC respectively a) and b) for a given one room environment and c) and d) for agiven tree rooms environment. R.T. for the context neurons have been set to 0.75. Histogram c) en d) show how are distributed PCs according to context neurons. For the one room environment, the first two cells are linked with around half of the PC population size. In the three room environment, more context neurons are recruited and consequently the distribution is more equally dispatched, but there are still three major context neurons. Histrograms b) and d) show that the majority of PCs are link with at most 2 context neurons.

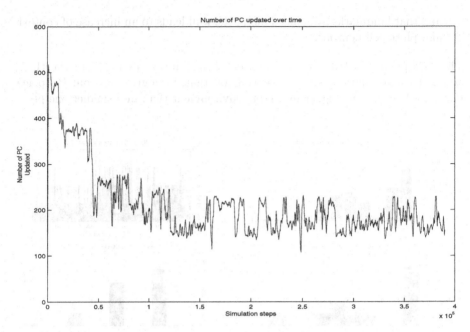

Fig. 5. Measure of the computational cost of the PC updating process for a three rooms environment. The computational cost is, in mean, reduced by a factor of two after the entire exploration of the environment (iteration 0.510^5).

Figure 4 shows detailed histograms for the number of PC linked to each context neuron and for the number of context neuron linked to each PC in two given experiments (one and three rooms environment). RT for the context layer has been fixed to 0.75 in these simulations.

Finally, we analysed the impact of this contextual bias on the computationnal cost of the PC activity update. Without contextual information, this cost is linear to the number of PC recruited in the whole environment. Exploiting a contextual information allows us to only update PCs (compute the equation) linked with the activated context neurons, since we know the other ones will have a very low activity. Figure 5 shows the number of place cells updated at each iteration over a period of 400000 iterations. This result shows an interesting decrease of the updating process cost. After, the exploration of the entire environment (iteration 0.510^5), this cost is really reduce, divided by two compared to the number of recruted PC (500).

4   Conclusion

The results described in this paper confirm the validity and the computational interest of our model of the interaction between place cells and contextual cells. However, this work is a very first step toward the constitution of a context. The model still need to be enhanced, for example by adding other information (like idiothetic information, other geometric or nongeometric information like sound). Nethertheless, even at this preliminary stage, this first model leads to several interesting properties, since contextual PCs can:

- increase the reliability of the PC code by limiting the range of the neurons competing for a given place. For example, a place cell a learned in a part of the environment coded by context neuron A can not pertubate the competition taking place in another location belonging to a context neuron B if no link exist between PC a and context neuron B.
- decrease the computatinal cost of the PC update process of the activity .
- provide a multiresolution localisation system to a more general navigation model.

We currently work on experimatal data (images and azimuth) acquired while a car perfomed a loop in the city near over 30km. We will certainly have to introduce path integration information and to introduce grid cell activity in order to discriminate ambiguous visual places when visual stimuli are ambiguous (long corridor for instance). We will also perform experiments on a real robot over a whole floor to validate these first findings.

We also plan to modify the context layer to obtain a distributed code of context, whereby context information is shared across a population of neurons. We think this kind of coding could take care of the remapping phenomenon observed in rodent hippocampus as recent finding suggest it [2]. Furthermore, this code could reduce the number of both needed PC and context neurons.

References

1. Anderson, M.I., Jeffery, K.J.: Heterogeneous modulation of place cell firing by changes in context. Journal of Neuroscience 23, 8827–8835 (2003)
2. Anderson, M.I., Killing, S., Morris, C., O'Donoghue, A., Onyiagha, D., Stevenson, R., Verriotis, M., Jeffery, K.J.: Behavioral correlates of the distributed coding of spatial context. Hippocampus 16, 730–742 (2006)
3. Cuperlier, N., Quoy, M., Gaussier, P.: Navigation adn planning in an unknown environment using vision and a cognitive map. In: Proc. of European robotic Symponium 2006, Palerme, pp. 129–142 (2006)
4. Cuperlier, N., Quoy, M., Gaussier, P.: Neurobiologically inspired mobile robot navigation and planning. Frontiers in NeuroRobotics 1(1) (2007)
5. Cuperlier, N., Quoy, M., Giovannangeli, C., Gaussier, P., Laroque, P.: Transition cells for navigation and planning in an unknown environment. In: Şahin, E., Spears, W.M., Winfield, A.F.T. (eds.) SAB 2006 Ws 2007. LNCS, vol. 4433, pp. 286–297. Springer, Heidelberg (2007)
6. Doboli, S., Minai, A.A., Best, P.J.: A computational model of the interaction between external and internal cues for the control of hippocampal place cells. Neurocomputing 52, 371–379 (2004)
7. Frank, L.M., Brown, E.N., Wilson, M.A.: Trajectory encoding in the hippocampus and entorhinal cortex. Neuron 27, 169–178 (2000)
8. Gaussier, P., Banquet, J.-P., Sargolini, F., Giovannangeli, C., Save, E., Poucet, B.: A model of grid cells involving extra hippocampal path integration and the hippocampal loop. Journal of Integrative Neuroscience 6, 447–476 (2007)

9. Giovannangeli, C., Gaussier, P., Banquet, J.-P.: Robustness of visual place cells in dynamic indoor and outdoor environment. International Journal of Advanced Robotic Systems 3(2), 115–124 (2006)
10. Hafting, T., Fyhn, M., Molden, S., Moser, M.-B., Moser, E.I.: Microstructure of a spatial map in the entorhinal cortex. Nature 436, 801–806 (2005)
11. Jung, M.W., McNaughton, B.L.: Spatial selectivity of unit activity in the hippocampal granular layer. Hippocampus 3, 165–182 (1993)
12. Kentros, C.G., Agnihotri, N.T., Streater, S., Hawkins, R.D., Kandel, E.R.: Increased attention to spatial context increases both place field stability and spatial memory. Neuron 42, 283–295 (2004)
13. Kubie, J., Ranck, J.: Sensory-behavioral correlates in individual hippocampus neurons in three situations: Space and context. In: Seifert, W. Neurobiology of the Hippocampus. Academic press, New York, pp. 433–447 (1983)
14. Muller, R.U., Kubie, J.L.: The effect of change in environment of the spatial firing of hippocampal complex-spike cells. Journal of Neuroscience 7, 1951–1958 (1987)
15. O'Keefe, J., Dostrovski, J.: The hippocampus as a spatial map. preliminary evidence from unit activity in the freely-moving rat. Brain Research 34, 171–175 (1971)
16. Quirk, G.J., Muller, R.U., Kubie, J.L., Ranck, J.B.: The positional firing properties of the medial enthorinal neurons: Description and comparison with hippocampal place cells. Journal of Neuroscience 10, 1945–1963 (1992)
17. Sharp, P.: Complementary roles for hippocampal versus subicular/enthorinal place cells in coding place, context, and events. Hippocampus 4, 432–443 (1999)
18. Sharp, P.E., Green, C.: Spatial correlates of firing patterns of single cells in the subiculum of the freely moving rat. J. Neurosci. 14(4), 2339–2356 (1994)

Linked Local Visual Navigation and Robustness to Motor Noise and Route Displacement

Lincoln Smith, Andrew Philippides, Paul Graham, and Phil Husbands

Centre for Computational Neuroscience and Robotics
University of Sussex BN1 9QG, United Kingdom
{lincolns,andrewop,paulgr,philh}@sussex.ac.uk

Abstract. This paper presents an investigation into the robustness to motor noise of an insect-inspired visual navigation method that links together local view-based navigation in a series of visual locales automatically defined by the method. The method is tested in the real world using specialist robotic equipment that allows a controllable level of motor noise to be used. Extensions to the method, which can improve its robustness to severe motor noise and to major disruptions such as being displaced along its route, are investigated.

1 Introduction

We recently introduced an insect-inspired method for navigating across relatively complex environments by linking together local view-based navigation using the across a series of automatically defined visual locales [1]. Within the Linked Local Navigation (LLN) framework, local homing was achieved using the Average Landmark Vector (ALV) model. The LLN method was demonstrated in a variety of simulated and real environments and was shown to be inherently robust to visual noise. In addition, many mobile robot platforms also suffer from high degrees of motor noise, as do insects while being buffeted by gusts of wind for example. Hence, in order to further evaluate the LLN's promise for real world applications and its potential use in biological modelling, this paper presents an investigation of the LLN method's robustness to motor noise. Using specialist robotic equipment we can systematically explore the effects of motor noise in controllable real world conditions, something not possible in previous route-based models [2,3]. The LLN method is shown to be robust to even high levels of motor noise.

We then investigate simple extensions of the LLN method, based on minimal 'place recognition', which can improve robustness to severe motor noise and to major disruptions such as displacement along the learned route. The ability to recognise a location depends on the representation of the visual scene and a range of representations have been used from Fourier components [4] to colour information [5]. Here we present an analysis of place recognition using only a two and three-dimensional Average Landmark Vector and discuss how place recognition can be used to augment our LLN method.

M. Asada et al. (Eds.): SAB 2008, LNAI 5040, pp. 179–188, 2008.

After briefly describing the main navigation algorithm and the robotic equipment used, the experimental results are presented.

2   Linked Local Navigation

This section describes the biologically inspired visual navigation algorithm under investigation. At its core is a computationally efficient view-based navigation method that performs well in locations close to the goal. View-based methods compare the current visual scene with a stored representation of the scene at the goal and derive a direction heading from the difference between the two. A series of local view-based navigation steps are chained together in a novel way to allow navigation over complex environments. For full details see [1].

While there are many algorithms capable of implementing local view-based navigation (for review see [6]), the most parsimonious is the ALV method which processes a view into a single vector. The ALV model requires little computation and memory, and has been shown to be effective for visual navigation in both simulation [7] and on autonomous mobile robots [8,1].

To calculate the ALV, features (landmarks) are selected from a 360 degree panoramic view. The ALV is simply the average of the unit vectors from the agent towards each landmark. For navigation, the agent is placed at a goal location and the ALV there (the goal ALV) stored. To return to the goal, the agent calculates the vector difference between the current ALV and the goal ALV and moves in that direction. Since the difference between the ALVs gives the approximate direction of goal, navigation is implemented by iterating this process [7].

Prerequisites for the ALV are therefore a $360°$ visual system, an ability to align views with an external reference (e.g., a compass direction) and a robust object detection system. Ants and bees have near spherical vision, both gain compass information from celestial cues [9] and it is assumed that they can reliably segregate objects from the background [10]. Thus the ALV method is biologically plausible and has been shown to be computable with simple artificial neural networks [11].

Taking inspiration from observations of ant navigation [12], the full linked local algorithm requires a training phase where a scaffolding behaviour dictates the route to be learnt. In training, the agent travels along a path from start to goal. If the number of landmarks currently seen is different to the number seen at the previous time-step, the ALV calculated at the previous time-step is stored as a waypoint; this simply requires that the agent can perceive the binary event of an appearance or disappearance of a landmark. When the agent is within 5cm of the goal, the goal ALV is calculated and stored as the final waypoint. In this way the environment is broken up into separate visual locales in which local view-based navigation will be effective.

The navigation phase begins with the agent at the start position with an ordered series of stored ALVs as waypoints. The agent then uses the local navigation method (ALV) to navigate towards the first waypoint. When the number of landmarks changes, the agent is assumed to have crossed a boundary into the next visual locale and the navigation system switches to using the ALV associated with the next waypoint in the list. This process continues until the agent reaches the final goal or times out.

While a movement calculation based purely on the difference between the current and goal (or waypoint) ALVs, is adequate for local navigation, it causes problems at the boundary between visual locales. As the difference becomes small so does the the resultant movement and the agent will very likely not reach the waypoint let alone cross the boundary. After investigating various methods to overcome this [1], we settled on a method which applies momentum to the heading of the agent together with an absolute step-size of $2cm$. The heading direction at time t, θ_t, is calculated as a weighted average of the angle dictated by the difference of the current and goal ALVs, ϕ_t, and the previous heading, θ_{t-1}, with the weight, ω_t, of the previous heading increasing with the value of $|\theta_{t-1} - \phi_t|$, as described by the following equations.

$$\theta_t = \omega_t \theta_{t-1} + (1 - \omega_t)\phi_t, \text{ where } \omega_t = \min\left(\frac{|\theta_{t-1} - \phi_t|}{0.5\pi}, 1\right) \qquad (1)$$

This method prevents large jumps in direction, ignoring ϕ_t altogether when $|\theta_{t-1} - \phi_t| \geq \frac{\pi}{2}$. Using egocentric polar coordinates, the movement vector is thus $\mathbf{r_t} = (\mathbf{2}, \boldsymbol{\theta_t})$.

3   The Sussex Gantry Robot and Visual Processing

All experiments reported in this paper were performed on a gantry robot: a large volume XYZ Cartesian robot (Figure 1) with an operating volume of $3 \times 2 \times 2m$. Black/dark-grey cardboard tubes of different diameters were placed within the environment to make high contrast landmarks against the white walls of the gantry. The gantry head can be moved with sub-millimetre precision which allows us, through software control, to effectively control the amount of motor noise experienced by the agent.

For the work presented here we used a panoramic camera mounted on the Z-axis. The camera, a VCAM 360, is shown in Figure 1B. The hemispherical mirror projects a 360° image of the environment on to the downward facing CCD video camera. The image was transformed from a circular reflection to a 1-dimensional image representing a 360- degree panorama (Figure 1C). The transformation was accomplished by taking eight 1-pixel-wide radial samples from the panoramic image. The radial positions of these annular samples are shown by the concentric circles in Figure 1C. Three hundred and sixty one-degree, grey-scale levels were calculated for each radial strip through interpolation. These are then averaged across the eight samples to give a 1 x 360 strip of mean grey-scales rounded to integers in the range [0, 255].

This one-dimensional strip is the raw visual input. At each time-step this is processed into landmarks (Figure 1D) from which the ALV is generated, as

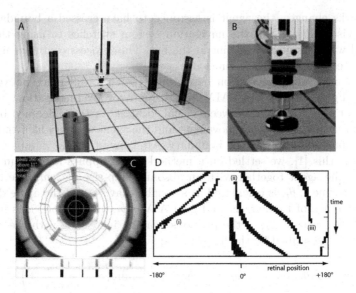

Fig. 1. The gantry robot. A: The gantry robot is an XYZ Cartesian robot, which can position a camera at any point in a 3mx2mx2m volume. B: The camera head is a catadioptric system that projects a 360° panoramic image of the world onto a CCD array. C: A frame capture from the video feed. The three concentric circles (outermost to innermost) indicate the sampled area's upper edge, horizon, and lower edge. The resulting strip, after unwrapping, is shown underneath along with a thresholded strip. D: A trace of the visual input experienced by the agent along a route. This trace demonstrates: (i) occlusion, (ii) appearance of landmarks as they come into perceptual range, and (iii) disappearance of landmarks as they leave perceptual range.

described briefly below. Note that due to occlusions and the limited perceptual range of the agent, the set of perceived landmarks will change during locomotion.

Landmark recognition is accomplished in several sequential stages, described in detail in [1]. Briefly, The raw visual input is first resolved into 90 panoramic facets resulting in an inter-facet angle of the same order as the inter-ommatidial angle of ants' eyes [13]. Each facet has a receptive field covering 8°, that is, itself and half of each of its neighbours. The activation within each facet is averaged and then thresholded to -1 or 1 depending on whether the output is less than 194. We take advantage of the robot's movement to implement two further processing steps based on lateral and temporal excitation/inhibition which serve to 'clean' the visual signal by ameliorating the problem of perceptual 'flickering' of landmarks; that is, single landmarks which are on the edge of the agents' perceptual range and occluding landmarks which are alternately perceived as one or two objects. Effectively, these steps mean that a new landmark appearing is not perceived until it is at least two facets in width. Similarly, once two landmarks have been perceived as one, they are not perceived as 2 landmarks until the gap between them is at least two facets.

Once this visual processing is complete, landmarks are defined as connected sets of active facets and the bearing of each landmark is calculated as the average of the angular position of the facets containing the landmark edges. Thus landmark bearings are accurate to $\pm 2^{\circ}$. These bearings are then used to generate the ALV and this, together with a signal determining whether the number of landmarks has changed, is passed to the main algorithm.

Landmark heights were extracted from the full panoramic image generated by the visual system. For each visible landmark (as detected by the image processing method detailed above) we find the point where the central facet of the landmark changes from 1 to -1, starting from the centre of the 2 concentric circles in Figure 1C and working downwards. As all our landmarks are equal height ($1m$, meaning the tops of visible landmarks are above the field of view of our camera) the elevation of the landmark base gives us a proxy for their perceived height. The visual processing is specialized to the environments used in the experiments, but could be generalized.

4   Results 1: Motor Noise

Controllable amounts of motor noise were added as follows. During the training phase (see Section 2), the agent travels along a path from start, $\mathbf{s}$, to goal, $\mathbf{g}$, in steps of $2cm$. When noise is added to the learning route, the learning step $\mathbf{l_t}$ is defined by:

$$\mathbf{l_t} = 2\frac{\mathbf{h_t}}{\|\mathbf{h_t}\|} + (\mathbf{N}\left(0, 2\mathbf{n_{pc}}\right), \mathbf{N}\left(0, 2\mathbf{n_{pc}}\right)) \qquad (2)$$

where $\mathbf{h_t} = \mathbf{g} - \mathbf{l_{t-1}}$ is the vector from current position to goal (so at the first step, $\mathbf{h_0} = \mathbf{g} - \mathbf{s}$) and $N\left(0, 2n_{pc}\right)$ is a Normally distributed random number drawn with mean 0 and standard deviation $2\,n_{pc}$, where $n_{pc} \in [0, 0.8]$ sets the standard deviation to a percentage of the step-size $2cm$. Thus noise is added independently to x and y dimensions independently and alters both size and direction of the step taken. At the highest noise levels, this can result in the step being taken in the opposite direction than was intended.

During the navigation phase, after the heading vector, $\mathbf{r_t}$ - as defined in Section 2, is transformed into Cartesian co-ordinates $\mathbf{u_t} = (2cos(\theta_t), 2sin(\theta_t))$ noise is added in the same way as for the learning step, to give a final step $\mathbf{v_t}$ of:

$$\mathbf{v_t} = \mathbf{u_t} + (\mathbf{N}\left(0, 2\mathbf{n_{pc}}\right), \mathbf{N}\left(0, 2\mathbf{n_{pc}}\right)) \qquad (3)$$

with parameters as defined above. Note that the momentum on the heading is calculated according to θ_t, the heading the agent 'thinks' it is taking, rather than the actual direction it takes. We feel this is most realistic in terms of an insect or robot's movements being moderated by environmental noise.

To assess the algorithm's robustness to motor noise, we initially ran the algorithm with 5 different noise levels applied to the navigation part of the algorithm. The learning run was performed with no noise so we could isolate the effects of noise on the navigation and learning phases of the algorithm independently. 10

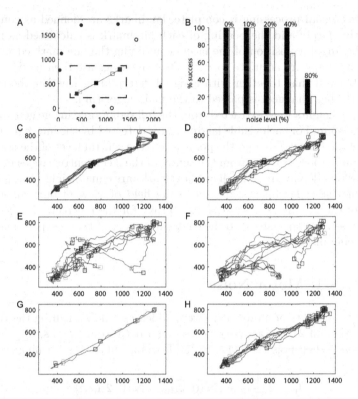

Fig. 2. Navigation with motor noise. A: Simple and complex environments with landmarks (circles), waypoints (squares) and learning run (solid line). Dashed line indicates area shown in C-H. Simple environment is as complex environment minus open symbols (2 waypoints and 2 landmarks). B: Navigation success over 10 runs in simple (black) and complex (white) environments for various noise levels (n_{pc}). C-F: 10 navigation runs in complex environment ($n_{pc} = 20, 40$ and 80%, C-E respectively) and simple environment ($n_{pc} = 80\%$, F). G: Learning run with 40% noise (straight line shows noiseless learning). H: Navigation after noisy learning run (complex environment, $n_{pc} = 40\%$).

runs were performed for each noise level in first a simple and then a complex environment. The results are shown in Figure 2. Failures were counted as any run in which the agent entered a visual locale which was not the next on its list of waypoints. In both environments, the algorithm was robust to high levels of motor noise.

In the simple environment, the algorithm was successful until the standard deviation of the Gaussian noise reached 1.6 cm, 80% of the step-size. As this noise is applied independently to x and y components of the movement vector, individual steps often took the agent in the opposite direction to which it was trying to home. Successful homing under these conditions demonstrates the benefits of view-based homing as opposed to odometric information. At each step, the direction to home is calculated anew and thus errors do not accumulate. The algorithm will therefore succeed unless by chance the noise on several sequential steps takes the agent in a consistent direction towards a new visual locale which is not the one the agnet 'expects' to encounter next. Thus, one can see that for the high noise level, clusters of failures appear at points where incorrect visual locales occur close to the paths taken by the agent.

In the complex environment, due to the increased number of waypoints and erroneous visual locales, the algorithm's performance dropped when the standard deviation of the Gaussian noise reached 40% of the step-size. Note however that two of the three failures ocurred near the end of the run where there are a succession of waypoints very close to each other (Figure 3D). For many of these 'failures', the algorithm did home successfully to the goal using the goal waypoint and ignoring subsequent boundaries experienced. While the chances of such 'lucky' successes are higher near the goal, success is not guaranteed and so we count these runs as failures.

Finally, we performed a further 10 runs with the standard deviation of the Gaussian noise at 40% on both the learning and navigation parts of the algorithm (Figure 3G-H). The agent homed successfully in all runs. This demonstrates that the algorithm is robust noise on the learning run. Moreover, the greater success with a noisy learning run shows firstly, there is no advantage to the straight-path learning runs we perform; and secondly, the lesser performance of the agent at 40% noise in the complex environment could be due to a 'difficult' set of waypoints to navigate.

5   Results 2: Route Recognition

We next wanted to investigate the effects of more severe noise. It is known that ants who are taken from the end of their route near the goal (nest/food) and placed at a new location somewhere near its usual route can successfully navigate back to the route and from there, to the goal. To reproduce this sort of robustness our agent needs, on displacement, to be able to assess which locale it is in, or, more pertinently, which waypoint it should navigate to. Can this be achieved using the minimal visual representation of the world that the agent has available to it?

As a first step to answering this question, we examined the runs with 40% noise and assessed which locale the agent 'thought' it was in at each step of each run. This was accomplished by calculating the distance (according to various metrics) of the (possibly parameterised) current view from the view from each of the stored waypoints. At each step, if the closest waypoint is the one associated with the locale the agent is in, it is assumed the agent can home successfully from that point, first to the winning waypoint and subsequently to the goal. We used three distance metrics based on different representations of the visual scene. The first is the sum square difference between a Cartesian representation of current and waypoint ALV. The second uses only the absolute angular difference between the headings of current and waypoint ALV. As a control, we also used the sum-square difference between the current visual scene and that at the waypoint, that is the

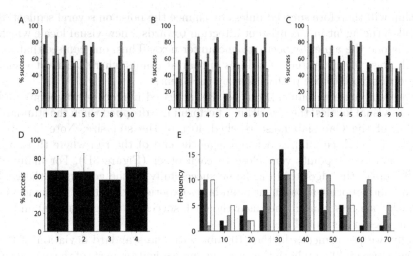

Fig. 3. A-C: Percentage route recognition success over 10 runs for Cartesian ALV (black), angular ALV (grey) and unprocessed vision (white). A: Simple environment with 40% noise. B: Complex environment with 40% noise. C: Different learning routes. D-E: Route recognition success over 66 horizontal and vertical transects through the complex environment. D: Mean and standard deviation of percentage success for (left to right) Cartesian ALV, angular ALV, unprocessed vision and 3D sum ALV. E: Distribution of successes of different routes for Cartesian ALV (black), angular ALV (dark grey), unprocessed vision (light grey) and 3D sum ALV (white).

visual input before it is processed into landmarks. Such raw visual input has been shown to be a good indentifier of location in regions close to the goal [14].

Results are shown in Figure 3A-B. In both simple and complex environments results show consistent patterns. The metric based on unprocessed vision is correct over less than half of the route ($48 \pm 9\%$), as would be expected from the results of [14]. In both environments the ALV-based metrics perform well in all runs (discounting run 6 of the complex environment, in which navigation failed early) with the angular ALV performing better ($69 \pm 13\%$ compared to $59 \pm 12\%$). All methods perform best close to the waypoints, on the side where the agent is approaching the waypoints. This is to be expected as the waypoints are points at which the visual scene changes significantly. This effect should be more evident for ALV rather than visual metrics. Indeed, examination of distances to waypoints along each route shows asymmetric profiles for the ALV methods, but characteristic symmetric funnels [14] for the unprocessed vision.

To assess the generality of this result, we examined place recognition over 10 learning runs from a fan of points around the complex environment. Results were fairly consistent with those achieved for the noisy runs (Figure 3C). The angular ALV was again the best ($69 \pm 17\%$), while the Cartesian ALV ($61 \pm 10\%$) and vision-based metric ($54 \pm 10\%$) were slightly more successful than previously. The angular ALV did however have a much greater variance than the other two methods with an almost bi-modal distribution of results. It is therefore worth noting that learning run 1 of these results - in which the angular ALV performs well - is the learning run used for the noisy runs in the complex environment. It is therefore possible that the angular ALV is suited to this particular run, and that its good performance in the noisy runs may well be an overestimate.

As a final test, we assessed route recognition for the whole environment using a database of images gathered at a grid of points with a spacing of $5cm$. To generate routes, we took either horizontal or vertical transects through the environment, with waypoints set down whenever the number of landmarks changed across the route. Averaged over all the 66 transects, differences between the ALV-based methods disappeared while the visual method remained at just over 50% (Figure 3D). To attempt to increase the success-rate we incorporated the height of the landmarks into a 3-dimensional ALV, and assessed all methods with a number of different metrics. Intriguingly, the best performing metric ($70 \pm 16\%$)was one which used the sum of the 3-dimensional vectors to each landmark (rather than an average which would result in a 3-D ALV). Presumably the success of this method is linked to the fact that it incorporates a sense of the number of visible landmarks. This latter factor is clearly important as it delineates visual locales, though is prone to aliasing. While differences between the best performing metrics were not statistically significant, the difference in distribution of success rates over the runs (Figure 3E) indicates that the methods are correct indifferent parts of the environment. It is therefore possible a combination of metrics could provide more robust route recognition.

6   Discussion

In this paper we have demonstrated that the linked local navigation framework is robust to large amounts of motor noise. Moreover, we have shown that using the ALV, or a representation derived from it, can provide a robust measure of place within a route should a more radical displacement occur.

We have also begun preliminary work to increase robustness further by incorporating route recognition within the navigational algorithm, with encouraging results. Briefly, the algorithm proceeds as usual, but should a change in the number of landmarks occur it does not automatically start using the next waypoint. Instead, the agent assesses whether the current view is associated with the visual locale it is expecting. If so, the algorithm proceeds to home to the next waypoint. If not, it continues homing with the current one. Our initial tests use the number of landmarks to assess whether the locale is correct or not. Future work will explore the use of the continuous encodings presented here, as well as more flexible ways of incorporating route recognition. Finally, we are investigating neural network-based implementations of route recognition.

Acknowledgments

This work was supported by EPSRC grant GR T08753 01. We thank Prof. Collett and members of the CCNR for useful discussion.

References

1. Smith, L., Philippides, A., Graham, P., Baddeley, B., Husbands, P.: Linked local navigation for visual route guidance. Adaptive Behavior 15(3), 257–271 (2007)
2. Vardy, A.: Long-range visual homing. In: Zha, H. (ed.) Proceedings of the IEEE International Conference on Robotics and Biomimetics, vol. 1, pp. 220–226 (2006)
3. Franz, M.O., Schölkopf, B., Mallot, H.A., Bülthoff, H.H.: Learning view graphs for robot navigation. Autonomous Robots 5, 111–125 (1998)
4. Menegatti, E., Maeda, T., Ishiguro, H.: Image-based memory for robot navigation using properties of the omnidirectional images. Robotics and Autonomous Systems 47(4), 251–267 (2004)
5. Gourichon, S.: Utilisation d'un compas visuel pour la navigation d'un robot mobile. PhD thesis, Université Pierre et Marie Curie, Paris, AnimatLab (2004)
6. Möller, R., Vardy, A.: Local visual homing by matched-filter descent in image distances. Biological Cybernetics 95(5), 413–430 (2006)
7. Lambrinos, D., Möller, R., Pfeifer, R., Wehner, R., Labhart, T.: A mobile robot employing insect strategies for navigation. Robotics and Autonomous Systems 30, 39–64 (2000)
8. Möller, R.: Modelling the landmark navigation behavior of the desert ant Cataglyphis. Technical Report IFI-AI-00.24, Artificial Intelligence Lab, Dept. Computer Science, University of Zurich (November 2000)
9. Wehner, R., Michel, B., Antonsen, P.: Visual navigation in insects: Coupling of egocentric and geocentric information. Journal of Experimental Biology 199, 129–140 (1996)
10. Möller, R.: Insects could exploit uv-green contrast for landmark navigation. Journal of Theoretical Biology 214(4), 619–631 (2002)
11. Hafner, V., Möller, R.: Learning of visual navigation strategies. In: Quoy, M., Gaussier, P., Wyatt, J. (eds.) Proceedings of the European Workshop on Learning Robots, vol. 1, pp. 47–56 (2001)
12. Collett, M., Collett, T.S.: How do insects use path integration for their navigation? Biological Cybernetics 83, 245–259 (2000)
13. Zollikofer, C., Wehner, R., Fukushi, T.: Optical scaling in conspecific cataglyphis ants. Journal of Experimental Biology 198, 1637–1646 (1995)
14. Zeil, J., Hofmann, M., Chahl, J.: Catchment areas of panoramic snapshots in outdoor scence. J. Opt. Soc. Am. A 20(3), 450–469 (2003)

Second Order Conditioning in the Sub-cortical Nuclei of the Limbic System

Adedoyin Maria Thompson[1], Bernd Porr[1], Christoph Kolodziejski[2], and Florentin Wörgötter[2]

[1] Department of Electronics & Electrical Engineering, University of Glasgow, Glasgow, G12 8LT, Scotland, United Kingdom
{mariat,b.porr}@elec.gla.ac.uk
[2] Bernstein Center for Computational Neuroscience, University of Göttingen, Bunsenstr. 10, D-37073 Göttingen, Germany
{kolo,worgott}@bccn-goettingen.de

Abstract. Three factor Isotropic sequence order (ISO3) learning is a form of differential Hebbian learning where a third factor switches on learning at relevant moments for example, after reward retreival. This switch enables learning only at specific moments and, thus, stablises the corresponding weights. The concept of using a third factor as a gating signal for learning at relevant moments has been extended in this paper to perform second order conditioning (SOC). We present a biological model of the sub-cortical nuclei of the limbic system that is capable of performing SOC in a food seeking task. The 3rd-factor is modelled by dopaminergic neurons of the VTA which are activated via a direct excitatory glutamatergic pathway, and an indirect dis-inhibitory GABAergic pathway. The latter generates an amplification in the number of tonically active DA neurons. This produces an increase in DA outside the event of a primary reward and enables SOC to be accomplished.

Keywords: Conditioning, Dopamine, Hebbian learning, three factor ISO Learning.

1 Introduction

In second-order conditioning (SOC), a conditioned stimulus (CS1) that becomes associated with a reward or unconditioned stimulus (US) can be used to develop further learning of behavioral associations for a second conditioned stimulus (CS2).

It has been established that Dopamine (DA) is involved in the reward system of the brain [1] and is released in short phasic bursts in response to primary rewards. Over time, dopaminergic neurons stop firing in the event of a reward and commence instead at the onset of the conditioned stimuli (CS) that predicts the reward [2]. The error signal in Temporal difference (TD) [3] learning has been a popular interpretation of the DA neuron activity. [4,5] suggest that weight change is achieved instead by using a modified form of differential Hebbian learning.

M. Asada et al. (Eds.): SAB 2008, LNAI 5040, pp. 189–198, 2008.

Here the DA activity corresponds to the 3rd-factor which when active, functions as a gating system and facilitates learning only at relevant moments. In SOC the relevant moments occur at the CS1 and CS2 onset. Anatomical findings suggest that there are two distinct methods that drive the spiking activity of DA neurons. These are glutamatergic excitation and GABAergic dis-inhibition [6] both of which result in an excitatory influence on the target neurons. We suggest that the glutamatergic excitation results in burst spiking at the moment of the primary reward while the GABAergic dis-inhibition is responsible for the burst at the onset of the conditioned stimulus (CS). [5] demonstrate how learning is achieved in event of the primary reward using the glutamatergic pathway. This paper illustrates an extended model in which an added dis-inhibitory pathway generates a second activity state of the DA neurons. These two activity states produce the 3rd factor in concert with the US and both CS1 and CS2 events.

Our model is based on the sub-cortical nuclei of the limbic system. It is capable of demonstrating a SOC behavior in a food seeking task. The model learns by implementing a form of differential Hebbian learning and a model of dopaminergic activity. A combination of these two systems is known as three factor Isotropic sequence order learning (ISO3) [7]. The model uses cues from its environment to eventually find a food reward. In the following section, we discuss the behavioral model and the environment. We show how certain nuclei form a network which is integrated into an agent. The agent utilises the signals obtained from the environment to perform the food seeking task.

2   The Model and the Simulated Environment and Agent

Figure 1A shows the agent searching for food in an octagonal playground comprising two compartments. There is a small opening in the wall dividing the playground at which a blue landmark is positioned for the agents navigation from one demarcation to the other. A green food disk is located in the second compartment and can only be detected when the agent enters the second compartment. The agent has left and right bump sensors for avoiding the walls and light detectors for detecting the landmark and food disk (Fig. 1B). It starts in the first compartment and explores in a straight line until it either bumps into the walls and changes direction, or it comes in contact with the landmark and food disk for which a natural reflex attraction reaction is exhibited. When the agent comes into contact with the food disk, it is returned to its starting location in the first compartment and the agent begins its search again. This process is repeated for a set period of time steps and the agent slowly reduces the time required to obtain the food disk by approaching the blue landmark and food disk from a distance. There are red lights embedded in both the landmark and food disk which are triggered when the agent comes into direct contact with either of the objects. The signal obtained from the red light is used to bootstrap learning and generates the stimulus that cause reflex approaching behavior towards the red light. Embedding the red lights in both the landmark and food disk ensure that a symmetry is maintained within the objects that interact with the naive agent.

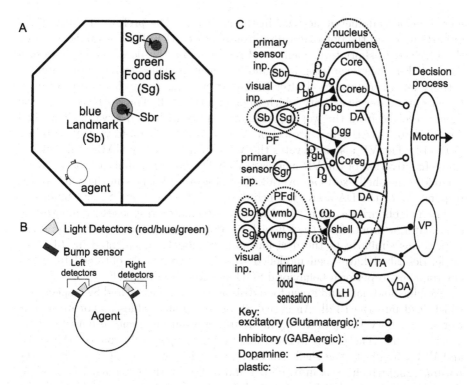

Fig. 1. A) A behavioral model of the agent in its environment in which the SOC food seeking task is performed. B) The agent containing both left and right bump sensors and red, blue and green light detectors. The agent uses the blue and green light detectors to detect the signals of the blue landmark (Sb) and green food disk (Sg) from a distance. The specific red lights enclosed in both the landmark and food disk respectively generate the Sbr and Sgr signals when the agent touches either landmark or food disk. C) The model of the limbic circuitry. The Sbr and Sgr signals activate the reflex towards the landmark and food disk respectively. The distal Sb and Sg signals converge onto the $core_b$ decision unit with plastic weights ρ_{bb}, ρ_{bg} and $core_g$ decision unit with plastic weights ρ_{gg}, ρ_{gb} The Sb and Sg signals also feed into the working memory neurons WM_b and WM_g of the PFdl which project to the shell with variable weights ω_b and ω_g respectively. Abbreviations: ventral tegmental area: VTA; lateral hypothalamus: LH; ventral pallidum: VP; dorsolateral prefrontal cortex: PFdl; Nucleus Accumbens: Nac.

Certain nuclei are modelled as shown in Fig. 1C to form a network as follows: The ventral tegmental area (VTA) afferents a variety of nuclei causing DA release at the target sites. This includes the nucleus accumbens (Nac) which is located in the Ventral striatum (VS) and can be sub-divided into the shell and core [8]. This doparminergic projection from the VTA to the VS is necessary for reward seeking behavior and motivation [9]. The lateral hypothalamus (LH) is known to be activated by natural reinforcers [10]. Its activation provides excitatory glutamatergic inputs to the shell and the VTA and also results in an increase of dopaminergic activity in the shell [11]. The prefrontal cortex (PF) in particular and the dorsolateral prefrontal cortex (PFdl) are established to be the involved in short term memory [12]. Therefore the dorsolateral prefrontal cortex (PFdl) has been integrated and has been modelled to maintain stimulus activity for a certain period after its initiation. This is another source of excitatory glutamatergic activity which feeds into the shell. The shell associates the inputs received from the working memory inputs with food rewards. The shell has inhibitory GABAergic projection neuron which innervate ventral pallidum (VP) [13]. The VP also innervates the VTA with inhibitory GABAergic projections. Inactivation of the VP increases VTA-DA neuron population activity [6]. Therefore the shell exerts an overall dis-inhibitory influence on the VTA and is responsible for the DA activity which develops at either the CS1 or CS2 onset.

The neighbouring core in the Nac develops motor responses by integrating glutamatergic inputs from the PF with the dopaminergic activity from the VTA. It performs similarly to the decision making model implemented in [14]. The nuclei described have been modelled to utilise the signals obtained from the environment to perform behavioral SOC in a food seeking task.

The blue landmark and green food disk produce signals Sb and Sg respectively which feed into the PFdl. An active Sb or Sg signal enables its corresponding working memory WM_b or WM_g input to maintain persitent activity for a set period of time. The plastic synapses ω_b and ω_g respectively connect the WM_b and WM_g inputs to the shell. An encounter with the food disk generates the *reward* signal which feeds into the LH and results in the DA burst in the VTA. Encountering the food disk results in shell activation via the LH and an association is made between the most recently activated working memory input (CS1) and the food reward.

The core comprises two decision making units namely $core_b$ and $core_g$ which control the agents approach to the blue landmark and green food disk respectively. The approach behavior is controlled by utilising a normalised difference between left and right light sensors. Originally in the naive agent, the approach occurs as a primary reflex reaction to the landmark or food disk. This behavior is driven by the cortical inputs labelled Sbr and Sgr which are triggered by direct encounter to the blue landmark and green food disk respectively i.e. when the agent activates the individual red lights. The Sbr and Sgr feed into the $core_b$ and $core_g$ units accordingly while Sb and Sg signals are both fed into each of the core units via plastic synapses ρ_b and ρ_g. After learning the plastic synapes are strengthened and the distal Sb and Sg signals enable anticipatory reactions towards the landmark or food disk from a distance.

The next section describes how the plastic synapses develop and learning occurs.

3   Plasticity in the Nac

The LH produces an excitatory glutamatergic activity to the shell and VTA and this activity is triggered upon contact with the food disk. Thus the LH can be represented as: $LH = h_{bp} * reward$ i.e. the *reward* signal with a transfer function h_{bp}. Neuronal systems are simulated by transforming the inputs with bandpass filters h_{bp} in which f is frequency of oscillation and q is the quality factor. The activity of the Nac shell is influenced by the weighted PFdl working memory input (WM_g or WM_b) and LH signals. While the activity of the core units are influenced by cortical inputs which signify distant and proximal contact with both the landmark and food disk as follows:

$$shell = LH + (WM_b \cdot \omega_b) + (WM_g \cdot \omega_g). \tag{1}$$

$$core_b = Sbr * h_{bp} + ((Sb * h_{bp}) \cdot \rho_{bb}) + ((Sg * h_{bp}) \cdot \rho_{bg}). \tag{2}$$

$$core_g = Sgr * h_{bp} + ((Sb * h_{bp}) \cdot \rho_{gb}) + ((Sg * h_{bp}) \cdot \rho_{gg}). \tag{3}$$

ρ and ω are the plastic synapses that change under the influence of the dopaminergic neuron activity of the VTA which is dis-inhibited by the shell via the VP.

$$VP = \theta(1 - shell \cdot \zeta) \tag{4}$$

$$VTA = lh \cdot \kappa - VP \cdot \upsilon \tag{5}$$

Where ζ, κ and υ represent the shell-VP, LH-VTA and VP-VTA pathways respectively and $\theta(x)$ is given by:

$$\theta(x) = \begin{cases} x, & \text{if } x > 0 \\ 0, & \text{otherwise} \end{cases} \tag{6}$$

We model the VTA DA burst spiking by passing the VTA signal through a highpass filter to give: $burst = \theta(VTA * h_{hp})$. Note that this burst can originate from excitation or from the dis-inhibition of the VTA which then generates the R burst or the CS burst.

The plastic synapses of the shell and core undergo an identical process of weight change. Therefore plastic weights in the Nac shell and core can be generalised with a symbol β as: $\beta \leftarrow \beta + \mu(u_j \cdot Nac' \cdot burst \cdot (1 - \beta))$ in which μ represents the learning rate.

4   Results

We demonstrate the performance of the model in simulations using an open dynamics engine (ODE) with the network of the model programmed in C++. The simulations were run on a computer running Ubuntu Linux. The behavioral environment and agent were programmed to perform the SOC food seeking task as described in the previous sections. This section details some results of the simulation which was run for a period of 500000 time steps. Figure 2 shows the simulation traces for the WM_b, WM_g and LH signals which correspond to the CS2, CS1 and US signals respectively. The VTA, resultant DA burst and shell weights development are included. The focus is on traces which occur within two specific time windows of the overall simulation. These are time steps between 8000 and 18000 and 43000 and 53000. Three relevant events numbered i, ii and iii

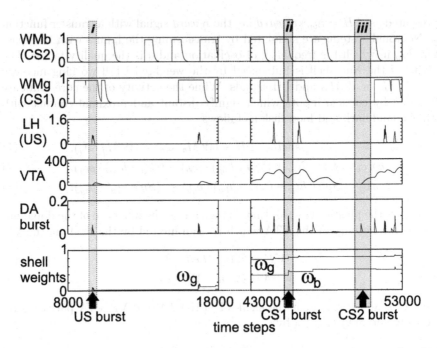

Fig. 2. The detailed simulation results showing two specific time intervals which demonstrate how the CS and US bursts indicated by the arrows are produced. At the beginning of simulation between time step 8000 and 10000, the WM_b/CS2 is triggered but no DA burst is generated. The first burst occurs when the agent encounters the reward and the LH becomes active as shown in the highlighted area i. Two bursts are highlighted between the timesteps 43000 and 53000 which occur at the onset of the WM_G/CS1 and WM_B/CS2 stimuli in the absence of LH activity (area ii and iii). Parameters: The bandpass filters (H_{bp}) were set with $f = 0.01$ and $q = 0.51$. The highpass filter producing a DA burst (H_{hp}) was set: $f = 0.1$, $q = 0.71$. The pathways are weighted: shell-VP (ζ) = -1, LH-VTA (κ) = 0.00001 and VP-VTA (v) = -1 The learning rates of the shell $\mu_{shell} = 0.5$ and the core $\mu_{core} = 0.02$.

have been highlighted within these time intervals and show the development of the DA burst activity which occur originally at the onset of the US highlighted i and later on at the CS1 and then CS2 events highlighted by ii and iii respectively. At the beginning of the simulation, all plastic weights were set to zero thus modelling a naive agent. The agent explored the environment until it encountered the food disk (US activation). The LH was triggered and the VTA was activated resulting in a DA burst which in turn enabled learning to occur at the relevant synapse. In this case, the working memory (WM_g) input from the green food disk to the shell maintained activity which coincided with the DA burst and facilitated the corresponding weight ω_g to develop. This is indicated by the arrow labelled US burst in the event i. The agent was returned to its starting location and commenced exploration and slowly developed a behavioral attraction to the food disk from a distance. The ω_g weights develop further as the agent repeatedly

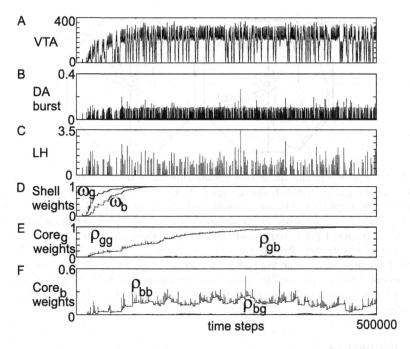

Fig. 3. An overview of the simulation results. A) The VTA activity B) The LH/US signal, C) The DA burst activity, D) The shell weights, E) The $core_b$ weights, F) The $core_g$ weights.

encounters the food disk and an association is made between the detection of the food disk form a distance (CS1) and the contact made with the food disk (US). The shell is activated when the agent detects the food disk from a distance and the WM_g signal is triggered. This resulted in a dis-inhibition of VTA and a DA release at the onset of the CS1. This is shown in the highlighted area ii in Fig. 2. The activity maintained by the working memory (WM_b) input from the blue landmark occurs in concert with this CS1 burst and causes its corresponding weight ω_b to grow. The agents behavioral activity throughout the course of the simulation can be seen in Fig. 3 which shows the (A) VTA, (B) DA burst, (C) LH, (D) shell weight and individual (E) $core_b$ and (F) $core_g$ units weight development. Figure 4 shows the paths made by the agent at (A) the first 10000 timesteps of the simulation and (B) the last 10000 time steps of the simulation. It can be seen that the agent originally explores along the walls until it comes in direct contact with the landmark or food disk. At the end of the simulation the agent has developed a behavioral reaction towards both the landmark and food disk. The agent makes five times more contacts with the food disk over the final 10000 time steps than it does during the first. It can be seen that after learning the agent heads directly for both the landmark and food disk.

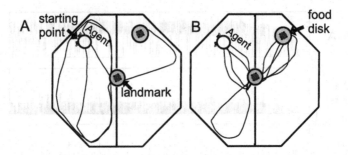

Fig. 4. An overview of the simulation results showing a trace of the paths the agent made when navigating through the playground. A) The trajectory of the agent during the first 10000 timesteps. The agent bumps into the wall and explores along the wall until it comes in contact with the landmark or food disk. The agent makes 1 contact with the food disk. B) The path made by the agent during the last 10000 timesteps. The agent makes its way directly to either the landmark or food disk when either is detected. On approach to the landmark, the agent bumped into the edges of the walls which forced it to change direction away from the landmark until the agent detects the landmark once again heads directly towards it. The agent makes 5 contacts with the food disk.

5   Discussion

In this paper, we have shown how a biologically inspired model has been developed which is capable of performing SOC in a food seeking task. The focus is on two pathways that result in the generation of the bursts in the VTA. Primary rewards facilitate bursts through the LH-VTA pathway while the predictive bursts are generated via CS activation of the shell and dis-inhibition of the VTA through the VP. These DA bursts enable three-factor learning to occur [15,4] in which the three interacting elements include the pre-synaptic activity triggered by the CS, the post-synaptic activity triggered by the US and the dopaminergic burst. Learning in this case is dependent on spike timing dependent plasticity STDP occuring locally at target sites.

Among a variety of limbic circuitry models are two that show certain similarities to the current model. [16,17] describe methods that emphasise two pathways which facilitate the production of DA bursts. While [17] provides a model based on the DA neurons of the substantia nigra compacta (SNc), similar nuclei are employed to generate a DA burst. The US and CS produce a burst through the LH and VS-VP pathways respectively. [16] accounts for the CS burst by implementing an additional nuclei known as the central nucleus of the amygdala (CNA).

The error signal which drives learning in TD methods incorporates predictions of future rewards to produce associations between both CS and reward related stimuli and CS and other CS in higher order conditioning. The earliest CS can eventually trigger a DA burst if and only if there is a series of uninterrupted chain of CS-CS and CS-US events. This system is biologically unrealistic. SOC is accomplished in the PVLV model by introducing a modification of the Rescorla-Wagner learning rules into two separate subsystems. One calculates the CS-US error while the other calculates the error between a second conditioned stimuli and the CS associated with the reward. SOC can be performed by both the current model and the PVLV model without the disadvantage of depending on a continuous series of CS representations. While the latter achieves this by using two systems as mentioned, our model uses CS traces which are maintained by working memory input.

Although the current model focusses only on a process of generating predictive bursts, [16,17] have also accommodated the eventual delayed inhibition of the bursts by incorporating added nuclei. The current model could also theoretically generate the inhibitory reward prediction by modifying the circuitry with an inhibitory GABAergic shell-LH pathway as confirmed in [18]. The PVLV model achieves this by implementing a direct Nac-VTA pathway. Strong evidence suggests that the major influence the Nac has on the VTA is through the indirect pathway via the VP which receives an abundant GABAergic projection from the Nac [19] and its inactivation results in an increase in the activity of the DA neuron population [6]. [8] state that the direct Nac-VTA or VS-VTA pathway implemented in the PVLV model is less prominent than the indirect pathway. This means that activation of the Nac would produce an overall excitatory effect on the VTA. In the TD and PVLV methods, omission is coded by a decrease in activity of the VTA dopaminergic neurons which results in long term depression (LTD). [20,21] have questioned the capability of the dip in DA activity to sufficiently code for negative prediction errors. [20] proposes cholinergic mechanisms of detecting the contrasts in DA neuron firing and [22] suggests that serotonin might play a role in coding the negative prediction errors. We posit that an increase in the number of tonically active dopaminergic neurons might be sufficient to code for these negative prediction errors. This paper emphasises that the dopaminergic signal which is distributed globally to various brain areas acts like a gating mechanism for learning where learning occurring locally at target sites depends on heterosynaptic STDP. It would be interesting to extend the behavioral model so that the agent is made to navigate in an environment comprising of additional landmarks that do not necessarily guide the agent towards the food disk. By implementing a mechanism by which STDP occurs locally, the landmarks can either be ignored or acknowledged depending on their relevance.

References

1. Robbins, T.W., Everitt, B.J.: Neurobehavioural mechanisms of reward and motivation. Curr. Opin. Neurobiol. 6(2), 228–236 (1996)
2. Schultz, W.: Dopamine neurons and their role in reward mechanisms. Curr. Opin. Neurobiol. 7(2), 191–197 (1997)
3. Sutton, R.S., Barto, A.: A temporal-difference model of classical conditioning. In: Proceedings of the Ninth Annual Conference of the Cognitive Science Society, Seattle, Washington, July 1987, pp. 355–378 (1987)

4. Porr, B., Wörgötter, F.: Learning with "relevance": Using a third factor to stabilise hebbian learning. Neural Computation (in press, 2007)
5. Thompson, A.M., Porr, B., Wörgötter, F.: Stabilising hebbian learning with a third factor in a food retrieval task. In: Nolfi, S., Baldassarre, G., Calabretta, R., Hallam, J.C.T., Marocco, D., Meyer, J.A., Miglino, O., Parisi, D. (eds.) SAB 2006. LNCS (LNAI), vol. 4095, pp. 313–322. Springer, Heidelberg (2006)
6. Floresco, S.B., West, A.R., Ash, B., Moore, H., Grace, A.A.: Afferent modulation of dopamine neuron firing differentially regulates tonic and phasic dopamine transmission. Nat. Neurosci. 6(9), 968–973 (2003)
7. Porr, B., Wörgötter, F.: Isotropic Sequence Order learning. Neural Comp. 15, 831–864 (2003)
8. Zahm, D.: An integrative neuroanatomical perspective on some subcortical substrates of adaptive responding with emphasis on the nucleus accumbens. Neurosci. Biobehav Rev. 24(1), 85–105 (2000)
9. Everitt, B.J., Robbins, T.W.: Neural systems of reinforcement for drug addiction: from actions to habits to compulsion. Nat. Neurosci. 8(11), 1481–1489 (2005)
10. Nakamura, K., Ono, T.: Lateral hypothalamus neuron involvement in integration of natural and artificial rewards and cue signals. J. Neurophysiol. 55(1), 163–181 (1986)
11. Maldonado-Irizarry, C., Swanson, C., Kelley, A.: Glutamate receptors in the nucleus accumbens shell control feeding behavior via the lateral hypothalamus. J. Neurosci. 15(10), 6779–6788 (1995)
12. Pribram, K.H., Mishkin, M., Rosvold, H.E., Kaplan, S.J.: Effects on delayed-response performance of lesions of dorsolateral and ventromedial frontal cortex of baboons. J Comp. Physiol. Psychol. 45(6), 565–575 (1952)
13. Lisman, J., Grace, A.: The Hippocampal-VTA loop: Controlling the entry of information into the long term memory. Neuron 46, 703–713 (2005)
14. Prescott, T.J., Gonzalez, F.M.M., Gurney, K., D., H.M., Redgrave, P.: A robot model of the basal ganglia: Behavior and intrinsic processing. Neural Networks 19(1), 31–61 (2006)
15. Kötter, R., Wickens, J.: Interactions of glutamate and dopamine in a computational model of the striatum. J. Comput. Neurosci. 2(3), 195–214 (1995)
16. O'Reilly, R., Frank, M., Hazy, T., Watz, B.: The primary value and learned value pavlovian learning algorithm. Behavioral Neuroscience 121(1), 31–49 (2007)
17. Brown, J., Bullock, D., Grossberg, S.: How the basal ganglia use parallel exitatory and inhibitory learning pathways to selectively respond to unexpected rewarding cues. J Neurosci. 19(23), 10502–10511 (1999)
18. Kelley, A.: Ventral striatal control of appetitive motivation: Role in ingestive behavior and reward-related learning. Neurosci Biobehav Rev. 27(8), 765–776 (2004)
19. Zahm, D.S., Heimer, L.: Two transpallidal pathways originating in the rat nucleus accumbens. J Comp. Neurol. 302(3), 437–446 (1990)
20. Cragg, S.: Meaningful silences: How dopamine listens to the ACh pause. Trends in Neuroscience 29(3), 125–131 (2006)
21. Daw, N.D., Doya, K.: The computational neurobiology of learning and reward. Curr. Opin. Neurobiol. 16(2), 199–204 (2006)
22. Daw, N.D., Kakade, S., Dayan, P.: Opponent interactions between serotonin and dopamine. Neural Netw. 15(4-6), 603–616 (2002)

Synthesising Novel Movements through Latent Space Modulation of Scalable Control Policies

Sebastian Bitzer, Ioannis Havoutis, and Sethu Vijayakumar

Institute of Perception, Action and Behaviour, University of Edinburgh, Edinburgh EH9 3JZ, UK
s.bitzer@ed.ac.uk, I.Havoutis@sms.ed.ac.uk,
sethu.vijayakumar@ed.ac.uk

Abstract. We propose a novel methodology for learning and synthesising whole classes of high dimensional movements from a limited set of demonstrated examples that satisfy some underlying 'latent' low dimensional task constraints. We employ non-linear dimensionality reduction to extract a canonical latent space that captures some of the essential topology of the unobserved task space. In this latent space, we identify suitable parametrisation of movements with control policies such that they are easily modulated to generate novel movements from the same class and are robust to perturbations. We evaluate our method on controlled simulation experiments with simple robots (reaching and periodic movement tasks) as well as on a data set of very high-dimensional human (punching) movements. We verify that we can generate a continuum of new movements from the demonstrated class from only a few examples in both robotic and human data.

1 Introduction

As we design robots to become more anthropomorphic with an aim for them to co-exist in human friendly environments, the number of degrees of freedom and consequently the variety of movements that they can execute have grown significantly. This raises many issues concerning the control and planning in these robots: Who defines such a large set of movements for every new robot? How do you make those movements look natural? How do you cope with the large degree of redundancy?

A promising way out of this dilemma is for the robot (student) to learn the desired movements from a teacher (e.g., human demonstrator) through imitation [1]. There are several approaches to this problem depending on the information available to the student. For example, Grimes et al. [2] observe the movement of a teacher in joint angles and learn a probabilistic model which entails a common latent space between teacher and student to produce a stable movement of the student. Peters and Schaal [3], on the other hand, observe an imprecise, supervised movement in the student's own joint space and then, improve on it with reinforcement learning (which needs additional feedback). Such approaches might solve the problems of producing naturally looking movements and appropriate resolution of redundancy, but being only able to imitate one particular movement is rather limiting. An interesting possibility would be to use the demonstrated examples as a basis for generation of more generalised movements from the same class.

Here, we assume that a set of demonstrated examples belong to the same class of movements, i.e., follows a consistent optimisation or redundancy resolution principle

M. Asada et al. (Eds.): SAB 2008, LNAI 5040, pp. 199–209, 2008.

in some lower dimensional (and common) *unobserved* task space. Additionally, we assume a rich repertoire of movements that achieve different task goals.

The problem of generating similar movements to a set of examples has been addressed in the computer graphics and animation communities. The aim there often is to generate natural looking human motion adapted to a certain situation given a database of recorded human motion. If the database is big enough and contains all the motions needed, it is often sufficient to use an efficient graph based search algorithm to generate desired movement sequences – however, we consider situations where extensive and exhaustive motion generation or capture is either expensive or infeasible. If two similar motions are available, linear interpolation between these works surprisingly well when they are represented as absolute positions and rotations of body parts in a global coordinate system [4]. Also linear combination of motion sequences has been shown to work reasonably well with the right representation [5]. These approaches, besides having to extrapolate movements in (usually) high dimensional movement space, have the problem of scalability and robustness under perturbation or goal modification because they generate an explicit, fixed movement plan indexed in time.

Ideally, we would want to represent and scale the movements in the corresponding task space, since such representations are very compact and interpretable. However, typically we only have access to the demonstrated movements in joint space. A potential solution to this problem is to find a low-dimensional space with similar properties as the task space by employing appropriate dimensionality reduction [6]. Tatani and Nakamura [7] apply autoassociative neural networks to find compact representations for motions from a humanoid robot, but they are missing a way to represent motion dynamics. While Wang et al. [8] incorporate dynamics in their dimensionality reduction to represent movements, this is not suitable for robotic applications, since it is not robust against perturbations and expensive to compute.

In this paper, we first investigate the qualitative relationship between latent spaces produced by the chosen dimensionality reduction technique and the task spaces of simple robotic setups. Then, we show that the resulting latent spaces can be used to encode and learn control policies which act as robust representations of the example movements and allow easy generalisation to new movements from the same class. Finally, we apply this methodology to human motion capture data to demonstrate its feasibility for complex, high-dimensional real world movement data.

2   Methodology

We adopt a 2-step approach, the schematic for which is laid out in Fig. 1. First, we explore a suitable latent space representation of the observed high dimensional movement data (e.g., in joint space) using appropriate dimensionality reduction techniques. Then, we formulate a representation of trajectories as control policies such that they are spatiotemporally scalable and robust against perturbations. In order to test the scalability of the methods, modulated control policies are then mapped back into the original movement space to generate novel target motion. While task space data (or constraints) are generally not accessible in real world demonstrated examples, we will exploit this formalism in artificial setups to test the viability of our methods against ground truth.

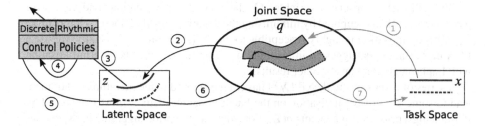

Fig. 1. Experimental methodology. The steps are: (1) inverse kinematics, (2) GPLVM learning, (3) control policy (CP) learning, (4) change of CP parameters, (5) generation of new trajectory with CP, (6) GPLVM mapping, (7) forward kinematics.

Next, we explore the two essential components of our method: a dimensionality reduction algorithm which possesses an inverse mapping and a robust 'control policy' representation that can be easily modulated.

2.1 Dimensionality Reduction

In general, joint and task spaces are nonlinearly related. Therefore, joint and latent spaces should be, too. Furthermore, we need a mapping from latent to joint spaces to generate new movements from modified trajectories in the latent space. Consequently, we have identified the Gaussian process latent variable model (GPLVM) as a promising candidate for our purposes, details of which are described below. An alternative method with similar properties is the Laplacian eigenmap latent variable model [9] – an extension of the spectral technique of Laplacian eigenmaps that adds continuous mappings between data and latent spaces. However, we do not follow up on this in this paper since our explorative experiments suggest that latent spaces recovered did not maintain a topology that was conducive to control policy modulation (see next subsection).

Gaussian Process Latent Variable Models. The Gaussian process latent variable model [10] is a nonlinear generalisation of probabilistic PCA. It is based on a generative model which uses Gaussian processes to map low-dimensional latent variables $\mathbf{z} \in \mathbb{R}^d$ to high-dimensional observed variables $\mathbf{q} \in \mathbb{R}^D$. The corresponding likelihood for a set of N latent variables $\mathbf{Z} = [\mathbf{z}_1, \ldots, \mathbf{z}_N]^\top$ can be written as

$$p(\mathbf{Q}|\mathbf{Z}, \beta) \sim \prod_{i=1}^{D} \exp\left[-\frac{1}{2}\hat{\mathbf{q}}_i^\top \mathbf{K}^{-1}\hat{\mathbf{q}}_i\right]$$

(D independent Gaussian processes) with $\hat{\mathbf{q}}_i = [q_i^1, \ldots, q_i^N]^\top$ the collection of all observed points in dimension i and $\mathbf{K}$ a covariance matrix dependent on $\mathbf{Z}$. Here we use the standard squared exponential covariance matrix with independent, identically distributed noise on the observed variables $\mathbf{q}$:

$$K_{mn} = \beta_1 \exp\left(-\frac{1}{2}\|(\mathbf{z}_m - \mathbf{z}_n)/\beta_2\|^2\right) + \delta_{mn}\beta_3$$

where $\delta_{mn} = 1$ for $m = n$ and 0 otherwise.

Given the set of observed variables $\mathbf{Q}$, their latent representations and values for parameters are then computed by minimising the negative log-likelihood ($-\log p(\mathbf{Q}|\mathbf{Z}, \beta)$). This optimisation is highly susceptible to the initialisation of $\mathbf{Z}$. Usually, we use a PCA initialisation as suggested in [10], but where indicated, we also use initialisation with Laplacian eigenmaps or other results.

A series of extensions to the GPLVM has been proposed in the literature. All of them lead to some kind of regularisation on the latent variables. This is mostly achieved by introducing a prior over the latents $p(\mathbf{Z})$. For example, the prior suggested in [8] defines dynamics on the latents.

2.2 Control Policies

We use discrete and periodic control policies (CPs) to represent goal-directed and periodic movements as attractors of nonlinear dynamical systems [11]. The advantages of this approach are robust representation of movements and easy modifiability of movement parameters such as amplitude, goal point and baseline of oscillations while shape of the CPs is maintained. Alternative ways of representing dynamics, for example with HMMs or linear Gaussian models, do not provide the same level of robustness, suffer from being either restricted to a fixed set of discrete states, only allowing linear dynamics, or expensive computations. In the following, we present our adaptation of the formulation in [11] such that we can explicitly incorporate modifiable start and end positions. Note that only motion in one dimension (e.g. joint) is represented. Consequently, for motion in d dimensions d control policies must be learnt.

Discrete. Discrete movements (e.g., reaching) are characterised by a starting state, z_0, some state trajectory and a goal state, g. The formalisation of such a system is shown in Table 1(left). Ignoring the details of the modulating function f, this is a linear, two-dimensional dynamical system with a single, attracting stable point at $[g, 0]$. f is used to shape the trajectory of the dynamical system between z_0 and g. It can be represented as a weighted sum of RBF basis functions which depend on the state, ξ, of a canonical system that converges to 0. The number of basis functions, n and their width and centres, h_i, c_i, are chosen a priori. Given a complete movement $[\mathbf{z}, \dot{\mathbf{z}}, \ddot{\mathbf{z}}]$, the weights, w_i, of the nonlinear component are learnt. Once the movement is learnt (or encoded as a CP with start state z_0^* and goal g^*), we can change the start state and goal to produce

Table 1. Definitions of discrete and periodic control policies. For discrete CPs the dynamic variable governing the nonlinearity converges to 0 while it monotonically increases for periodic CPs.

discrete	periodic
$\frac{1}{\tau}\dot{v} = \alpha_v(\beta_v(g-z)-v) + \dfrac{g-z_0}{g^*-z_0^*}f(\xi)$	$\frac{1}{\tau}\dot{v} = \alpha_v(\beta_v(z_m-z)-v) + Af(\phi)$
$\frac{1}{\tau}\dot{z} = v \qquad \frac{1}{\tau}\dot{\xi} = -\alpha_\xi\xi$	$\frac{1}{\tau}\dot{z} = v \qquad \frac{1}{\tau}\dot{\phi} = \omega$
$f(\xi) = \xi\dfrac{\sum_{i=1}^{n}\Psi_i(\xi)w_i}{\sum_i \Psi_i(\xi)}$	$f(\phi) = \dfrac{\sum_{i=1}^{n}\Psi_i(\phi)w_i}{\sum_i \Psi_i(\phi)}$
$\Psi_i(\xi) = \exp\left(-h_i(\xi-c_i)^2\right)$	$\Psi_i(\phi) = \exp\left(-h_i(1-\cos(\phi-c_i))\right)$

a qualitatively equivalent dynamics of motion in different parts of the state space of z (which can either be a joint angle, or a dimension in our latent space).

Periodic. Periodic control policies work similarly, as shown in Table 1(right). Instead of a goal state, we have a baseline of oscillation, z_m. The nonlinearity, f, is now governed by a periodic, canonical system with phase velocity ω. Once the weights are learnt to fit a given periodic movement, we can adapt the amplitude, A, of that movement and move it around in state space by changing the baseline, z_m, without losing the shape of the CP. In our implementations, we choose the mean of a data set as an approximation for the initial baseline of the oscillation.

3   Experiments

We use the robotics toolbox for Matlab [1] to implement simulations of two different robots. Our first simulation features a 3 Degree of Freedom (DoF) **planar robot arm** that has a shoulder joint and 2 elbow joints, with the end effector constraint to move in a 2D plane. We resolve the redundancy in the inverse kinematics by choosing the joint space configuration, q, closest to a default pose, q^*, for which the task space constraints are fulfilled. In other words, we minimise $\|q - q^*\|^2$ subject to $k(q) - x = 0$ with $k(q)$ being the forward kinematics. The second platform that we use is the **PUMA-560 robot arm** with 6 DoFs joints (3 translational plus 3 rotational). However, we fix the rotation of the end-effector to a default value in our simulation. For the PUMA-560 robot, there are always 8 alternative joint angle configurations which all correspond to the same translation and rotation of the end-effector. Of these alternatives, we choose the solution which is right handed, has elbow up and non-flipped wrist.

3.1   Task Space vs. Latent Space

In our first experiment, we explore the relationship between the task space used to produce the example movements and the latent space resulting from nonlinear dimensionality reduction on such data. To begin with, we use a uniform grid data in task space to verify that the important properties of the task space are recovered. In particular, we sample 256 data points regularly spaced from a 2D task space. For the planar arm, the data points are spaced at 0.1m (see Fig. 2, left, blue +) while for the Puma arm[2] the points are separated by 0.027m (see Fig. 2, left, green +). For each of the 256 points in task space, we obtain a corresponding robot configuration in joint space using inverse kinematics and run the GPLVM on them to find a latent space configuration.

If for the same robot, a different inverse kinematic solution is chosen, i.e. existing redundancies are resolved in a different way, the data in joint space corresponding to the original task space points will change. Ideally, we would like the dimensionality reduction technique to show some sort of invariance to this source of variability. We investigate resulting latent spaces for 3 different simulations: we use the planar arm with the inverse kinematics as described above as well as one where the deviation from

---

[1] http://www.petercorke.com/Robotics%20Toolbox.html

[2] The Puma's workspace is 3D, here the data points lie in the X-Y plane with Z=0.

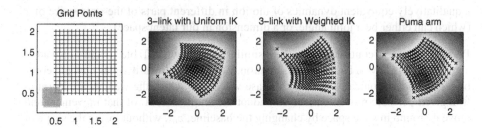

Fig. 2. GPLVM results on simulated robot poses with end-effector positions spanning a 2D grid. (a) The grid in task space (end-effector positions). GPLVM result on: 3DOF planar arm data with (b) standard inverse kinematics (c) weighted inverse kinematics; (d) 6 DOF PUMA arm.

default for the first joint is weighted four times higher. In the third simulation, we use the Puma arm as described above.

Fig. 2 shows the resulting GPLVM latent spaces. Compared to the original grids, we see that the grids in latent space are nonlinearly distorted. However, the spatial topology of the original task space grids are maintained in the latent space. This suggests that interpolation between neighbouring points in latent space has a direct correspondence to modulation in the underlying task space. As expected, the GPLVM is sensitive to the exact choice of redundancy resolution (e.g., inverse kinematics) – the nonlinear distortions are subtly different in all three examples. However, the properties of all resulting latent spaces allow that a continuous trajectory in task space can be represented as a continuous trajectory in latent space, i.e., recovering a structure topologically similar to the unobserved task space is possible from joint data only.

3.2   Reaching and Periodic Movements with Control Policies

Next, we investigate reaching movements which are constrained in specific ways in the task space. The aim of this investigation is three fold. Firstly, we would like to verify that topology is maintained in the extracted latent space. Secondly, we would like to investigate whether modulations of the CPs in the latent space recovers the same class of task (and joint) space movements that was used to train the GPLVM. Thirdly, we want to assess the level of generalisation to novel, unseen movements.

The following experiments are done with the simulated Puma robot which has more degrees of freedom than the planar arm. We start with a family of straight line, minimum jerk, reaching movement data in task space. To test whether we can reliably reconstruct a movement in latent (and task) space that was not used to produce the latent space, we leave one movement out when training the GPLVM (Step 2, Fig. 1).

After we obtain the latent space, we fit discrete control policies (Step 3, Fig. 1) to a single representative trajectory in latent space. We then generate new latent space movements through modulating the CPs (Step 4, Fig. 1) by reparametrising the start state and goal to match those of the remaining desired movements in latent space. Importantly, to test the generalisation ability, we generate a movement that was not used in the GPLVM training by interpolating the start and goal state of two neighbouring movements. We

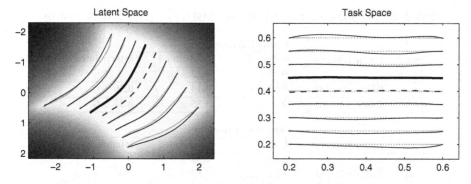

Fig. 3. Fitted (bold), modulated (thin) and interpolated (dashed) discrete CPs for the parallel trajectories of the PUMA arm in latent and task space

then evaluate movements generated by the CPs against the original one in latent, joint and task space.

Parallel Trajectories. We begin by considering parallel task space trajectories depicted by grey dots in Fig. 3(right) with the resulting trained latent space shown on the left. The shading visualises the probability that the GPLVM puts on a corresponding point in joint space. In both panels, the grey dots are the data points available in that space. The bold lines in latent space represent the fitted control policy while the thin lines are the result of CP modulation. The trajectories in task space result from mapping the latent space trajectories through joint space to task space (Steps 6-7, Fig. 1).

The deviation of the trajectories from the data points in task space has two possible sources: (i) discrepancy between the latent space data and CP modulated trajectories (thin lines); (ii) reconstruction errors of the GPLVM(i.e, Steps 2 & 6, Fig. 1). Statistics of the trajectory errors in various spaces are summarised in Table 2. We find that the GPLVM reconstruction error is negligible (see first column, Table 2). Consequently most of the deviation in task space is due to deviation of the CP modulation from latent space data exemplars. Overall, however, the generated movements fit the original task space and joint space movements exceptionally well.

One can note that, as expected, the fitted control policy has smaller trajectory errors than the result of modulation of the CPs to other movements. This can be attributed to the slightly varying shapes that the representations of the movements have in latent space. Also, the topological relationship is preserved as can be seen by the fact that movements close by in latent space have similar 'shapes' - lending itself to better CP modulation. Indeed, that explains the very low error of the interpolated CP (being near the original fitted CP).

Star Trajectories. Next, we test whether these findings transfer to reaching movements where the task constraints are slightly more complex. Our data consists of 10 minimum jerk trajectories in task space where the start and end points are distributed along a quarter circle with radius 0.5 and 2, respectively (see Fig. 4(top right)).

We find comparable results to the earlier discussion – both qualitatively (see Fig. 4(top)) and quantitatively (the statistics of the error, which is very similar to Table 2, is left out

Table 2. Reconstruction and Trajectory Errors (nMSE$\times 10^{-3}$ and standard deviation)

Parallel Trajectory

space	reconstruction	fitted	modulated	interpolated
latent	–	**0.14** ± 0.16	**4.12** ± 4.70	–
joint	**0.006** ± 0.009	**0.13** ± 0.17	**4.14** ± 5.19	**0.36** ± 0.41
task	**0.009** ± 0.013	**0.16** ± 0.21	**3.85** ± 4.95	**0.31** ± 0.42

Figure-8 Trajectory

space	reconstruction	fitted	modulated	interpolated
latent	–	**2.14** ± 5.44	**2.52** ± 2.58	–
joint	**0.000** ± 0.000	**1.81** ± 3.76	**2.83** ± 3.37	**7.96** ± 15.71
task	**0.000** ± 0.000	**0.61** ± 1.36	**3.17** ± 3.18	**4.98** ± 7.66

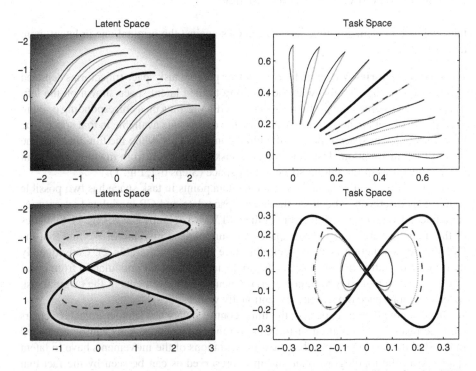

Fig. 4. Fitted (bold), modulated (thin) and interpolated (dashed) Control Policies for the (TOP row) star; (BOTTOM row) figure-8 trajectories of the PUMA arm in latent and task space

in the interest of space). Consequently, we expect our method to be applicable to a wide range of reaching movements with diverse task constraints and orientation.

Periodic Movements. Having explored discrete, point-to-point movements, the natural question is whether the method extends to periodic movements? Again we utilise the Puma platform and simulate figure-8 movements in a 2D task space. We now fit, modulate and interpolate periodic control policies. The task space trajectories: $x(t) =$

$A\sin(\pi t), y(t) = A\sin(2\pi t)$ are generated and then, translated and rotated to fit the Puma workspace. In the latent space, we fit CPs on the figure 8 with $A = 0.3$, modulate with $A = 0.1$ and interpolate for $A = 0.2$.

Results are shown in Fig. 4(bottom) and Table 2. Again, the generated movements follow the figure 8s in task space. However, movement shapes show larger variation in latent space, resulting in slightly higher error rates in task space. It is remarkable, though, that we can generate a continuum of complicated task space movements from just two examples.

3.3 Human Motion Capture

The simulation experiments are useful to compare our results to known ground truth, but compared to what we want to achieve the problem setting in these experiments is still easy with very regular movements in only a few degrees of freedom. A realistic setting is provided by real human data recorded with motion capture.

Here, we apply our method to 3 different punching motions from the same person. The 3 movements all have the same style of punch, but differ in the height that the punch hand (right) is travelling. In particular there is a high, a low and a very low punch (see Fig. 5, top right). The recorded data has 63 dimensions (60 angles plus the root offset).

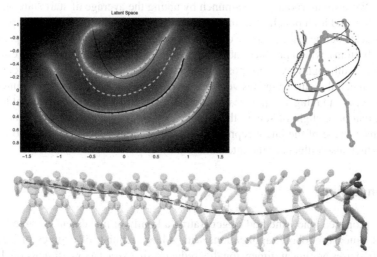

space	reconstruction	fitted	modulated	interpolated
latent	–	0.44 ± 0.81	72.75 ± 65.56	–
joint	0.001 ± 0.002	2.78 ± 5.00	351.55 ± 510.42	–
task	0.000 ± 0.000	0.67 ± 1.07	399.42 ± 657.52	–

Fig. 5. Fitted (bold), modulated (thin) and interpolated (dashed) discrete Control Policies for the human punching motion in latent and end effector space. MIDDLE: Very low punch with path of right hand for original punch (red triangles) and result from CP modulation (solid line). BOTTOM: Errors for human movements (nMSE$\times 10^{-3}$ together with its standard deviation).

First, we note that linear dimensionality reduction like PCA does not work for this data set. For a 2D PCA latent space the reconstruction error of the data in joint space is very high (nMSE: 4780.6e-3). Even for a 10D latent space this error is still significantly higher than for the GPLVM with a 2D latent space (86.8e-3 versus 0.001e-3).

Although a standard GPLVM with 2D latent space has virtually no problem reconstructing the data used to train it, the resulting latent space is not useful for learning control policies, because spatiotemporal topological relation is not well maintained, e.g. data points belonging to single punch sequences are broken up and spread discontinuously. Adding a dynamics prior on the data sequences as suggested in [8] improves results a little bit, but not sufficiently.

We find that a suitable initialisation of the latent space is of key importance. We carried out Laplacian Eigenmaps (LE) on a subset of the movement data that only contains motion of the punch arm and then trained a GPLVM on the same data, initialising with the LE result. This gives a good latent space in which topological invariance is maintained, but in this form, it did not provide a mapping to the full body; and furthermore, low and very low punches were switched in order in latent space. We overcome these problems by recomputing a GPLVM on the full data while using an initialisation based on the previous result which we bias towards correct order of the movements.

The resulting latent space is shown in Fig. 5(top left). We learn discrete CPs on the low punch and adapt their start state and goal to fit those of the high and very low punches. We also interpolate a new punch by taking the average of start state and goal between low and high punches. Using the position of the right hand (also compare Fig. 5, top right) to define the task space of these movements, we report nMSEs as presented in Fig. 5(bottom), which produce satisfying results. The learnt punch closely resembles the original, although it was not possible to modulate the learnt control policies such that they match the other punches very precisely – especially , for the high punch scenario. However, playing a sequence of the generated movement creates natural looking punches that have slight offsets in the joint space. This is a consequence of the different intrinsic shape of the latent representation of the high punch compared to the low punch, which also influences the interpolation.

4   Conclusion

We have proposed a new method of generating a family of movements from examples which is suited for robotic applications with a large number of degrees of freedom. The method uses nonlinear dimensionality reduction to extract a low-dimensional space which captures the essence of the task space constraints and then, learns control policies on the resulting compact representations. New movements are generated by adapting parameters of the learnt control policies in the low-dimensional space and mapping the result back to the original joint space. We have demonstrated this approach in simulated experiments with simple robots and have shown its feasibility for more complicated movements with human motion capture data. In future work, we will investigate how to iteratively use feedback from the mapping to bias the dimensionality reduction such that representations of movements in the resulting latent spaces share stronger shape similarity and hence, allow better interpolation of new movements.

Acknowledgement. This work has been carried out within the SENSOPAC project which is supported by the European Commission through the Sixth Framework Programme for Research and Development.

References

1. Schaal, S., Ijspeert, A., Billard, A.: Computational approaches to motor learning by imitation. Phil. Trans. Royal Soc. London B Biol. Sci. 358(1431), 537–547 (2003)
2. Grimes, D.B., Chalodhorn, R., Rao, R.P.N.: Dynamic imitation in a humanoid robot through nonparametric probabilistic inference. In: Proc. Robotics: Science & Systems (2006)
3. Peters, J., Schaal, S.: Reinforcement learning for parameterized motor primitives. In: 2006 International Joint Conference on Neural Networks, IJCNN, pp. 73–80 (2006)
4. Wiley, D.J., Hahn, J.K.: Interpolation synthesis of articulated figure motion. IEEE Computer Graphics and Applications 17(6), 39–45 (1997)
5. Giese, M.A., Poggio, T.: Morphable models for the analysis and synthesis of complex motion patterns. International Journal of Computer Vision 38(1), 59–73 (2000)
6. Grochow, K., Martin, S.L., Hertzmann, A., Popovic, Z.: Style-based inverse kinematics. In: ACM Transactions on Graphics (Proceedings of SIGGRAPH) (2004)
7. Tatani, K., Nakamura, Y.: Dimensionality reduction and reproduction with hierarchical NLPCA neural networks - extracting common space of multiple humanoid motion patterns. In: Proc. IEEE Intl. Conf. on Robotics and Automation, pp. 1927–1932. ICRA (2003)
8. Wang, J.M., Fleet, D.J., Hertzmann, A.: Gaussian process dynamical models for human motion. IEEE Trans. on Pattern Analysis and Machine Intelligence 30(2), 283–298 (2008)
9. Carreira-Perpinan, M.A., Lu, Z.: The laplacian eigenmaps latent variable model. In: Proc. of the 11th Intl. Conference on Artificial Intelligence and Statistics. AISTATS (2007)
10. Lawrence, N.: Probabilistic non-linear principal component analysis with gaussian process latent variable models. Journal of Machine Learning Research 6, 1783–1816 (2005)
11. Ijspeert, A.J., Nakanishi, J., Schaal, S.: Learning attractor landscapes for learning motor primitives. In: Advances in Neural Information Processing Systems, vol. 15, pp. 1523–1530 (2003)

Incremental Evolution of Animats' Behaviors as a Multi-objective Optimization

Jean-Baptiste Mouret[1] and Stéphane Doncieux[1]

ISIR, Université Pierre et Marie Curie-Paris6, CNRS FRE2507, Paris, F-75005
mouret@isir.fr, doncieux@isir.fr

Abstract. Evolutionary algorithms have been successfully used to create controllers for many animats. However, intuitive fitness functions like the survival time of the animat, often do not lead to interesting results because of the bootstrap problem, arguably one of the main challenges in evolutionary robotics: if all the individuals perform equally poorly, the evolutionary process cannot start. To overcome this problem, many authors defined ordered sub-tasks to bootstrap the process, leading to an incremental evolution scheme. Published methods require a deep knowledge of the underlying structure of the analyzed task, which is often not available to the experimenter. In this paper, we propose a new incremental scheme based on multi-objective evolution. This process is able to automatically switch between each sub-task resolution and does not require to order them. The proposed method has been successfully tested on the evolution of a neuro-controller for a complex-light seeking simulated robot, involving 8 sub-tasks.

1 Introduction

One of the main goal of animat research is to design controllers so that the "intelligence" of the animat emerges from its interactions with its environment, avoiding the biases inducted by human analysis and, hopefully, leading to solutions as smart as the ones found by nature [1]. Evolutionary methods have been widely used to that aim (see [2] for an overview), succeeding in creating controllers for complex behaviors like the combination of locomotion, obstacle avoidance and gradient following in an insect-like robot [3].

However, many early researches in evolutionary robotics showed that rewarding the efficiency of the final behavior was not enough to obtain working controllers because of the bootstrap problem: if the objective is so hard that all the individuals in the first generations perform equally poorly, evolution cannot start and no functioning controllers can be found. For instance, it has been found hard to evolve a light-seeking behavior in a complex arena without having before evolved an obstacle-avoidance reflex [4].

In consequence, researchers emphasized the need to help the evolutionary process by adding some kind of reward for intermediate steps. More precisely, we will call an *incremental task* a task such that:

M. Asada et al. (Eds.): SAB 2008, LNAI 5040, pp. 210–219, 2008.
© Springer-Verlag Berlin Heidelberg 2008

- no fitness gradient exists on most part of the search space, i.e. the fitness of all random individuals for the goal task is minimum;
- the experimenter is able to define some simpler sub-tasks whose completion is useful to the completion of the whole task.

Evolutionary processes specially designed for such tasks will be referred as *incremental evolution*. A lot of experiments have been reported to be more efficiently solved with such approaches than with direct evolution [5,6,3,7].

Despite these successes, published incremental evolution rely on some assumptions that require an accurate knowledge of the problem to solve and can lead the evolutionary algorithm to a local extremum. Most of them, for instance, require to precisely order the different sub-tasks or to determine when to switch from a sub-task to another one. These biases prevent published methods to scale-up well to more complex or more open tasks; noticeably, most of them have been tried with only two or three sub-tasks.

In this article, we propose an original view of incremental evolution based on multi-objective optimization [8], each sub-task defining an objective. Recent multi-objective evolutionary algorithms can then be employed to efficiently solve the bootstrap problem while automatically switching between sub-tasks and not requiring to order the sub-tasks or even to exploit all of them before solving the goal task. This work is a milestone on the way towards the generation of complex behaviors by evolutionary algorithms; nevertheless, to focus ourselves on the evolutionary process, the benchmark task and the genotype have been chosen as simple as possible. To that aim, we evolved feed-forward neuro-controllers for a simulated robot that has to avoid obstacles and turn on a target light, a task that requires to successively switch on at least four different lights in a predefined order. The task has been designed to be extensible and easily modifiable.

This paper is organized in four parts. In the first part, we briefly review the literature concerning incremental evolution. Next, we introduce a way to view incremental evolution as multi-objective optimization. In the third part, we show how multi-objective evolutionary algorithms can be used on incremental tasks. The fourth part describes a simulated robotics experiment using the presented method. A short discussion concludes this paper.

2   Previous Work

Papers dealing with incremental evolution can be categorized in four main approaches: staged evolution, environmental complexification, fitness shaping and behavioral decomposition.

In staged evolution, the main task is split into ordered sub-tasks, each with a corresponding fitness function. Individuals are first selected using the first fitness. Once a convergence criterion has been reached for a stage, for instance when a "good enough" fitness is obtained by the best individual or when the best fitness doesn't change for some generations, the experimenter switches to the next sub-task. This technique has first been successfully employed by Harvey [5] to evolve a vision-based target location task. Three stages were used, from locating a large immobile target to tracking a moving smaller one. Many other examples of staged evolution can be found in the literature [3,9,10,11,7,12].

Environmental complexification allows to more continuously change the complexity of the task through the tuning of dedicated parameters. Gomez et al. [6] thus worked on a prey-capture task parameterized with the prey speed and the delay before starting the pursuit. Ten ordered sub-tasks were thus defined by specific values of these parameters and their use proved to lead to more efficient solutions.

In behavioral decomposition, the robot controller is divided into sub-controllers, each one evolved separately to solve a sub-task. The sub-controllers are then combined together using a second evolutionary process. This approach led to controllers for a composite behavior where a Lego robot had to push a movable light into a designated goal area [13]. A variation of this concept has been recently used to evolve a position controller for an autonomous helicopter [14].

Fitness shaping is sometimes used to help bootstrapping an evolutionary process, though it is not often seen as an incremental evolution scheme. The fitness is defined as an aggregation (a weighted sum or a product) of different evaluation criteria in order to create a followable fitness gradient. Using this idea, Nolfi et al. [15] successfully evolved a feed-forward neuro-controller for a robot to locate, recognize and grasp a target object. The fitness was increased if the individual was close to the target object, if the target was in front of the robot, if the robot tried to pick-up the object, if the robot had the object in the gripper and if the robot released the object outside the area.

3   Incremental Evolution as a Multi-objective Optimization

While the four previously described approaches allowed the evolution of substantially complex behaviors, they imply an accurate knowledge of the global task and of the defined sub-tasks. Staged evolution methods, for instance, imply a manual switch between the sub-tasks. Such an approach can be represented in the fitness space by a "L" shape (in 2D, figure 1 (a)): the first sub-task T_s is optimized first, while the second T_g is not and, after the switch, the performance on the T_s remains constant – it is at least expected not to decrease either because a good behavior on T_s is required to solve T_g or because T_s controller is frozen – while the performance on T_g starts to climb. This method requires to choose when to trigger the switch: if too early, the first sub-task won't be completely solved and if too late, generated solutions might be over-specialized and make optimizing T_g even more difficult (figure 1 (b)). More generally, viewing incremental tasks using this figure suggests that a trade-off can exist between T_g and T_s. For instance, a controller could use all the available ressources to obtain an optimal fitness with regard to T_s, leaving nothing left to solve the goal-task.

The order of the sub-tasks is another important bias of current incremental methods. Let us consider the sub-tasks a typical rodent should manage to survive:

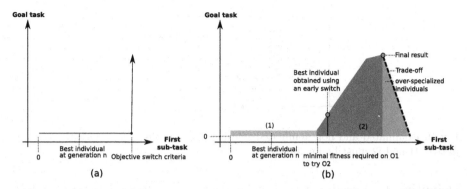

Fig. 1. (a) Staged evolution: a goal task depends on the minimal completion of a sub-task. The first sub-task is optimized and, at a given generation, the experimenter changes the fitness to optimize the final task. (b) Typical fitness space for an incremental task with one sub-task. (1) the first sub-task is optimized; (2) the minimal fitness required to try the goal task is reached, therefore it is possible to have a non-minimum fitness for the goal task; a trade-off can exists between the goal-task and the sub-task, leading to over-specialized individuals.

- walking;
- running;
- avoiding obstacles;
- avoiding predators, implying being able to perceive them and to know an escape strategy;
- eating, implying to be able to explore its environment and locate the food;
- resting, when necessary;
- finding a sexual partner;
- procreating (final goal).

How to successfully learn all these sub-tasks? All of them share complex dependencies but organizing them in a single incremental learning scheme is a challenging problem which could have no solution, even for this simplified rodent.

Consequently, a good incremental algorithm should explore all the sub-tasks and the goal task at the same time, continuing to improve on each tasks at each moment and trying to switch automatically to other tasks when possible. This is exactly what multi-objective evolutionary algorithms do if each task is viewed as an objective.

4   Multi-objective Evolutionary Algorithms

Recent research in evolutionary computation proposed numerous algorithms to simultaneously optimize several objectives [8]; most of them rely on the concept of domination and generate the so-called Pareto Front (figure 2):

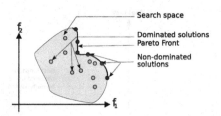

Fig. 2. Dominated and non-dominated solutions for a problem for which the two objectives f1 and f2 must be maximized

Definition 1. *A solution* $\mathbf{x}^{(1)}$ *is said to dominate another solution* $\mathbf{x}^{(2)}$, *if both conditions 1 and 2 are true:*

1. *the solution* $\mathbf{x}^{(1)}$ *is not worse than* $\mathbf{x}^{(2)}$ *with respect to all objectives;*
2. *the solution* $\mathbf{x}^{(1)}$ *is strictly better than* $\mathbf{x}^{(2)}$ *with respect to at least one objective.*

The non-dominated set of the entire feasible search space is the globally Pareto-optimal set (Pareto front).

A typical multi-objective algorithm [8] sorts individuals with respect to domination. Non-dominated individuals may, for instance, be ranked 1, making them the most suitable for reproduction. Individuals which are only dominated by non-dominated ones may be ranked 2, and so on.

Such an algorithm will first try all sub-tasks and the goal-task simultaneously. If the task is incremental, only a small subset of corresponding fitnesses should be different from the minimum. As the fitness on other tasks will be almost equal, the most simple sub-tasks only differentiate individuals and will then be the support of selection. They will be automatically considered as the first sub-tasks to solve. Once individuals are able to test their skills on other sub-tasks or on the goal-task, the algorithm will automatically sort individuals using theses objectives, maintaining the complete set of optimal trade-offs, from the best individuals for each sub-task to the best one for the goal-task.

The aggregation of objectives, as used by Nolfi et al. [15], is a simple way to perform a multi-objective optimization. However, at least three aspects differentiate their approach to the one presented here:

- depending on the shape of the Pareto front, some optimal trade-offs cannot be found using an aggregation approach [8];
- weights have to be chosen using a time-consuming and often not theoretically justified trial-and-error approach;
- best individuals on sub-tasks are not kept, making impossible the simultaneous exploration of different paths potentially leading to more efficient results; for instance, an aggregation approach can be trapped in area (3) of figure 1 or lead to over-specialized individuals;

5   Example: A Complex Light-Seeking Robot

5.1   Light-Seeking Robot

To benchmark the proposed approach on a typical incremental task, we designed a variant of the light-seeking task which involves eight sub-tasks with complex dependencies. A simulated wheeled robot is placed in an arena with some obstacles and seven different colored lights (figure 3(a)). Each light is mounted on a button switch which, when pressed, turns on one or more lights in the arena. Once a light is on, it remains in the same state during the whole experiment. The main goal is to turn on a particular light. The connection between lights and switches, a simple explicit dependency between tasks, is to be discovered by the evolutionary algorithm. The only knowledge used to bootstrap the process is that turning on a light should help to turn on the goal-light. As we consider separately each light, turning on a particular light is considered a particular sub-task[1]. To benchmark future incremental algorithms, the task can be easily complexified by adding lights or changing the dependencies between lights.

The underlying structure of this incremental task is depicted on figure 3 (a). The first light turns on two other lights, creating two paths to accomplish the goal-task. A part of the population may choose to learn to turn on lights $\{0,1,2,3,6\}$ and another one may learn sequence $\{0,4,5,6\}$. This task can therefore be learned in different ways. Moreover, all robots have to avoid obstacles. Despite the simplicity of this problem and of the different sub-tasks, it involves at least five sub-tasks (avoiding obstacles, turn on the good lights

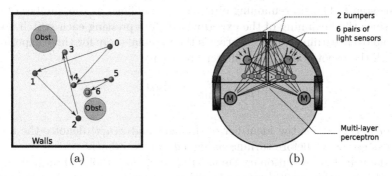

(a)   (b)

Fig. 3. (a) Overview of the benchmark task. Seven colored lights and two obstacles are placed in a arena delimited by four walls. Each light is mounted on a button switch which turns on some other lights, the light circuit (unknown to the algorithm) being represented by arrows. The goal-task is to turn on the sixth light. (b) The simulated robot is equiped with two bumpers and six pairs of binary light sensors, each one sensing a different light. The controller is a multi-layer perceptron with 14 inputs, 14 hidden nodes and 2 outputs (the speed of the motors).

---

[1] This corresponds to seven sub-tasks, the eighth being obstacle avoidance.

of the path having the least number of sub-tasks), making it one of the most complex composite tasks explored with incremental evolution at this time.

The simulated robot (figure 3(b)) is equiped with two bumpers, to avoid obstacles, and six pairs of light-sensors, each one sensing a different light color. Light sensors have a 90 degrees field of view and a binary output (1 if the light is in the field of view, 0 otherwise). Robots are controlled by a simple feed-forward neural network with 14 inputs, 2 outputs and 14 hidden nodes. Neurons use the following activation function, based on $\tanh(x)$, to easily allow the inhibition of a neuron in the case of negative inputs:

$$f(x) = \begin{cases} \tanh(x) \text{ if } x > 0 \\ 0 \text{ otherwise} \end{cases}$$

Cross-over is not used and a Gaussian mutation is applied to weights.

Runs of a direct evolution algorithm designed to minimize the number of time-steps to turn-on the last light, did not find any working controller, all individuals obtaining the minimum fitness. An incremental method seems therefore required to bootstrap the process.

To use the proposed approach, eight objectives are defined. The first one evaluates the obstacle avoidance skills by rewarding how much the robot moves during the whole experiment, thus punishing robots blocked by obstacles[2]:

$$F_0 = \frac{1}{T} \sum_{t=0}^{t=T} \sqrt{\left(x(t) - x(t-1)\right)^2 + \left(y(t) - y(t-1)\right)^2}$$

where T is the length of an experiment and $(x(t), y(t))$ the position of the robot at time-step t. The six remaining objectives are defined as the number of time-steps since the beginning of the experiment before pressing each switch button. To avoid over-learning, each objective is the minimal score for three experiments in which the robot starts from different positions:

$$F_i = \min_{n=1,2,3} \frac{-\varphi(n,i)}{T}$$

where $i = 1, 2, ..., 7$ is the identifier of the light and $\varphi(n, i)$ denotes the number of time-steps spent before turning on light i, for experiment n.

We launched 10 evolutionary runs, with a population of 150 individuals and the multi-objective evolutionary algorithm ε-MOEA [16]. 60 new individuals were generated at each generation.

5.2   Results

About 300 generations are needed to obtain efficient neuro-controllers able to reach the sixth light. The best individuals found by evolution are those that

---

[2] In the simplified simulation we used, a robot touching an obstacle is systematically stopped until it goes backward.

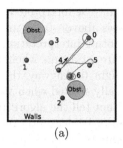
(a)

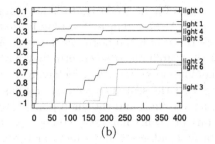
(b)

Fig. 4. (a) Typical trajectory of an evolved simulated robot. Lights 0, 4, 5 and 6 are successfully turned on. (b) Best fitness obtained for each fitness with regards to generation. Lights 0, 1 and 4 are first optimized; they are followed by light 5 (gen. 50), 2 (gen. 70), 3 (gen. 130) and 6 (gen 200).

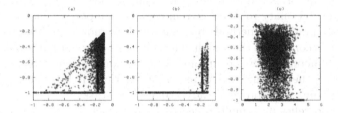

Fig. 5. Typical patterns observed in fitness space. (a) light 4 (y axis) versus light 0 (x-axis); (b) light 5 versus light 0. (c) light 0 versus obstacle avoidance.

used the shortest path, i.e. $\{0, 4, 5, 6\}$. They can be found by looking at the best individuals on the "light 6" criterion. Figure 4(a) depicts a typical trajectory followed by an evolved robot which uses its sensors to avoid obstacles and to locate the successive lights. When the last light has been turned on, it draws an infinite circle around it.

Figure 4(b) shows the best fitness obtained on each sub-task with regards to the generation number. The successive steps followed by the evolutionary process are visible: good controllers are obtained in the first generations for lights 0, 1 and 4; a controller able to reach the fifth light is found around generation 50 and the sixth at generation 230. This switch between sub-tasks was automatic and allowed the process to continue to optimize the first objectives.

Figure 5 shows the explored search space projected on two dimensions. Analyzing these representations may help in finding the dependencies between the criteria, but their interpretation must be careful as they also include historical aspects of the search process[3]. The triangular shape of light 4 versus light 0 criterion (figure 5 (a)) shows the linear dependency of this criterion on light 0 – it is an example of part (2) of figure 1(b). Equivalent figures with light n relative to its predecessor in the path representing the whole task (figure 3(a)) show a similar structure, although it may be shifted or scaled down. The "L" shape of

---

[3] Actually, these sets are only subsets of the whole search space.

the graph of light 5 relative to light 0 (figure 5 (b)) may be interpreted in different ways. At first glance, it may indicate that a fitness above -0.3 is mandatory before the search can begin on these criteria. But another interpretation may be that the search on these criteria may have started late, at a time when the only remaining individuals were behaving very well on light 0. Anyway, this clearly shows that the search on this criteria has automatically started when some conditions were met and the main point is that we did not tell the algorithm when to start or even what these conditions were. Graph of light criteria relative to obstacle avoidance show a more or less densely populated square (figure 5 (c)) indicating that the two objectives are relatively independent.

6 Discussion and Conclusion

In this paper, we proposed an original method to solve the bootstrap problem observed in evolutionary robotics. It removes some important biases present in the previously published methods by requiring less knowledge from the experimenter and less constrains the search. The sub-tasks does not have to be ordered and actually different orders are automatically explored by the evolutionary process. Moreover, the system behaves as an automatic switch between the stages while not stopping to improve already explored sub-tasks.

This method relies on a multi-objective view of incremental evolution which could lead to powerful future algorithms by exploiting the particular shape of the fitness space. One of the difficulties for the proposed method to scale up is the high number of objectives involved. Current multi-objective algorithms are designed to handle a few objectives (typically up to four) and are not as efficient in presence of more objectives. However, the particular shape of the fitness space suggests that the number of dimensions could be reduced during the evolutionary process, for instance using principal component analysis methods. This kind of dimension reduction could allow to neglect the evaluation of sub-tasks during some generations and possibly to reduce the computational cost induced by the simultaneous evaluation of the sub-tasks. Indeed, in the light-seeking task, an agent is observed during a certain amount of time and some fitness functions are derived. Consequently, adding new sub-tasks does not require new evaluations; but some other problems may require sub-tasks to be evaluated in different contexts, thus adding a significant computational cost.

The cross-over operator, not used in this work, could benefit from the proposed method and should be investigated in future work. Since the process maintains specialists in each sub-task, the cross-over operator could allow the evolutionary process to combine sub-parts of each specialist to create more efficient individuals. Such investigations probably requires a modular encoding for neural-networks, as ModNet [17]. Increasing the complexity of the neural-network during evolution also seems to be a key element. All this pleads for considering both selection algorithms and controller encodings as a whole in future work on incremental evolution.

References

1. Meyer, J.A.: From natural to artificial life: Biomimetic mechanisms in animat design. Robotics and Autonomous Systems 22(3-21) (1997)
2. Meyer, J.A., Husbands, P., Harvey, I.: Evolutionary robotics: a survey of applications and problems. In: Husbands, P., Meyer, J.A. (eds.) Proceedings of The First European Workshop on Evolutionary Robotics - EvoRobot 1998. Springer, Heidelberg (1998)
3. Kodjabachian, J., Meyer, J.A.: Evolution and development of neural networks controlling locomotion, gradient-following, and obstacle-avoidance in artificial insects. IEEE Transactions on Neural Networks 9, 796–812 (1997)
4. Urzelai, J., Floreano, D., Dorigo, M., Colombetti, M.: Incremental robot shaping. Connection Science Journal 10(384), 341–360 (1998)
5. Harvey, I., Husbands, P., Cliff, D.: Seeing the light: artificial evolution; real vision. In: Cliff, D., Husbands, P., Meyer, J.A., Wilson, S. (eds.) From Animals to Animats 3, Proceedings of the third international conference on Simulation of Adaptive Behavior. MIT Press/Bradford Books (1994)
6. Gomez, F., Miikkulainen, R.: Incremental evolution of complex general behavior. Adaptive Behavior 5, 317–342 (1997)
7. Urzelai, J., Floreano, D.: Incremental Evolution with Minimal Resources. In: Proceedings of IKW 1999 (1999)
8. Deb, K.: Multi-objectives optimization using evolutionnary algorithms. Wiley, Chichester (2001)
9. Winkeler, J., Manjunath, B.: Incremental evolution in genetic programming. Genetic Programming, 403–411 (1998)
10. Walker, M.: Comparing the performance of incremental evolution to direct evolution. In: 2nd International Conference on Autonomous Robots and Agents (2004)
11. Parker, G.: The Incremental Evolution of Gaits for Hexapod Robots. In: Proceedings of the Genetic and Evolutionary Computation Conference (GECCO 2001), pp. 1114–1121 (2001)
12. Mouret, J.B., Doncieux, S., Meyer, J.A.: Incremental evolution of target-following neuro-controllers for flapping-wing animats. In: From Animals to Animats: Proceedings of the 9th International Conference on the Simulation of Adaptive Behavior (SAB), pp. 606–618 (2006)
13. Larsen, T., Hansen, S.: Evolving composite robot behaviour-a modular architecture. In: Proceedings of the Fifth International Workshop. Robot Motion and Control 2005, pp. 271–276 (2005)
14. De Nardi, R., Togelius, J., Holland, O., Lucas, S.: Evolution of Neural Networks for Helicopter Control: Why Modularity Matters. In: IEEE Congress on Evolutionary Computation, 2006. CEC 2006, pp. 1799–1806 (2006)
15. Nolfi, S., Paris, D.: Evolving non-Trivial Behaviors on Real Robots: an Autonomous Robot that Picks up Objects. In: Topics in Artificial Intelligence: Proceedings of 4th Conference of the Italian Association for Artificial Intelligence, AI* IA 1995, Florence, Italy, October 11-13 (1995)
16. Deb, K., Mohan, M., Mishra, S.: Evaluating the ε-domination based multi-objective evolutionary algorithm for a quick computation of pareto-optimal solutions. Evolutionary Computatition 13(4), 501–525 (2005)
17. Doncieux, S., Meyer, J.A.: Evolving PID-like neurocontrollers for non-linear control problems. International Journal of Control and Intelligent Systems (IJCIS) 33(1), 55–62 (2005); Special Issue on nonlinear adaptive PID control

Integrating Epistemic Action (Active Vision) and Pragmatic Action (Reaching): A Neural Architecture for Camera-Arm Robots

Dimitri Ognibene[1,2], Christian Balkenius[3], and Gianluca Baldassarre[1,*]

[1] LARAL-ISTC-CNR, Lab. of Autonomous Robotics and Artificial Life, Istituto di Scienze e Tecnologie della Cognizione, Consiglio Nazionale delle Ricerche, Via San Martino della Battaglia 44 - 00185 Roma, Italy
[2] DIST, Dip. di Informatica Sistemistica e Telematica, Universita' di Genova, Via all'Opera Pia 13 - 16145 Genova, Italy
[3] Lund University Cognitive Science, Kungshuset, Lundagård SE - 222 22 Lund, Sweden
dimitri.ognibene@istc.cnr.it, christian.balkenius@lucs.lu.se, gianluca.baldassarre@istc.cnr.it

Abstract. The active vision and attention-for-action frameworks propose that in organisms attention and perception are closely integrated with action and learning. This work proposes a novel bio-inspired integrated neural-network architecture that on one side uses attention to guide and furnish the parameters to action, and on the other side uses the effects of action to train the task-oriented top-down attention components of the system. The architecture is tested both with a simulated and a real camera-arm robot engaged in a reaching task. The results highlight the computational opportunities and difficulties deriving from a close integration of attention, action and learning.

1 Introduction

Consider a primate exposed to a new environment scattered with bushes carrying red fruits. It might initially look at the bright green foliage and trunks of bushes popping out of the scene, and try to interact with them without any useful result. Then it might look at a fruit and then pick and taste it. Now that it understands that fruits are useful, how can it find more of them? As its gaze often focuses on the bushes' foliage, it should learn to look away from them, and below them, as the fruits of these bushes hang below their leaves. It should also learn to trigger reaching actions on the basis of the fruits' sight and to shape actions on the basis of the gaze direction.

This example shows typical interactions between attention, perception, action and learning processes taking place in an organism acting in a natural context. These interactions have often been overlooked by the *information-processing*

---

* This research was supported by the EU Projects *ICEA*, contract no. FP6-IST-027819-IP, and *MindRACES*, contract no. FP6-511931-STREP.

M. Asada et al. (Eds.): SAB 2008, LNAI 5040, pp. 220–229, 2008.

framework widely used in machine vision, initiated with Marr's theory of vision [1]. This framework views attention and vision as processes directed to construct "objective" detailed general-purpose representations of the environment later used to guide action and learning [2]. Computationally, building representations totally detached from the embodiment and the specific needs of the system tends to produce scene representations containing an overwhelming amount of non-needed information and hence often computationally heavy or intractable.

The *active vision approach* [3] introduced action in visual processes to allow a high-sensitive *fovea* to scan the scene and perform heavy computations only on portions of it relevant for the task in hand, similarly to what happens in human attention [4,5]. Moreover, it proposed to exploit gaze motion to simplify representations and learning processes, for example by using "deictic representations" encoding information with respect to the current state of sensors, or by applying object or feature recognition processes only to the foveated points [3].

As it emphasizes the importance of action in perception, the active vision perspective has been fully embraced by *evolutionary robotics* [6]. This has proposed systems that fully integrate actions directed to gather information (*epistemic actions*) and actions directed to accomplish the systems' goals in the environment (*pragmatic actions*) [7]. In general, with respect to active vision, evolutionary algorithms have the advantage of co-evolving complementary fovea movements and feature detectors [8,9]. Moreover, they are not affected by the *perceptual aliasing problem* [10] introduced by the fovea's partial view of scenes as reinforcement learning algorithms are. In this respect, an important feature of the architecture proposed here is that, while it uses reinforcement learning to exploit the advantages of on-line adaptation, it ameliorates the aliasing problem by using a *potential action map (PAM)* that stores information on past percepts in the form of potential actions (*memory* is a typical solution to aliasing).

Interestingly, within psychology, Allport [11] proposed a new perspective on attention that, in line with the ideas of active vision, claims that attention serves primarily to guide organisms' action, for example by directly setting some of its parameters [12]. Within the modeling literature, Balkenius [13] echoes this view and specifies the *attention-for-action perspective* with four basic principles: (1) inhibition can be used to disengage the focus of attention from the current location; (2) attentive (epistemic) actions can be computationally treated as other (pragmatic) actions; (3) focussing processes can lead to select targets for (pragmatic) action; (4) gaze direction can be used to produce implicit arguments for action. These principles not only emphasize that attention and action are closely coupled, but they also stress that learning principles generally used to acquire pragmatic actions can also be used to learn attentive actions (principle (2)). Indeed, in the past several systems have been proposed that, contrary to processes that detect information on the basis of intrinsic salience of images' features (*bottom-up attention*; e.g. [14]), exploit *reinforcement learning* algorithms to learn to detect task-relevant information (*top-down attention*; e.g. [15]).

This work proposes a novel neural-network architecture where perception, attention (bottom-up and top-down), action, and learning are integrated to an extent that goes well beyond what is done in existing models. The architecture is tested both with a simulated and with a real camera-arm robot engaged in a reaching task. As we shall see, Sect. 2 on methods and Sect. 3 on results show that the principles of attention-for-action proposed in [13] are fully integrated in the system, either by design or as features emerging from the learning processes. The overall value of this research resides not only in the mechanisms that are proposed to implement the aforementioned integration, but also in the analysis of the system that shows the computational advantages that derive from it.

2   Methods

This section first overviews the system and then explains in detail its components' functioning. The system (Fig. 1b) integrates two previous models: (1) a bottom-up and top-down attention model [16]; (2) an arm control model [17]. These models are based on common computational principles: population codes (here 2D neural maps) to represent sensorimotor information and probability distributions of variables controlling eye/arm behavior [18,19]; dynamic neural-field networks to integrate information and select actions through biased competition mechanisms [20,21]; a progressive developmental of skills of the neural components (cf. [17]). These principles were chosen for their biological plausibility.

The simulated/hardware experimental setup is formed by a down-looking webcam set above a robotic arm (Fig. 1a). The arm's working plane is a CRT monitor. A host computer grabs the camera images, runs the robot's controller, issues motor commands to the arm, and controls the images of the monitor (task). A moving sub-image (*input image*) is extracted from the camera image to simulate eye movements. The input image is used by a *periphery map* that implements bottom-up attention. The central part of input image (*fovea*) is the

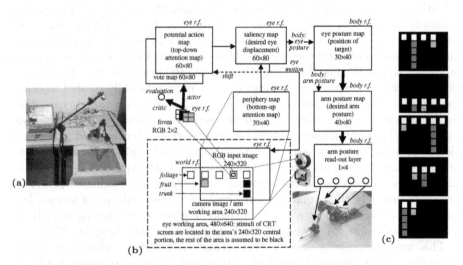

Fig. 1. (a) Robotic setup. (b) Model's architecture. (c) Examples of input images.

input of *reinforcement-learning actor-critic* component that learns to predict the spatial position of the rewarded arm targets with respect to the foveated cues (top-down attention). A *potential action map (PAM)* accumulates this evidence while the fovea explores various cues. A *saliency map* integrates information from the periphery map and the PAM to select the next eye movement using a biased competition. Each fixation point, encoded in a *eye posture map*, suggests a potential arm target to a *arm posture map*: when the eye fixates a location for long, the arm posture map triggers a related action on the basis of a biased competition. If the reached target is a "fruit", the system gets rewarded otherwise it gets slightly punished (energy consumption).

2.1 Robotic Setup and Task

In the following, with the exception of some weight matrices, the bold symbols of mathematical notations represent column vectors.

Camera. A low cost Spacecam 150 Live webcam (by Trust; see Fig. 1) was used to acquire visual information from the environment. The webcam grabs ten RGB images per second with a 240×320 pixel definition (24 bit/pixel). The webcam is set above the robotic arm with its view field covering exactly the monitor's screen forming the arm's working plane. The webcam is connected to the host computer via a USB port and is interfaced with software built with Java Multimedia Framework libraries (by SUN). In the tests running in simulation, the camera is simulated by directly using the task's monitor images.

Robotic Arm. The robotic arm was built using low cost components (e.g. by Lynxmotion). The arm (Fig. 1) is composed of a base and 3 segments (upper arm $15.9cm$, forearm $17.5cm$, and "hand" $9.5cm$). The arm has four degrees of freedom: two at the shoulder (planar rotation, $15° - 165°$, and vertical rotation, $20° - 140°$), one at the elbow ($35° - 145°$), and one at the wrist ($110° - 220°$). Each joint is powered by one digital servo (by Hitech) with the exception of the shoulder vertical-rotation joint having two servos. The servos are controlled by a servo controller SSC32 (by Lynxmotion) connected to the host computer via a serial port. The simulated version of the arm is a simplified kinematic plant with segments' size and degrees of freedom like those of the real arm.

Environment and Task. The horizontal working plane of the arm is a CRT monitor screen ($37 \times 28cm$) connected to the host computer. The monitor generates the images of the task used to test the system. These images are formed by red, green and blue squares set on the vertexes of a 5×5 grid covering the whole screen. In particular, the images form stylized "trees" (see Fig. 1) with 2-5 green blocks representing the foliage (100% luminosity), 1-4 blue blocks representing the trunk (80% luminosity), and 1 red block representing a fruit (80% luminosity). After each reaching action, the system gets a reward of 1 if it touches a fruit and a punishment of -0.05 otherwise. Saccades are not directly rewarded or punished. A new tree randomly structured and positioned in the image is generated after the execution of each reaching action.

2.2 Attention Control Components

Preprocessing Filters. The 240 × 320 pixel RGB image of the webcam is noisy and contains reflections, so it is first subtracted by an image grabbed by the camera with the screen switched off (black image), and then it is filtered into main-color components. The resulting image in copied into the centre of a bigger 480 × 640 pixel black image, and a 240 × 320 pixel image is extracted from this to simulate the system's *input image* grabbed by a moving eye.

Periphery Map (Bottom-Up Attention). The 30 × 40 periphery map **pm** is activated with a grayscale image: first the input image is divided into 30 × 40 blocks of 8 × 8 pixels each, then the RGB color values of the pixels of each block are averaged to obtain a gray value. A more sophisticated bottom-up saliency (e.g. as in [14]) is not needed as this research focuses on top-down attention.

Actor-Critic Component (Top-Down Attention). The fovea is simulated with an image **f** of 2 × 2 RGB pixels taken from the input image centre. This image is fed into two feedforward neural networks forming a reinforcement-learning actor-critic architecture [22]. The *critic* is a network with a linear output unit v_t which learns to evaluate the current state on the basis of the expected future discounted rewards. The system gets a reward r_t after the execution of a reaching action, and this, together with v_t, is used to compute the *surprise signal* s_t [22] used to update both the critic's weights $\mathbf{w}^c$ and the actor's weights $\mathbf{W}^a$. The *actor* is a network whose output layer is a *vote map* **vm** of 60 × 80 sigmoid neurons which signal to the PAM the possible positions of rewarded targets with respect to the currently foveated visual cue ($\gamma = 0.9$; T is the transpose operator):

$$v_t = \mathbf{w}^{cT}\mathbf{f} \quad s_t = (r_t + \gamma\, v_t) - v_{t-1} \quad \mathbf{vm} = g\left[\mathbf{W}^a\mathbf{f}\right] \quad g[x] = 1/(1 + e^{-x}) \quad (1)$$

The critic is trained on the basis of s_t, used as error signal, and the input signal **f** [22]. The actor is trained with a Hebb rule involving the activation of the saliency map $\mathbf{sm}_t$ (encoding the last eye displacement, see below) and the input signal **f** so as to increase or decrease the probability of doing the same saccadic movement again on the basis of the surprise signal s_t [16] ($\eta^c = 10^{-7}$, $\eta^a = 10^{-5}$):

$$\mathbf{w}^c_{t+1} = \mathbf{w}^c_t + \eta^c\, s_t\, \mathbf{f}_t \quad \mathbf{W}^a_{t+1} = \mathbf{W}^a_t + \eta^a\, s_t\, \mathbf{sm}_t \bullet (\mathbf{vm}_t \bullet (1 - \mathbf{vm}_t))\, \mathbf{f}^T_t \quad (2)$$

where • is the entrywise product operator.

Potential Action Map (Top-Down Attention Memory). The PAM **pam** is formed by 60 × 80 leaky neurons and accumulates evidence, furnished by the vote map **vm** via topological connections, on the possible positions of rewarded targets. During each saccade the map's activation is shifted in the direction opposite to the eye's motion to maintain eye-centred representations (as it might happen in real organisms [23]). The PAM is reset each time the tree image from the camera changes (also this might happen in real organisms [24]).

Saliency Map. The 60 × 80 saliency map **sm** selects saccade movements on the basis of the sum of the topological input signals **pm** and **pam**. The saccade movement is selected by first identifying the unit with the maximum activation and then by activating the map with a Gaussian population code centred on it ($\sigma = 1$). The eye movement is the winning neurons' preferred eye displacement ($\Delta x, \Delta y$). This selection mechanism is a computationally fast approximation of a biased dynamic competition process as the one reported in [21] (cf. Eq. 3).

2.3   Arm Control Components

Eye Posture Map. This 30×40 neuron map encodes the current eye posture as a Gaussian population code **emp** ($\sigma = 0.3$).

Arm Posture Map. This 40×40 map *apm* is the output layer of a neural network pre-trained with a Kohonen algorithm (see below and [17]) and encodes arm postures in the 2D map space. During the tests reported in Sect. 3, a neural biased competition [21] takes place in the map (similarly to what happens in real organisms [20]) in order to select a target for reaching actions when any neuron achieves a threshold th ($th = 0.3$; $\delta = 0.1$):

$$\mathbf{apm}_{t+1} = max\left[(1 - \delta)\,\mathbf{apm}_t + \mathbf{W}^{apm\,l}\mathbf{apm}_t + \mathbf{W}^{apm}\mathbf{epm}_t, 0\right] \qquad (3)$$

where $\mathbf{W}^{apm\,l}$ are the weights of *lateral* close-excitatory far-inhibitory connections having a Gaussian distribution dependent on the distance between neurons (see [17] for details), and $\mathbf{W}^{apm}$ are the weights from the eye posture map.

Arm Posture Readout Layer. This is a layer of four sigmoid neurons **aprl** that encode the desired arm joint angles issued to the arm real/simulated servos. The map is activated by the arm posture map through the weights $\mathbf{W}^{aprl}$.

Training. The weights $\mathbf{W}^{aprl}$ and $\mathbf{W}^{apm}$ are trained using the simulated arm to avoid stressing the hardware robot. Training is composed of three succeeding learning phases based on random movements of the arm (*motor babbling*). To avoid redundancy problems during training, the hand segment is kept parallel to the working plane at a fixed distance from it (see Fig. 1a; see [25] for a version of the model addressing redundancy issues). In these phases the system (see [17] for details): (a) performs a vector quantization of postures, within the arm posture map, on the basis of a Kohonen algorithm; (b) learns the inverse kinematic mapping ($\mathbf{W}^{apm}$) between the gaze directions corresponding to the seen hand (**epm**) and the corresponding arm posture (**apm**) with supervised learning; (c) trains the arm posture readout map ($\mathbf{W}^{aprl}$) with supervised learning.

3   Results

This section analyses the behavior of the system tested in the simulated and real robot and the functioning of its neural components emerged with learning. Fig. 2.a shows the reward received by the robot for every reaching action during learning. After the first reward, the performance of both the simulated and real robot increases rapidly and soon reaches a near-optimal steady state. Fig. 2.b shows the average distance between reaching actions' targets and the τth saccade's target executed before such actions during learning. For $\tau = 1$ the distance

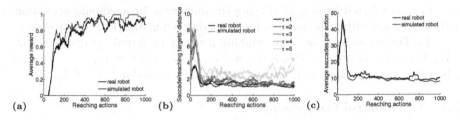

Fig. 2. Learning dynamics during 1000 reaching actions of the simulated and real robot. (a) Moving average of reward. (b) Moving average of the distance between the targets of the last five saccades (τ) and the reaching targets. (c) Moving average of saccades per reaching action. All moving averages have a 50-step window.

initially increases from $2cm$ to $4cm$ and then goes back to $2cm$, for $\tau = 2, 3, 4$ the distance goes from $4 - 8cm$ to $2cm$, for $\tau = 5$ the distance decreases from $6 - 8cm$ to $3cm$ and then goes to $3 - 4cm$. This data indicate that initially only the last saccade ($\tau = 1$) is related to the reaching target but with learning the attention-action coordination increases until the last four/five saccades are focussed on the target. Fig. 2.c shows how the average number of saccades per reaching action evolves during learning. Initially this number is about 20 but then increases to 50 in correspondence to the maximum learning progress in reaching (Fig. 2.a), and finally stabilises at about 10. This dynamics is due to the fact that the system initially tends to trigger reaching actions directed to bushes' foliage (which has a high salience) or trunk, then it learns to inhibit these actions so that eye exploration increases, and finally it learns to anticipate the position of fruits so that saccades become very efficient in localising fruits and in triggering correct reaching actions.

This interpretation is corroborated by data reported in Fig. 3 that shows the 20 most frequent sequences of objects foveated by the trained simulated and real robots in 1000 reaching actions. In both cases, the two most frequent sequences start with a saccade on foliage followed by 4-5 saccades on fruit that trigger action: the system has learned to suitably inhibit the high-saliency foliage cues and to stay on the fruit once found. Other sequences focus only on fruit: these are the "lucky" cases where the eye is already on the fruit in the new tree image. Finally, other sequences are those that start with a foliage saccade followed by a trunk saccade and then a fruit: they indicate that the PAM retains information on the first "ambiguous" saccade target and integrates it with the information from the second saccade target, so in part solving the partial observability problem caused by the limited view of the fovea (see Sect. 1). Note that the most frequent sequences are quite similar for the simulated and real robot. However, the number of total sequences in general is higher for the real robot (53) than for the simulated one (391) due to a higher noise which in the latter case tends to lead saccades to the background.

Fig. 4 shows the activation of the vote map, the PAM and the saliency map in a sequence of three saccades targeted respectively to the foliage, trunk and fruit. While foveating the foliage (or the trunk), the vote map activates as follows

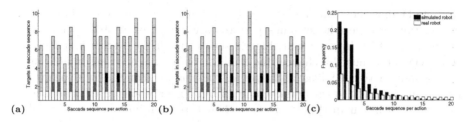

Fig. 3. Most frequent sequences of targets foveated in 1000 reaching actions by the simulated (a) and real robot (b), and corresponding frequencies (c).

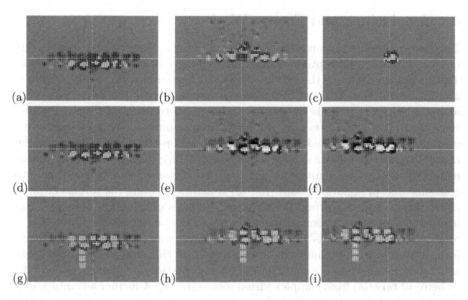

Fig. 4. Activation of the vote map (a-c), PAM (d-f), and saliency map (g-i) in a sequence of three saccades targeted to the foliage (a, d, g), trunk (b, e, h), and fruit (c, f, i) of the tree of the top graph of Fig. 1.c. White and black dots indicate neurons activated respectively above and below 0.5 (a-c) or above and below 0 (d-i).

(Fig. 4a-c): (a) a cluster of neurons activates below 0.5 in correspondence to the whole row of foliage elements (or the column of trunk elements): this biases the eye to move away from them and constitutes an *emergent* form of self-tuned *inhibition of return* related to visual cues (as in organisms [24]) and not to spatial locations, as in the *hardwired* implementations of it of previous models (e.g. [14,16]); (b) a cluster of neurons activates above 0.5 in correspondence to the row of elements below the foliage (or left and right to the trunk): this biases the eye to move there and captures the spatial relations existing between the foliage (or trunk) and the fruit. While foveating the fruit, the vote map exhibits a high-contrast Mexican-hat-shaped activation formed by a cluster of neurons activated above 0.5, surrounded by neurons activated below 0.5, in correspondence to the centre: this bias the eye to stay on the target. The activations of the PAM

(Fig. 4d-f) show how this memory plays an important function in integrating information in time (the system's performance decreases of 50% without this memory, see [17]). In particular, Fig. 4e shows that if the system first foveates the foliage and then, by chance, the trunk it maintains a strong inhibition in correspondence to the foliage and sums up the bias to go on the fruit (below the foliage and laterally to the trunk) coming from both the foliage and trunk cues. Last, the saliency map (Fig. 4g-i) shows how top-down information is suitably integrated with bottom-up information in order to select the most promising locations. For example, notice the strong activation in correspondence to the fruit, laterally to the trunk, in Fig. 4h compared to Fig. 4e.

4   Conclusions and Future Work

This paper presented an architecture for controlling a camera-arm robot that integrates attention, perception, action and learning well beyond existing models. The integrated nature of the system allows it to instantiate the four principles of attention-for-action [13], and this gives the system several interesting properties and strengths: (1) it leads the system to learn self-tuned object-related inhibitions that allow it to disengage attention from scanned or non-relevant visual cues: this can be considered as an emergent inhibition-of-return mechanism commonly hardwired in attentional systems (e.g. [14,16]); (2) it allows using similar neural structures and algorithms, such as reinforcement learning, to train both epistemic and pragmatic actions [13]; (3) it allows selecting the targets of pragmatic action, and triggering the latter, on the basis of attention processes: this lead to a strong integration of the *decision* and *parametrisation* of actions, as observed in real organisms' brains [20]; (4) it allows using the direction of gaze to furnish an implicit parameter to reaching actions: this simplifies computations as it allows extracting a simple and clean information for guiding action (the "where" of targets) from complex visual scenes [11,12]. A further advantage produced by the integration is that the architecture does not need to be furnished the representation of "target objects" to which associate a reward signal, as it usually happens (cf. [8]) in other top-down attention learning systems (e.g., [15]). In fact. the reward produced by behaviour allows the system to autonomously build representations of objects that should trigger actions.

Notwithstanding its strengths, the model has various limits: a simplified feature extraction component, based on simple colour-detection, a simplified bottom-up attention component, based only on luminosity (see the components used in [14]), and a hardwired reset of the PAM memory when the scene changes (cf. [24]). However these limits, which will be tackled in future work, concern the specific system's components used here and not its overall architecture.

References

1. Marr, D.: Vision: A Computational Investigation into the Human Representation and Processing of Visual Information. W. H. Freeman, New York (1982)
2. Fermuller, C., Aloimonos, Y.: Vision and action. Image Vision Comput. 13(10), 725–744 (1995)

3. Ballard, D.: Animate vision. Artif. Intell. 48, 57–86 (1991)
4. Posner, M.I.: Orienting of attention. Q J. Exp. Psychol. 32(1), 3–25 (1980)
5. Treisman, A.M., Gelade, G.: A feature-integration theory of attention. Cognit Psychol. 12(1), 97–136 (1980)
6. Nolfi, S., Floreano, D.: Evolutionary Robotics: The Biology, Intelligence, and Technology. MIT Press, Cambridge (2000)
7. Floreano, D., Kato, T., Marocco, D., Sauser, E.: Coevolution of active vision and feature selection. Biol. Cybern. 90(3), 218–228 (2004)
8. Cliff, D., Noble, J.: Knowledge-based vision and simple visual machines. Philos. T Roy Soc. B 352(1358), 1165–1175 (1997)
9. de Croon, G., Postma, E.: Sensory-motor coordination in object detection. In: IEEE Symp. ALIFE 2007, pp. 147–154 (2007)
10. Whitehead, S.D., Ballard, D.H.: Learning to perceive and act by trial and error. Mach. Learn. 7(1), 45–83 (1991)
11. Allport, D.: Selection for action: Some behavioral and neurophysiological considerations of attention and action. In: Perspectives on perception and action, vol. 15, pp. 395–419. Erlbaum, Hillsdale (1987)
12. Neumann, O.: Direct parameter specification and the concept of perception. Psychol. Res. 52(2-3), 207–215 (1990)
13. Balkenius, C.: Attention, habituation and conditioning: Toward a computational model. Cogn. Sci.Quart. 1(2), 171–204 (2000)
14. Itti, L., Koch, C.: Computational modelling of visual attention. Nat. Rev. Neurosci. 2(3), 194–203 (2001)
15. Schmidhuber, J., Huber, R.: Learning to generate artificial fovea trajectories for target detection. Int. J. Neural Syst. 2(1-2), 135–141 (1991)
16. Ognibene, D., Balkenius, C., Baldassarre, G.: A reinforcement-learning model of top-down attention based on a potential-action map. In: The Anticipatory Approach. Springer, Berlin (2008)
17. Ognibene, D., Rega, A., Baldassarre, G.: A model of reaching that integrates reinforcement learning and population encoding of postures. In: 9th Int. Conf. Simul. Adapt. Behav., September 2006, pp. 381–393. Springer, Heidelberg (2006)
18. Pouget, A., Ducom, J.C., Torri, J., Bavelier, D.: Multisensory spatial representations in eye-centered coordinates for reaching. Cognition 83(1), B1–11 (2002)
19. Pouget, A., Zhang, K., Deneve, S., Latham, P.E.: Statistically efficient estimation using population coding. Neural Comput. 10(2), 373–401 (1998)
20. Cisek, P.: Integrated neural processes for defining potential actions and deciding between them: a computational model. J. Neurosci. 26(38), 9761–9770 (2006)
21. Erlhagen, W., Schöner, G.: Dynamic field theory of movement preparation. Psychol. Rev. 109(3), 545–572 (2002)
22. Sutton, R., Barto, A.: Reinforcement Learning. MIT Press, Cambridge (1998)
23. Dominey, P.F., Arbib, M.A.: A cortico-subcortical model for generation of spatially accurate sequential saccades. Cereb Cortex 2(2), 153–175 (1992)
24. Klein: Inhibition of return. Trends Cogn. Sci. 4(4), 138–147 (2000)
25. Herbort, O., Ognibene, D., Butz, M.V., Baldassarre, G.: Learning to select targets within targets in reaching tasks. In: IEEE 6th Intern. Conf. Development Learning, July 2007, pp. 7–12 (2007)

Neural Coding in the Dorsal Visual Stream

Eris Chinellato and Angel P. del Pobil

Robotic Intelligence Lab, Jaume I University, 12071 Castellón, Spain
{eris,pobil}@icc.uji.es

Abstract. The information flow along the dorsal visual stream of the primate brain is being thoroughly studied in neuroscience, and this research is being used in artificial intelligence applications. The knowledge regarding one of its most critical stages though, the posterior intraparietal area CIP, remains relatively undeveloped. This paper offers new computational descriptions of the tasks performed by CIP as a fundamental relay station between the visual cortex and the visuomotor areas downstream. Analytical expressions of the transfer functions realized by surface and axes orientation selective neurons (SOS and AOS) of CIP are derived and discussed.

1   Introduction

The visual system of human and other primates is composed of two main information streams [1]. The ventral stream is devoted to perceptual analysis of the visual input, such as in recognition, categorization, assessment tasks. The dorsal stream is instead concerned with providing the subject the ability of interacting with its environment in a fast, effective and reliable way. In our laboratory, we are developing a model involving various areas of the dorsal stream and their interaction with the ventral stream in vision-based grasping actions [2]. The goal of the project is to validate neuroscience findings through implementation on a real robotic setup. At the same time, we expect to endow our robotic system with advanced grasping capabilities typical of primates.

In this paper, we deal with an area of the primate cortex which plays a fundamental role in the sequence of transformations performed along the dorsal stream. The caudal intraparietal area, CIP, constitutes the central node of a spatial analysis process which endows the subject with the ability of interacting with her/his surrounding proximal environment. Neuroscience studies both on monkeys and humans have depicted a reasonably clear image of the sort of processing performed by CIP. At the computational level, though, this area has been rather neglected compared to its downstream neighbor AIP, more directly related to grasping actions. The goal of this paper is to fill this gap by providing a detailed analytical interpretation of CIP tasks which takes into account both the computational and the neurophysiological point of view. The focus is put on two kinds of neurons that have been described in the literature, that is, axis orientation selective and surface orientation selective neurons. We provide plausible transfer functions for these neurons, and discuss the consequences of the proposal for neuroscience research and computational implementations.

M. Asada et al. (Eds.): SAB 2008, LNAI 5040, pp. 230–239, 2008.

2   The Dorsal Visual Stream of the Primate Cortex

The basic visual areas of the primate brain, V1 and V2, perform the most fundamental processing on the input coming from the retina. They send their output,
consisting of simple but consistent visual features, such as edges, corners, textures, to visual area V3. This area has orientation-selective cells disposed in
columns of similar disparity tuning properties, and such layout is ideal for sim-
plifying the processing of extracting more complex properties, such as disparity
gradients. Area V3A is the first which belongs exclusively to the dorsal stream,
and it is in charge of computing gradients, thus refining the job of detecting the
pose of visual features began in V3.

Following the dorsal stream, the posterior parietal cortex (PPC) is largely
recognized as the main associative area of our brain dedicated to the coordination
between sensory information and motor response. The intraparietal sulcus (IPS)
separates the superior and inferior lobes of the PPC. Several areas within and
close to the IPS are dedicated to visuomotor transformations [3,4,5].

AIP (anterior intraparietal) is the most anterior zone of the IPS, and is largely
recognized as the area of the primate brain (humans included) dedicated to the
planning and monitoring of grasping actions. In AIP the visuomotor transformations necessary to map visual stimuli onto hand configurations suitable for
grasping target objects are performed [6].

The posterior lateral part of the IPS (usually called cIPS or CIP, caudal intraparietal) is dedicated to 3D shape and orientation processing, and is also active
during visually guided grasping. CIP has a central role in the dorsal stream, as
it connects visual area V3A with grasping area AIP. CIP neurons are strongly
selective for the orientation of visual stimuli, represented in a viewer-centered
way. Exhaustive studies [7,8] showed that selectivity toward disparity based orientation cues is predominant, but many neurons also respond (some exclusively)
to perspective based disparity cues. Indeed, it seems that cue integration for
obtaining better estimates of orientation is performed in this area. This sort of
processing performed by CIP neurons is the logical continuation of the simpler
orientation responsiveness found in V3 and V3A [5,9].

Two main neuronal populations have been distinguished in CIP: surface orientation selective and axis orientation selective neurons. Surface orientation selective (SOS) neurons [10] respond to a 2D shape in different orientations, but
extract the signal of 3D surface orientation from a 2D contour viewed in a linear
perspective: i.e., these neurons interpret the stimuli as the silhouette of a square
plate slanted in depth. Experiments with different proportions of the visual features showed that the responsiveness is maximum for "square" shapes, in which
the two major dimensions are similar, whilst elongation in either width or length
inhibits the response. The third, minor dimension seems not affect the response
up to a certain thickness threshold. Above this, a clear decrease in responsiveness
can be noted. We hypothesize that such threshold could represent the graspability of the feature, as it appears close to the size of the hand. The second class of
CIP neurons, axis orientation selective (AOS) neurons, represent the 3D orientation of the longitudinal axes of elongated objects [11]. Their response increases

with decreasing thickness (the two minor dimensions) and with increasing length (the major dimension), showing a complementarity with SOS neurons. It is not clear from the provided data if the reduced responsiveness with thicker objects is only due to the relative proportion between the object dimension or also by some comparison with the hand size. We rather promote this last possibility, for consistency with the role of CIP in providing AIP information regarding graspable features. Some AOS neurons are also shape selective, distinguishing for example between cylinders and square columns of similar length and thickness. This supports the view that disparity gradients are used in CIP to detect also the curvature of objects.

Overall, a population of mixed CIP neurons, including different types of SOS and AOS is able to provide full information about 3D proportion and orientation of a target shape. It is most likely this sort of information that is forwarded to AIP, where 3D orientation and shape can be coded as a unique combined feature and possible grip configurations generated.

3   Modeling the Caudal Intraparietal Sulcus

The research presented in this work is part of a detailed model of the information flow through the dorsal stream of the primate brain [2]. In the situation to face a simple object (possibly box-like, or cylindrical) lies on a table, slanted about a vertical axes. The goal is to generate, using only binocular visual information, possible grips on the object, emulating as much as possible the data flow connecting V3/V3A - CIP - AIP. In particular, we focus here on the tasks performed by the caudal intraparietal area, which can be schematized as in Fig. 1. The module on the left of the image, already developed [12], integrates stereoscopic and perspective visual information to estimate distance, orientation and size of simple 3D objects. Its fidelity to the findings described in the neuroscience literature has been proved through experimental validation on a simulated environment. The module has also been implemented on a robotic setup to work on real world situations (submitted research).

The following step (right side of Fig. 1) requires an action-based point of view, to assess the intermediate level object features with the purpose of evaluating their suitability for grasping. Orientation, relative and absolute size of the major axes of the object are thus compared and the response is synthesized in the SOS and AOS neurons output. The activation of these two kinds of neurons will depend on the general proportion of the object (two similar dimensions, a smaller one: prevailing SOS activation; two similar dimensions, a bigger one:

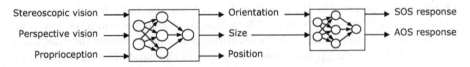

Fig. 1. Elaboration of visual data in the posterior intraparietal sulcus (CIP)

prevailing AOS activation). As a convention, from now on the three dimensions will be named a, b and c, where a and b are close in size, and c is the bigger (AOS) or smaller one (SOS).

3.1 Understanding and Interpreting the Available Data

Despite the recent efforts and encouraging advancements [13], the most important insights regarding the nature of 3D object representation by CIP neurons date back to the second half of the last decade [10,11]. The basic concepts were clear from the beginning, such as the distinction between SOS and AOS neurons and their responsiveness trend as a function of an object relative dimensions. The number and variety of different experiments is nevertheless reduced, and their characterization remains at most qualitative. Our goal is to analyze such experiments with modeling purposes, and possibly hazard new interpretation hypothesis deriving from a pragmatic point of view.

Fig. 2(a) reproduces the response of an AOS neuron to the view of a slanted elongated object as a function of object width [11]. The authors of the original study briefly comment on it suggesting that neuronal response and object width are inversely proportional. As an alternative, we propose a sigmoidal response function, which fits better with the observed data, as can be observed in Fig. 2(b), where two differently parameterized sigmoidals are superposed to the data of Fig. 2(a). Moreover, the sigmoidal is a transfer function very commonly found in brain mechanisms, especially when some threshold effects have to be taken into account. Indeed, in our case there is a very important threshold to consider, that is, the size of the grasping hand.

We suggest that the cut-off value for the sigmoid is the dimension of the open hand or even better the extension of a comfortable grip. For the monkey performing the experiment of Fig. 2(a) this value is reasonably around 12-15cm. Such hypothesis is supported by the fact that many CIP neurons are sensitive not only to relative object dimensions (and thus shape) but also to its absolute

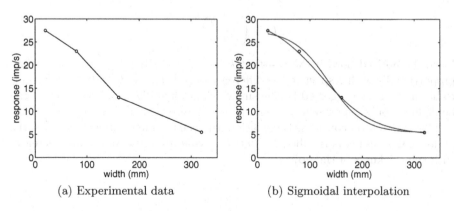

(a) Experimental data   (b) Sigmoidal interpolation

Fig. 2. Response of an AOS neuron as a function of object width. Experimental data (adapted from [11]) and interpolation with sigmoidal functions.

size. Thus, if the size of a potentially graspable object has to be represented in the brain, hand size is a very useful and convenient unit of measure to use. In the next section, we further develop and exploit this principle for defining the analytic expressions which model the function of SOS and AOS neurons.

Overall, CIP is responsive for all the following features of an object: relative size of main axes, absolute size, orientation in 3D, local curvature. Studies reported in the literature describe SOS and AOS neurons that are selective only for width and not for thickness, or only for relative and not for absolute size. Indeed, just a minority of CIP neurons are selective for all the features at the same time, but globally, at a population level, all relevant information regarding object shape in relation to potential grasping actions is processed by the posterior intraparietal area [5]. The transfer functions we propose take into account dimensional aspects at a neural population level, leaving aside for the moment orientation and curvature.

3.2   SOS Neurons Transfer Function

As a general principle, SOS neurons respond preferentially when two dimensions of the object are similar, while the third is sensibly smaller: $a > b \gg c$. Experiments performed varying the width and the thickness of the object gave the results reproduced in Fig. 3 [10]. These graphs and the comments of the authors, together with the principles previously introduced, allowed us to define a transfer function for modeling the behavior of a population of SOS neurons. The transfer function we propose takes into account three main aspects represented by three penalty, or inhibition terms. In a hypothetical ideal situation, all inhibition terms would be zero and activation maximal.

We call the first component of the transfer function I_s, *symmetry inhibition term*. It takes into account the difference between the two major dimensions of the object a and b: responsiveness is maximal, and inhibition minimal, for equal major axes. Asymmetrical situations are given higher penalties. The value of I_s is 0 when the major dimensions are equal, and increases with their difference:

$$I_s = \left(\frac{a-b}{a+b}\right)^{k_s} \tag{1}$$

It can be noticed how, if the constant k_s is bigger than 1, I_s grows more than proportionally with respect to the difference between a and b. Indeed, this seems to be the case, as suggested by the single-cell studies. The exact value of k_s can be deduced only experimentally, and is not necessarily stable across conditions.

The second term considers the relation between the minor dimension c and the major ones a and b. It is called I_f, *flatness inhibition term*, and it just increases with the thickness of the object:

$$I_f = \frac{c}{a+b} \tag{2}$$

The two previous terms are independent from the absolute size of the object. As discussed in the previous section, it is though likely that the hand size is

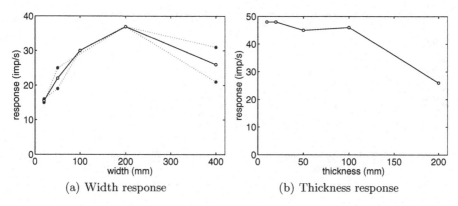

(a) Width response   (b) Thickness response

Fig. 3. Response of an SOS neuron as a function of object width (average of two major sizes response, dotted lines) and thickness. Experimental data adapted from [10].

playing an important role in determining the global responsiveness of CIP to a given target object. We thus introduced I_g, the *graspability inhibition term*. As anticipated, it is expressed as a sigmoidal function. I_g decreases when increasing the graspable dimension c, and its symmetry point is the limit of a comfortable hand opening, called H:

$$I_g = \sigma(c, H) = \frac{1}{1 + e^{-k_g(c-H)}} \tag{3}$$

Constant k_g affects in this case the non-linearity of the equation: the larger k_g, the steepest the slope of the sigmoid function, and thus the importance of hand size H.

The global response R_{SOS} of a population of SOS neurons is thus estimated detracting the inhibitory quantities, appropriately weighted, from the theoretical 100% activation:

$$R_{SOS} = 1 - w_s * I_s - w_f * I_f - w_g * I_g \tag{4}$$

The given equation is still undetermined, as the two parameters k_s and k_g and the three weights w have not been assigned any value yet. Starting with the symmetry term alone, we used least squares fitting to compute the value of k_s and w_s that best fits (1) to the data corresponding to Fig. 3(a). This gave us $k_s = 1.948$, and $w_s = 1.059$. It looks reasonable to simplify setting $k_s = 2$ and $w_s = 1$. In this way, I_s is the square of the fraction $(a - b)/(a + b)$ and its weight can be omitted. Similarly, we fitted (3) to the data of Fig. 3(b) and obtained 0.042 for the estimation of k_g and 0.458 for w_g. With a little approximation, $k_g = 0.04$ and $w_g = 0.5$. Finally, the only remaining coefficient w_f was estimated through least squares fitting of (2) to the data of Fig. 3(b) (taking into account the contribution of (3)). The final result was $w_f = 0.030$. After substituting all

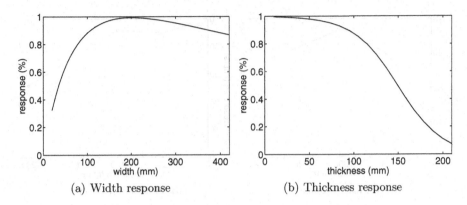

(a) Width response   (b) Thickness response

Fig. 4. Response of an SOS neuron as a function of object width and thickness. Simulated data obtained with (5).

these values in the corresponding formulas, the response of (4) remains more explicitely defined as:

$$R_{SOS} = 1 - \left(\frac{a-b}{a+b}\right)^2 - 0.03\frac{c}{a+b} - 0.5\frac{1}{1+e^{-0.04(c-H)}} \tag{5}$$

We calculated the global SOS response according to 5, as a function of object width and thickness. The results depicted in Fig. 4 show how the proposed model, properly parameterized, nicely fits to the experimental data ($H = 150mm$).

3.3   AOS Neurons Transfer Function

Axis orientation selective (AOS) neurons activate when one of the three dimensions of the object is quite larger than the other two, which are closer in size: $c \gg a > b$. Compared to SOS, less numerical results are though available in the literature, and the main source of information is Fig. 2(a) with the description of the corresponding experiments [11].

SOS and AOS neurons are intermixed in CIP, and it is thus plausible to assume that their response functions are similar. We thus decided to compose the hypothetical transfer function of AOS neurons starting from the same three inhibition terms introduced above.

AOS *symmetry inhibition term* is equal to 0 when the two similar dimensions of the object a and b are equal, and increases proportionally with their difference, exactly as in (1):

$$I_s = \left(\frac{a-b}{a+b}\right)^{k_s} \tag{6}$$

No experiments explicitly design to verify the effect of differences between the two minor dimensions have been carried out for AOS neurons. This effect is probably not very strong, but it can be reasonably assumed that a strong asymmetry would indeed affect the perception of the elongated object. Such reduced influence of the fraction $(a - b)/(a + b)$ on the total response can be obtained with a smaller constant k_s.

Similarly to (2), the next term compares the major and minor dimensions of the object. This time, we call it *length inhibition term*, as it decreases with increasing the major dimension c of the object:

$$I_l = \frac{a}{c} \tag{7}$$

The graspable dimension, a in this case, is again the numerator of the fraction, as was c in (2). In this case the numerator could also be $(a + b)/2$, but if a and b are quite similar this would likely be a pointless calculation.

The *graspability inhibition term* is again a sigmoidal function decreasing with the increasing of the minor dimension a, having as symmetry point the limit of a comfortable hand opening H.

$$I_g = \sigma(a, H) = \frac{1}{1 + e^{-k_g(a-H)}} \tag{8}$$

Again, the activation of a population of AOS neurons is estimated detracting the inhibition quantities from the theoretical 100% activation:

$$R_{AOS} = 1 - w_s * I_s - w_l * I_l - w_g * I_g \tag{9}$$

Due to the limited availability of data, a bigger extrapolation effort is needed in the AOS case to estimate appropriate values for parameters and coefficients. The case of the symmetry term is the most critical, as there is no published numerical data which can help in determining the values of k_s and w_s. This second coefficient can be set to the same value as for SOS neurons, $w_s = 1$, whilst k_s should be assigned a value such that the influence of the term on the overall response is reduced with respect to the SOS case. The easiest solution, but certainly no the only possible one, is to set $k_s = 1$, and leave only the fraction component. Response would thus linearly increase when reducing the difference between a and b.

Regarding graspability, there are no reasons to believe that parameter k_g and weight w_g should be much different from the SOS case. Least squares fitting of (8) to the data of Fig. 2(a) gives values included in $[0.02, 0.05]$ for k_g and in $[0.5, 0.8]$ for w_g, depending on the initial conditions. It seems thus reasonable, for symmetry and ecological reasons, to set $k_g = 0.04$ and $w_g = 0.5$, as in (5). Sakata and colleagues [11] state that: "discharge rate of the AOS neurons increased monotonically with increasing length of the stimulus". The authors did not provide further information on this issue, but this comment describes how to generate additional data which could help us in fitting our functions. We thus prepared a small dataset of 6 points in which response linearly increases with c. We used the newly generated dataset to fit (7) and thus set the value of w_l. We obtained values between 0.2 and 1 using different graspable sizes of a. There is no reason why the value of w_l should not changed dynamically, but for the

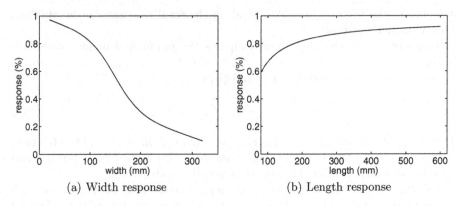

(a) Width response   (b) Length response

Fig. 5. Response of an AOS neurons as a function of object width and length. Simulated data obtained with (10).

moment we chose an intermediate value of $w_l = 0.374$, obtained for $a = 80$. The overall formula for AOS response is thus defined as:

$$R_{AOS} = 1 - \frac{a-b}{a+b} - 0.37\frac{a}{c} - 0.5\frac{1}{1 + e^{-0.04(a-H)}} \quad (10)$$

The behavior of (10) with changing thickness and length of the object is shown in Fig. 5 ($H = 150mm$). Fig. 5(a) tries to reproduce the effect depicted in Fig. 2(a), whilst Fig. 5(b) shows how the response grows when increasing c. Again, the effects described in the neurosciense literature are well reproduced.

4   Discussion and Future Developments

The model we put forth offers some solutions to the problem of identifying the transfer functions of the different areas of the dorsal stream, but opens at least as many questions.

More experiments are needed to validate the proposal. The actual importance of hand size on SOS and AOS activation should be explicitly analyzed, through experimental protocols designed to distinguish the effect of relative and absolute size of features. For example, no experiments are described in the CIP literature regarding non graspable (or strangely shaped) objects, and these are definitely required at this point. Similarly, there is the need to disambiguate the influence of shape and size on neuronal response. This can be done changing gradually the proportion and size of objects, and analyzing the response as a function of only one driving variable at a time. It is very likely that the equations we propose will need to be updated and suited to the new findings, but they constitute a helpful tool for orienting the future studies on the subject.

The second development regards the transformation from object description to hand configuration. This is performed in AIP, and various studies can help in defining what sort of operations are done on the data coming from CIP in order to generate suitable hand configurations. Taking as a reference the responsiveness patterns of AIP neurons described by Murata and colleagues [6], a preliminary study we have performed suggests that visuomotor AIP neurons could elaborate a simple linear combination of SOS and AOS activation. Moreover, although hand shape is strongly dependent on hand size, it is not at all unlikely that flexion of proximal phalanges is more associated to SOS activation (flat objects, opposing grips), whilst AOS activation would affect more the distal phalanges (elongated objects, involving grips).

References

1. Goodale, M.A, Milner, A.D.: Sight Unseen. Oxford University Press, Oxford (2004)
2. Chinellato, E., Demiris, Y., del Pobil, A.P.: Studying the human visual cortex for achieving action-perception coordination with robots. In: IASTED ASC Intl. Conf. on Artificial Intelligence and Soft Computing (2006)
3. Culham, J.C., Cavina-Pratesi, C., Singhal, A.: The role of parietal cortex in visuo-motor control: what have we learned from neuroimaging? Neuropsychologia 44(13), 2668–2684 (2006)
4. Castiello, U.: The neuroscience of grasping. Nat. Rev. Neurosci. 6(9), 726–736 (2005)
5. Sakata, H., Tsutsui, K.-I., Taira, M.: Toward an understanding of the neural processing for 3D shape perception. Neuropsychologia 43(2), 151–161 (2005)
6. Murata, A., Gallese, V., Luppino, G., Kaseda, M., Sakata, H.: Selectivity for the shape, size, and orientation of objects for grasping in neurons of monkey parietal area AIP. J. Neurophysiol. 83(5), 2580–2601 (2000)
7. Taira, M., Tsutsui, K.I., Jiang, M., Yara, K., Sakata, H.: Parietal neurons represent surface orientation from the gradient of binocular disparity. J. Neurophysiol. 83(5), 3140–3146 (2000)
8. Tsutsui, K., Jiang, M., Yara, K., Sakata, H., Taira, M.: Integration of perspective and disparity cues in surface-orientation-selective neurons of area CIP. J. Neurophysiol. 86(6), 2856–2867 (2001)
9. Shikata, E., Hamzei, F., Glauche, V., Knab, R., Dettmers, C., Weiller, C., Büchel, C.: Surface orientation discrimination activates caudal and anterior intraparietal sulcus in humans: an event-related fMRI study. J. Neurophysiol. 85(3), 1309–1314 (2001)
10. Shikata, E., Tanaka, Y., Nakamura, H., Taira, M., Sakata, H.: Selectivity of the parietal visual neurones in 3D orientation of surface of stereoscopic stimuli. Neuroreport 7(14), 2389–2394 (1996)
11. Sakata, H., Taira, M., Kusunoki, M., Murata, A., Tanaka, Y., Tsutsui, K.: Neural coding of 3D features of objects for hand action in the parietal cortex of the monkey. Philos. Trans. R. Soc. Lond. B: Biol. Sci. 353(1373), 1363–1373 (1998)
12. Chinellato, E., del Pobil, A.P.: Integration of stereoscopic and perspective cues for slant estimation in natural and artificial systems. In: Mira, J., Alvarez, J.R. (eds.) Nature Inspired Problem-Solving Methods in Knowledge Engineering, pp. 399–408. Springer, Heidelberg (2007)
13. Naganuma, T., Nose, I., Inoue, K., Takemoto, A., Katsuyama, N., Taira, M.: Information processing of geometrical features of a surface based on binocular disparity cues: an fMRI study. Neurosci. Res. 51(2), 147–155 (2005)

Modeling the Bat LSO Tonotopical Map Refinement during Development

Bertrand Fontaine and Herbert Peremans

Universiteit Antwerpen
13, Prinsstraat 2018 Antwerpen, Belgium
bertrand.fontaine@ua.ac.be

Abstract. The Lateral Superior Olive (LSO) codes for interaural intensity difference (IID), a cue used for sound localization. Between birth and maturation, the LSO undergoes plasticity driven by input neurons activity. During this developmental phase, a number of inputs are pruned out leading to a refinement of the frequency tuning. The goal of this paper is to show that, using a physiologically plausible network architecture and neuronal model, the activity dependent plasticity of the LSO can be modeled using Spike-Timing Dependent Plasticity (STDP). In particular, we show that the time properties of STDP coupled with the fact that the frequency axis in the LSO can be considered as a delay axis leads to the observed tonotopical map refinement. The response of both the individual neurons as well as population are shown to be in accordance with data taken from physiological analysis.

1   Introduction

It is believed that mammals use predominantly Interaural Intensity Differences (IID) to determine the azimuth of an acoustic target [1,2]. Due to the shadowing effects of the head and the pinna, the intensity perceived by the ear ipsilateral to the sound will be higher than the one on the contralateral side. This shadowing effect changes along with the sound source position allowing the IID to code for the latter.

In the Lateral Superior Olive (LSO), the neurons show a firing rate dependent on IID [3] by integrating excitatory inputs from the ipsilateral ear and inhibitory inputs from the contralateral one. The rate is maximum for IID values corresponding with the sound coming from the ipsilateral side and decreases as the sound becomes more centered to reach zero at a particular IID value. Neurophysiological experiments on bats [3] have shown that response of the neurons in the LSO to transient stimuli (short sounds) can be explained by a latency hypothesis: the relative arrival times of the inputs from the two ears differ among cells.

Before it reaches maturity, activity-driven synaptic plasticity shapes the LSO network [1,4]. This plasticity, which involves the strenghtening of certain input synapses and the pruning of others, is driven by the activity of the input neurons. The decrease in the number of input synapses leads to a refinement of the

M. Asada et al. (Eds.): SAB 2008, LNAI 5040, pp. 240–249, 2008.

frequency tuning as the inputs come from a more restricted area of the tonotopic map.

Else where we have shown [5] that individual LSO neurons can be tuned to particular IID's using STDP by presenting teacher stimuli from a fixed azimuth. This paper extends this concept by considering population learning with teacher stimuli taken from a gaussian distribution centered around azimuth zero. The STDP based learning mechanism described here could be used to implement developmental systems [12] that allow real robots to learn, in a bottom-up way, models of their own sensory systems.

We choose an FM-bat, the Eptesicus fuscus [2], as case-study because its received stimulus is mainly transient and thus the response can be explained by a pure latency hypothesis. However, most of the parameters of our model are taken from neurophysiological studies of gerbils and rats for the LSO and of guinea pigs for the peripheral system. The assumption, that those data can be used to model the bat auditory system is consistent with the numerous similarities encountered in the development of this part of the auditory system among mammals.

2   The Lateral Superior Olive

2.1   Mature LSO Response Mechanisms

Most of the LSO neurons receive their excitatory inputs from the ipsilateral ear through the cochlear nucleus (CN) and inhibitory inputs from the contralateral ear. The contralateral pathway passes first through the contralateral CN and then through the ipsilateral Medial Nucleus of the Trapezoid Body (MNTB) which finally projects to the LSO. The different connections can be seen in Fig.1.

Neurons in the LSO have firing behaviours which depend on IID. By convention we express the IID as the difference between ipsilateral intensity and contralateral intensity. As shown in [3], every neuron in a LSO population has a different IID-firing rate curve but most of them have the same sigmoid shape

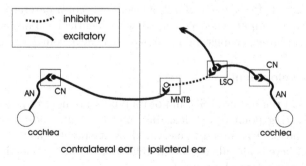

Fig. 1. Auditory pathway until the LSO. AN: Auditory Nerve, CN: Cochlear Nucleus, MNTB: Medial Nucleus of the Trapezoidal Body, LSO: Lateral Superior Olive.

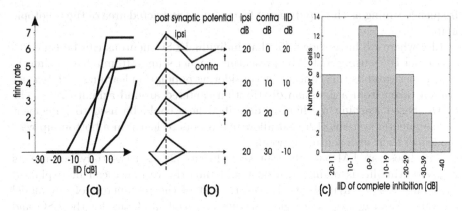

Fig. 2. (a): Firing response of different LSO neurons when the IID is varied. (b): Mechanisms explaining the firing rate curve of an LSO neuron with complete inhibition at 0dB IID (adapted from[3]). (c):Histogram of IID of complete inhibition of a population of 50 LSO neurons (adapted from[3]).

shifted along the abscissa as sketched in Fig.2a. A standard response feature allowing the characterisation of these IID-firing rate curves is the IID of Complete Inhibition (IID_{CI}). IID_{CI} denotes the IID value for which the neuron passes from a firing to a completely silent state.

The response of a population of LSO neurons is presented in Fig.2c as the histogram of the IID_{CI} of 50 neurons. From these results it can be concluded that the population is mostly sensitive to IIDs around zero ranging from 20dB (excitatory ear more intense) to -40dB (inhibitory ear more intense). This range of IID selectivities corresponds well with the range of IIDs that this species would normally encounter in the free field [3].

For the LSO cells considered in the present analysis, the IID-firing rate curve can be explained by the so-called latency hypothesis [3]. This hypothesis is based on two features of the input spike trains: i. the relative arrival times of the inputs from the two ears differ among LSO cells, and ii. changes in the intensity of the stimuli at the ear shift the latencies of the inputs [7]. For example, as shown in Fig.2b, if, for a particular LSO cell, the contralateral input arrives after the ipsilateral input when presented with identical stimuli at both ears, the contralateral intensity must be increased to achieve coincidence.

2.2   LSO Development

The properties of the mature LSO and of its inputs are not yet present at birth. We take as starting point of our learning model the first postnatal day (P1). At this time the LSO is already innervated by its inputs thanks to molecular tracers [4] and there is already a good matching of frequency selectivity between ipsilateral and contralateral inputs [1].

Between birth and LSO maturity, the ipsilateral and contralateral input pathways are refined by the pruning of the afferent axon terminals of CN and MNTB neurons within the LSO. The spread of the inputs over the frequency axis decreases leading to a tonotopical refinement. The morphological remodeling of these pathways seems to be activity dependent, because it is prevented by reducing neural activity of the inputs [4]. The amplitude of the post-synaptic potential due to an individual synapse changes during development [1] suggesting that the plasticity can be modeled as a change in the synaptic efficiencies or weights.

This evidence suggests that a learning algorithm using neural activity as driver could be feasible. Due to the immaturity of the peripheral system, the initial input available to the LSO is spontaneous activity (noise) until the auditory onset (P12 for the gerbil). After hearing onset, the inputs have mature temporal properties, thus reflecting the cues present in the acoustic stimuli.

3   Models

The peripheral auditory system model takes as input an acoustic signal and gives as output a stream of spikes in a collection of Auditory Nerve (AN) fibers. The basilar membrane, the inner hair cells and the spike generation in the AN are modeled as in [8]. This model, which is based on the guinea pig data, is applied to the bat peripheral auditory system by changing the parameters of the filterbank to values known to apply for bats .

The AN fibers project to neurons in the Cochlear Nucleus (CN). We model CN neurons using single-variable Leaky Integrate and Fire (LIF) neurons. We chose the membrane time constant so that the modeled neurons behave as coincidence detectors [9]. This peripheral system model yields a intensity-dependent latency as shown in Fig.3.

The LSO neurons are modeled as conductance based neurons, taking into account the synapses but not the ion channels. This simple model is precise enough as we use transient stimuli and thus need not model the steady-state

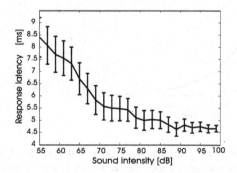

Fig. 3. Intensity-dependent mean output latency function (length error bars is equal to 2σ) of a modeled CN neuron

behaviour of those neurons. The membrane voltage V_m follows the differential
equation

$$C_m(t)\frac{dV_m(t)}{dt} = -\frac{V_m(t) - E_m}{R_m(t)} - \sum_{i=1}^{N_{ipsi}} g_i^{ipsi}(t)\left(V_m(t) - E_{rev}^{ipsi(i)}(t)\right)$$

$$- \sum_{i=1}^{N_{contra}} g_i^{contra}(t)\left(V_m(t) - E_{rev}^{contra(i)}(t)\right) \quad (1)$$

where the synaptic conductance $g_i(t)$ is given by

$$g_i(t) = w_i(t)\exp\left(-(t - t_i^{in} - d_i)/\tau_i\right)u(t - t_i^{in} - d_i) \quad (2)$$

$u(t)$ is the step function. C_m is the membrane capacitance, R_m is the membrane
resistance. The membrane time constant is $\tau_m = R_m C_m$. $E_m = 55mV$ is the
membrane resting potential. N_{ipsi} is the number of ipsilateral inputs. N_{contra} is
the number of contralateral inputs. E_{rev} is the reversal potential of a synapse.
w_i is the weight of the synapse i and d_i the corresponding synaptic delay. t_i^{in}
is the arrival time of an input neuron spike at the synapse i. τ_i is the synaptic
time constant, and is computed using the value of maximum duration of the
corresponding synapse (see Fig.4b).

Output LSO spikes are formal events characterized by a firing time t^{out}. t^{out} is
defined by the time when V_m crosses a threshold voltage ($V_{threshold} = -45mV$)
coming from lower values. Immediately after t^{out}, the potential is reset to a new
value $V_{reset} = -60mV$ and the neuron cannot fire for a refractory period of
$0.75ms$.

Some neuronal parameters also change over time independently of activity to
reach maturation after a certain period [4]. Of great importance for us, using time

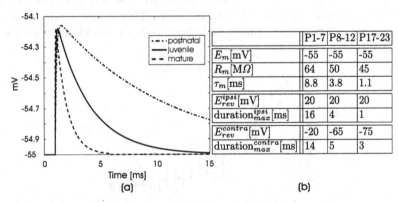

	P1-7	P8-12	P17-23
E_m[mV]	-55	-55	-55
R_m[MΩ]	64	50	45
τ_m[ms]	8.8	3.8	1.1
E_{rev}^{ipsi}[mV]	20	20	20
duration$_{max}^{ipsi}$[ms]	16	4	1
E_{rev}^{contra}[mV]	-20	-65	-75
duration$_{max}^{contra}$[ms]	14	5	3

(a)   (b)

Fig. 4. (a): LSO postsynaptic potentials at three different developmental stages: the
temporal precision increases over time. (b): LSO neurons parameters used in the simulation [4]. Duration$_{max}$ is the maximum duration of an PSP. The other parameters
are explained in the text.

as principal cue, is the temporal precision of the neurons and synapses. Indeed, both the time constant of the membrane and the time constant of the input synapses decrease during development. This leads to a more precisely defined spike time by increasing the slope of the rising edge of a PSP and by shortening its duration. In Fig.4a, we can see an excitatory PSP (EPSP) of an LSO neuron at different stages of development.

The contralateral connections also change properties during development. In the case of the rat, from P1 to P4, the contralateral side sends excitatory inputs to the LSO whereas at P8 this pathway is entirely inhibitory [4]. We model the switch from excitatory to inhibitory of the glycinergic synapses by changing over time their reverse potential, being above the membrane resting potential (-20mV) when excitatory and under the membrane resting potential (-70mV) when inhibitory [4].

The prelearning LSO network consists of 40 LSO neurons fully connected with 50 CN neurons and 50 MNTB neurons (Fig.5). The input neurons are uniformly distributed over the frequency axis between 40kHz and 60kHz. This range does not span the entire frequency range of interest to the bat in order to take into account the gross tonotopical convergence already present at birth.

The input synapses differ systematically in their delay. Indeed, the finite speed of the traveling wave on the BM causes delays in the displacement onset of the latter[6]. In our model, the response onset occurs after a delay of 2ms for the highest frequency increasing linearly to reach 8ms for the lowest frequency. A random term drawn from a normal distribution centered around zero with standard deviation of 0.5ms ($N(0, 0.5)$) is added to take into account synaptic delay differences due to, for example, differences between nerve fibers in conductance, length or diameter. The synaptic delay d_i (see Eq.2) of an individual synapse of an LSO neuron having center frequency at f [kHz] is then

$$d_i = -0.3(f - 40) + 8 + N(0, 0.5) \text{ [ms]} \tag{3}$$

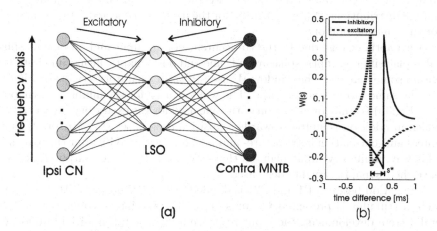

Fig. 5. (a): Prelearning architecture of the network. (b): Windows used for the STDP.

4   Learning

During development, the contralateral inputs become inhibitory but are still plastic. Applying STDP to training inhibitory networks is not straightforward. STDP follows a modified Hebb's rule, such that weights are increased if the activity between pre- and postsynaptic neurons is correlated and the weights are decreased if this activity is uncorrelated. Unlike excitatory synapses, the activity of an inhibitory presynaptic neuron is negatively correlated with the activity of its postsynaptic partner. We propose a modified learning window to take this into account.

The learning process is as follows: starting from an initial value, the synaptic weights $w_i(t)$ are updated at the end of every learning period T_L (20ms) chosen to contain the interval between stimulus onset and the last LSO output spike elicited by this stimulus. The learning rule we used for the modification of the synaptic weights $w_i^{\pm}$ of an LSO neuron consists of three parts [10]: (1) the arrival of an input spike from a synapse i changes w_i by a certain amount c_{in}, (2) when an output spike is triggered, all the weights change by a certain amount c_{out} and (3) time differences between all pairs of input and output spikes influence a third change of weights:

$$\Delta w_i^{\pm} = \eta^{\pm} \left[\sum_{t_i^{in}}{}' c_{in}^{\pm} + \sum_{t^{out}}{}' c_{out}^{\pm} + \sum_{t_i^{in},t^{out}}{}' W^{\pm}(t_i^{in} - t^{out}) \right]. \tag{4}$$

The $\pm$ sign denotes excitatory and inhibitory synapses. $\eta^{\pm}$ is the learning rate. η, which can be seen as the efficiency of the plasticity, decreases linearly with time to reach zero at maturation. The prime indicates that only firing times t_i^{in} and t^{out} in the time interval $[t, t + T_L]$ are to be taken into account.

To avoid unlimited growth we impose an upper and lower bound on the weights: $w_i^{\pm} \in [0, w_{max}]$. w_{max} is set to 1.5e-8 for the ipsilateral inputs and to 6e-8 for the contralateral ones. The weights w_i are initialized randomly from normal distributions: $N(1e^{-8}, 0.2e^{-8})$ for the ipsilateral one and $N(3e^{-8}, 0.8e^{-8})$ for the contralateral one.

A synaptic change due to the co-occurence of an input and an output spike takes place if presynaptic spike arrival time and postsynaptic firing time both fall within some window. The window W used to train excitatory synapses (Fig.6b) has the same formulation as in [5]. The window used for the inhibitory synapses W^{-}(Fig.6b) is a mirrored version of the one used for the excitatory synapses W^{+} which in addition is shifted towards a positive time offset s^{*-}. Mirroring guarantees anti-correlation between pre- and postsynaptic activity after the learning. This way, inhibitory input neurons that have fired before an output spike will have their synaptic weights decreased.

The period between P1 and P21 is divided in 10000 time steps. At every time step, a stimulus is presented to our system. To take into account the peripheral system development, the input response latency is taken initialy uniformly

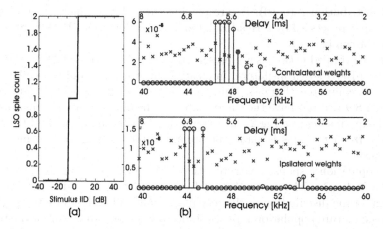

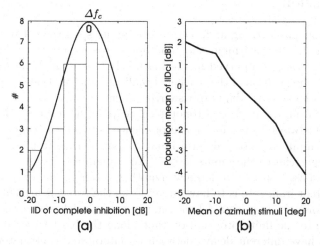

Fig. 6. (a): Firing rate when neuron is presented with different IID's. (b): Initial weights (x) and final weights (o) of both contralateral and ipsilateral inputs to one LSO neuron.

Fig. 7. (a): Histogram of the IID_{ci}'s of the resulting LSO population. The curve is the probability density function of the teacher stimuli azimuth. (b): IID_{CI}'s mean distribution of the population in function of the teaching stimuli distribution mean.

distributed between 2ms and 8ms for each ear. This interval decreases linearly to reach at P10 the randomness encountered in the mature peripheral system (Fig.3). After P10, the spike trains are elicited by simulated acoustic stimuli using the mature peripheral system model presented before. The acoustic stimuli are downward FM-sweeps from 80kHz to 20kHz of 2ms. The sound intensity of an individual stimulus is taken randomly between 50dB and 80dB (uniform distribution) and the azimuth of an individual sound stimulus is drawn from a normal distribution $N(0, 10)$. Before passing through the inner ear model, the sound is first processed by a Head Related Transfer Function (HRTF) measured on a big brown bat [2].

Next, the trained networks are presented with stimuli whose IID's range from -30dB to +15dB (with ipsilateral intensity kept constant) and the spike count response is computed (Fig.6a). Note that the resulting responses of individual modeled LSO neurons have the same shapes as the one measured in neurophysiological experiments (see Fig.2). In Fig.6b, the distribution of the input neurons along the frequency axis is shown before and after learning for one individual LSO neuron.

The population response characteristics, plotted as the histogram of IID_{CI}'s, is shown in Fig.7a. The mean IID_{CI} of the population is 0.8dB, not far from the mean of the distribution of the teaching stimuli. This shows that experience shapes the mature population reponse. Indeed, if we vary the mean of the teaching stimuli distribution, the population mean response changes as well (Fig.7b).

5   Conclusion

Whereas in our pre-learning architecture the inputs span the entire considered frequency band, after learning, the tonotopical connectivity is refined corresponding with what is observed in the mature LSO system. Moreover, experience, through the position of the teacher stimuli, shapes the resulting population response, i.e. the system adapts to its environment.

Differently from the measured tonotopical map [1], however, the learned one here does not always results in a correspondence between the center frequency of ipsilateral input and contralateral input as can be seen from the difference in center frequencies of the inputs (Fig.7). To explain this result, we note that our model is based on a pure latency hypothesis. Hence, the only way to get IID_{CI}'s different from zero is to have inputs from the two ears arriving at different times. Moreover, in our model the only systematic delay difference between inputs is due to the finite propagation time along the basilar membrane. Consequently, to have different delays between ipsilateral and contralateral inputs, an LSO neuron has to receive inputs from different frequency ranges from both sides.

This discrepancy could be remedied by replacing the pure latency hypothesis with a weak latency hypothesis [11], i.e. also modelling the difference in thresholds among input neurons, would result in neurons having non-zero IID_{CI} but still a correspondence between the frequency spread of ipsilateral input and contralateral input. It would therefore restrict the number of neurons in the population not having correspondence between frequency spread from both sides.

Acknowledgments. This work is sponsored by the EU as part of the CILIA project. The authors would like to thank A. Murat and C. Moss for having made available their Big Brown bat HRTF measurements.

References

1. Sanes, D.H., Rubel, E.W.: The ontogeny of inhibition and excitation in the Gerbil Lateral Superior Olive. J. Neurosci. 8(2), 682–700 (1988)
2. Aytekin, M., Grassi, E., Sahota, M., Moss, C.F.: The bat head-related transfer function reveals binaural cues for sound localization in azimuth and elevation. J. Acoust. Soc. Am. 116(6), 3594–3605 (2004)
3. Park, T.J., Grothe, B., Pollak, G.D., Schuller, G., Koch, U.: Neural Delays Shape Selectivity to Interaural Intensity Differences in the Lateral Superior Olive. J. Neurosci. 16(20), 6554–6566 (1996)
4. Kandler, K., Friauf, E.: Development of glycinergic and glutamatergic synaptic transmission in the auditory brainstem of perinatal rats. J. Neurosci. 15(10), 6890–6904 (1995)
5. Fontaine, B., Peremans, H.: Tuning bat LSO neurons to interaural intensity differences through spike-timing dependent plasticity. Biol. Cyber. 8(3), 332–338 (2007)
6. Ruggero, M.A., Rich, N.C.: Timing of spike initiation on cochlear afferents: dependence on site of innervation. J. Neurophysiol. 53(8), 379–403 (1987)
7. Krishna, B.S.: An unified mechanism for spontaneous-rate and first-spike timing in the auditory nerve. J. Comput. Neurosci. 13(2), 71–91 (2002)
8. Sumner, C.J., O'Mard, L.P., Lopez-Poveda, E.A., Meddis, R.: A non-linear filterbank model of the guinea-pig cochlear nerve: Rates responses. J. Acoust. Soc. Am. 113(6), 3264–3274 (2003)
9. Joris, P.X., Smith, P.H., Yin, T.C.: Enhancement of neural synchronization in the anteroventral cochlear nucleus. II. Responses in the tuning curve tail. J. Neurophysiol. 71(3), 1037–1051 (1994)
10. Kempter, R., Gerstner, W., van Hemmen, J.L.: Hebbian learning and spiking neurons. Phys. Rev. E 59(4), 4498–4514 (1999)
11. Park, T.J., Monsivais, P., Pollak, G.D.: Processing of interaural intensity differences in the LSO: role of interaural threshold differences. J. Neurophysi. 77(6), 2863–2878 (1997)
12. Olsson, L., Nehaniv, C.L., Polani, D.: Development via Information From Unknown Sensors and Actuators to Actions Grounded in Sensorimotor Perceptions. Connection Science; Special Issue On Developmental Robotics (Bank, D., Meeden, L., guest (eds.)), vol.18(2), pp.121-144 (2006)

A Reinforcement Learning Technique with an Adaptive Action Generator for a Multi-robot System

Toshiyuki Yasuda and Kazuhiro Ohkura

Graduate School of Engineering, Hiroshima University
Kagamiyama 1-4-1, Higashi-Hiroshima, Hiroshima 739-8527, Japan
{yasu,kohkura}@hiroshima-u.ac.jp
http://www.ohk.hiroshima-u.ac.jp

Abstract. We have developed a new reinforcement learning (RL) technique called Bayesian-discrimination-function-based reinforcement learning (BRL). BRL is unique, in that it does not have state and action spaces designed by a human designer, but adaptively segments them through the learning process. Compared to other standard RL algorithms, BRL has been proven to be more effective in handling problems encountered by multi-robot systems (MRS), which operate in a learning environment that is naturally dynamic. Furthermore, we have developed an extended form of BRL in order to improve the learning efficiency. Instead of generating a random action when a robot functioning within the framework of the standard BRL encounters an unknown situation, the extended BRL generates an action determined by linear interpolation among the rules that have high similarity to the current sensory input. In this study, we investigate the robustness of the extended BRL through further experiments. In both physical experiments and computer simulations, the extended BRL shows higher robustness and relearning ability against an environmental change as compared to the standard BRL.

Keywords: Multi-Robot System, Reinforcement Learning, Autonomous Specialization, Action Search.

1 Introduction

A robust instance-based reinforcement learning (RL) approach for controlling autonomous multi-robot systems (MRS) is introduced in this paper. Although RL has been proven to be an effective approach for behavior acquisition for an autonomous robot, it generates considerably sensitive results for the segmentation of the state and action spaces. This problem can yield severe results with increase in the complexity of the system. When segmentation is inappropriate, RL often fails. Even if RL obtains successful results, the achieved behavior might not be sufficiently robust. In conventional RL, human designers segment the state and action spaces by using implicit knowledge based on their personal experience, because there are no guidelines for segmenting the state and action spaces.

M. Asada et al. (Eds.): SAB 2008, LNAI 5040, pp. 250–259, 2008.

Two main approaches for solving the abovementioned problem and for learning in a continuous space have been discussed. One of the methods applies function-approximation techniques such as artificial neural networks to the Q-function. Sutton [1] used CMAC and Morimoto and Doya [2] used Gaussian softmax basis functions for function approximation. Lin represented the Q-function by using multi-layer neural networks called *Q-net* [3]. However, these techniques have the inherent difficulty that a human designer must properly design their neural networks before executing RL. The other method involves the adaptive segmentation of the continuous state space according to the robots' experiences. Asada *et al.* proposed a state clustering method based on the Mahalanobis distance [4]. Takahashi *et al.* used the nearest-neighbor method [5]. However, these methods generally require large learning costs for tasks such as the continuous update of data classifications every time new data arrives.

Our research group has proposed an instance-based RL method called the continuous space classifier generator (CSCG), which proves to be effective for behavior acquisition [6]. We have also developed a second instance-based RL method called Bayesian-discrimination-function-based reinforcement learning (BRL) [7]. Our preliminary experiments proved that BRL, by means of adaptive segmentation of state and action spaces, exhibits better performance as compared to CSCG.

BRL has an extended form that accelerates the learning speed [8]. Our focal point for the extension is the process of action searching. In a standard BRL, a robot performs a random action and stores an input-output pair as a new rule when it encounters a new situation. This random action sometimes produces one novel situation after another, which results in unstable behavior. In order to overcome this problem, we added a function that performs an action on the basis of acquired experiences. Our previous study demonstrated that MRS that employ the extended BRL learn behaviors faster as compared to those that employ the standard BRL. In this study, we conduct further experiments in which a robot in an MRS is initialized after successful learning, and thus we investigate the robustness and relearning ability of the extended BRL.

The remainder of this paper is organized as follows. The target problem is introduced in Section 2. Our design concept and the controller details are explained in Section 3. The results of our experiments are provided in Section 4. The conclusions are provided in Section 5.

2 Task: Cooperative Carrying Problem

Our target problem is a simple MRS composed of three autonomous robots, as shown in Fig. 1. This problem is called the *cooperative carrying problem* (CCP), and it involves requiring the MRS to carry a triangular board from the start position to the goal position. A robot is connected to the different corners of the load so that it can rotate freely. A potentiometer measures the angle between the load and the robot's direction θ. A robot can perceive the potentiometer measurements of the other robots as well as its own. All three robots have

Fig. 1. Cooperative carrying problem

the same specifications; each robot possesses two distance sensors d and three light sensors l. The larger the value of d / l, the shorter is the distance to an obstacle or a light source. The resolutions of the distance sensor, light sensor, and potentiometer are 350, 400, and 100, respectively. Each robot possesses two motors for rotating two omnidirectional wheels. The resolution of the motor signal is 16. A wheel provides a powered drive along the direction in which it points and a passive coasting along an orthogonal direction, simultaneously.

The difficulties involved in executing this task can be summarized as follows:

- The robots have to cooperate with each other to move around.
- They begin with no predefined behavior rule sets or roles.
- They have no explicit communication functions.
- They cannot perceive the other robots through the distance sensors because the sensors do not have sufficient range.
- Each robot can perceive the goal (the location of the light source) only when the light is within the range of its light sensors.
- Passive coasting of the omnidirectional wheels causes a dynamic and uncertain state transition.

3   APPROACH

3.1   BRL: RL in Continuous Learning Space

Our approach, called BRL, adaptively updates classifications on the basis of interval estimation, only when such an update is required. In BRL, the state space is covered by multivariate normal distributions, each of which represents a rule cluster, C_i. A set of production rules is defined by Bayesian discrimination. This method can assign an input, x, to the cluster, C_i, which has the largest posterior probability, $\max \Pr(C_i|x)$. Here, $\Pr(C_i|x)$ indicates the probability (calculated by Bayes' formula) that a cluster C_i holds the observed input x. Therefore, by using this technique, a robot can select a rule that is most similar to the current sensory input. In this RL, production rules are associated with clusters segmented by Bayes boundaries. Each rule contains a state vector $\boldsymbol{v}$, an action vector $\boldsymbol{a}$, a utility u, and parameters for calculating the posterior probability, i.e., a prior probability f, a covariance matrix $\boldsymbol{\Sigma}$, and a sample set $\boldsymbol{\Phi}$.

The learning procedure is as follows:

(1) A robot perceives the current sensory input $\boldsymbol{x}$.
(2) By means of Bayesian discrimination, the robot selects the most similar rule from a rule set. If a rule is selected, the robot executes the corresponding action $\boldsymbol{a}$; otherwise, it performs a random action.
(3) The robot transfers to the next state and receives a reward r.
(4) All rule utilities are updated according to r. Rules with utility below a certain threshold are removed.
(5) When the robot performs a random action, the robot produces a new rule by combining the current sensory input and the executed action. This executed new rule is memorized in the rule table.
(6) If the robot receives no penalty, an interval estimation technique updates the parameters of all the rules. Otherwise, the robot updates only the parameters of the selected rule.
(7) Go to (1).

Action Selection and Rule Production. In BRL, a rule in the rule set is selected to minimize g, i.e., the risk of misclassification of the current input. We obtain g on the basis of the posterior probability $\Pr(C_i|\boldsymbol{x})$. $\Pr(C_i|\boldsymbol{x})$ is calculated as an indicator of classification for each cluster by using Bayes' Theorem:

$$\Pr(C_i|\boldsymbol{x}) = \frac{\Pr(C_i)\Pr(\boldsymbol{x}|C_i)}{\Pr(\boldsymbol{x})}. \tag{1}$$

A rule cluster of i-th rule, C_i, is represented by a $\boldsymbol{v}_i$-centered Gaussian with co-variance $\boldsymbol{\Sigma}_i$. Therefore, the probability density function of the i-th rule's cluster is represented by

$$\Pr(\boldsymbol{x}|C_i) = \frac{1}{(2\pi)^{\frac{n_s}{2}}|\boldsymbol{\Sigma}_i|^{\frac{1}{2}}} \cdot \exp\left\{\frac{-1}{2}(\boldsymbol{x}-\boldsymbol{v}_i)^{\mathrm{T}}\boldsymbol{\Sigma}_i^{-1}(\boldsymbol{x}-\boldsymbol{v}_i)\right\}. \tag{2}$$

A robot requires g_i instead of calculating $\Pr(C_i|\boldsymbol{x})^1$, because no one can correctly estimate $\Pr(\boldsymbol{x})$ in Eq.(1). A robot must select a rule on the basis of only the numerator. The value of g_i is calculated as

$$\begin{aligned}
g_i &= -\log(f_i \cdot \Pr(\boldsymbol{x}|C_i)) \\
&= \frac{1}{2}(\boldsymbol{x}-\boldsymbol{v}_i)^{\mathrm{T}}\boldsymbol{\Sigma}_i^{-1}(\boldsymbol{x}-\boldsymbol{v}_i) - \log\left\{\frac{1}{(2\pi)^{\frac{n_s}{2}}|\boldsymbol{\Sigma}_i|^{\frac{1}{2}}}\right\} - \log f_i,
\end{aligned} \tag{3}$$

where f_i is synonymous with $\Pr(C_i)$.

---

[1] The higher the value of $\Pr(C_i|\boldsymbol{x})$, the lower is the value of g_i.

After calculating g for all the rules, the winner rl_w is selected as that with the minimal value of g_i. As mentioned in the learning procedure in Sec. 3.1, the action in rl_w is performed if g_w is lower than a threshold $g_{th} = -\log(f_0 \cdot P_{th})$, where f_0 and P_{th} are predefined constants. Otherwise, a random action is performed.

3.2   Extended BRL

Basic Concept. We have some RL approaches that provide learning in continuous action spaces. An actor-critic algorithm built with neural networks has a continuous learning space and modifies actions adaptively [9]. This algorithm modifies policies based on TD-error at every time step. Theoretically, the REINFORCE algorithm requires immediate rewards [10]. These approaches are not useful for tasks such as the navigation problem shown in Sec. 2, because the robot gets a reward only when it reaches the goal. However, BRL proves to be robust against a delayed reward.

In the standard BRL, a robot performs a random search in its action space; such random actions often resulted in instability in the global behavior of MRS in our preliminary experiments. Therefore, reducing the chance of random actions may accelerate behavior acquisition and provide a more robust behavior. Instead of performing a random action, BRL requires a function that determines actions on the basis of acquired knowledge.

BRL with an Adaptive Action Generator. In order to accelerate learning, in this study, we introduce an extended BRL by modifying the learning procedure, Step (2) in Sec. 3.1. In this extension, instead of a random action, the robot performs a knowledge-based action when it encounters a new environment. Therefore, we set a new threshold, $P'_{th}(< P_{th})$, and provide three cases for rule selection in Step (2), as follows:

- $g_w < g_{th}$: The robot selects the rule with g_w and executes its corresponding action a_w.
- $g_{th} \leq g_w < g'_{th}$: The robot executes an action with parameters determined based on rl_w and other rules with misclassification risks within this range, as follows:

$$a' = \sum_{l=1}^{n_r} (\frac{u_l}{\sum_{k=1}^{n_r} u_k} \cdot a_l) + N(0, \sigma), \tag{4}$$

where n_r denotes the number of referred rules, and $N(0, \sigma)$ is a zero-centered Gaussian noise with variance σ. This utility-weighted-average action is regarded as an interpolation of previously-acquired knowledge.
- $g'_{th} \leq g_w$: The robot generates a random action.

In this rule selection, the first and third cases are the same as the standard BRL.

4   Experiments

4.1   Settings

Figure 2 shows the general view of the experimental environments for the simulation and physical experiments. In the simulation runs, the field is a square surrounded by a wall. The real robots are situated in a pathway with length and width 3.6 m and 2.4 m, respectively. The task for the MRS is to move from the start position to the goal position (light source). All the robots get a reward when one of them reaches the goal ($l_0 > thr_{goal} \lor l_1 > thr_{goal} \lor l_2 > thr_{goal}$). A robot gets a punishment when it collides with a wall ($d_0^i > thr_d \lor d_1^i > thr_d$). We represent a unit of time as a *step*. A *step* is a sequence that allows the three robots to obtain their own input information, make decisions by themselves, and execute their actions independently. When the MRS reaches the goal, or when it cannot reach the goal within 200 steps in the simulations and 100 steps in the physical experiments, it is returned to the start position. This time span is called an *episode*.

The robot controller comprises a prediction mechanism and a behavior learning algorithm. The settings for these two mechanisms are as follows.

Prediction Mechanism (NN). In our previous study [7], we verified BRL to be a successful approach to CCP by introducing reformations such that the state space was constructed by using sensory information and predictions of the posture of the other robots in the subsequent time step in order to decrease the learning problem dynamics.

The prediction mechanism attached is a three-layered feed-forward neural network that performs back propagation. The hidden layer has eight nodes. The input of the i-th robot is a short history of sensory information, $I^i = \{ \cos\theta_{t-2}^i, \sin\theta_{t-2}^i, \cos\psi_{t-2}^i, \sin\psi_{t-2}^i, \cos\theta_{t-1}^i, \sin\theta_{t-1}^i, \cos\psi_{t-1}^i, \sin\psi_{t-1}^i, \cos\theta_t^i, \sin\theta_t^i, \cos\psi_t^i, \sin\psi_t^i\}$, where $\psi_t^i = (\theta_t^j + \theta_t^k)/2$ ($i \neq j \neq k$, j and k indicate the IDs of the neighboring robots). The output is a prediction of the posture of the other robots in the subsequent time step $O^i = \{\cos\psi_{t+1}^i, \sin\psi_{t+1}^i\}$. The behavior learning mechanism utilizes O^i as a part of sensory information input.

(a) Simulation

(b) Real robots

Fig. 2. Experimental environment

Behavior Learning Mechanism (BRL). The input of the i-th robot is $\boldsymbol{x}^i = \{ \cos\theta_t^i, \sin\theta_t^i, \cos\psi_{t+1}^i, \sin\psi_{t+1}^i, d_0^i, d_1^i, l_0^i, l_1^i, l_2^i \}$. The output is $\boldsymbol{a}^i = \{m_{rud}^i, m_{th}^i\}$, where m_{rud}^i and m_{th}^i are the motor commands for the rudder and the throttle respectively. The value of σ in Eq.(4) is 0.05. For the standard BRL, $P_{th} = 0.012$. For the extended BRL, $P_{th} = 0.012$ and $P'_{th} = 0.01$. The other parameters are the same as the values recommended in our journal [7].

We introduce a change in an environment by initializing one of the three robots. This may correspond to a situation in which a robot is replaced with a new one. Such changes occur when the MRS continuously reaches the goal for 100 consecutive episodes in the simulations and for 25 consecutive episodes in the physical experiments.

4.2   Result: Simulations

We have investigated the improved performance of the extended BRL by means of three-/four-/five-robot CCP simulations in which robots must learn cooperative behavior from scratch [8]. In these experiments, we observed that robots always achieve cooperative behavior by developing team play organized by a leader, a sub-leader, and a follower. This implies that acquiring cooperative behavior always involves *autonomous specialization*.

The experiments in this section are conducted to observe the robustness of BRLs against a change in an environment. The MRS is disturbed in such a manner that one of the three robots is initialized immediately after a globally stable behavior is observed. Then, we count the number of episodes required for the MRS to relearn a new, stable behavior.

Figure 3 shows the average and the deviations in the number of episodes for 10 independent runs. The difficulty in relearning is apparently different for each case. The most difficult cases are those in which the initialized robot is the leader of the team (Fig. 3(a)). If a leader robot is initialized, the robots require a large number of episodes to relearn a new, stable behavior; however, such cases show the largest difference among those employing BRLs. The extended

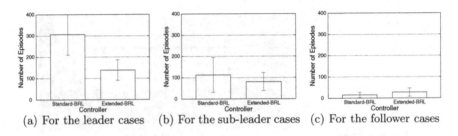

(a) For the leader cases   (b) For the sub-leader cases   (c) For the follower cases

Fig. 3. Numbers of episodes required to relearn a behavior after an environmental change

BRL generates 50% better results as compared to the standard BRL. Since the acquired cooperative behavior possesses slight instability and the robots must coordinate their behaviors, particularly in a case in which a follower is initialized, the extended BRL provides a slightly worse result. The improvement can be observed from the graphs for our proposed extensions. This implies that in terms of learning speed, the extended BRL outperforms the standard BRL.

4.3 Result: Physical Experiments

We conducted five independent experimental runs for each case employing the BRL. The standard BRL provided two successful results and the extended BRL provided four successful results from scratch [8].

Figures 4–6 illustrate the learning results after one of the robots is initialized by using the best results in [8] for the standard and extended BRL. Before an environmental change, Robot1, Robot2, and Robot3 are the leader, sub-leader, and follower, respectively, in the experiments for both the BRLs. These figures illustrate the number of steps and punishments in each episode. Comparing these results shows that the extended BRL requires fewer episodes to newly develop a globally stable behavior. Similar to the simulation results, the case where a leader robot is initialized demonstrates the most significant difference. In this case, the standard BRL could not achieve a globally stable behavior and hence resulted in failure. In the other cases, the extended BRL required smaller number of episodes to relearn cooperative behavior. Further, the extended BRL is more stable than the standard BRL because the MRS with the standard BRL gets several punishments.

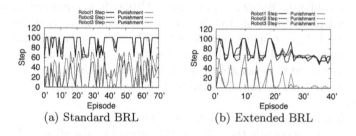

(a) Standard BRL	(b) Extended BRL

Fig. 4. Learning history after a leader is initialized

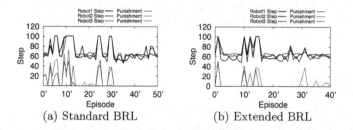

(a) Standard BRL	(b) Extended BRL

Fig. 5. Learning history after a sub-leader is initialized

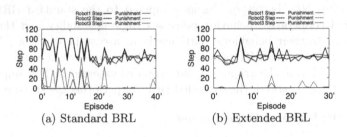

(a) Standard BRL   (b) Extended BRL

Fig. 6. Learning history after a follower is initialized

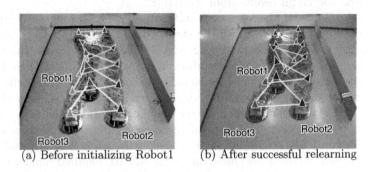

(a) Before initializing Robot1   (b) After successful relearning

Fig. 7. Acquired behavior: extended BRL

Figure 7 shows examples of the stable behaviors acquired by the extended BRL, before and after Robot1 is initialized. Although an environmental change occurred for Robot2 and Robot3, the robots achieved a globally stable behavior similar to the behavior before initialization. The robots trooped right, left and right, and then reached the goal. By observing the rule parameters, we found that Robot1 learned to be another type of a leader and the other robots utilized some rules stored before initialization and the newly generated rules based on our extension.

Although parameters that are more refined might provide better performance, parameter tuning is outside the scope of this study because BRL is designed for acquiring a reasonable behavior as quickly as possible, rather than the optimal behavior. In other words, the focal point of our MRS controller is not optimality but versatility. In fact, we obtain similar experimental results through experiments with an arm-type MRS, similar to that in [6], by using the same parameter settings.

5 Conclusions

We investigated an RL approach for the behavior acquisition of an autonomous MRS. Our proposed RL technique, BRL, has a mechanism for the autonomous segmentation of the continuous learning space, and it proves to be effective for an MRS through autonomous specialization. For improving the robustness of an MRS, we proposed an extension of BRL by adding a function to generate interpolated actions based on previously acquired rules. The results of the simulations and physical experiments demonstrated that the MRS with the extended BRL relearns behaviors faster than that with the standard BRL, after an environmental change.

In the future, we plan to analyze the learning process in detail. We also plan to increase the number of sensors and adopt other expensive sensors such as an omnidirectional camera that will allow a robot to incorporate a variety of information, and thereby acquire more sophisticated cooperative behavior in more complex environments.

References

1. Sutton, R.S.: Generalization in Reinforcement Learning: Successful Examples Using Sparse Coarse Coding. In: Advances in Neural Information Processing Systems, vol. 8, pp. 1038–1044. MIT Press, Cambridge (1996)
2. Morimoto, J., Doya, K.: Acquisition of Stand-Up Behavior by a Real Robot using Hierarchical Reinforcement Learning for Motion Learning: Learning "Stand Up" Trajectories. In: Intl. Conf. on Machine Learning, pp. 623–630 (2000)
3. Lin, L.J.: Scaling Up Reinforcement Learning for Robot Control. In: the 10th Intl Conf. on Machine Learning, pp. 182–189 (1993)
4. Asada, M., Noda, S., Hosoda, K.: Action-Based Sensor Space Categorization for Robot Learning. In: IEEE/RSJ Intl. Conf. on Intelligent Robots and Systems, pp. 1502–1509 (1996)
5. Takahashi, Y., Asada, M., Hosoda, K.: Reasonable Performance in Less Learning Time by Real Robot Based on Incremental State Space Segmentation. In: IEEE/RSJ Intl. Conf. on Intelligent Robots and Systems, pp. 1518–1524 (1996)
6. Svinin, M., Kojima, F., Katada, Y., Ueda, K.: Initial Experiments on Reinforcement Learning Control of Cooperative Manipulations. In: IEEE/RSJ Intl. Conf. on Intelligent Robots and Systems, pp. 416–422 (2000)
7. Yasuda, T., Ohkura, K.: Autonomous Role Assignment in Homogeneous Multi-Robot Systems. Journal of Robotics and Mechatronics 17(5), 596–604 (2005)
8. Yasuda, T., Ohkura, K.: Improving Search Efficiency in the Action Space of an Instance-Based Reinforcement Learning. In: e Costa, F.A., Rocha, L.M., Costa, E., Harvey, I., Colutinho, A. (eds.) ECAL 2007. LNCS (LNAI), vol. 4648, pp. 325–334. Springer, Heidelberg (2007)
9. Doya, K.: Reinforcement Learning in Continuous Time and Space. Neural Computation 12, 219–245 (2000)
10. Williams, R.J.: Simple Statistical Gradient-Following Algorithms for Connectionist Reinforcement Learning. Machine Learning 8, 229–256 (1992)

A Multi-cellular Developmental System in Continuous Space Using Cell Migration

Nicolas Bredeche

TAO / LRI ; Univ Paris-Sud, CNRS, INRIA Futurs
Université de Paris-Sud - LRI, Bat. 490 - F-91405 Orsay Cedex, France
bredeche@lri.fr
http://www.lri.fr/~bredeche

Abstract. This paper introduces a novel multi-cellular developmental system where cells are placed in a continuous space. Cells communicate by diffusing and perceiving substances in the environment and are able to migrate around following affinities with substance gradients. The optimization process is performed using Echo State neural networks on the problem of minimizing tile size variations in the context of a tiling problem. Experimental results show that problem complexity only impacts the number of substances used, rather than the number of cells, which implies some sort of scalability with regards to the size of the phenotype. Symmetry breaking and robustness are addressed by adding noise as an intrinsic property of the model. A (positive) side effect is that the resulting model produces very robust solutions with efficient self-healing behavior in the presence of perturbations never met before.

1   Introduction

Evolutionary Design deals with the optimization of structures such as objects, buildings or robot morphologies (to name a few). A key problem in this context is the search space: it is now widely accepted that indirect encoding approaches [16, 10, 2] (ie. representing a construction plan) may yield to better results in some cases than direct encoding (ie. representing the construction itself). However, these approaches failed to address scalability issues since the size of the construction plan still grows with the size of the construction itself. Recent works on developmental systems have taken a step further by optimizing the very developmental process rather than the construction plan. This new approach is often referred to as multi-cellular Artificial Ontogeny [4, 17] and addresses both scalibility and robustness issues.

A multi-cellular developmental system is defined as a dynamical system with a group of interacting cells. These cells communicate by emitting and receiving substances in the environment (e.g. concentration of chemicals) and may perform a given number of operations depending on substances concentration. The most basic operation is cell differentiation, where a cell appearance may change to a given state (e.g. color, size or shape of a cell). At each step of development, all cells are updated synchronously until a halting criterion is met (e.g. maximum

M. Asada et al. (Eds.): SAB 2008, LNAI 5040, pp. 260–269, 2008.

number of development steps, stability measure, etc.). When development stops, the state of all cells is considered as the resulting phenotype. The important point is that all cells share the same duplicated controller, which results in different output values depending on each cell current context.

Some few works have addressed evolutionary design problems using a multi-cellular approach, ranging from evolution of neural networks [11] to artificial creatures morphologies [4] and gene regulatory networks [3]. However, these setups make it difficult to study the internal dynamics of such systems and most recent works have focused on simpler problem. The basic setup for studying such developmental systems, proposed by [13], is to consider a matrix of pixels (ie. an image) where each pixel contains one cell. Each of these cells may emit and receive substance concentration to neighboring cells, usually in either a von neumann or moore neighborhood fashion. Evaluation of a developmental model is usually performed by optimizing the cell controller so that the resulting phenotype matches a given image pattern.

This paper proposes an extension of multi-cellular developmental systems in the context of a continuous environment. In this scope, development is considered both through time and space: cells communicate with one another by diffusing substances in the environment and are either attracted or repulsed along the gradient of the diffused substances. Each cell perceives substance concentration at its position in a totalistic fashion so that the search space is dependant only on the number of substances in the environment rather than the number of cells. As a result, a simple yet large phenotype may be build with very few substances. Another key issue of our model is that symmetry breaking and robustness towards noise are both addressed by featuring noise as an intrinsic property of the environment. Such a model makes possible to address problem where cell positionning is necessary such as group behaviors in swarm robotics, mobile sensor optimal positionning, light sources positionning in an architectural building, etc.

In section 2, a model for multi-cellular developmental system in the continuous space is described. Cell migration, symmetry breaking and controller issues are also addressed. Section 3 presents an implementation of the model and describes the benchmark problem used for evaluation. Section 4 gives the experimental settings as well as results regarding both optimization performance and self healing capabilities. Then, section 5 provides a comparaison between our model and multi-cellular developmental models from the litterature and establish some links with some application problems such as mobile sensor deployment and diffusion of a swarm of robots

2   Development in a Continuous Space

Our model for multi-cellular development addresses the problem of development of a group of moving cells in a continuous environment. In this setup, cells are placed in a continuous environment and are free to move around (ie. cell migration). Cell communication is performed through emitting and perceiving substance concentrations from the neighborhood. Cell substance emission is performed through radial diffusion. Substance concentration decreases depending on the distance to the emitting cell. As a result, a cell neighborhood is defined in a given radius and neighboring cells influences depend on distance to the target cell. Substance concentrations is thus perceived in a totalistic fashion, ie. the number of inputs for a cell depends on the number of substances only.

2.1   Cell Migration

Cell migration refers indirectly to a coordinate system. In this model, cell migration is performed depending on available substances in the environment, limited to the cell perception radius. A cell controller determines current *affinities* with substance concentrations in the environment. Affinities range between -1.0 (ie. repulsed) and $+1.0$ (ie. attracted), if affinity is close to 0.0, the cell maintains its current position. By considering each substance affinities with substance actual concentrations nearby the cell at hand, it is possible to compute a migration vector (direction and (bounded) force) for each cell. Given N substance concentrations perceived by a cell, the cell migration vector $\overrightarrow{\Delta^{c_0}}$ for one specific cell c_0 at time t is computed as follow:

$$\overrightarrow{\Delta^{c_0}} = f_{max}(\sum_{i=1}^{N_s}(aff_{c_0}^i * \sum_{k=1}^{N_c}(\overrightarrow{(x_{c_k}^{N_d} - x_{c_0}^{N_d})} * f_{\theta_{c_0}^{rec}(i)}(\theta_{c_k}^{emit}(i), d(x_{c_0}^{N_d}, x_{c_k}^{N_d})))))$$

Roughly, this means that the migration vector for a given cell is computed at each time step from the cell affinities with substances produced by neighboring cells within a predefined radius, given that perceived substance concentrations depend both on distance and neighboring cells production. Notations are as follow: c_0 is the target cell, Δ_{c_0} is the localization update for the current iteration, N_d is the number of space dimensions, N_s is the number of substances used for communication, N_c is the number of cells in the (possibly limited) neighborhood of c_0, $x_{c_k}^{N_d}$ is the coordinate of cell c_k in the N_d-dimensional space, and $d(x_{c_1}^{N_d}, x_{c_2}^{N_d})$ is the euclidian distance between cell 1 and cell 2. $aff_{c_0}^i$ state affinity of cell c_0 with substance i and f_{max} is a function to limit the norm of the resulting migration vector such as, for example, a simple maximum norm threshold. $\theta_{c_k}^{emit}(i)$ is the concentration value of substance i produced by cell c_k and $f_{\theta_{c_0}^{rec}(i)}$ is the corresponding concentration value *perceived* by cell c_0. In this setup, $\overrightarrow{\Delta}_{c_0}$ determines the instant speed to update the position of cell c_0[1]. Step 2 of figure 1 illustrates the cell migration operation in a 2D world.

A key feature of our model is that the coordinate system is generated by the developing system itself as there is no reference to any external coordinate system - for example: given three cells and three substances, these cells may form an equilateral triangle by each emitting one specific substance (through radial diffusion) and migrating so that perceived concentrations of the two other substances is equal. As a consequence, this triangle of cells result in a non-ambiguous two dimensional space where each point in the 2D space is defined by a unique triplet of substance concentrations.

---

[1] However it should be noted that $\overrightarrow{\Delta}_{c_0}$ could be used to determine acceleration, for example.

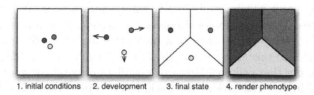

1. initial conditions   2. development   3. final state   4. render phenotype

Fig. 1. Example of development of an embryo to produce a voronoi diagram. Starting with an initial condition where cells are placed in an environment (step 1), cells migrate by communicating with each other (step 2) so as to reach a stable state (step 3). At this point, the resulting embryo may be rendered as a phenotype which can be evaluated (step 4 - an example of voronoi diagram rendering).

However, two (or more) cells may be located at the exact same position. In this case, cells potentially receive exactly the same inputs and may produce exactly the same outputs thereafter. This is what happens if all cells start at the same initial position: at each step, all cells behave in a mimetic fashion. In order to avoid this, it is necessary to introduce a random perturbation during cell migration, ie. each cell position is updated at each step according to its migration affinities and migration noise.

Because substance concentrations are perceived in a totalistic fashion, a cell may have N_s inputs (perceived substance concentrations) and $2 * N_s$ outputs (diffused substance concentrations + substance affinities for migration), with N_s the number of substances in the environment. If cell differentiation is enabled (which is not the case in the scope of this paper), additional outputs may be required (e.g. cell color/size/shape state).

3   Optimization of Tiling Problems

The objective function is defined as follow: given N_c cells in an environement, the goal is to position the cell in the environment such as each cell determines a voronoi site. The phenotype is then evaluated as the voronoi diagram built from these voronoi sites and the goal is to minimize the standard deviation σ of the average of the resulting voronoi regions. More formaly, the objective function is the following:

$$fitness(x) = \sqrt{\frac{1}{N_c} * \sum_i^{N_c} (surface_i - \overline{surface})^2}$$

With the following notations: x is the genotype to be optimized; N_c is the number of cells (ie. voronoi sites); $surface_i$ the surface covered by the ith voronoi region and $\overline{surface}$ the average surface of all voronoi regions. This benchmark problem is referred to as the "tiling problem" since the goal is to tile space with possibly repeating patterns, or at least with patterns of the same size (if not of the same shape). Figure 1 shows an example of development from the initial starting point which results in a voronoi diagram where cell positions act as voronoi sites. Moreover, the environment is considered as a non-bounded 2D world (ie. a toroidal world) in order to avoid interfering with cell migrations. This makes it possible to take into account the fact that the very coordinate system of our model implies that cell positions are dependant only on one another (i.e. not related to the image translation/orientation as it is perceived after rendering).

Evaluation is performed as follow: development starts from an initial state where N_c cells are placed in the center of the image. As stated in the previous section, cells move away from one another because of migration noise, then each cell is able to behave in a different fashion as it is experiencing a different environmental context. Development stops when a termination criterion is met. In this particular setup, development stops after a maximum number of iterations, as proposed in other works [9,13,14]. Then, the resulting phenotype is evaluated by rendering the corresponding Voronoi diagram and computing the aforementionned fitness value. Voronoi sites are located in a continuous environment and Voronoi regions are rendered (and evaluated) in a discrete environment. Both migration and rendering assumes that the environment is a toroidal world.

In practical, voronoi regions are rendered in a $32x32$ or $128x128$ RGB image (depending on the number of cells) using cell coordinates in the environment. At each iteration it, cell position update (δ) results from the combination of random noise perturbation and cell migration vector as follows:

$$\overrightarrow{\delta}_{it}^{co} = \frac{max_{it} - it}{max_{it}} * \overrightarrow{perturbation} + \overrightarrow{\Delta_{it}^{co}}$$

Cell migration maximum speed ($\overrightarrow{\Delta^{co}}$) is bounded and scaled so that a cell may not move faster than one pixel per iteration. Cell random noise perturbation ($\overrightarrow{perturbation}$) starts with a maximum bounded at 1.0 pixel-equivalent size and decreases linearly over time so as to reach zero influence when development stops (max_{it} is the number of development iterations).

4   Experiments

In order to evaluate our model, a set of experiments with a different number of cells have been conducted. The number of cells is either 3 or 16 cells (corresponding to tiling the environment with 3 or 16 voronoi regions) with only one substance for communication. The perception radius for each cell is set to a little more than twice (= 2.15) the radius of the optimal surface voronoi regions should cover. It also ensures that each cell is still able to communicate with its neighbors. Finally, this means that some cells may possibly not be able to directly communicate with one another depending on their respective position.

The experiments presented hereafter evaluate the relevance (or not) of an additional internal substance which is different to each cell and is set with a unique read-only value. This makes it possible to number each cell in a unique fashion, so that the controller has access to a cell identification number.

An echo state network with a reservoir size of 10 (with a connection density of 0.2 and damping factor set to 0.8) is used as controller[2]. The state-of-the-art CMA Evolution Strategy algorithm with restart feature [1] is used to optimize the set of neural network weights. The genotype size depends on the number of inputs and outputs as well as the size and connectivity of the hidden layer and is set here to 20 (ESN with or without cell identification - while number of inputs varied, only weights from the reservoir to the output nodes are considered for optimization). In preliminary experiments, classic multi-layered perceptron (MLP) was also considered (results not shown here). MLP used 6 hidden nodes so that the number of dimensions of the optimization problem remains the same as with ESN. Results were shown to be comparable in the end of the evolution process with that obtained with ESN (which implies that temporal capability is not crucial), but convergence tends to be slower than with ESN.

The number of iterations for the development process of one genotype is set to 256 (note that less than 32 iterations is enough for one cell to travel accross the whole environment) and noise perturbation is set as stated before. In order to compute an accurate performance, evaluation is averaged from 2 tries so as to smooth the effect of random perturbation, which is particularly useful in the first steps of optimization where genotypes either fail to feature robustness wrt. developmental perturbations or suffer from the discretization process during rendering[3]. All experiments in the next section are limited to 1000 evaluations (approx. 60 generations). CMA-ES automatically selects population size (initial population size is 12) and default restart parameters are used.

4.1 Result, Robustness Towards Perturbations and Scaling Issues

All results are shown in Figure 2. All figures show results of the best individuals from 11 independant runs (median best, best of best, worst of best and 25% and 75% quartiles for each generation) - the y-axis features the fitness values and the x-axis features the number of evaluations (rather than the number of generations). Columns: either 3 or 16 cells problems ; Rows: without or with Cell Identifier. Random development figures are given as a naive baseline, and are always quickly outperformed. From these results, two main considerations emerge: (1) a cell unique identifier does not provide any advantage, which is no surprise in this context since the cell population may be homogeneous and there is no clear gain as to identify each celll explicitly ; (2) Problem complexity seems not to be related with the number of cells as the 3 cells problem even

---

[2] ESN can be defined as a discrete-time recurrent neural network based on the reservoir computing framework known to perform very well in the context of temporal data. The reader may refer to [12] for a full description of ESN and to [7] for application in control problem related to multi-cellular developmental systems.

[3] Fitness values for the same phenotype may vary slightly since Voronoi diagram rendering leads to slightly different phenotypes depending on cell positions because of the discretization process that results in a 32x32 or 128x128 image. Evaluation for the 16 cells problem is performed in 128x128 so as to provide comparable fitness figures with that of the 3 cells problem.

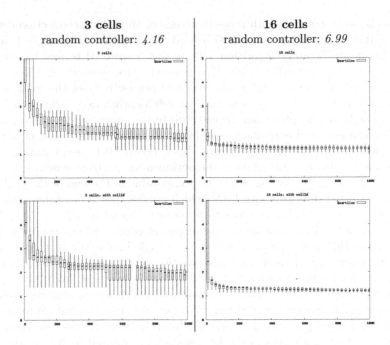

Fig. 2. Summary of results (see text for details)

show slightly slower convergence, which may be explained by the fact that each cell must cover a wider region (implying more freedom of migration).

In order to evaluate robustness towards noise, best individuals from all the runs were put to the test by applying two kind of perturbations: development is started with either (1) all cells positionned at the same location, as used during optimization (ie. intrinsic random perturbation may lead to a different outcome); or (2) cells are randomly placed in the environment. In both case, the resulting phenotype is evaluated and fitness values are compared to the predicted fitness values recorded during evolution. Development in case 1 shows robustness toward noise for all runs, which is no surprise since this setup is very close to what was used during optimization (which confirms that individuals are able to cope with the noisy fitness function). Case 2 leads to similar results - which is more difficult since robustness wrt. random starting positions was not considered during optimization. In this latter case, all individuals showed great robustness and converged towards phenotypes that displayed the awaited fitness, both in the 3 and 16 cells setup, whatever the starting initial cell positions. Figure 3 shows an example monitoring the development from an optimized genotype for the 16 cells problem. The instant fitness value at each of the development step is traced in both case and the development course for the two experimental conditions can be compared: in case 2, convergence is not as straight-forward as in case 1 at first, but the cell population displays an efficient self-healing behavior as it quickly recovers from the initial localization pertubation and end up with the

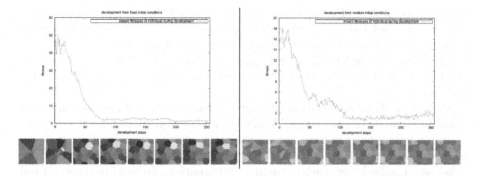

Fig. 3. Example of robustness towards development for the 16 cells problem (best individual). Left: development with centered voronoi sites (ie. all cells start from the same position). Right: development from random starting positions (all cells start from a different position). the y-axis gives the instant fitness value (translation/rotation invariant) of the phenotype and the x-axis gives the number of development steps so far for this individual. Voronoi region colors are randomly assigned and are used only for visualization.

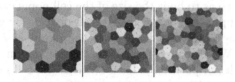

Fig. 4. Scaling up to 32, 64 and 96 cells (best controller from the 16 cells setup)

same fitness value as expected. In both case, the best fitness value is also reached before the end of development.

Lastly, the scaling problem was considered. Figure 4 shows the final stage of development of the best individual from the 16 cells problem but in the context of 32, 64 and then 96 cells. Indeed, optimized controllers produced behaviors very close to a simple repulse behavior (where each cell flies away from substance concentration) combined with a threshold limit so as to maintain a stable distance between cells, which is very simple yet efficient solution for this problem and scales up easily. It is interesting to note that resulting phenotype closely resemble a regular tiling, which is the optimal solution whenever it is possible (i.e. when all regions have either 3,4 or 6 borders, which may not be the case for the problems at hand).

5   Discussion and Conclusions

This paper introduces a new model for multi-cellular artificial embryogeny in a continuous space for optimization. The proposed model relies on cell migration based on substance affinities rather that explicit cell neighborhood and noise so as to break possible symmetries. Results showed that problem difficulty in our model is related to the number of substances rather than the number of cells, which is a key feature with regards to scalability since the size of the cell population is independant from problem complexity. It was also shown to produce a robust and, to some extent, scalable behavior at least for the problem at hand.

Other models have already addressed development of patterns as benchmark problem for artificial embryogeny. While the main feature of our model is that we consider development in a continuous environment with cell migration, it shares some similarities with other models. Firstly, Gordon and Bentley's model [9] also relies on a totalistic approach to compute information perceived by a specific cell as neighboring cell states are summed. However, this implies breaking symmetry by introducing some bias as initial starting condition. This differs from our model because symmetry breaking is automaticaly handled through the combination of additional random positional noise and cell relative positionning through selected affinities. The model presented here is able to automaticaly build from scratch a coordinate system. A direct advantage of such a system is that the number of substances is directly related to the problem complexity, and *not* to the number of cells (ie. scalability wrt. size). Secondly, Miller's model [13] already considered spatial development so as to limit the size of the cell population wrt. to the task at hand. In this setup, a cell would trigger cell division by growing a new cell on a nearby free site targeted in an explicit coordinate system (in this setup: North, South, East, West). As a consequence, the population of cells would grow spatially even if each cell position is fixed at birth (no cell migration involves). While it is possible to position a cell with a specific state at one precise location, a possibly large population may be needed. Thirdly, Miller [13], Federici [8] as well as our previous work in [6] addressed the issue of self-healing, ie. robustness of development toward noise. In [6], we showed that impressive results could be achieved by adding a strong penalization to individuals that could not reach a stable state[4] *before* the maximum number of development steps allowed. This feature provided very reliable controllers which were able to completely recover even from a 100% noise perturbation. In the current model, stability is also achieved but in a different fashion: positional noise perturbation is intrinsic to the environment and robustness is thus forced from the beginning and results in very robust phenotypes even if perturbation occurs.

The work presented here has also some connections with the problem of dispersion of a swarm of robots (or, more generally, particles). Swarm dispersion is a distributed coverage control problem defined by a clustering of the environment where each agent applies a local strategy based on information from its neighborhood. In this context, both homogeneous and heterogeneous dispersions have been addressed in simulation [5] and with a real-world robot swarm [15]. To some extent, our model can be related to these approaches and is indeed more general in the sense that it relies only on raw information and does not require

---

[4] In this previous work, a stable state is reached when all neural network controller internal activity does not change for a given number of iterations.

explicit identification of cells/particles. Of course some fundamental differences still exist since our model does not rely on *physical* cells that are subject to environmental dynamics when moving around and that noise is artificially added to the model. However, it is considered as a promising direction for future works.

References

1. Auger, A., Hansen, N.: A restart cma evolution strategy with increasing population size. In: Proceedings of the IEEE Congress on Evolutionary Computation (CEC 2005), pp. 1769–1776 (2005)
2. Bentley, P., Kumar, S.: Three ways to grow designs: A comparison of embryogenies for an evolutionary design problem. In: Banzhaf, W., et al. (eds.) GECCO 1999, pp. 35–43. Morgan Kaufmann, San Francisco (1999)
3. Bongard, J.: Evolving modular genetic regulatory networks. In: Fogel, D.B., et al. (eds.) CEC 2002, pp. 1872–1877. IEEE Press, Los Alamitos (2002)
4. Bongard, J., Pfeifer, R.: Evolving complete agents using artificial ontogeny. In: Morpho-functional Machines: The New Species (Designing Embodied Intelligence), pp. 237–258. Springer, Heidelberg (2003)
5. Cortes, J., Martinez, S., Karatas, T., Bullo, F.: Coverage control for mobile sensing networks. IEEE Transactions on Robotics and Automation 20(2), 243–255 (2004)
6. Devert, A., Bredeche, N., Schoenauer, M.: Robust multi-cellular developmental design. In: Thierens, D., et al. (eds.) GECCO 2007. ACM Press, New York (2007)
7. Devert, A., Bredeche, N., Schoenauer, M.: Unsupervised learning of echo state networks: A case study in artificial embryogeny. In: 8th International Conference on Artificial Evolution (Evolution Artificielle - EA 2007) (2007)
8. Federici, D., Ziemke, T.: Why are evolved developing organisms also fault-tolerant? In: SAB 2006, pp. 449–460 (2006)
9. Gordon, T.G.W., Bentley, P.J.: Bias and scalability in evolutionary development. In: GECCO 2005, pp. 83–90. ACM Press, New York (2005)
10. Gruau, F.: Genetic synthesis of modular neural networks. In: ICGA-1993, pp. 318–325. Morgan Kaufmann, San Francisco (1993)
11. Hotz, P.E.: Evolving morphologies of simulated 3d organisms based on differential gene expression. In: Husbands, P., Harvey, I. (eds.) ECAL 1997. MIT Press, Cambridge (1997)
12. Jaeger, H.: Tutorial on training recurrent neural networks. Technical report, GMD Report 159, Fraunhofer Institute AIS (2002)
13. Miller, J.F.: Evolving a self-repairing, self-regulating, french flag organism. In: GECCO, pp. 129–139 (2004)
14. Roggen, D., Federici, D.: Multi-cellular development: Is there scalability and robustness to gain? In: Yao, X., Burke, E.K., Lozano, J.A., Smith, J., Merelo-Guervós, J.J., Bullinaria, J.A., Rowe, J.E., Tiňo, P., Kabán, A., Schwefel, H.-P. (eds.) PPSN 2004. LNCS, vol. 3242, pp. 391–400. Springer, Heidelberg (2004)
15. Schwager, M., McLurkin, J., Rus, D.: Distributed coverage control with sensory feedback for networked robots. In: Proceedings of Robotics: Science and Systems, Philadelphia, USA (August 2006)
16. Sims, K.: Evolving virtual creatures. In: SIGGRAPH 1994, July 1994, pp. 15–22. ACM Press, New York (1994)
17. Stanley, K.O., Miikkulainen, R.: A taxonomy for artificial embryogeny. Artificial Life 9(2), 93–130 (2003)

Toward a Theory of Embodied Statistical Learning

Daniel Burfoot[1], Max Lungarella[2], and Yasuo Kuniyoshi[1]

[1] The University of Tokyo, 7-3-1 Hongo, Bunkyo-ku, Tokyo 113-8656, Japan
{burfoot,kuniyoshi}@isi.imi.i.u-tokyo.ac.jp
[2] Artificial Intelligence Laboratory, Andreasstrasse 15, 8050 Zurich, Switzerland
lunga@ifi.uzh.ch

Abstract. The purpose of this paper is to outline a new formulation of statistical learning that will be more useful and relevant to the field of robotics. The primary motivation for this new perspective is the mismatch between the form of data assumed by current statistical learning algorithms, and the form of data that is actually generated by robotic systems. Specifically, robotic systems generate a vast unlabeled data stream, while most current algorithms are designed to handle limited numbers of discrete, labeled, independent and identically distributed samples. We argue that there is only one meaningful unsupervised learning process that can be applied to a vast data stream: adaptive compression. The compression rate can be used to compare different techniques, and statistical models obtained through adaptive compression should also be useful for other tasks.

1 Introduction

One striking characteristic of human competence is that it requires many years of learning to develop. Learning can be regarded as a form of *statistical adaptation* in which the brain adjusts to data flowing into it from the senses. Recently, researchers in the field of statistical learning have made important progress in understanding the nature of learning and the conditions under which learning can occur. This understanding supports the definition of several powerful learning algorithms [1,2].

The field of embodied artificial intelligence is also deeply concerned with the issue of adaptation, and has recently made several important conceptual advances [3]. One such advance is the realization that in many cases good performance can be achieved without advanced information processing, by relying on techniques such as reactivity, self-organization, and exploitation of body dynamics [4,5]. Another achievement is the identification of a set of design principles to guide the construction of robotic systems [3].

Unfortunately, there is not much communication between these two disciplines. In particular, it is difficult for roboticists to apply the strong results of statistical learning theory to embodied agents research. This difficulty is caused by a mismatch between the form of data assumed by current statistical learning

M. Asada et al. (Eds.): SAB 2008, LNAI 5040, pp. 270–279, 2008.
© Springer-Verlag Berlin Heidelberg 2008

algorithms and the form of data available to embodied agents. The purpose of this paper is to argue for a new formulation of statistical learning that can be applied to the vast unlabeled data stream generated by robotic systems. Furthermore, we argue that given this type of input, there is only one meaningful learning process that can be applied: adaptive compression. An important advantage of the view of learning as compression is that it provides a rigorous and highly practical research methodology within which to proceed. We refer to the hybrid field resulting from a combination of ideas from statistical learning and embodied artificial intelligence as "Embodied Statistical Learning" (EStL).

2   Background

2.1   Statistical Learning: The Current Formulation

The goal of the field of statistical learning is to discover algorithms which build statistical models from data. This field has developed an impressive mathematical theory [2,6,7] and has demonstrated strong results on various applications, such as face detection, handwritten digit recognition, and machine translation. The basic problem statement of statistical learning, in its current form, is given in the first sentence of the first chapter of the great work by Vapnik [6]:

> In this book we consider the learning problem as a problem of finding a desired dependence using a *limited* number of observations.

Several important ideas are contained in this statement. First, a critical aspect of this type of learning is the limitations on the amount of available data. Second, the goal is to find dependencies - for example, finding a rule that can assign a label to an image (e.g. "face" or "no face"). Third, the data is assumed to be partitioned into a number of distinct "observations". Because of this assumption of partitionability, it is then natural to assume that the samples are independent and identically distributed (IID). A critical piece of the VC theory is a set of probabilistic bounds on the difference between the real and empirical performance of a model class, in terms of the complexity of the class and the number of observations. These bounds are obtained using the assumption of IID samples, and their purpose is to describe when it is possible to generalize from limited data [6]. Thus, the above assumptions are essential to the theory. Other formulations of statistical learning are mostly similar to Vapnik's; we refer to these collectively as CStL.

2.2   Embodied Artificial Intelligence

The subfield known as embodied artificial intelligence originated with the work of Brooks [8,9]. Writing in the early 1990s, Brooks was reacting to what he saw as an unhealthy overemphasis on the *physical symbol system hypothesis*, which was a major influence on AI at the time. In this view, the key role of intelligence was to use formal logic and symbolic manipulation to construct plans from logical propositions about the world, which would be delivered by an unspecified perceptual system. The typical strategy for finding a plan was to conduct a heuristic search in a large action space. Brooks made two important criticisms of this approach. First, he noted that perception was a major problem in itself, and it was naïve to assume that the "vision guys in white hats down the corridor" would be able to obtain the necessary world descriptions [9]. Second, he argued that search was not the right tool for intelligence. As a counter-point to the symbol system hypothesis, Brooks offered his own *physical grounding hypothesis*: "to build a system that is intelligent it is necessary to have its representations grounded in the physical world" [9]. This view motivated Brooks' research into embodied agents, i.e. real robots operating in the real world.

The ideas of Brooks were pursued vigorously by later researchers [3,4,5,10]. An important part of this work is the development of a set of design principles to guide the construction of embodied agents [3]. The principles were obtained through extensive experience with robotic systems and from detailed study of biological organisms. These ideas were primarily targeted toward the physical construction of robots, but they have important implications for the design of learning algorithms as well. In particular, the *Complete Agent Principle* instructs designers to build agents that are "autonomous, self-sufficient, embodied, and situated". Another important idea is the *Principle of Information Self-Structuring*, which states that the agent should take advantage of statistical regularities induced by body-environment interactions, and should actively attempt to seek out such regularities [5,10,11]. As we discuss below, it is difficult to reconcile these principles with the current formulation of statistical learning.

3   The Setting of the Embodied Learning Problem

3.1   Two Types of Learning

To motivate the following discussion, we postulate a rough separation of learning into two types: perceptual and behavioral. The former allows the agent to understand the world, while the latter guides the agent's choice of actions. A necessary component of behavioral learning is reinforcement. Agents are assumed to receive a reward signal from the environment that instructs them to behave in an adaptive way: actions that produce positive rewards are strengthened, while actions that produce negative rewards are weakened. Reinforcement learning is an active area of research [12], and the fundamental principles are well understood.

Our view is that reinforcement is sufficient to explain behavioral learning but not perceptual learning. Simply stated, the information from the reinforcement signal is not sufficient to determine the huge complexity of the brain, which has on the order of 10^{12} synapses. For example, it is difficult to believe that differential reinforcement can be used to tune the synaptic weights in the lower levels of the visual cortex.

One can imagine constructing a Complete Agent and equipping it with two learning mechanisms: one for behavioral learning, and one for perceptual learning. For these two learning tasks, one might reasonably choose a reinforcement

Table 1. Summary of differences between problem formulation in Embodied Statistical Learning (EStL) and the current formulation of statistical learning (CStL)

Aspect of problem	EStL	CStL	Section
Form of data	stream	discrete samples	3.2
Supervisory signal	scarce	frequent	3.3
Volume of data	vast	limited	3.4
Key problem	prediction	recognition	3.6
Agent contribution	actively structures data	passively observes data	3.7

learning algorithm for the behavioral component, and a CStL algorithm for the perceptual component. The problem with this approach is that the current formulation of statistical learning is not well suited to the type of data encountered by robots. We now outline the setting of the problem of EStL, and contrast it to the CStL formulation. The constrast is summarized in Table 1.

3.2   Data Is a Stream

In the real world, for both robots and organisms, data arrives in the form of a stream. No obvious method exists for partitioning the stream into samples that can satisfy the assumptions of CStL. Any such partitioning of the stream will destroy the IID property. If one partitions a stream of visual images into frames, then each frame is strongly dependent on the previous frames.

The data is a stream, but it is not necessary to treat the stream simply as a sequence of bits. For example, if the stream is a sequence of images, it is reasonable to assume that the dimensions of the images are known. Or, if the stream is a sequence of video, audio, and sonar data, then it is reasonable to assume knowledge of which bits correspond to each sensory modality. In return for giving up the assumption of discrete sample data we get a "consolation prize": the temporal structure of the data stream, which can and should be exploited.

3.3   Labels and Reinforcement Events Are Scarce

The Complete Agent Principle instructs us to build agents that are "autonomous, self-sufficient, embodied, and situated" [3]. When applied to learning in robotic systems, this principle requires that the agents should learn in an unsupervised or self-supervised way. The amount of supervisory information provided to the agents, in the form of labeled training data and reinforcement signals, should be strictly limited.

In order to perform a pattern recognition task in CStL, one typically assumes a set of data points (e.g., images) and associated labels. Usually there are as many labels as data points. The labeling process may require a substantial amount of human labor, and is often error-prone. For example, in an image annotation task one must label each image with a set of words describing the objects and activities displayed in the image. However, different people might use different sets of words to describe an image. Also, this model of learning is fundamentally limited to the labels: a vision system trained using labeled data to recognize car models will not be able to determine if the car is parked or in use.

Reinforcement learning for perception faces similar limitations. Imagine we have built a humanoid robot and want to train it to fetch coffee in an office environment. The simplest method would be to give the robot a big reward when it arrives with the coffee. However, this simple scheme will require the robot to explore for years before it happens to retrieve the coffee and get the reward. By adding some complexity to the reward signal, we can potentially improve performance. Maybe we give the robot a small reward for obtaining the coffee, and a larger reward for delivering it. We could then go further, defining rewards for entering the hallway, pouring the coffee into the cup, adding sugar and cream, and so on. We could also define negative rewards for spilling the coffee or bumping into people. However, this violates the Complete Agent Principle, because it requires us to provide greater and greater levels of supervision to the robot, in the form of defining complex reward signals.

In the view of EStL we are developing, the agent is able to learn in an unsupervised and autonomous manner. However, this learning should be thought of as *preparation*, so that when supervisory information arrives, the agent can adapt to it as quickly as possible. When the user gives the coffee fetching robot a command, it should not have to perform a lengthy learning process; it should already know enough about the world to execute the command. Thus, if the robot is a humanoid, it must already have complex knowledge of grasping, walking, the visual stimuli corresponding to coffee cups, and so on.

3.4   The Stream Is Vast

Above we argued that data should be thought of as a stream. We now point out that it is an *enormous* stream. Robotic systems can obtain data from cameras, microphones, laser range finders, odometers, gyroscopes, and many other devices.

One of the great insights of statistical learning theory [6,7] is that when the data is limited, the model employed must be simple. There are various ways to calculate the model complexity, but the idea is the same. A basic rule of thumb is that the complexity of the model cannot exceed the information content of the data being modeled.

Consider the problem of classifying handwritten digits, which can be thought of as a "typical problem" of CStL. We wish to learn a rule that gives a good estimate of the probability distribution $p(Y|X)$ where Y is the label and X is the image. In this case the data being modeled is the set of labels, each of which has an information content of $\log_2 10 \approx 3.2$ bits. Assuming there are 10000 samples, the information content of the entire set of labels is about 32000 bits. Thus we cannot use models that have complexity of greater than 32000 bits. The essence of learning in this low-data regime is to find low complexity models that have high explanatory power. The success of the Support Vector Machines can be attributed to the fact that only a small number of parameters corresponding to the support vectors need to be specified, and the data is separated using the optimal separating hyperplane.

The data available to robots is enormously more vast than the 32000 bits available in the handwritten digit recognition problem. This multiple order of magnitude difference means that the learning problem must be thought about in an entirely new way. In particular, the vast amount of data available justifies the use of highly complex models. To sum up, the basic problem of CStL is: how can one generalize well from a limited amount of data? In contrast, the basic problem of EStL is: how can one efficiently exploit the huge amount of data to build a complex model of the complex world?

3.5 Fast Is Better Than Slow

This point is a combination of three distinct ideas, all of which emphasize speed in different ways. The first idea is that an agent must react rapidly to supervisory information when it appears, as was illustrated by the example of the coffee fetching robot. Similarly, when considering biological situations, strong negative reinforcement signals often relate to life-threatening events (e.g. a rabbit eating a poisonous plant), so the agent must adapt to those signals rapidly. To allow the behavioral learning component to adapt quickly, the perceptual learning component must provide it with meaningful abstractions.

The second reason for emphasizing speed is a consequence of our emphasis on learning vast data. If the computational architecture cannot process the data efficiently, it will choke on the vast size of the stream. In CStL, the learning bottleneck is the limited amount of data; in EStL it will likely be the computational complexity of learning.

The third idea is that the learning process should be primarily online. Learning should begin immediately once the sensor data stream starts flowing, and should proceed in a continuous fashion thereafter. Ideally, at each step the learning machine should update itself to reflect the new piece of data that has arrived.

3.6 Prediction Is Critical

The claim of this section is that the ability to predict is necessary and sufficient for intelligent behavior in the sense of optimizing future reward. To see that prediction is sufficient for reward-optimizing behavior, consider the following reinforcement learning strategy. We assume that the agent has experienced a large amount of data with sensor, motor, and reward components. It has built a model which interleaves these data types, allowing it to predict the future reward from the sensor stimulation and motor actions. Then it predicts the future reward given the current stimulus and a variety of action plans, and chooses the plan corresponding to the highest predicted reward.

To see that prediction is necessary, notice that real agents must be able to predict that an action like jumping off a cliff will be harmful without actually experimenting with it. This is fairly obvious, but standard reinforcement learning algorithms do not provide a mechanism to avoid bad states without actually visiting those states.

Note that other problems of classification, recognition, and so forth can be thought of as subproblems of prediction. If one can recognize the numbers and letters written on a business card, one can predict the identity of the person who will answer the phone when the number is dialled.

Because of its relationship to reward-optimization and classification, prediction can be thought of as a fundamental cognitive task. Thus, if a powerful and general purpose prediction method can be achieved, it will bring us much closer to the goal of intelligent machines. The critical role of prediction was recently articulated by Hawkins [13].

3.7   Agents Influence Their Own Learning

An agent learns by adapting to the vast stream of data entering its experience. However, some types of data are better suited to the learning process than others. For example, it is probably not very useful to observe visual data from a television tuned to a dead channel. On the other hand, data that has a certain kind of statistical regularity may be especially helpful in guiding the learning process. We say that this type of data is *structured.*

The Principle of Information Self-Structuring discussed in Section 2.2 states that an agent should actively attempt to induce structure in the data entering its experience. If this can be done successfully, the learning machine will be able to adapt more rapidly to the environment.

One mechanism of information self-structuring is the idea of intrinsic rewards for learning, referred to as the "autotelic drive" by Steels [11]. In order to implement an autotelic drive, the learning machine reports a signal that characterizes the degree of information structure in the incoming stimuli. It should also reflect the extent to which the information structure is useful in *improving* the performance of the learning machine: even if a certain pattern is highly structured, it may not be useful to observe it repeatedly after it has been thoroughly learned. This intrinsic reward is combined with external reward to guide behavior.

Another mechanism of information self-structuring is morphological computation, which is the idea that the body can act as a computational device to reduce the cognitive burden on the brain [14]. For example, it can be shown that a fly's eye is morphologically suited to the problem of detecting motion, because of the curvature of the lens [15]. The lens preprocesses the incoming sensory data in such a way as to simplify the computational problem of motion detection.

The Principle of Information Self-Structuring does not fit easily into the CStL paradigm. With the important exception of research in active learning [16], most CStL algorithms assume that the agent itself plays no part in the selection of data points used for training. This is required because if the robot uses information from the first $\frac{N}{2}$ data points to decide how to select the next $\frac{N}{2}$ samples, the IID assumption breaks down. Thus, the CStL theory does not provide insight regarding how to implement an autotelic drive, or how to design agent bodies to facilitate fast learning.

4   Synthesis: Adaptive Compression

Historically, there have been two main paradigms in statistical learning. The first is that of learning as induction, described by the Vapnik quote above. The second is the view of learning as compression, which has its roots in the idea of Minimum Description Length modeling [7]. There is a deep relationship between induction and compression [6,7].

In the above discussion, we described what we consider to be the proper inputs to the learning algorithm. We are now faced with the question of what the learning algorithm should actually *do* with the input. We claim that the only meaningful learning process that can be performed on the basis of a vast stream of unlabeled data is adaptive compression. Specifically, the learning algorithm should incrementally update a statistical model so as to reduce the bit rate per unit time required to represent the incoming data stream. This view connects directly to the idea of *redundancy reduction* which has been proposed as a fundamental principle explaining the function of the cortex [17].

Compared to the compression view, the induction view may seem more attractive for practical reasons. A program that can determine if a face is present in an image may seem more useful than a program that can compress images with faces in them. Thus, to further justify the goal of compression, we propose the following hypothesis: *statistical models obtained through the adaptive compression process will be useful for other applications*. There is a variety of evidence for this hypothesis. In recent work by Hinton *et al.*, it is demonstrated that building a generative model of handwritten digit images is useful in recognizing their labels [2]. Also, in the field of statistical natural language processing, improvements in the language model immediately yield improvements in applications such as speech recognition and machine translation [18]. In both cases, the model is obtained by finding a set of parameters that minimizes the log-likelihood of the original data (text or digit images); this process is basically equivalent to compression.

Note also the strong link between compression and prediction. If one can predict a data stream, then one can compress it. Thus while we use the compression rate for comparison purposes because it is a hard number, what we are really measuring is an algorithm's ability to *predict*.

It is important to note that this view is agnostic with regard to the choice of computational approach (e.g. dynamical systems or physical symbol systems) underlying the learning process. Given a computational model, it is easy to construct a compression algorithm on top of it. Thus the compression rate can be used by advocates of various perspectives on cognition to provide strong quantitative evidence for their views. One simply constructs a compression algorithm inspired by a particular idea about cognition and applies it to some large dataset (ideally, a benchmark dataset). If the new algorithm achieves a significant reduction in compressed data size, this provides strong quantitative evidence for the cognitive model. Thus, the view of learning as compression supports a rigorous methodology, and we consider this to be one of the major arguments in favor of it. This rationale for using compression rates to quantify progress in artificial intelligence research was recently articulated by Mahoney [19].

As an example of how the compression methodology can benefit embodied agents research, consider the work of Tani and Nolfi [20], which describes a method for hierarchical learning of different categories of sensory-motor data generated by a mobile robot. The authors show that the recurrent neural network modules self-organize such that each module becomes an expert at one type of data. This is an interesting result, but it is difficult to compare the method to other possible techniques. If the paper reported the compression rate achieved by the system on the sensory-motor data, it would be a much more powerful vindication of the method. In work using a similar experimental setup, but a very different modeling scheme, we showed how the compression rate can be used as a performance measure in cases where there is no obvious task to perform (i.e., the robot is simply exploring without a specific goal) [21]. The important result here is that the statistical model obtained in this process is useful for other tasks such as localization, thus supporting the hypothesis given above. However, our modeling method is fairly simplistic; it is likely that other methods (such as the one proposed by Tani and Nolfi) will provide better performance. The compression rate should allow us to select the best general method.

5   Conclusion

The goal of Embodied Statistical Learning is to fuse together the strong mathematical theory of statistical learning with the design principles of Embodied AI. This requires a new setting for the learning problem, because of the mismatch between the type of data available to embodied agents, and the type of data assumed by the current theories. In the new formulation, the input data is a vast unlabeled stream which is actively structured by the agent. We argued that the only meaningful learning process that can be applied to a vast unlabeled data stream is adaptive compression. Compression is equivalent to prediction, and allows for rigorous comparisons of results. We also hypothesize that the statistical model obtained through compression will be useful for other applications.

Compared to CStL, we consider EStL to be a more realistic setting of the learning problem. It may also be an *easier* setting, for the following reasons. First, the agent can exploit the temporal structure of the data stream. Second, the agent can perform information self-structuring. But the most important reason is that the amount of data available is enormous. The exploitation of this vast data resource may allow us to construct models of complexity comparable to the human brain.

References

1. Vapnik, V., Golowich, S.E., Smola, A.J.: Support vector method for function approximation, regression estimation and signal processing. In: Advances in Neural Information Processing Systems, pp. 281–287 (1996)

2. Hinton, G., Osindero, S., Teh, Y.: A fast learning algorithm for deep belief nets. Neural Computation 18, 1527–1554 (2006)
3. Pfeifer, R., Scheier, C.: Understanding Intelligence. MIT Press, Cambridge (1999)
4. Pfeifer, R., Lungarella, M., Iida, F.: Self-organization, embodiment, and biologically inspired robotics. Science 318, 1088–1093 (2007)
5. Kuniyoshi, Y., Yorozu, Y., Suzuki, S., Sangawa, S., Ohmura, Y., Terada, K., Nagakubo, A.: Emergence and development of embodied cognition: A constructivist approach using robots. In: von Hofsten, C., Rosander, K. (eds.) From Action to Cognition, Progress in Brain Research, vol. 164, pp. 435–455. Elsevier, Amsterdam (2007)
6. Vapnik, V.: The Nature of Statistical Learning Theory. Springer, Heidelberg (1998)
7. Rissanen, J.: Modeling by shortest data description. Automatica 14, 465–471 (1978)
8. Brooks, R.A.: Intelligence without representation. Artificial Intelligence 47, 139–160 (1991)
9. Brooks, R.A.: Elephants don't play chess. Robotics and Autonomous Systems 6, 3–15 (1990)
10. Lungarella, M., Sporns, O.: Information self-structuring: Key principle for learning and development. In: International Conference on Development and Learning (2005)
11. Steels, L.: The autotelic principle. In: Embodied Artificial Intelligence, pp. 231–242 (2003)
12. Sutton, R., Barto, A.: Reinforcement Learning: An Introduction. MIT Press, Cambridge (1997)
13. Hawkins, J., Blakeslee, S.: On Intelligence. Times Books, New York (2004)
14. Pfeifer, R., Iida, F., Gomez, G.: Morphological computation for adaptive behavior and cognition. International Congress Series 1291, 22–29 (2006)
15. Franceschini, N., Pichon, J., Blanes, C.: From insect vision to robot vision. Philosophical Transactions: Biological Sciences 337(1281), 283–294 (1992)
16. Tong, S., Koller, D.: Support vector machine active learning with applications to text classification. Journal of Machine Learning Research 2, 45–66 (2001)
17. Atick, J.: Could information theory provide an ecological theory of sensory processing? Network 3, 213–251 (1992)
18. Rosenfeld, R.: A maximum entropy approach to adaptive statistical language modeling. Computer, Speech and Language 10, 187–228 (1996)
19. Mahoney, M.: Rationale for a large text compression benchmark (2006), http://www.cs.fit.edu/~mmahoney/compression/rationale.html
20. Tani, J., Nolfi, S.: Learning to perceive the world as articulated: an approach for hierarchical learning in sensory-motor systems. Neural Networks 12(7-8), 1131–1141 (1999)
21. Burfoot, D., Kuniyoshi, Y.: Maximum entropy statistical modeling of sensor data from robotic systems. In: Proceedings of the Symposium on Language and Robots, pp. 107–116 (2007)

Closing the Sensory-Motor Loop on Dopamine Signalled Reinforcement Learning

Paul Chorley and Anil K. Seth

Dept of Informatics, University of Sussex, Brighton, BN1 9QJ, UK
p.chorley@sussex.ac.uk

Abstract. It has been shown recently that dopamine signalled modulation of spike timing-dependent synaptic plasticity (DA-STDP) can enable reinforcement learning of delayed stimulus-reward associations when both stimulus and reward are delivered at precisely timed intervals. Here, we test whether a similar model can support learning in an embodied context, in which timing of both sensory input and delivery of reward depend on the agent's behaviour. We show that effective reinforcement learning is indeed possible, but only when stimuli are gated so as to occur as near-synchronous patterns of neural activity and when neuroanatomical constraints are imposed which predispose agents to explorative behaviours. Extinction of learned responses in this model is subsequently shown to result from agent-environment interactions and not directly from any specific neural mechanism.

1   Introduction

There is much evidence to suggest that the activity of midbrain dopamine neurons is correlated with both prediction and receipt of reward under reinforcement learning paradigms, for example in the monkey *Macaca fascicularis* [10]. Dopamine levels are known to increase both in anticipation of expected reward and on the occurrence of novel reward, as well as to decrease when previously expected rewards are omitted. Further evidence that changes in dopamine levels affect synaptic efficacy (see Schultz [9] for a review) suggests that dopamine signalled modulation of synaptic spike-timing dependent plasticity (DA-STDP) might act as a neurobiological mechanism for reinforcement learning.

One recent computational model of DA-STDP [6] has demonstrated how such a mechanism might solve what is known as the distal reward problem, in which reinforcing reward signals arrive some time after corresponding stimuli, using a method similar to TD(λ) learning [14]. In this model, synapse-specific records of correlated neural activity (eligibility traces) enable associations between past events to persist over time, allowing rewarded responses to be appropriately reinforced when reward eventually arrives. An extension of this model, to include topological representation of cortical projections to the midbrain dopaminergic system [6], further demonstrated temporal shifts in reward predicting responses similar to those found *in vivo*.

M. Asada et al. (Eds.): SAB 2008, LNAI 5040, pp. 280–290, 2008.

In this paper we analyse a simple agent-based computational model incorporating DA-STDP on its capacity to implement reinforcement learning in an embodied context in which the precise timing of sensory stimuli and reward signals result directly from agent-environment interactions. We show that it is possible for the DA-STDP mechanism to support reinforcement learning in this context, but only when constraints are imposed on both neuroanatomy and on the encoding of sensory input. We also find that feedback from the agent's environment, occurring in response to changing patterns of rewarded behaviour, allow for the extinction of conditioned behaviour under conditions in which the DA-STDP mechanism alone does not.

2   Dopamine Signalling in a Simple Foraging Task

The agent-based model presented consists of a simulated low-inertia wheeled robot [11] controlled by a spiking neural network implementing the Izhikevich model of DA-STDP [5] [6], detailed in Appendix A. The circular agent (Figure 1a) has a radius of 2 units and is tasked with navigating a 200×200 unit toroidal environment in which two types of resource (green and blue) are present.

Five instances of each resource type (each having a radius of 4 units), are randomly distributed in the environment at the beginning of each trial, along with the agent itself, which faces in a random direction. Agents are able to move freely around the environment and can collect resources by making contact with them. Reward is signalled to the agent in the form of a burst of dopamine to its neural controller upon collection of just one of the two types of resource. Collection of a resource results in another instance of the same resource type immediately being created at some other random location in the environment.

Each agent is provided with 2 arrays (left and right) of 100 evenly distributed ray sensors for each type of resource (i.e. 4 arrays and a total of 400 ray sensors) and 2 motors (one per wheel). Each array extends over an angle of 2.5 radians with the arrays positioned such that they overlap slightly in front of the agent. Each sensor innervates a single input neuron in the network and signals the exact distance (up to a maximum of 100 units) at which the associated ray intercepts the nearest resources of the corresponding type. The agent's motors are each driven by the activity of a further 200 (output) neurons in the neural controller.

In each trial agents roam their environment continuously for a total of 6 simulated hours, taking just over 30 minutes of real time on a 2.66Ghz PC. In the first 2 hours reward is delivered on collection of green resources. In the second 2 hours reward is shifted to blue resources. In the final 2 hours neither resource is rewarded.

2.1   Neural Controller

The agent's neural controller consists of 800 excitatory and 200 inhibitory neurons (Figure 1b), initially having 10% random synaptic connectivity (i.e. 100 synapses per neuron projecting uniformly throughout the network). Synaptic

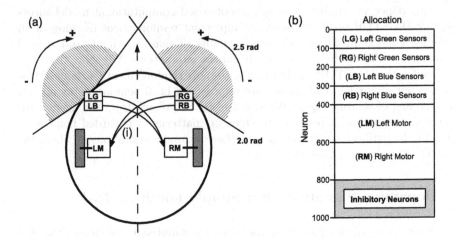

Fig. 1. Agent morphology (a) and neuron function (b). Sensors are connected such that objects placed directly in front of the agent appear as stimulus to higher indexed neurons in left hand clusters and to lower indexed neurons in right hand clusters. When imposed, anatomical constraints (i) predispose a simple taxis behaviour.

conductance delays are uniformly distributed in the (integer) range [1,10] and STDP parameters are equal for every synapse with a 50% negative asymmetry ($A^+ = 0.1$, $A^- = 0.15$, $\tau_{+/-} = 0.02s$). A small random excitatory input is provided to each neuron such that it spikes with an average frequency of 1Hz when not receiving external stimulation.

3   Experiment 1: Constraint on Sensory Input

3.1   Experimental Setup

Embodied simulations generate a continuous stream of sensory input, yet many cognitive functions such as visual processing appear to function too rapidly for information to be coded simply in mean spike firing rates [15]. This, among other things, suggests that sensory information may instead be encoded by temporal patterns of activity, distributed across a population of neurons [3] [8]. In our first experiment we therefore investigate two alternative ways of encoding this input stream, classical 'rate coding' and a novel method we refer to as 'stimulus gating', on their ability to support reinforcement learning. Our 'stimulus gating' mechanism yields volleys of sensory activity that are reminiscent of the 'spike waves' described by VanRullen and Thorpe [16]. In particular, this procedure implements a reset mechanism of the kind hypothesised by these authors to ensure separate processing of successive inputs; or in our case, separate processing of inputs from different stimuli. Examples of typical firing patterns induced by each method of stimulus composition are shown in Figure 2.

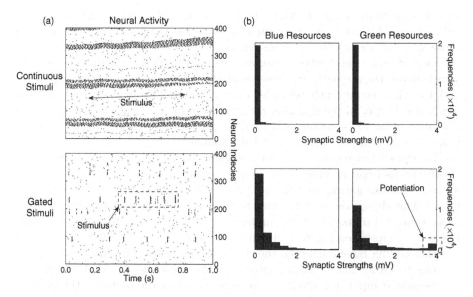

Fig. 2. (a) Sample neural activity and (b) synaptic strength distributions after 2 hours rewarding green resource collection, with and without stimulus gating. Histograms show the strength of synapses leading from input neurons associated with blue and green resources. Without gating, stimuli result in a continuous stream of neural activity and no significant long-term synaptic potentiation occurs. However, stimulus gating results in near-synchronous pulses of activity and a small proportion of those synapses associated with the green type (highlighted) are potentiated.

Rate coded input to the neural controller is implemented by the constant application of a small current I to each sensory neuron, proportional to the strength of the stimulus. This induces Poisson distributed spike trains in the sensory neurons with spike rates varying with the strength of stimulation.

Stimulus gating is applied by modulating the sensory input with a Poissonian distributed delta function at a frequency of 10Hz. This causes input neurons to fire near synchronously in short bursts rather than independently in continual streams. Each of the 4 clusters of input neurons are gated separately to ensure that stimuli of the same type (i.e. from the same bank of sensors) occur as synchronous patterns of activity, whilst those from different types occur asynchronously. The information content of the stimulus presented to the network is therefore available not only in the spike rates of individual neurons, but also in the spike patterns of different groups of neurons firing at similar times [12].

3.2   Results

Responses from ten randomly initialised agents produced similar results in which a marked difference is seen in both neuronal activity the corresponding pattern of synaptic potentiation, between the two coding regimes (Figure 2). When stimuli are rate coded, synapses leading from sensory to motor neurons do not undergo significant potentiation and, due to the asymmetry of the STDP window, instead fall to near minimal values very rapidly. With gated stimuli however many of those same sensory-motor synapses remain above their minimum values for the duration of each trial. Furthermore, with gated stimuli a clear difference can be seen between the strengths of those synapses projecting from input neurons associated with the currently rewarded object type and those associated with the unrewarded type. In the stimulus gating trials a higher frequency of strong potentiation is found for sensorimotor synapses associated with rewarded resource, than those associated with unrewarded resource.

The agent's behaviour is of course affected by these changes, resulting in a significant increase in mean collection frequency from $1.65/\text{min}$ ($\sigma^2 = 1.82$) without stimulus gating, to $2.62/\text{min}$ ($\sigma^2 = 1.31$) with ($p \ll 10^{-3}$ by a two-tail t-test). This change can be attributed to an increase in the average velocity of the agent from 8.6 units/sec to 9.2 units/sec, enabled by the greater potentiation found at synapses leading to motor neurons. However, little difference is found in either coding regime between the strengths of synapses leading from sensory neurons to either left or right motor neurons, which means that the motors receive similar input in response to any sensory stimulation, driving the agent in a straight line regardless of the location of sensed resources. Consequently there is no increase in the relative rate at which rewarded resources are collected.

4   Experiment 2: Neuroanatomical Constraints

In our second experiment the connectivity of the agent's neural network is constrained so as to predispose it toward learning generic approach behaviours. In this way, any changes in the relative potentiation of synapses occurring in response to reward should be reflected in a greater tendency for the agent to approach one or other type of resource, thus facilitating a more effective exploration of possibly rewarding behaviours. If DA-STDP is able to reinforce only those synapses associated with one particular behaviour (i.e. approaching the rewarded resource), then the agent will gain greater rewarding feedback from the environment and further reinforcement should result.

4.1   Experimental Setup

The agent's neuroanatomy is constrained so as to cause simple taxis behaviour [2] whenever synapses leading from neurons associated with a particular type of input stimuli are potentiated (Figure 1). Specifically, projections from left hand sensory neurons terminate on right hand motor neurons and *vice-versa*. Synaptic potentiation may therefore lead only to approach behaviours, although the designation of which resource types are to be approached is deferred for selection by the DA-STDP mechanism. The distribution of synapses connected to inhibitory neurons is left unchanged as these are thought to function mainly in modulating the overall excitation in the network [7].

4.2   Results

The results of this experiment show a marked difference from Experiment 1. Figure 3 shows a representative trial in which green resources are rewarded first. After approximately 30 minutes the agent begins to collect a significant number of those resources. Whilst there is a slight increase in the number of blue resources collected (from ≈1/min to ≈4/min), the increase in green is significantly greater (up to ≈13/min). When reward is switched to blue resources the agent immediately begins to alter its behaviour. Within 15 minutes collection frequencies are completely reversed, with more blue (≈13/min) than green (≈4/min) resources being collected. After reward is removed from either resource type, instead of extinguishing, the agent's blue-selective behaviour is maintained up to the end of the trial. This behavioural persistence is reflected in continued potentiation of the corresponding synapses (Figure 3, top right).

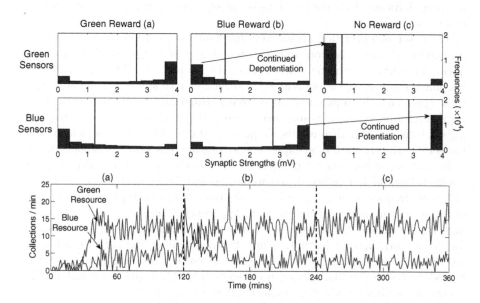

Fig. 3. Synaptic strength distributions (top, vertical bars denote mean synaptic strengths) and resource collection frequencies (bottom) with anatomical constraint whilst rewarding green resources (a), blue resources (b) and after reward has been removed from either type (c). Synaptic strengths were measured at the end of each time period, whilst collection frequencies were measured in non-overlapping 1 minute windows. Within the first hour agents learn to collect significantly more green (rewarded) resource. When reward is switched to blue resources (a) this preference is reversed. At the end of each phase a significant proportion of those synapses projecting from input neurons corresponding to the currently rewarded type are potentiated to near maximal values. When reward is finally removed from either type (b) the previous pattern of (de)potentiation is maintained and agents continue preferentially to collect blue resources.

The distribution of synaptic strengths at the end of each phase reveals that synapses leading from input neurons corresponding to the rewarded resource type undergo significantly more potentiation than in the previous experiment. Rather than showing a slight increase in the frequency of strongly potentiated synapses there is now a complete redistribution, such that the *majority* of synapses leading from such neurons are potentiated to near maximal values (Figure 3, top).

The results also show the agent switching almost entirely from collecting green to blue resources after reward is reallocated part way through the trial. As well as seeing an increase in the collection frequency of blue resources, we also find that the number of green resources collected is significantly reduced. This effect cannot simply be explained by the fact that collection of one instance of resource (usually) negates the simultaneous collection of another. If this were the case we would expect to see an equal yet smaller number of each resource type being collected after the switch. Instead, the observation of a clear preference for blue resources indicates that the previous behaviour has undergone almost complete extinction.

The observed extinction presumably functions by means of an increase in the proportion of motor neuron activity occurring in response to (now more frequent) interaction with the newly rewarded resource type. This increased proportion of (post-synaptic) activity is uncorrelated with any continued activity of those (pre-synaptic) input neurons receiving stimulation from the previously rewarded resource type and is sufficient to cause depotentiation of the corresponding synaptic pathway through the asymmetry of the STDP window, which tends to decrease the strength of synapses connecting neurons that fire independently.

In contrast, when reward is removed from both types of resource no alternative behaviour is reinforced and we see no significant decrease in either the collection frequency of the previously rewarded resource type or the strength of the corresponding synaptic pathway. Apparently, there is no inherent mechanism for the depotentiation of a functional synaptic pathway (i.e. no mechanism for extinction), beyond learning a new behaviour, once the corresponding behaviour has been established. If the agent does not begin to acquire a competing behaviour (as in the final phase) no change in post-synaptic neuronal activity occurs and extinction via depotentiation does not result.

5   Discussion

Although the mechanisms of DA-STDP have previously been shown to support reinforcement learning in abstract network models [6], it has thus far remained unclear whether it can do so in an embodied context in which the precise timing of sensory input and reward signals is contingent upon agent behaviour. We have shown here that embodied reinforcement learning via DA-STDP is possible, however in our model it was necessary both to modulate sensory input so as to induce near-synchronous patterns of neural activity, as well as to impose constraints on the agent's neuroanatomy which predispose generic foraging behaviours. The results we report have several additional implications which we discuss below.

Firstly, our experiments on the effects of stimulus encoding are clearly related to the growing body of work suggesting the importance of temporal patterning in neuronal signalling [3]. Of interest in this context is the old but recurring idea that perception might occur in discrete 'frames' [4]. A key component of this idea is that cortical rhythms modulate neuronal excitability so as to implement a 'shutter' separating successive perceptual frames. In this light our results show a functional benefit of such a mechanism in terms of facilitating embodied reinforcement learning. Our findings invite further work testing the functional benefits of sensory gating (perceptual framing) for embodied cognition.

Secondly, the need for constraints upon the agent's neuroanatomy in our model demonstrates a further extension to the DA-STDP model useful in embodied contexts. When exploration of the environment is not under the direct control of an experimenter, the agent must be predisposed to some form of exploratory behaviour in order to attain reward at an adequate frequency. Without such a predisposition there is little chance that random neural activity will consistently generate the behaviour necessary to bootstrap the reinforcement learning process. We have shown here that constraining the agent's neuroanatomy to predispose generic foraging behaviours can be sufficient to enable this process. Further work will address how sufficient exploratory behaviours can emerge autonomously, in embodied models not incorporating such task-specific constraints.

Finally, we observed that feedback from the agent's environment was critical for behavioural extinction, as mediated by synaptic depotentiation occurring in response to increased rates of uncorrelated neuronal activity. This result suggests that changes in sensory stimulation which result from engaging in a novel behaviour may have more of an active role in extinction than previously recognised [13]. In our experiments, because uncorrelated neuronal activity results from ongoing agent behaviour as well as from intrinsic network activity, an interaction between environmental feedback and synaptic depotentiation is implicated, along with removal of reinforcement, in extinction. This finding therefore invites new conditioning experiments with real organisms in which ongoing behaviour and environmental feedback are explicitly kept to a minimum after the removal of reinforcement, so that the effect of these factors upon extinction might be investigated. More generally however, our findings lend support to the embodied approach to computational neuroscience undertaken here and encourage continued investigation under this paradigm.

References

1. Bi, G., Poo, M.: Synaptic modifications in cultured hippocampal neurons: dependence on spike timing, synaptic strength, and postsynaptic cell type. Journal of Neuroscience 18, 10464–10472 (1998)
2. Braitenberg, V.: Vehicles: Experiments in Synthetic Psychology. MIT Press, Cambridge (1984)
3. deCharms, R.C., Zador, A.: Neural representation and the cortical code. Annual Review of Neuroscience 23, 613–646 (2000)

4. Gho, M., Varela, F.J.: A quantitative assessment of the dependency of the visual temporal frame upon the cortical rhythm. Journal of Physiology 83(2), 95–101 (1988)
5. Izhikevich, E.M.: Simple model of spiking neurons. IEEE Transactions on Neural Networks 14, 1569–1572 (2003)
6. Izhikevich, E.M.: Solving the distal reward problem through linkage of stdp and dopamine signaling. Cerebral Cortex 17, 2443–2452 (2007)
7. Izhikevich, E.M., Gally, J.A., Edelman, G.M.: Spike-timing dynamics of neuronal groups. Cerebral Cortex 14, 933–944 (2004)
8. Rieke, F., Warland, D., de Ruyter van Steveninck, R.R., Bialek, W.: Spikes: exploring the neural code. MIT Press, Cambridge (1997)
9. Schultz, W.: Predictive reward signal of dopamine neurons. Journal of Neurophysiology 80, 1–27 (1998)
10. Schultz, W., Apicella, P., Ljungberg, T.: Responses of monkey dopamine neurons to reward and conditioned stimuli during successive steps of learning a delayed response task. Journal of Neuroscience 13, 900–913 (1993)
11. Seth, A.K.: Evolving action selection and selective attention without actions, attention, or selection. In: Pfeifer, R., Blumberg, B., Meyer, J.-A., Wilson, S.W. (eds.) From Animals to Animats 5: Proceedings of the Fifth International Conference on the Simulation of Adaptive Behaviour, pp. 139–147. MIT Press, Cambridge (1998)
12. Seth, A.K., McKinstry, J.L., Edelman, G.M., Krichmar, J.L.: Visual binding through reentrant connectivity and dynamic synchronization in a brain-based device. Cerebral Cortex 14, 1185–1199 (2004)
13. Skinner, B.F.: The Behavior of Organisms. Appleton-Century-Crofts, New York (1938)
14. Sutton, R.S., Barto, A.G.: Reinforcement Learning: An Introduction. MIT Press, Cambridge (1998)
15. VanRullen, R., Thorpe, S.J.: Rate coding vs temporal order coding: what the retinal ganglion cells tell the visual cortex. Neural Computation 13(6), 1255–1283 (2001)
16. Van Rullen, R., Thorpe, S.J.: Surfing a spike wave down the ventral stream. Vision Research 42(23), 2593–2615 (2002)

Appendix A: Neural Model

The agent's neural controller was implemented using the Izhikevich model of spiking neurons [5] with axonal conductance delays, synaptic spike-timing dependent plasticity [1] and dopamine neuromodulation [6].

According to this method, neurons are modelled by the two differential equations

$$v' = 0.04v^2 + 5v + 140 - u + I \tag{1}$$

and

$$u' = a(bv - u) \tag{2}$$

which calculate the membrane potential (v) of the neuron and a membrane recovery variable (u). Variable I represents synaptic current injected into each neuron, whilst parameters a, b, c and d define the type of neuron being modelled. If $v \geq 30mV$, neurons emit a spike and are reset according to

$$v \leftarrow c \tag{3}$$

and

$$u \leftarrow u + d \tag{4}$$

In the experiments implemented here excitatory neurons are regular spiking ($a = 0.02$, $b = 0.2$, $c = -65$, $d = 8$) and inhibitory neurons are fast spiking ($a = 0.1$, $b = 0.2$, $c = -65$, $d = 2$) [5].

Input to the network is implemented by scaling the value of each sensor by a gain of 2 before clipping to ≤ 1. Corresponding input neurons are subsequently innervated by a direct current of magnitude $I = 26S$, in which I is the applied current and S is the sensor reading after gain and clipping. Conversely, motor output is generated via a leaky integrator whose value is incremented by a value of 0.1 whenever an output neuron spikes, otherwise decaying exponentially with a time constant of 0.02s. The value of the integrator at each time step is converted directly to wheel velocities after linearly scaling by a motor gain of $M_G = 20$, enabling agents to reach a maximum speed of approximately 15 units/s under normal conditions. Finally, neural noise is introduced into the network by perturbing I by a random number in the range $[-6.5, 6.5]$ at each time-step.

Synaptic conductance delays are modelled separately for each neuron in the range [1,10ms] with excitatory neurons projecting plastic synapses with strengths in the range [0,4mV]. In all experiments excitatory synaptic strengths were initially set to $s = 0.1mV$, whilst inhibitory neurons projected non-plastic synapses with constant strength $s = -0.1mV$.

Following Izhikevich [7], synaptic spike-timing dependent plasticity is implemented via the derivative of synaptic strength, sd, such that only the rate of change in synaptic strength is affected directly by the relative timings of pre- and post-synaptic spikes. Using an earliest-neighbour method, the firing of a post-synaptic neuron i at time t increases the value of sd by $A^+ e^{\frac{t_j-t}{T^+}}$, where t_j is the time of arrival (after axonal conductance delay) of the last spike of each pre-synaptic neuron j. Similarly, when a pre-synaptic neuron fires and emits a spike (again, after axonal conductance delay) at time t the value of sd is reduced by $A^- e^{\frac{t_i-t}{T^-}}$, where t_i is the time of last spike of each post-synaptic neuron i. The variable sd otherwise decays exponentially with time constant $\tau_{sd} = 1s$. The parameters $A^\pm$ and $T^\pm$ therefore determine the relative size of the STDP window for both causal and anti-causal firings.

Dopamine modulation of synaptic plasticity is subsequently implemented (following Izhikevich [6]) in the calculation of s from sd, where

$$s' = d \times sd \tag{5}$$

Here the variable d corresponds to the current level of extracellular dopamine and regulates the rate at which synaptic strength changes with respect to the value of sd. The value of d decays exponentially with time constant $\tau_d = 0.2s$. A 2nM baseline concentration of dopamine was maintained by the tonic release of $0.01\mu M/s$, allowing synaptic plasticity to occur at a slow rate at all times. Whenever reward was received however, d was step increased by a value of $0.5\mu M$ and therefore increases synaptic plasticity significantly for a period of around 1s. In all experiments, membrane potentials were integrated by the Euler method with an time-step of 0.5ms, whilst synaptic plasticity was integrated at 1ms and updated every 10ms.

Mutual Development of Behavior Acquisition and Recognition Based on Value System

Yasutake Takahashi, Yoshihiro Tamura, and Minoru Asada

Graduate School of Engineering, Osaka University
Yamadaoka 2-1, Suita, Osaka, 565-0871, Japan
{yasutake,yoshihiro.tamura,asada}@ams.eng.osaka-u.ac.jp
http://www.er.ams.eng.osaka-u.ac.jp/

Abstract. Both self-learning architecture (embedded structure) and explicit/implicit teaching from other agents (environmental design issue) are necessary not only for one behavior learning but more seriously for life-time behavior learning. This paper presents a method for a robot to understand unfamiliar behavior shown by others through the collaboration between behavior acquisition and recognition of observed behavior, where the state value has an important role not simply for behavior acquisition (reinforcement learning) but also for behavior recognition (observation). That is, the state value updates can be accelerated by observation without real trials and errors while the learned values enrich the recognition system since it is based on estimation of the state value of the observed behavior. The validity of the proposed method is shown by applying it to a dynamic environment where two robots play soccer.

Keywords: Reinforcement Learning, Behavior Recognition, Value system, Learning by Observation.

1   Introduction

Reinforcement learning has been studied well for motor skill learning and robot behavior acquisition in both single and multi-agent environments. Especially, in the multi-agent environment, observation of others make the behavior learning rapid and therefore much more efficient [1,2,3]. Actually, it is desirable to acquire various unfamiliar behavior with some instructions from others in real environment because of huge exploration space and enormous learning time to learn. Therefore, behavior learning through observation has been more important. Understanding observed behavior does not mean simply following the trajectory of an end-effector or joints of demonstrator. It means reading his/her intention, that is, the goal of the observed behavior and finding a way how to achieve the goal by oneself regardless of the difference of the trajectory. From a viewpoint of the reinforcement learning framework, this means reading rewards of the observed behavior and estimating sequence of the value through the observation.

M. Asada et al. (Eds.): SAB 2008, LNAI 5040, pp. 291–300, 2008.

Takahashi et al.[4] proposed a method of not only to learn and execute a variety of behaviors but also to recognize behavior of others supposing that the observer has already acquired the values of all kinds of behaviors the observed agent can do. The recognition means, in this paper, that the robot categorizes the observed behavior to a set of its own behaviors acquired beforehand. The method seamlessly combines behavior acquisition and recognition based on "state value" in reinforcement learning scheme. Reinforcement learning generates not only an appropriate behavior (a map from states to actions) to accomplish a given task but also an utility of the behavior, an estimated discounted sum of rewards that will be received in future while the robot is taking an appropriate policy. This estimated discounted sum of reward is called "state value." This value roughly indicates closeness to the goal state of the given task if the robot receives a positive reward when it reaches the goal and zero else, that is, if the agent is getting closer to the goal, the value becomes higher. This suggests that the observer may recognize which goal the observed agent likes to achieve if the value of the corresponding task is going higher.

This paper proposes a novel method that enhances behavior acquisition and recognition based on interaction between learning and observation of behaviors. A robot learns its behaviors through not only trials and errors but also reading rewards of the observed behaviors of others (including robots and humans). Fig.1 shows a rough idea of our proposed method. $V(s)$ and $\hat{V}(s)$ are the state value updated by oneself and the state value estimated though observation, respectively. Takahashi et al. [4] showed the capability of the proposed method mainly in case that the observer has already acquired a number of behaviors to be recognized beforehand. Their case study showed how this system recognizes observed behaviors based on the state value functions of self-behaviors. This paper shows how the estimated state value of observed behavior, $\hat{V}(s)$, gives feedback to learning and understanding unfamiliar observed behaviors and this feedback loop enhances the performance of observed behavior recognition. The validity of the proposed method is shown by applying it to a dynamic environment where two robots play soccer.

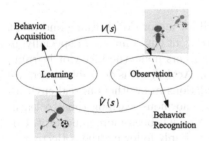

Fig. 1. Interaction between Learning and Observation of Behavior

Fig. 2. Robots with a human player in a Soccer Field

2   Experimental Setup and an Assumption

Fig.2 shows two robots, a human player and color-coded objects, e.g., an orange ball, and a goal. The robot has an omni-directional camera on top. A simple color image processing is applied in order to detect the color-coded objects and players in real-time. The mobile platform is based on an omni-directional vehicle. These two robots and the human play soccer such as dribbling a ball, kicking it to a goal, passing a ball to the other, and so on. While playing with objects, they watch each other, try to understand observed behaviors of the other, and emulate them. In this paper, all experiments are done in computer simulation environment due to space limitation.

A learning/recognizing robot assumes that all robots and even the human player share reward models of the behaviors. For example, all robots and the human player receive a positive reward when the ball is kicked into the goal. This assumption is very natural as we assume that we share "value" with colleagues, friends, or our family in our daily life.

3   Outline of the Mechanisms

3.1   Behavior Learning Based on Reinforcement Learning

An agent can discriminate a set S of distinct world states. The world is modeled as a Markov process, making stochastic transitions based on its current state and the action taken by the agent based on a policy π. The agent receives reward r_t at each step t. State value V^π, the discounted sum of the reward received over time under execution of policy π, will be calculated as follows:

$$V^\pi = \sum_{t=0}^{\infty} \gamma^t r_t \ .$$

(1)

In case that the agent receives a positive reward if it reaches a specified goal and zero else, then, the state value increases if the agent follows a good policy π. The agent updates its policy through trials and errors in order to receive higher positive rewards in future. Analogously, as animals get closer to former action sequences that led to goals, they are more likely to retry it. For further details, please refer to the textbook of Sutton and Barto [5] or a survey of robot learning [6].

Here we introduce model-based reinforcement learning method. A learning module has a forward model which represents the state transition model and a behavior learner which estimates the state-action value function based on the forward model in a reinforcement learning manner. Each learning module has its own state transition model. This model estimates the state transition probability $\hat{\mathcal{P}}^a_{ss'}$ for the triplet of state s, action a, and next state s':

$$\hat{\mathcal{P}}^a_{ss'} = Pr\{s_{t+1} = s' | s_t = s, a_t = a\}$$

(2)

Each module has a reward model $\hat{\mathcal{R}}_s$, too:

$$\hat{\mathcal{R}}(s) = E\{r_t|s_t = s\} \tag{3}$$

All experiences (sequences of state-action-next state and reward) are simply stored to estimate these models. Now we have the estimated state transition probability $\hat{\mathcal{P}}_{ss'}^a$ and the expected reward $\hat{\mathcal{R}}_s$, then, an approximated state-action value function $Q(s, a)$ for a state action pair s and a is given by

$$Q(s,a) = \sum_{s'} \hat{\mathcal{P}}_{ss'}^a \left[\hat{\mathcal{R}}(s') + \gamma V(s') \right] \tag{4}$$

$$V(s) = \max_a Q(s, a), \tag{5}$$

where γ is a discount factor.

3.2   Modular Learning System

In order to observe/learn/execute a number of behaviors in parallel, we adopt a modular learning system. Many modular architectures have been proposed so far (for example [6]). Each module is responsible for learning to achieve a single goal. One arbiter or a gate module is responsible for merging information from the individual modules in order to derive a single action performed by the robot.

We prepare a number of behavior modules each of which adopts the behavior learning method described in 3.1. The module is assigned to one goal-oriented behavior and estimates one action value function $Q(s, a)$. A module receives a positive reward when it accomplishes the assigned behavior or zero reward else. The behavior module has a controller that generates predictions of next state values, selecting the action with the maximum value. The gating module will then select one output from the inputs of the different behavior modules according to the player's intention.

The same behavior modules are used for the behavior recognition. Each behavior module estimates the state value based on the estimated state of the observed demonstrator[1] and calculates reliability of observed behavior, that is, how likely the demonstrator is taking the behavior of the module. The details are described in following sections.

3.3   Behavior Recognition Based on Estimated Values

Each behavior module can estimate a state value of observed behavior at an arbitrary time t to accomplish the specified task. An observer watches a demonstrator's behavior and maps the sensory information from an observer viewpoint to a demonstrator's one with a simple mapping of state variables. Fig.3 shows a simple example of this transformation. It detects color-coded objects on the

---

[1] For reasons of consistency, the term "demonstrator" is used to describe any agent from which an observer can learn, even if the demonstrator does not have an intention to show its behavior to the observer.

Table 1. List of behaviors learned by self and state variables for each behavior

Behavior	State variables
Approaching a ball	d_b
Approaching a goal	d_g
Approaching the teammate	d_r
Shooting a ball	d_b, d_g, θ_{bg}
Passing a ball	d_b, d_r, θ_{br}

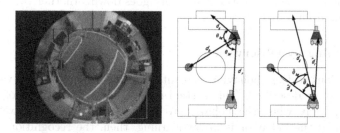

Fig. 3. Estimation of view of the demonstrator. Left : a captured image the of observer, Center : object detection and state variables for self, Right : estimation of view of the demonstrator.

omni-directional image, calculates distances and directions of the objects in the world coordinate of the observer, and shifts the axes so that the position of the demonstrator comes to center of the demonstrator's coordinate. Then it roughly estimates the state information in the egocentric coordinate and the state of the demonstrator. Every behavior module estimates a sequence of its state value from the estimated state of the observed demonstrator and the system selects modules which values are increasing. The learner tries to acquire a number of behaviors shown in Table 1. The table also describes necessary state variables shown in Fig.3 for each behavior. Each state variable is divided into 11 in order to construct quantized state space. 4 actions are prepared to be selected by the learning modules: Approaching the goal, approaching the teammate, going in front of the ball while watching the goal, and going in front of the ball while watching the teammate.

While an observer watches a demonstrator's behavior, it uses the same behavior modules for recognition of observed behavior as shown in Fig.3. Each behavior module estimates the state value based on the estimated state of the observed demonstrator and sends it to the selector. The selector watches the sequence of the state values and selects a set of possible behavior modules of which state values are going up as a set of behaviors the demonstrator is currently taking. As mentioned before, if the state value goes up during a behavior, it means that the module is valid for explaining the behavior. The observed behavior is recognized by a set of behaviors whose modules' values are increasing.

Here we define reliability g that indicates how much the observed behavior would be reasonable to be recognized as a behavior

$$g = \begin{cases} g + \beta & \text{if } V_t - V_{t-1} > 0 \text{ and } g < 1 \\ g & \text{if } V_t - V_{t-1} = 0 \\ g - \beta & \text{if } V_t - V_{t-1} < 0 \text{ and } g > 0 \ , \end{cases}$$

where β is an update parameter, and 0.1 in this paper. This equation indicates that the reliability g will become large if the estimated utility rises up and it will become low when the estimated utility goes down. Another condition is to keep g value from 0 to 1.

3.4   Learning by Observation

In the previous section, behavior recognition system based on state value of its own behavior is described. This system shows robust recognition of observed behavior [7] only when the behavior to be recognized has been well-learned beforehand. If the behavior is under learning, then, the recognition system is not able to show good recognition performance at beginning. The trajectory of the observed behavior can be a bias for learning behavior and might enhance the behavior learning based on the trajectory. The observer cannot watch actions of observed behavior directly and can only estimate the sequence of the state of the observed robot. Let s_t^o be the estimated state of the observed robot at time t. Then, the estimated state value $\hat{V}^o$ of the observed behavior can be calculated as below:

$$\hat{V}^o(s) = \sum_{s'} \hat{\mathcal{P}}_{ss'}^o \left[\hat{\mathcal{R}}(s') + \gamma \hat{V}^o(s') \right] \tag{6}$$

where $\hat{\mathcal{P}}_{ss'}^o$ is state transition probability estimated from the behavior observation. This state value function $\hat{V}^o$ can be used for can be used as a bias of the state value function of the learner V. The learner updates its state-action value function $Q(s, a)$ during trials and errors based on the estimated state value of observed behavior $\hat{V}^o$ as below:

$$Q(s, a) = \sum_{s'} \hat{\mathcal{P}}_{ss'}^a \left[\hat{\mathcal{R}}(s') + \gamma V'(s') \right] \tag{7}$$

while

$$V'(s) = \begin{cases} V(s) & \text{if } V(s) > \hat{V}^o(s) \\ \hat{V}^o(s) & \text{else} \end{cases}$$

This is a normal update equation as shown in (4) except using $V'(s)$. The update system switches the state value of the next state s' between the state value of own learning behavior $V(s')$ and the one of the observed behavior $\hat{V}^o(s')$. It takes $V(s')$ if the state value of own learning behavior $V(s')$ is bigger than the one of the observed behavior $\hat{V}^o(s')$, $\hat{V}^o(s')$ else. This means the state value update system takes $\hat{V}^o(s')$ if the learner does not estimate the state value $V(s')$ because of lack of experience at the state s' from which it reaches to the goal of the behavior. $\hat{V}^o(s')$ becomes a bias for reinforcing the action a from the state s even though the state value of its own behavior $V(s')$ is small so that it leads the learner to explore the space near to the goal state of the behavior effectively.

A demonstrator is supposed to show a number of behaviors which are not informed directly to the observer. In order to update the estimate values of the behavior the demonstrator is taking, the observer has to estimate which behavior the demonstrator is taking correctly. If the observer waits to learn some specific behavior by observation until it becomes able to recognize the observed behavior well, bootstrap of leaning unfamiliar behaviors by observation cannot be expected. Therefore, two strategies of updating value functions of observed behaivors here:

- update all value functions of observed behaviors ($\hat{V}^o(s')$) based on all observed trajectories
- update value functions of observed behaviors with high reliability using history of the observed trajectories

The former strategy contributes to propagate values/rewards to the neiboring state. Even if the observed behavior does not match the expected behavior, the state transition through the observation gives rough hints on distances between states that help to develop rough state values of behaviors. The latter strategy enhances to estimate appropriate values of the observed behavior. This directly contributes to bootstrap learning of the behavior. The former strategy sometimes produces wrong state value estimation and the latter strategy corrects the estimation based on appropriate state transition of the behaviors.

4   Behavior Learning by Observation

4.1   Experimental Setup

In order to validate the effect of interaction between acquisition and recognition of behaviors through observation, two experiments are set up. One is that the learner does not observe the behavior of other but tries to acquire shooting/passing behaviors by itself. The other is that the learner observes the behavior of other and enhances the learning of the behavior based on the estimated state value of the observed behavior. In former experiment, the learner follows the learning procedure:

1. 10 episodes for behavior learning by itself
2. evaluation of self-behavior performance
3. evaluation of behavior recognition performance
4. goto 1.

On the other hand, the later experiment, it follows :

1. 10 episodes for observation of the behavior of the other
2. 10 episodes for behavior learning by self-trials with observed experience

3. evaluation of self-behavior performance
4. evaluation of behavior recognition performance
5. goto 1.

The both learners attempt to acquire behaviors listed in Table 1. The demonstrator shows the behavior one by one but the observer does not know which behavior the demonstrator is taking. In both experiments, the learner follows ϵ-greedy method; it follows the greedy policy with 80% probabilty and takes a random action else. Performance of the behaviors execution and recognition of observed behavior during the learning time is evaluated every 10 learning episodes. The performance of the behavior execution is success rate of the behavior while the learner, the ball, and the teammate are placed at a set of pre-defined positions. The one of the behavior recognition is average length of period in which the recognition reliability of the right behavior is larger than 70% during the observation. The soccer field area is divided 3 by 3 and the center of the each area is a candidate of the position of the ball, the learner, or the teammate. The performances are evaluated in all possible combinations of the positions.

4.2 Recognition of Observed Behaviors

Before evaluating the performance of the behavior execution and behavior recognition of other during learning the behavior, we briefly review how this system estimates the values of behaviors and recognizes the observed behavior after the observer has learned behaviors. When the observer watches a behavior of the other, it recognizes the observed behavior based on repertoire of its own behaviors. Figs.4 (a) and (b) show sequences of estimated values and reliabilities of the behaviors, respectively. The line that indicates the passing behavior keeps tendency of increasing value during the behavior in this figures. This behavior is composed of behaviors of approaching a ball and approaching the teammate again, then, the line of approaching a ball goes up at the earlier stage and the line of approaching the teammate goes up at the later stage in Fig.4(a). All

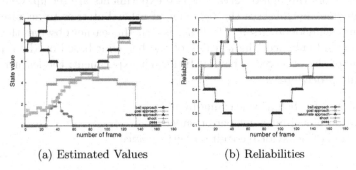

(a) Estimated Values   (b) Reliabilities

Fig. 4. Sequence of estimated values and reliabilities during a behavior of pushing a ball to the magenta player, red line : approaching a ball, green line : approaching the goal, light blue line : passing, blue line : approaching the other, magenta line : shooting

reliabilities start from 0.5 and increase if the value goes up and decrease else. Even when the value stays low, if it is increasing with small value, the reliability of the behavior increases rapidly. The reliability of the behavior of pushing a ball into the teammate reaches 1.0 at middle stage of the observed behavior. The performance of observed behavior recognition is 85% here ,that means, the period in which the reliability of passing behavior is over 70% is 85% during the observation.

4.3   Performance of Behavior Learning and Recognition

In this section, performances of the behavior execution and behavior recognition during learning the behavior are shown. Fig.5 shows success rates of the behaviors and their variances during learning in cases of learning with/without value update through observation. The success rates with value update of all kinds of behaviors grows more rapidly than the one without observation feedback. Rapid learning is one of the most important aspect for a real robot application. The success rate without value update through observation sometimes could not reach the goal of the behavior at the beginning of the learning because there is no bias to lead the robot to learn appropriate actions. This is the reason why the variances of the rate is big. On the other hand, the system with value update through observation utilizes the observation to bootstrap the learning even though it cannot read exact actions of observed behavior.

Recognition rates for observed behaviors and their variances shown in Fig.6 indicate a similar aspect with the ones of success rates. The performance of the behavior recognition depends on the learning performance. If the learning system has not acquired data enough to estimate state value of the behavior, it cannot perform well. The learning system with value update with observed behavior rapidly enables to recognize the behavior while the system without value update based on the observation has to wait to realize a good recognition performance until it estimates good state value of the behavior by its own trials and errors.

Those figures show the importance of learning through interaction between behavior acquisition and recognition of observed behaviors.

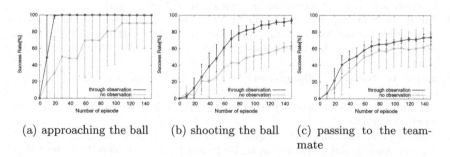

(a) approaching the ball   (b) shooting the ball   (c) passing to the team-
                                                         mate

Fig. 5. Success rate of the behaviors during learning with/without observation of demonstrator's behavior

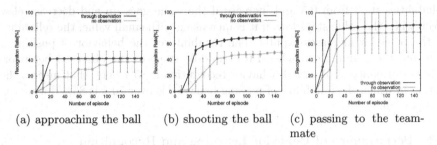

(a) approaching the ball   (b) shooting the ball   (c) passing to the team-
                                                           mate

Fig. 6. Recognition rate of the behaviors during learning with/without observation of
demonstrator's behavior

5   Conclusion

Above, values are defined as behaviors, which are defined by the achieved goals.
The observer uses its own value functions to recognize what the demonstrator will
do. Preliminary investigations in a similar context have been done by Takahashi
et al. [7] and they showed much better robustness of behavior recognition than
a typical method. In this paper, unknown behaviors are also understood in term
of one's own value function through learning based on the estimated values
derived from the observed behaviors. Furthermore, value update through the
observation enhances not only the performance of behavior learning but also the
one of recognition of the observed behavior effectively.

References

1. Whitehead, S.D.: Complexity and cooperation in q-learning. In: Proceedings Eighth
   International Workshop on Machine Learning (ML 1991), pp. 363–367 (1991)
2. Price, B., Boutilier, C.: Accelerating reinforcement learning through implicit imita-
   tione. Journal of Articial Intelligence Research (2003)
3. Bentivegna, D.C., Atkeson, C.G., Chenga, G.: Learning tasks from observation and
   practice. Robotics and Autonomous Systems 47, 163–169 (2004)
4. Takahashi, Y., Kawamata, T., Asada, M., Negrello, M.: Emulation and behavior understanding through shared values. In: Proceedings of the 2007 IEEE/RSJ International Conference on Intelligent Robots and Systems, October 2007, pp. 3950–3955
   (2007)
5. Sutton, R., Barto, A.: Reinforcement Learning: An Introduction. MIT Press, Cambridge (1998)
6. Connell, J.H., Mahadevan, S.: ROBOT LEARNING. Kluwer Academic Publishers,
   Dordrecht (1993)
7. Takahashi, Y., Kawamata, T., Asada, M.: Learning utility for behavior acquisition
   and intention inference of other agent. In: Proceedings of the 2006 IEEE/RSJ IROS
   2006 Workshop on Multi-objective Robotics, October 2006, pp. 25–31 (2006)

Improving Situated Agents Adaptability Using Interruption Theory of Emotions

Clément Raïevsky and François Michaud

Department of Electrical Engineering and Computer Engineering
Université de Sherbrooke, Québec Canada J1K 2R1

Abstract. Emotions play several important roles in the cognition of human beings and other life forms, and are therefore a legitimate inspiration to provide adaptability and autonomy to situated agents. However, there is no unified theory of emotions and many discoveries are yet to be made in the applicability of emotions to situated agents. This paper investigates the feasibility and utility of an artificial model of anger and fear based on Interruption Theory of Emotions. This model detects and highlights situations for which an agent's decision-making mechanism is no longer pertinent. These situations are detected by analyzing discrepancies between the agent's actions and its intentions, making this model independent from the agent's environment and tasks. Collective foraging simulations are used to characterize the influence of the model. Results show that the model improves the adaptability of a group of agents by simultaneously optimizing multiple performance criterion.

Introduction

In spite of significant evidence that emotion plays several crucial roles in cognitive processes [1][2][3][4], no consensus currently exists about a unified theory from which an artificial model can be derived. Therefore, to enhance our understanding of existing theories and to appreciate their usage and effects, it is still appropriate to implement them on artificial systems. Among research related to process models of emotions [5] which are applied to situated agents, we can find emotional mechanisms aimed at enhancing interaction quality between humans and synthetic agents [6], increasing synthetic agents learning abilities [7], and improving coordination among situated agents [8][9]. In these previous works, emotions are generated in two ways : by detecting specific features in the environment [6][7], or monitoring specific task progress variables without taking into account the agent's intentions [8][9]. Once generated, emotions either directly modify the agent's behavior [8][9][6], or influence other cognitive process of the agent decision-making architecture [7]. These models of emotions are limited in their versatility because they are specific either to environmental conditions for emotions generation or to mission objectives for emotional responses. However, emotions should be derived from a generic model to capture the fact that different situations can lead to the same emotions, and that the same situation can lead to different emotions. To our knowledge, no environment-independent and

M. Asada et al. (Eds.): SAB 2008, LNAI 5040, pp. 301–310, 2008.

task-independent artificial model of emotions has yet been validated, and that is the purpose of our research.

This paper presents an artificial model of anger and fear that reproduces functions of emotions identified by the Interruption Theory of Emotions (ITE) [2][3]. This theory has not yet been implemented in situated agents. ITE states that emotions are elicited when the current decision-making process of an individual is not adapted to the experienced situation [2][3][1].Our model detects these conditions and generates emotions by monitoring temporal models of the agent's intentions. These models are independent of the agent's environment. The main function of emotions identified by ITE is to highlight the cause of the current emotional state [1][2][3]. In our model, the cause of the current emotional arousal is determined by an analysis of the agent's intentions. Identifying this cause allows the agent's motivations to adapt the agent's intentions. This signaling process is independent of the agent's tasks.

Section 1 presents ITE, followed by the description of our model in Section 2. To demonstrate and evaluate this model, it has been implemented in a behavior-based cognitive architecture and applied to a collective foraging task. Section 3 presents the experiments carried out and the results, illustrating that our emotional process improves the adaptability of a group of agents.

1   Interruption Theory of Emotions

ITE has been primarily developed by Hebb [2] and Mandler [3]. It states that interruptions of ongoing cognitive or behavioral activity trigger the *arousal* of the sympathetic nervous system which is the beginning of an emotional experience. This is also supported by the Affect Control Theory and the Self-Discrepancy Theory [10]. ITE identifies three main sources of interruption:

1. Experiencing an unexpected effect of a behavior. This occurs when conditions hinder or prevent a behavior from carrying out its function.
2. Experiencing conflicting intentions. This occurs when different decisional processes generate incompatible intentions, i.e. intentions that cannot be carried out simultaneously by actions.
3. Experiencing an unexpected situation, not anticipated by a predictive model of the world.

According to ITE, the main function of the arousal triggered by the occurrence of such interruptions is to signal to the individual that events in the environment require attention and adjustment. This generic arousal is followed by the orientation of the individual's attention toward the cause of the arousal. This highlighting process allows the individual to focus on the cause of the emotion and take the appropriate actions accordingly. Unlike other cognitive theories of emotions ITE focuses on elicitation of emotion rather than on elicitation of the different emotions. It is thus not a complete theory of emotions and a model based on this theory should be extended by models of other aspects of human emotions such as appraisal [4] and stimulus analyses to get the full range of functions associated with human emotions.

2   Artificial Model of Emotions

The artificial model of emotions we have developed aims at detecting and highlighting interruptions of cognitive or behavioral activities in order to trigger an adaptive reaction when 'normal' decision-making is no longer pertinent. This relates to ITE's first two sources of interruption. Figure 1 illustrates the hypotheses we make about the decision-making architecture of an agent to design the process which implement our model of emotions. Actions of the agent are derived from concurrent processes (i.e., Behaviors) which are activated by an Action Selection mechanism according to the agent's Intentions. The agent must have cognitive processes (i.e., Motivations) responsible for generating Intentions and determining their desirabilities. Intentions are data structures which represent particular activations and configurations of one or several Behaviors. The information regarding which Intentions are realized by the agent's actions must be available to the decision-making processes.

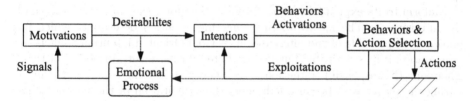

Fig. 1. Cognitive processes and concepts required by our model of emotions

The emotional process we have developed unfolds as follows: when a situation needing adaptation occurs, an interruption is detected by the appearance of a discrepancy between the agent's intentions and the way they are satisfied by its actions (exploited). Once an interruption is detected, its cause is identified by an analysis of the current agent's intentions. The occurrence of an interruption and its cause are then signaled to the agent's motivations, which can change the agent's intentions accordingly.

Coherence between intentions and actions is checked through the monitoring of temporal models of intentions' exploitation. These models depend on the type of intentions: Goal-Oriented intentions are related to behaviors which make the agent accomplish actions aimed at fulfilling it's goals and Safety-Oriented intentions are there to keep the agent away from problematic situations. Therefore, a Goal-Oriented intention has an exploitation model of being exploited when desirable and, conversely, a Safety-Oriented intention conform to its exploitation model when not exploited.

The accumulated time $a_I(t)$ during which intention I does not conform to its exploitation model at time t is expressed by (1) and (2):

$$a_I(t) = \int_{-W_I}^{0} b_I(t)dt \tag{1}$$

$$b_I(t) = \begin{cases} 0 \text{ if intention } I \text{ conform to its exploitation model} \\ \\ 1 \text{ otherwise} \end{cases} \tag{2}$$

where W_I is the length of the sliding time window over which intention I is monitored. An interruption is detected when $a_I(t)$ becomes greater than a time threshold.

Anger is elicited by interruptions involving Goal-Oriented intentions, and conversely Fear is elicited by interruptions involving Safety-Oriented intentions. These emotions are used to modify the agent's behavior through its motivations, but the associated results are beyond the scope of this article.

Detecting and highlighting the cause of the interruption is carried out by an analysis of the agent's current intentions. It is important to distinguish the intention which *triggers* the interruption (by not conforming to its exploitation model) from the intention which is the cause of this interruption (i.e., the "responsible intention"). The responsible intention prevents the triggering intention to conform to its exploitation model and is, therefore, the subjective source of the interruption. The nature of the triggering intention determines which intention is responsible of the interruption. A Goal-Oriented intention triggers an interruption if it is desirable but not exploited during a certain period of time. Therefore, the responsible intention is the one which is being exploited the most during the recent past because it hinders the exploitation of the triggering intention. Conversely, a Safety-Oriented intention triggers an interruption because it has been exploited during a certain period of time. The responsible intention is, in this case, the triggering one. Once identified, the responsible intention is signaled to the agent's motivations as the cause of the interruption. The agent's motivations are then responsible for the adaptation of the agent behavior, keeping the model independent of particular adaptive reaction and of the agent's mission.

Interestingly, the exploitation models of intentions used by the interruption detection process are independent of the way intentions are carried out by actions; only the intentions' nature is taken into account and this is independent of the expected specific effects of the intention on the environment. Furthermore, these models are independent of the situation experienced by the agent because its perceptions are not taken into account. However, a model of emotions cannot be completely disembodied and independent from reality because emotions are not pure cognition. The emotional process we have developed is grounded in agent's reality for two reasons: first, the model parameters are time periods (i.e., a time window length and a time threshold) and capture the 'normal' operation of the agent. They therefore are bounded to the agent reality (and can then be determined by a designer through observation and measurement instead of a tedious trial and error process). Second, the emotional process originates in the intentions' exploitations which is the result of the interaction between the agent's action selection process and its environment.

The key strength of the emotional process described here is its independence from both specific effects of intentions on the environment and from reactions triggered by interruptions. This independence implies that our model is not coupled with the way the agent's behaviors carry out its intentions and ensure its applicability to other behaviors and other missions.

3   Experiments and Results

Multi-agent foraging is a widely used task with clear metrics to evaluate performance (e.g., physical interferences, traveled distance, time to complete). It is therefore suitable to illustrate our emotional mechanism. Fig. 2 illustrates the simulated environment (implemented in Stage [11]) used for the foraging experiments. The simulated agents are Pioneer 2 DXs in a pen of 6×10 meters. Six agents have to collect 12 pucks and take them one-by-one to the home region. Each agent is given two simulated sensors: one laser range finder with an 8 meter range and 180° of field of view, and one fiducial finder which returns the identifier and relative position of objects with a fiducial tag, in a range of 5 meters and a 180° field of view. Each agent has an unique fiducial identifier which allow them to perceive others' relative positions. Home flags and pucks have also fiducial ids. Agents are considered to be homogeneous, i.e., they all have the same physical and decisional capabilities. They can communicate with each other using broadcast mode (through network link).

To apply our emotional process to this mission, we integrated it in a modular decision-making architecture called Motivated Behavioral Architecture (MBA) [12]. In MBA, the Behaviors are independent modules issuing commands based on the agent's perception and Intentions. Behaviors issue commands only if they are activated. Their activations and parameters are derived by a *Selection* module from the agent's Intentions. These intentions are generated by the *Motivational Modules* (MM) and are stored in the *Dynamic Task Workspace* (DTW). They are organized in a tree-like structure according to their interdependencies, from high-level/abstract intentions to primitive/behavior-related intentions. MMs are asynchronous, independent modules that can add Intentions, modify or monitor their parameters, and give activation recommendations

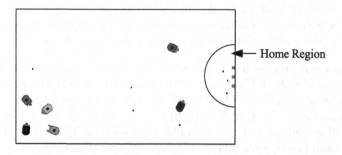

Fig. 2. Experimental setup for multi-agent foraging

about them. These recommendations correspond to the desirabilities of Intentions according to MMs and can take three different values: positive, negative and undetermined. The *Selection* module applies a policy to these recommendations to determine the behaviors' activations; a behavior is activated if its corresponding intention has at least one positive and no negative recommendation. The associations between Intentions and Behaviors is implemented in a *System Know-How* (SNOW) module. This module is also in charge of updating information about exploited Intentions, i.e., Intentions which are being carried out by the agent's actions.

For this mission, the five behaviors, arbitrated using subsumption, are (in order of priority):

- **Escape** makes the agent turn on itself to find a safe passage to leave the current location.
- **Obey** makes the agent execute a particular action such as stopping or turning left, according to a parameter associated with the agent's intentions.
- **Avoid** makes the agent move safely in the environment by avoiding obstacles using the laser range finder readings. Only obstacles within a 0.9 meter radius of the agent are taken into account.
- **Forage** tracks pucks, collects them one at a time and takes them back to the home region.
- **Move Forward** gives the agent a constant linear velocity.

The **motivational modules** and intentions they manipulate are:

- **Survive** ensures the security of the agent by adding and recommending the high-level *Stay Safe* intention, and specifying it by adding *Avoid* intention or *Escape* intention as its child. These two intentions are directly associated with behaviors.
- **Curiosity** makes the agent explore its environment by recommending the *Explore* intention, associated with **Move Forward**.
- **Forage** manages the foraging task by recommending the *Forage* intention (which is associated with the **Forage** behavior) and inhibiting the *Stay Safe* intention when a puck is about to be collected by the agent.
- **Social** carries out the group coordination strategy which is based on a dominance hierarchy. When an agent perceives a higher-ranked agent in a range of 1.5 meter in front of it, it stops (through the activation of the **Obey** behavior with its parameter set to stop). This distributed strategy aims at avoiding physical interference while minimizing distance traveled.

The emotional process has been implemented in a separate module, called the *Emotional Module* (EM). To detect interruptions, the EM monitors the *Stay Safe* and *Forage* intentions. The *Forage* intention has a time window length of 70 seconds and and time threshold of 60 seconds. *Stay Safe* has a time window length of 140 seconds and a time threshold of 120 seconds. These parameters have been fixed from pre-experiments trials by observation of intentions exploitations during normal situations.

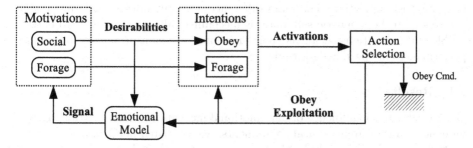

Fig. 3. Adaptation Process Example

These models have shown useful in detecting and adapting to two typical kinds of situation. The first typical situation occurs when the agent is in a high obstacle density area. In this situation, the *Stay Safe* intention is used often enough to trigger an interruption from the under-exploitation of the *Forage* intention or an over-exploitation of the *Stay Safe* intention. In both cases, the responsible intention is *Stay Safe* and is signaled as the cause of the interruption to the MM. The *Survive* MM uses this signal to adapt the avoidance strategy by switching the children intention of *Stay Safe* in the DTW from *Avoid* to *Escape*.

The second typical situation occurs when an agent stops because it perceives a superior agent which is experiencing some kind of failure. Fig. 3 presents the concepts involved in the adaptation mechanism triggered by this kind of situation. This adaptation process is triggered by the prolonged exploitation of *Obey* intention which prevents the *Forage* intention from being exploited. The *Forage* intention then generates an interruption after not being exploited during 60 seconds over the last 70 seconds. Because *Forage* is a goal-oriented intention, the EM looks for the most exploited intention as the cause of the interruption and find **Obey**. This intention is then signaled to the agent's motivations as the cause of an interruption. This intention has been added by the Social MM to enforce the social rules of the group. The Social MM is therefore responsible to adapt this intention in response to the emotional signal. To do so, it triggers an update of the dominance link established between the superior and the inferior agent. The result of this update depends on the emotional state of the superior agent : if its dominant emotion is fear (meaning it is experiencing an unwanted situation involving its security) the agent is lowered in the hierarchy. Conversely, if it is experiencing anger, the agent keeps its rank in the hierarchy. This update only changes the relationship between the two involved agents.

Work by Murphy *et al.* [9] and Parker [8] are closely related to ours as they trigger behavioral adaptation from affective evaluation of task progression. However, in their work, emotion related variables are generated from an analysis of the agent's perceptions and from dedicated social messages. Their models are therefore tightly coupled with the agent's environment and with its task. Furthermore, adaptations triggered by their emotional mechanisms are dedicated to group task allocation. The main difference with our work is that our model is intended to be a generic self-analysis mechanism allowing an agent to detect problematic situations, whereas their emotional mechanism aim at improving performances of specific algorithms.

3.1   Results

To characterize the influence of our emotional mechanism while neutralizing the influence of other architectural components, we compared performances of four controllers which differ only by the presence or absence of the Social MM and of the EM: 1) the Control Group (CG) does not use the Social MM nor the EM; 2) EM refers to the controller only using the EM without the Social MM, and is used to characterize the influence of the EM without social coordination; 3) CG-So is a controller using the Social MM without the EM, and is used as a reference to evaluate the coordination strategy performance during the foraging task; 4) EM-So has both the EM and the Social MM, and is used to characterize the combined influence of the EM on the avoidance behavior and on the coordination strategy. Each controller has been used on the same series of forty randomly generated initial positions of agents and pucks to eliminate the influence of the initial conditions on the group's performance. Table 1 summarizes the observed results in terms of the following metrics. Success Rate is the ratio of failed trials (determined when pucks remain to be collected after 30 min) over the total number of trials. This metrics captures the ability of the group to recover from situations that cause it to fail and that we have not anticipated reflecting the adaptability of the group. Physical Interference Ratio is the part of time spent at a distance of 0.7 m from other agents. This metric represents the risk of collision between agents. Completion Time is the time spent to take all pucks to the home region. Traveled Distance is the total distance traveled by the agents during one trial. All these metrics, except Success Rate, take only successful experiments into account.

Comparing controllers both with the EM (i.e. EM and EM-So) and without (i.e. CG and CG-So) shows that the presence of the EM has improved the Success Rate of the group with and without coordination. This therefore suggests that our emotional process improves both the adaptability of the avoidance behavior and of the coordination strategy. As expected, the introduction of the Social MM has reduced the physical interferences between agents, making them safer, and

Table 1. Experiments results

Control Type	CG	EM	CG-So	EM-So
Success Rate	90 %	95 %	54.8 %	87.5 %
Physical Interference Ratio	26.4 %	24.1 %	15.6 %	16.8 %
Completion Time	439	462	467	429
Traveled Distance	375	404	338	327

the total distance they have traveled to complete the task, making them more efficient. However, without the EM, the coordination strategy has dramatically reduced the Success Rate of the group. This can be explained by the recurrent occurrence of an endless situation, illustrated in Fig. 2. In this situation, an agent (the black one in the bottom left corner), carrying a puck, is surrounded by inferior agents which are thus blocking it. Adding the EM to the Social MM has balanced the effects of this typical situation, bringing the Success Rate, Physical Interference Ratio, Completion Time, and Traveled Distance metrics to an optimum.

These results also show that emotions we have generated can be used to partly replicate the structuring function of emotions in some human groups [13]. It does so by keeping hierarchy relations between individuals coherent with the situation experienced by the group allowing it to adapt to the situation.

4   Conclusion

Situated agents adaptability ultimately depends on the detection of the situations for which their decision-making is not pertinent and which require a behavioral or cognitive reaction. This detection is a key problem for situated agents because their environment is dynamic, continuous and unpredictable. Psychologists have identified that one of the human emotions' functions is to highlight this kind of situation, allowing other cognitive processes to address them. We have developed an emotional mechanism which allows situated agents to detect this kind of situation by using temporal models of intentions. One of the key strength of this mechanism is that it increases the agent's adaptability without introducing either specific knowledge about the environment or about the tasks. Results from simulation experiments show that agents can recover from the malfunction of two specific algorithms (i.e., an avoidance behavior and a coordination strategy) through the use of our emotional mechanism. This has been achieved without relying either on specific knowledge about these algorithms or about specific features of the environment. The independence between our model of emotions and these algorithms guarantees the applicability of our emotional process to other algorithms and by extension to other applications. Our emotional process can be extended by adding other kind of intentions analyses such as observation of intentions resulting status, desirabilities oscillations (eliciting confusion) or lack of change in intentions (eliciting boredom) for example. We believe this versatility allows us to see our emotional process as the basis of a generic self-analysis mechanism allowing situated agents to detect and then adapt to situations for which their actions or decisions are not pertinent. Such a generic mechanism will be useful to autonomously trigger modification of decisional processes which are needed by situations or environments not anticipated by human designers and therefore, bringing artificial systems closer to complete autonomy.

Acknowledgments

The authors gratefully acknowledge the contribution of the Canada Research Chair, the Natural Sciences and Engineering Research Council of Canada and the Canadian Foundation for Innovation, in the support of this work.

References

1. Frijda, N.H.: The emotions. Studies in Emotion and Social Interaction. Cambridge University Press, Cambridge (1986)
2. Hebb, D.O.: The organization of behavior: a neuropsychological theory. A Wiley book in clinical psychology. Wiley, New York (1949)
3. Mandler, G.: Mind and Body: Psychology of Emotion and Stress. W. W. Norton, New York and London (1984)
4. Scherer, K., Schorr, A., Johnstone, T.: Appraisal theories of emotions: Theories, methods, research. Oxford University Press, New York (2001)
5. Scheutz, M.: How to determine the utility of emotions. In: Proceedings of AAAI Spring Symposium (2004)
6. Velásquez, J.: A computational framework for emotion-based control. In: Proceedings of Fifth International Conference on Simulation of Adaptive Behaviors (SAB 1998) (1998)
7. Gadanho, S.C.: Emotional and cognitive adaptation in real environments. In: Symposium ACE 2002 of the 16th European Meeting on Cybernetics and Systems Research, Vienna, Austria (2002)
8. Parker, L.E.: ALLIANCE: An Architecture for Fault Tolerant Multi-Robot Cooperation. IEEE Transactions on Robotics and Automation 2(14) (1998)
9. Murphy, R., Lisetti, C., Tardif, R., Irish, L., Gage, A.: Emotion-based control of cooperating heterogeneous mobile robots. IEEE Transactions on Robotics and Automation 18(5), 744–757 (2002)
10. Stets, J.E.: Emotions and Sentiments. In: De Lamater, J.D. (ed.) Handbook of Social Psychology. Springer, New York (2003)
11. Vaughan, R.T., Gerkey, B.P., Howard, A.: On device abstractions for portable, reusable robot code. In: Proceedings IEEE/RSJ International Conference on Intelligent Robots and Systems, pp. 2421–2427 (2003)
12. Michaud, F., Côté, C., Létourneau, D., Brosseau, Y., Valin, J.M., Beaudry, E., Raïevsky, C., Ponchon, A., Moisan, P., Lepage, P., Morin, Y., Gagnon, F., Giguère, P., Roux, M.A., Caron, S., Frenette, P., Kabanza, F.: Spartacus attending the 2005 AAAI Conference. In: Autonomous Robots 2005; Special Issue on the AAAI Mobile Robot Competitions and Exhibition (2007)
13. Plutchik, R.: A general psychoevolutionary theory of emotion. In: Plutchik, R., Kellermann, H. (eds.) Emotion; Theory, Research and Experience, vol. 1, pp. 3–33. Academic Press, London (1980)

Dynamical Systems Account for Meta-level Cognition

Michail Maniadakis and Jun Tani

Laboratory for Behavior and Dynamic Cognition,
Brain Science Institute (BSI), RIKEN, Japan
{mmaniada,tani}@brain.riken.jp

Abstract. The current paper studies possible neuronal mechanisms for meta-level cognition of rule switching. In contrast to the conventional approach of hand-designing the cognitive functions, our study employs evolutional processes to search for neuronal mechanisms accounting for meta-level cognitive functions required in the investigated robotic tasks. Our repeated simulation experiments showed that the different rules are embedded in separate self-organized attractors, while rule switching is enabled by the transitions among attractors. Furthermore, the results showed that although certain segregation between the lower sensory-motor level and the higher cognitive level enhance the task performance, meta-level cognition is significantly supported by the embodiment and the lower level sensory-motor properties.

1 Introduction

Both animals and humans, use meta-level cognition in order to adapt their action strategies to unpredictable environment situations. Three decades ago, Gregory Bateson [1] demonstrated this ability in dolphins that learned both behavior scheme rules and additionally how to manipulate them in a meta-level, considering that rules have to repeatedly switch in order for the overall behavior to be successful. He argued that dolphins were manipulating rules at a meta-level, but how humans and other animals acquire a meta-rules, has yet to be explained. What are the neural mechanisms that enable meta-level phenomena?

To better understand how animals and humans execute rule switching, neuroscientists have relied on tests like the Wisconsin Card Sorting (WCS). The WSC task creates a situation in which the subject needs to have meta-level awareness of the rules to accomplish card sorting. A large number of experiments with humans and monkeys have shown that impairments to the prefrontal lobe adversely affect performance in WCS tests, suggesting that this region is important for meta-level cognitive behaviors.

In order to explain rule switching in meta-level, modelling researchers describe possible brain mechanisms from a human-specified cognitive science perspective. In a typical explanation, a rule is retrieved from long-term memory to working memory in order to be applied to the current situation. If the rule matches the situation (i.e. it supports gaining a reward) it is kept in the working memory.

M. Asada et al. (Eds.): SAB 2008, LNAI 5040, pp. 311–320, 2008.

Otherwise, the rule is switched. Such predefined cognitive mechanisms have been implemented in neural network models of prefrontal functionality [2,3]. The models describe rule switching mechanisms by combining discrete cognitive modules responsible for memory retrieval, matching, working memory, long-term memory and rule alternation.

However, these hardwired explanations seem to lack generality because of their potential arbitrariness. Therefore, the question now is, can we explain meta-level cognition only by combining these predefined cognitive modules? Is there any alternative approach that accounts for meta-level cognition? In order to investigate this issue, we should model meta-cognitive phenomena from the lower grain size level, without assuming apriori functions. Then, we might find novel and more natural mechanisms for explaining the same phenomena. The current paper pursues this type of approach for investigating meta-rule processing.

The motivation for our experiments is to examine possible neuronal mechanisms accounting for rule switching, by providing self-organization pressure on simple neural network models. Our research methodology is based on the combination of evolutionary robotics [4] and dynamic neural networks [5,6]. Specifically, we conduct robotics simulation experiments without abstracting predefined processes at different cognitive levels. The tasks used in our study integrate meta-level rule manipulation with sensory-motor interaction, highlighting the inseparable nature of these processes. In summary, our experiments and the analysis of the obtained results elucidate that:

- Attractor switching could be a universal mechanism accounting for rule switching in meta-level.
- Some kind of information processing segregation between higher and lower levels enhances the overall network performance. This findings may account for the higher level role of prefrontal lobe in tasks.
- Embodiment is essential even for meta-level cognitive processes because sensory-motor dynamics can support the rule switching mechanisms.

2   Behavioral Tasks

We have designed a robotic task that resembles Wisconsin Card Sorting test, but additionally emphasizes dynamic body-environment interaction. Following this approach, we consider both meta-level cognition and sensory-motor coupling as inseparable parts of a complex behavioral problem.

The task goes as follows. Let's assume that a simulated robotic agent is located in the lower part of a T-maze, and a light sample appears at its left or right side [7]. The agent has to move to the end of the corridor making a $90°$ left or right turning, depending on the side of the light sample. Two different response rules are defined (see Fig 1) similar to [8]. According to the Same-Side (SS) rule, the agent should turn left if the light source appeared at its left side, and it should turn right if the light source appeared at its right side. The complementary response rule named Opposite-Side (OS), implies that the robot should turn to the opposite direction of the light source sides. For both rules, when the agent

Same Side (SS) Rule   Opposite Side (OS) Rule

Fig. 1. A schematic representation of the delayed response rules. Light samples are represented by double circles. Target locations are represented by ×, while reward corresponds to the gray area. The behavioral task asks for controllers capable of switching between the two rules.

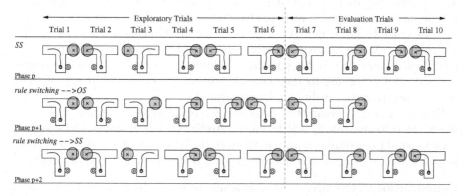

Fig. 2. A schematic demonstration of the task. For the current example, we assume that $T_p = 10, T_{p+1} = 8, T_{p+2} = 10$.

responds correctly it receives positive reinforcement indicating it is following the correct rule.

The task trials are repeated with resetting robot position to the start position but without resetting the neural state of the robot controller (see below). Because the rules are switched from one to the other unpredictably after several trials, the robot has to adapt its response strategy to the newly adopted rules by monitoring the reward signal. The details of the experimental procedure are described below.

2.1 Task Setup

The task is separated to $P \in \{1...10\}$ phases, each one including T_p trials. The number of trials $T_p \in \{8, 10, 12, 14\}$ is randomly specified, so that the agent can not predict the end of a phase. During phase p, the agent has to follow the same response rule for all T_p trials. Let's assume for example that it should follow the SS rule. Each trial tests the response of the robot after light sample appearance at its left or right side (their order is randomly chosen). When a trial starts, the robot is sensing the light and then it moves to the end of the corridor where it makes a turn choice. According to the SS rule, the response is correct when the robot turns towards the side of light sample. If the robot makes the correct choice, it drives to a reward area receiving positive reinforcement. In case that the robot turning is not correct, it will drive to an area that no reward exists, indicating that the currently adopted rule is not correct and it should be switched. During phase p, the robot is given six free exploratory trials to discover what is the correct rule. In the remaining $T_p - 6$ trials the performance of the robotic agent is evaluated in terms of following the desired response rule.

If phase p is completed successfully, the robot moves to phase $p + 1$, where the response rule is switched. This means that the reward signals have changed places and -for our example- they are now positioned according to the OS rule. However, the agent is not informed that the rule has been switched and thus it will continue responding according to the previous rule. In that case, the agent will be unable to get more reward, that indicates it is not following the correct rule. In order to get a new reward, the robot must reconsider its rule choice, switching to OS. In phase $p + 1$, the robot is given again six free exploratory trials to discover rule switching. In the remaining $T_{p+1} - 6$ trials agent's responses are evaluated according to the currently correct response rule. If any of these trials is incorrect, the evaluation is interrupted.

If phase $p + 1$ is completed successfully, the robot moves to phase $p + 2$, where the response rule is switched again -to SS, for our example- and a similar experimental procedure is repeated. Overall, the task evaluates agent's switching behavior for a maximum of P phases (if all of them are completed successfully).

Computational Details. At the beginning of trials the robot is located at a predefined starting position with its direction randomly specified in the range $[85° - 95°]$ degrees. The robot is kept in the same initial position for five simulation steps, and then it is allowed to navigate freely in the environment for 165 more simulation steps. After the completion of one trial the simulated robot is automatically transferred to the initial position having a new random direction, in order to experiment for the next trial.

3   CTRNN Model and Input-Output Connectivity

We use Continuous Time Recurrent Neural Network (CTRNN) models to investigate how meta-level mechanisms self-organize in neuronal dynamics under the given task pressure. Interestingly, in CTRNNs contextual memory is implicitly represented by internal neurons dynamics. In our experimental setup, the neuronal state is initialized only once in the beginning of the first trial, and then neuronal dynamics continues across trials and phases without resetting. In this manner, we speculate that dynamical states will emerge for representing the rule stored in working memory, and additionally, these dynamical states might switch to one another according to the currently adopted rule.

Additionally, both bottleneck (BN) [6] and fully connected CTRNN configurations are explored (see Fig 3), investigating what kind of network structure is essential for achieving meta-level functions. As shown in Fig 3(a), a CTRNN is squeezed in the middle with BN neurons (i.e. the upper and lower parts can interact only through BN neurons). The lower part perceives the sensory flow and outputs motor flow and the higher part receives reward stimuli. The BN

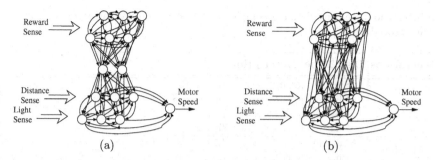

Fig. 3. Schematic representation of (a) the bottleneck CTRNN and (b) the fully connected CTRNN

is supposed to segregate information processing in different levels, maintaining minimum interactions between them. In contrast, in the fully connected case (see Fig 3(b)) information processing levels can hardly differentiate. Overall, the CTRNN consists of 15 neurons for the case of the bottleneck structure and 13 neurons for the case of fully connected structure. All neurons are governed by the standard leaky integrator equations described in previous studies [5,6].

In order to investigate embodied rule switching in meta-level we employ a two wheeled simulated robotic agent equipped with 8 uniformly distributed distance, light and reward sensors. The robotic platform is based on YAKS environment[1], which simulates motion dynamics of the Khepera robot.

4   Evolutionary Procedure

A Genetic Algorithm[2] (GA) is utilized to explore how the desired meta-level switching behavior can self-organize in the CTRNNs.

Incremental Evolution. In order to facilitate successful convergence of the evolutionary process we have used an incremental approach investigating gradually more complex versions of the problem. In the first 60 generations (see Table 1) the evolutionary process asks for robot controllers capable of adopting both SS and OS response rules. Two different tasks are used to evaluate robot controllers. Each task consists of only one phase. The accomplishment of $Task1$ implies that the robot can adopt SS rule, while the accomplishment of $Task2$ implies that the robot can adopt OS rule. At the beginning of each task the states of all CTRNN neurons are reset to zero, which means that the robot is in a neutral state without following any rule. The robot explores the environment to find out which is the rule that should be adopted in order to get rewards.

During generations 61-140, the tasks are getting more complex asking for controllers capable of switching between rules. At this version of the tasks only

---

[1] The simulator has been slightly modified for the needs of the present study.

[2] The current evolutionary procedure does not mean to represent an artificial counterpart of biological evolution. It only serves our study as a consistent mechanism to explore the domain of solutions for our problem.

Table 1. The incrementally more complex tasks solved in different parts of the evolutionary procedure

Generations	Task Type	Description
1-60	Single Phase	CTRNN reset - *Task 1*: SS CTRNN reset - *Task 2*: OS
61-140	Two Phase	CTRNN reset - *Task 1*: SS → OS CTRNN reset - *Task 2*: OS → SS
141-300	Multiple Phase	CTRNN reset - *Task 1*: SS → OS → SS → OS ... SS → OS CTRNN reset - *Task 2*: OS → SS → OS → SS ... OS → SS

one switching step is explored, therefore tasks consist of two phases. Properly positioned reward signals indicate the desired response strategy in each phase. The $Task1$ examines agent's ability to adopt SS and then switch to OS. In a similar way, the $Task2$ examines robot's ability to first adopt OS and then switch to SS. At the beginning of each task the CTRNN state is reset to zero, but then it is kept continuous implying that special memory pathways have to develop facilitating rule switching from SS to OS and visa versa.

Finally, in generations 141-300 we ask for controllers capable of repeatedly switching between rules. Both $Task1$ and $Task2$ are now described by ten phases. Similarly to previous generations CTRNN is reset to zero at the beginning of each task, and then keeps continuous memory state when passing from one phase to the other (i.e. continuously switching between SS and OS rules).

Task Evaluation. The accomplishment of tasks is evaluated based on the goal positions of each trial. The goal positions are specified according to (i) the current rule, and (ii) the side of the light sample (see Fig 1). For each response of the robot the minimum distance $d_{min} \in [0, D]$ between the goal and the robot route, is used to measure the success of robot turning choice (D is the distance between the starting position and the goal). For a task i evaluating the behavior of the robot for p phases, the success on rule switching is given by:

$$E_i = \sum_{q=1}^{p} \left(\sum_{t=7}^{T_q} \left(1 - \frac{d_{min}}{D} \right) \right) \tag{1}$$

The evaluation starts from trial $t = 7$ because the first six trials of each phase are exploratory and they are not considered in evaluation. The higher the value of E_i the more rule switches the agent has accomplished.

Fitness Measure. The individuals encoding CTRNN controllers are tested on $Task1$ and $Task2$ described above. The accomplishment of each task is evaluated separately according to eq (1). The total fitness of the individual is then estimated by:

$$fit = E_{Task1} \cdot E_{Task2} \tag{2}$$

SS response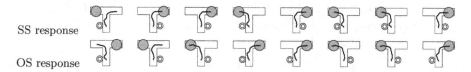

OS response

Fig. 4. The response of the agent in two consecutive phases

5   Results

For each network topology (either bottleneck or fully connected) we have conducted ten independent evolutionary runs to acquire CTRNN controllers capable of rule switching in meta-level. Specifically, a standard GA with mutation, but without crossover, is employed to evolve populations of 500 encoded CTRNN solutions. In the case of the bottleneck network eight out of the ten runs converged successfully, producing controllers capable of switching between SS and OS rules. However, for the fully connected networks, the success of evolutionary runs was reduced to five out of ten. This significant difference of the success rate implies that bottleneck structures are more appropriate to address meta-level cognitive process. This is because the bottleneck adequately segregates the composite system facilitating self-organization of distinct partial functionalities. For the rest of the paper we will concentrate on the results of bottleneck networks.

The performance of the agent for one representative CTRNN is demonstrated in Fig 4. Despite the fact that evolutionary procedures have been statistically independent, they have all produced CTRNNs with similar internal dynamics. The consistently similar characteristics of neural networks supports the validity of our results in terms of considering their internal dynamics as a valuable alternative scenario for rule switching in meta-level.

Additionally, we have investigated neural activity in the higher and lower levels of the CTRNN network observing different qualitative characteristics in their firing (see Fig 5), implying that layers have acquired different roles in the functionality of the overall system. In particular, the activity of higher level neurons varies mostly in (i) the beginning of trials, indicating they are involved in deciding the response of the agent, and (ii) the end of trials, indicating they are involved in assessing agent's response in terms of accomplishing the expected reward (Fig 5 (a),(b) top two lines). In contrast, variance in the activity of lower level neurons is observed during the whole trial indicating its involvement in the execution of higher level plans taking also into account environmental interaction issues, e.g. wall avoidance (Fig 5 (a),(b) lower two lines). It is worth emphasizing that the design procedure does not artificially force CTRNN to develop different roles in the higher and lower levels. This property is an emergent result of evolutionary self-organization, and appears consistently in all CTRNNs capable of rule switching in meta-level.

Comparing higher level activity for all four possible cases (left and right response for either the SS or the OS rule), we observe that the two rules are separated by the activation of neurons at the beginning and end of trials. Specifically,

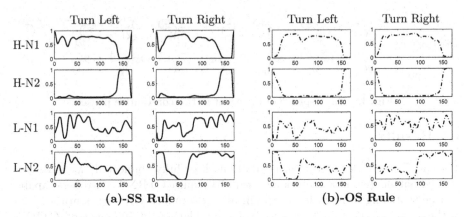

Fig. 5. The activation of two higher (H-N1, H-N2) and two lower (L-N1, L-N2) level neurons when the agent follows the SS rule and the OS rule. SS is demonstrated with a solid line while OS is demonstrated with a dashed line.

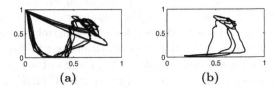

Fig. 6. Phase plot of higher level neural activity when the agent follows (a) the SS rule and (b) the OS rule. Neural activities stabilize to attractors having distinct shapes for each case.

both left and right SS rule turnings start and end with maximum activity at H-N1, and minimum activity for H-N2 (see Fig 5(a)). In contrast, these neurons have a complementary activation pattern for OS rule, that is common for both the left and the right turnings (see Fig 5(b)). The fact that the neurons have similar activities for both turnings of a rule facilitates positioning agents behavior in the same context when the underlying rule is adopted. At the same time, the fact that neural activity is very different when different rules are followed facilitates separating agent's behavioral context between OS and SS rules.

The higher level differences between the two rules can be clearly demonstrated by conducting the attractor analysis taking the phase plots of neuron activities in the higher level. Specifically, for both rules, we ask the agent to perform 30 random turning trials (either left or right) after randomly resetting the neurons in the higher level (Fig 6). For each rule we observed the same shape of attractors to appear in the plot, regardless of the randomness in the initial state. Therefore, each plot represents a distinct invariant set for the corresponding rule. The neuronal state always converge to one of the two invariant sets depending on the reward stimuli condition that specify the currently correct rule. The generation of distinct attractors for each rule is observed in all successful evolutionary runs, implying that this might be a general mechanism for rule encoding.

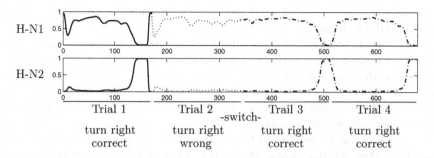

Fig. 7. Higher level neural activity during rule switching. In the trial 1 the agent follows the SS rule giving a successful right response (demonstrated with a solid line). In the next trial the rule has unexpectedly changed to OS, therefore the agent gives an erroneous response (demonstrated with dotted line). In the trial 3 the agent tries the OS rule that is correct (demonstrated with dashed line), and therefore it continues with the same rule in the next trial (demonstrated again with a dashed line).

The activity of neurons for the case of rule switching is demonstrated in Fig 7. Specifically, when the agent makes a wrong response (due to an unexpected rule change) higher level neurons are unable to get values that confirm the current rule is correct, providing instability to the whole system. This unstable condition facilitates changing the currently adopted rule to the new one (i.e making a transition from the one attractor to the other). We note that according to the current experimental setup the robot is provided 6 free trials to consider rule switching. However, in the majority of cases, the attractor state can transit immediately after a single wrong response, and rarely two or three unsuccessful trials are necessary.

6   Discussion

The current study investigates possible mechanism capable of accomplishing rule switching in meta-level. In contrast to traditional approaches using cognitive science terminology to explain rule switching, our results showed that new mechanisms based on dynamical systems principles can also accomplish meta-level manipulation of behavioral rules.

In our results, the self-organized internal dynamics encode each rule to a distinct attractor, while switching of rules takes place by accompanying state transition from one attractor to another. This mechanism seems to be widely valid because the same internal dynamics was observed in the most of our experiments and additionally because the assumptions made in our neural network model are minimal. Our only constraint is the bottleneck connectivity in the network [6]. In that case it is shown that partial segregation of information processing between the meta-level and the sensory-motor level enhances the performance of the global network. This may account why meta-level cognition has developed in a semi-autonomous higher level module (i.e. prefrontal cortex) in mammals and especially in humans.

It is noted that the agent follows different trajectories depending on the rules adopted. In other words, sensory-motor level dynamics have considerable correlations with meta-level processing. This means that distinct attractors are generated not only by abstract meta-level processing but with the active participation of sensory-motor dynamics. Therefore, we argue that the embodiment plays an important role in meta-level cognition.

Our results can be related to the study on emergence of turn-taking, conducted by Ikegami and Iizuka [9]. In this work, a couple of evolved agents alternates predator-pray roles acquiring autonomous, self-generated behavior switching mechanisms without external triggering. A similar phenomenon of self-generated rule switching was observed in our study, when the reward signal is completely removed for all trials. In that case, the agent repeatedly switches the adopted rule from one trial to the other whenever the robot enters the location where reward "used to be".

Our future research includes more complex cases with having more than two rules. Our preliminary result showed that 3 rules case can be evolved but with slightly lower success rate. The rule switching dynamics becomes much more complex compared to the two rule case. The detail analysis of these results is left for a future study.

References

1. Bateson, G. (ed.): Steps to an Ecology of Mind: Collected Essays in Anthropology, Psychiatry, Evolution, and Epistemology. Chandler Publishing Company (1972)
2. Kaplan, G., Sengr, N., Grvit, H., Gen, I., Gzelis, C.: A composite neural network model for perseveration and distractibility in the wisconsin card sorting test. Neural Networks 19(4), 375–387 (2006)
3. Stemme, A., Deco, G., Busch, A.: The neuronal dynamics underlying cognitive flexibility in set shifting tasks. J. of Comp. Neuroscience 23(3), 313–331 (2007)
4. Nolfi, S., Floreano, D.: Evolutionary Robotics: The Biology, Intelligence, and Technology of Self-Organizing Machines. MIT Press/Bradford Books (2000)
5. Yamauchi, B.M., Beer, R.D.: Spatial learning for navigation in dynamic environment. IEEE Trans. Syst. Man Cybern. 26(3) (1996)
6. Paine, R., Tani, J.: How hierarchical control self-organizes in artificial adaptive systems. Adaptive Behavior 13(3), 211–225 (2005)
7. Ziemke, T., Thieme, M.: Neuromodulation of Reactive Sensorimotor Mappings as Short-Term Memory Mechanism in Delayed Response Tasks. Adaptive Behavior 10(3/4), 185–199 (2002)
8. Maniadakis, M., Trahanias, P.: Hierarchical cooperative coevolution facilitates the redesign of agent-based systems. In: Proc. 9th Int. Conf. on the Simulation of Adaptive Behavior (SAB 2006), pp. 582–593 (2006)
9. Ikegami, T., Iizuka, H.: Joint attention and dynamics repertoire in coupled dynamical recognizers. In: Proc. of 2nd International Symposium on Imitation in Animals and Artifacts (AISB 2003), pp. 125–130 (2003)

A Computational Model of the Amygdala Nuclei's Role in Second Order Conditioning[*]

Francesco Mannella, Stefano Zappacosta,
Marco Mirolli, and Gianluca Baldassarre

Laboratory of Autonomous Robotics and Artificial Life,
Istituto di Scienze e Tecnologie della Cognizione,
Consiglio Nazionale delle Ricerche (LARAL-ISTC-CNR)
Via San Martino della Battaglia 44, I-00185 Roma, Italy
{francesco.mannella,stefano.zappacosta,marco.mirolli,
gianluca.baldassare}@istc.cnr.it

Abstract. The mechanisms underlying learning in classical conditioning experiments play a key role in many learning processes of real organisms. This paper presents a novel computational model that incorporates a biologically plausible hypothesis on the functions that the main nuclei of the amygdala might play in first and second order classical conditioning tasks. The model proposes that in these experiments the first and second order conditioned stimuli (CS) are associated both (a) with the unconditioned stimuli (US) within the basolateral amygdala (BLA), and (b) directly with the unconditioned responses (UR) through the connections linking the lateral amygdala (LA) to the central nucleus of amygdala (CeA). The model, embodied in a simulated robotic rat, is validated by reproducing the results of first and second order conditioning experiments of both sham-lesioned and BLA-lesioned real rats.

1 Introduction

Individual learning plays a fundamental role in adaptive behavior of organisms, especially in most sophisticated ones like mammals. Some of the most important mechanisms underlying learning are those studied in classical (Pavlovian) conditioning experiments. In these experiments an animal experiences a systematic association between a neutral stimulus, for example a light (the "conditioned stimulus" or "CS"), and a biologically salient stimulus, for example food (the "unconditioned stimulus" or "US"), to which it tends to react with an innate set of responses appropriate for the US, for example orienting and approaching (the "unconditioned responses" or "UR"). After repeated exposure to couples of CS-US the animal produces the UR even if CS are presented alone.

Since Pavlov's pioneering works [1], a lot of research has addressed classical conditioning phenomena producing a huge amount of behavioral and neural data

---

[*] This research was supported by the EU Project *ICEA - Integrating Cognition, Emotion and Autonomy*, contract no. FP6-IST-IP-027819.

M. Asada et al. (Eds.): SAB 2008, LNAI 5040, pp. 321–330, 2008.

[2]. However, we still lack a comprehensive theory able to explain the full range of these empirical data. Trying to build detailed *biologically plausible* computational models is a necessary step to overcome this knowledge gap. The current most influential models on classical conditioning, those based on "temporal-difference reward prediction error" [3,4] , suffer of several limitations. The main reason is that they have been developed within the machine learning framework with the aim of building artificial machines capable of autonomously learning to perform actions useful for the user. For this reason they are suitable to investigate *instrumental conditioning* phenomena – a type of associative learning based on stimulus-actions associations – but less adequate to explain Pavlovian phenomena mainly based on stimulus-stimulus associations [5,6].

A crucial question on classical conditioning regards the nature of the acquired association between the CS and the UR: is this association direct (CS-UR), as Pavlov himself seemed to claim [1], or does it pass through the unconditioned stimuli (CS-US-UR), as Hull [7] suggested? The long-lasting debate on this topic [2] seems now settled in favor of both hypotheses: in fact, there is now strong empirical evidence supporting the co-existence of both CS-UR and CS-US associations [5,8]. However, a clear understanding of the neural substrates which might be responsible for these two kinds of associations has yet to be gained. In particular, none of the computational models of classical conditioning based on the temporal-difference mechanisms, nor the models which have been proposed as alternatives to them [5,6,9,10], make any significant claim on this point.

Within the empirical literature, Cardinal et al. [8] formulated an interesting hypothesis on the neural basis of stimulus-stimulus and stimulus-response Pavlovian associations. According to this hypothesis, the basolateral amygdala (BLA) stores the CS-US associations, whereas the central nucleus of amygdala (CeA) receives or stores the CS-UR associations (CS-UR associations encoded in the cerebellum [11] are not considered here).

This paper presents an original computational model implementing that general hypothesis. In particular, it represents the first working model specifying the different functions played by the main sub-nuclei of amygdala in classical conditioning. The model, embodied in a simulated robotic rat, is validated by reproducing the results obtained with some first and second order conditioning experiments conducted with sham and BLA-lesioned real rats [12].

Sect. 2 presents the target experiment and the simulated experimental setup. Sect. 3 describes the model's general functioning and the biological constraints taken into account. The mathematical details of the model are presented in the Appendix. Sect. 4 shows the results of the tests of the model and compares them with those obtained with real rats. Finally, Sect. 5 concludes the paper.

2 The Target Experiment and the Simulated Environment

The model is validated by reproducing second-order conditioning experiments on real rats (reported as experiment 1a in [12]). The real experiment was conducted with 19 BLA-lesioned rats and 27 sham-lesioned rats, measuring the behaviours of walking, orienting and "food-cup" (insertion of head in the food dispenser). Namely, in the first phase both groups were trained for 8 sessions lasting 64 min each to acquire a first order conditioned behaviour. Each session was formed by a sequence of trials. In each trial a 10 sec light stimulus was presented, followed by the delivery of Noyes pellets (food) in the food dispenser. Recordings showed that both sham and lesioned rats were able to acquire first order conditioned behaviours. In the second phase the same rats were trained for 3 sessions of 64 min each to acquire a second order conditioned behaviour. A tone stimulus was presented for 10 sec followed by the light stimulus; every 3 trials a "reminder" of the light-food association was presented. The key result was that only sham rats acquired the second order CS-UR association. In accordance with other empirical evidences (see [8] for a review), these experiments suggest that BLA plays a fundamental role in the formation of the association between the CS and the incentive value of the US, and that this association plays a key role in the acquisition of the CS-UR association in second order conditioning.

The real experiment was simulated through a robotic rat ("ICEAsim") developed within the EU project ICEA on the basis of the physics 3D simulator Webots^{TM}. The model was written in Matlab^{TM} and was interfaced with ICEAsim through a TCP/IP connection. The robotic setup is shown in Fig. 1. The environment is formed by a gray-walled chamber, and the stimuli are expressed by 3 panels (vision is used, as no sound is supported by Webots): food delivering in the dispenser occurs when the green panel turns on, light when the yellow one is on, and tone when the red one is on. When one of those stimuli elicits an orienting response within the controller (see Sect. 3), the rat turns toward the panel and then approaches it (these behaviors are hardwired). This behavioural sequence terminates when the rat touches the food-dispenser (that is assumed to correspond to a food-cup behaviour).Although the "degree of embodiment and situatedness" of the setup is rather limited, nevertheless a robotic test was used because in the future we plan to scale the model to more realistic scenarios (for example, the random-lasting time intervals elapsing between rats' orienting and food deliver already started to challenge the robustness of the associative learning algorithms used).

3   The Model

This section presents a general description of the functioning of the model and the biological constraints that it satisfies, while a detailed mathematical description of it (included all the equations) is reported in the Appendix. A key feature of the model (Fig. 1) is the explicit representation of the three major anatomical components of the amygdala [13]: the lateral amygdala (LA), the basolateral amygdala (BLA), and the central nucleus of amygdala (CeA). The model assumes that these components form two functional sub-systems: (1) the LA-CeA sub-system, which forms S-R associations, and (2) the BLA sub-system, which forms S-S associations. Note that in the following "neurons" have to be

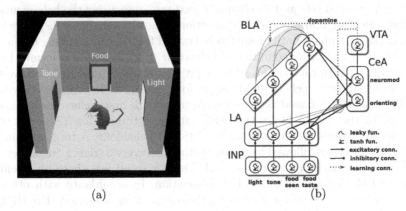

(a)   (b)

Fig. 1. (a) A snapshot of the simulator, showing the simulated rat at the centre of the experimental chamber, the food dispenser (at the rat's right hand side), the light panel (behind the rat) and the tone panel (in front of the rat). (b) The architecture of the model: bold and plain arrows indicate innate and trained connections, respectively.

intended as units whose functioning abstracts the collective functioning of whole assemblies of real neurons.

The Stimulus-Response Associator (LA-CeA). The LA is the main input of the amygdala system. It receives afferent connections from various sensory and associative areas of cortex, from thalamus, and from deeper regions within the brain-stem, and it sends efferent connections both to BLA and to CeA. The model has an input layer (INP) of four leaky neurons (**inp**) activated by four binary sensors (**s**) which encode the presence/absence of four stimuli: light (s_{li}), tone (s_{to}), food sight (s_{fs}) and food taste (s_{ft}) (Eq. (1)). LA (**la**) is formed by four leaky neurons receiving one-to-one afferent connections from INP (Eq. (2)).

The CeA is one of the main output gates of amygdala. Its efferent connections innervate regions of the brainstem controlling mainly: (1) body and behavioral reactions through the hypothalamus and periaqueductal gray [14]; (2) the release of basic neuromodulators through the ventral tegmental area (dopamine), the locus coeruleus (norepinephrine), and the raphe nuclei (serotonin) [13,15,16]. These neuromodulators play a fundamental role in learning processes but for simplicity this model considers only dopamine [17] (in particular it ignores the role that norepinephrine plays in AMG learning [18]). In the model CeA (**cea**) is formed by two leaky neurons, one (cea_{or}) encoding the rat's orienting behavior, and one (cea_{da}) connected to the ventral tegmental area (VTA) to produce the dopamine signal (da) (Eqs. (4) and (5)).

In the model, all LA neurons are connected to the orienting neuron of CeA (cea_{or}), whereas only the food taste neuron (la_{ft}) is connected to the neuromodulator neuron of CeA (cea_{da}). These connectivity allows stimuli representations of LA to be associated with the orienting behaviour in CeA but not with the dopamine neuromodulation. This is a key assumption to explain why LA-CeA associations can learn first order CS-US associations but not second order ones: conditioned stimuli cannot access the incentive value of rewarding stimuli.

The connections from LA to CeA are trained on the basis of a Hebb rule. In particular, the strengthening of connections takes place in the presence of three conditions (Eq. (6)): (1) a high value of the trace of the LA activation onset (la_tr): the use of the *onset* makes learning happen only when LA neurons' activation precedes CeA neurons' activation, while the use of the *trace* allows overcoming the time gap between CS and UR; (2) a high activation of CeA neurons (cea_{or} and da); (3) a dopamine level (da) over its threshold (th_{da}).

The Stimulus-Stimulus Associator (BLA). The BLA has afferent connections from LA and efferent connections to CeA [19,20]. BLA is also interconnected with the orbitofrontal cortex and hippocampus, and sends efferent connections to the nucleus accumbens: all these connections are ignored here (see [21] for a model where BLA-nucleus accumbens connections play a key role).

In the model, BLA (**bla**) is formed by four leaky units which receive one-to-one connections from LA (**la**) and have all-to-all lateral connections (Eq. (7)). Only the neuron encoding food taste (bla_{ft}) is connected to CeA neurons. This implies that all neurons of BLA representing stimuli different from the US (bla_{ft}) can exert effects on the CeA output neurons only via lateral stimulus-stimulus connections with the BLA's US neuron.

Learning of BLA lateral connections is based on a time-dependent Hebb algorithm. The key aspect of the algorithm is that it allows both the onset and the offset of BLA neurons preceding the onset of other BLA neurons to increase the connection from the former to the latter, provided that dopamine overcomes its threshold (Eqs. (8), (9), (10)). The sensitivity to the offset of stimuli was necessary due to the long duration of the CS stimuli, see Sect. 2 (cf. [21] for a simpler version of the algorithm using only the onset of presynaptic neurons).

4   Results

Figure 2a compares the percentage of times the tone elicits an orienting behaviour in real [12] and simulated rats after the second order conditioning phase. The main result of the experiment has been qualitatively reproduced by the model: in both real and simulated rats a BLA lesion prevents second order conditioning to take place. The analysis of the detailed functioning of the model provides an explanation for this result. Figure 2b shows the activations of some key neurons of: (1) a simulated sham rat during the first order conditioning phase with the light-food contingency; (2) the same sham rat during the second order conditioning phase with the noise-light contingency; (3) a simulated BLA-lesioned rat during the second order conditioning phase.

Figure 2b, first block, shows the mechanisms underlying first order conditioning in a sham simulated rat. At the beginning of the first trial, the appearance of light activates the light-related BLA neuron (bla_{li}). After a while, the appearance of food activates the food-sight BLA neuron (bla_{fs}). The bla_{fs} pre-activates the BLA food-taste neuron (bla_{ft}) before the rat actually reaches the food thanks

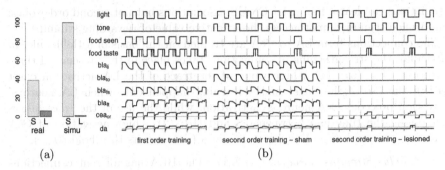

(a)   (b)

Fig. 2. (a) Percentage of orienting behaviours of sham (S) and lesioned (L) rats in response to the tone after second order conditioning: data from real (first two bars) and simulated rats (last two bars). (b) Stimuli, activations of key neurons, and dopamine release in 3 conditions: first-order and second-order conditioning phases of a sham rat (first and second block, respectively), and second-order conditioning phase of a BLA-lesioned rat. Trials are separated by short vertical dotted lines; thresholds (for orienting behavior and dopamine learning) are represented as gray horizontal dotted lines.

to a bla_{fs}-bla_{ft} excitatory connection which is assumed to be learned before the conditioning training (see the Appendix). In turn, the bla_{ft} triggers both the orienting behavior via the orienting CeA neuron (cea_{or}) and the release of dopamine (da) by the VTA via the CeA neuromodulation neuron. The release of above-threshold dopamine triggers the learning of both the connection between the light neuron in LA and the orienting neuron in CeA (implementing the CS-UR association) and the connections linking the light neuron with the food sight and food taste neurons in BLA (implementing the CS-US association). The result is that after a very few trials the bla_{fs} and bla_{ft} neurons start to be pre-activated as soon as the light is perceived. This results in an early activation of both CeA neurons and, consequently, in an early dopamine release and an early orienting response to the light.

As in the target experiment, during the second order conditioning phase the rats are exposed to sequences of four trials composed by three tone-light presentations and one light-food "reminder". Thanks to the CS-US BLA association acquired during the first phase, in sham rats (Fig. 2b, second block) the presentation of light immediately triggers both orienting behavior and dopamine release. This ability of light to trigger dopamine release permits the acquisition of the second-order association between the tone and the URs (orienting response and dopamine release) in a manner which is completely analogous to what happens in the first-order conditioning with respect to light.

On the other hand, second order conditioning cannot take place in BLA-lesioned rats (Fig. 2b, third block). The reason is that in this case light can trigger only the orienting response via the connection linking the light representation in LA with the orienting neuron in CeA (the direct CS-UR association), but not the dopamine release, which requires the activation of the food-taste representation in either BLA (which is lesioned) or LA (which is activated only when food is effectively eaten). As a result, since synaptic modification depends on dopamine, no learning can takes place during second-order conditioning.

5   Conclusions

This paper presented an original computational model of the basic brain mechanisms underlying classical conditioning phenomena. The architecture and functioning of the model was constrained on the basis of neural empirical data on the amygdala. The fundamental assumption underlying the model is that the association between conditioned stimuli (CS) and unconditioned responses (UR) formed in classical conditioning experiments is due to two related but distinct mechanisms: (1) stimulus-stimulus associations (CS-US-UR) involving unconditioned stimuli (US) stored in the BLA; (2) direct stimulus-response associations (CS-UR) stored in the LA-CeA neural pathway.

The model was embedded in a simulated robotic rat and was validated by reproducing the behaviours exhibited by both sham and BLA-lesioned rats in first and second order conditioning experiments. In particular, as in real rats, while after training the simulated sham rats react with UR (orienting) to both first and second order CS, BLA-lesioned simulated rats associate UR only to first order CS, but not to second order CS. The model is able to reproduce and explain these results thanks to the fundamental aforementioned assumption. During first order conditioning sham rats acquire both the direct CS-UR and the indirect CS-US-UR association. It is the first order CS-US association within BLA which permits the acquisition of the second order association as it allows the CS to reactivate the appetitive value of the US even when the US is absent. In contrast, BLA-lesioned rats can acquire direct first order CS-UR associations stored in the LA-CeA neural pathway but they cannot acquire the second order association because the first order CS has not access to the appetitive value of the US. To the best of the authors knowledge, this is the first model to propose such a specific computational hypothesis regarding the double association CS-US and CS-UR in classical conditioning.

Notwithstanding its strengths, the model suffers at least two significant limitations. First, the whole behavioral sequence triggered by the activation of the orienting neuron in CeA (orienting, approaching, and food-cup) is fully hardwired. For this reason, the model cannot reproduce the results on CeA-lesioned rats which are reported in the same article of the experiment targeted here [12]. Second, in contrast to most existing models of classical conditioning [5,6,9], the current model does not implement any mechanism for reproducing the exact timing of dopamine release observed in real animals. For this reason the model cannot reproduce another fundamental aspect of classical conditioning, that is extinction (the ability to re-learn not to respond to the CS if it stops to be followed by the US). We are currently working on improved versions of the present model for tackling both these limits.

References

1. Pavlov, I.P.: Conditioned Reflexes: An Investigation of the Physiological Activity of the Cerebral Cortex. Oxford University Press, Oxford (1927)
2. Lieberman, D.A.: Learning, behaviour and cognition. Brooks/Cole (1993)
3. Schultz, W., Dayan, P., Montague, P.: A neural substrate of prediction and reward. Science 275, 1593–1599 (1997)
4. Sutton, R.S., Barto, A.G.: Reinforcement Learning: An Introduction. MIT Press, Cambridge (1998)
5. Dayan, P., Balleine, B.: Reward, motivation and reinforcement learning. Neuron 36, 285–298 (2002)
6. O'Reilly, R.C., Frank, M.J., Hazy, T.E., Watz, B.: PVLV: the primary value and learned value Pavlovian learning algorithm. Behav. Neurosci. 121(1), 31–49 (2007)
7. Hull, C.L.: Principles of behavior. Appleton-century-crofts (1943)
8. Cardinal, R., Parkinson, J., Hall, J., Everitt, B.: Emotion and motivation: the role of the amygdala, ventral striatum, and prefrontal cortex. Behav. Cogn. Neurosci. Rev. 26(3), 321–352 (2002)
9. Balkenius, C., Morèn, J.: Dynamics of a classical conditioning model. Auton. Robot. 7(1), 41–56 (1999)
10. Morèn, J., Balkenius, C.: A computational model of emotional learning in the amygdala. In: Meyer, J.A., Berthoz, A., Floreano, D., Roitblat, H.L., Wilson, S.W. (eds.) From Animals to Animats 6: Proceedings of the 6th International Conference on the Simulation of Adaptive Behaviour. The MIT Press, Cambridge (2000)
11. Thompson, R.F., Swain, R., Clark, R., Shinkman, P.: Intracerebellar conditioning–Brogden and Gantt revisited. Behav. Brain Res. 110(1-2), 3–11 (2000)
12. Hatfield, T., Han, J.S., Conley, M., Gallagher, M., Holland, P.: Neurotoxic lesions of basolateral, but not central, amygdala interfere with Pavlovian second-order conditioning and reinforcer devaluation effects. J. Neurosci. 16(16), 5256–5265 (1996)
13. Pitkänen, A., Jolkkonen, E., Kemppainen, S.: Anatomic heterogeneity of the rat amygdaloid complex. Folia Morphol. 59(1), 1–23 (2000)
14. Phelps, E.A., LeDoux, J.E.: Contributions of the amygdala to emotion processing: from animal models to human behavior. Neuron 48(2), 175–187 (2005)
15. Rosen, J.B.: The neurobiology of conditioned and unconditioned fear: a neurobehavioral system analysis of the amygdala. Behav. Cogn. Neurosci. Rev. 3(1), 23–41 (2004)
16. Fudge, J.L., Emiliano, A.B.: The extended amygdala and the dopamine system: another piece of the dopamine puzzle. J. Neuropsych. Clin. N. 15(3), 306–316 (2003)
17. LaLumiere, R.T., Nawar, E.M., McGaugh, J.L.: Modulation of memory consolidation by the basolateral amygdala or nucleus accumbens shell requires concurrent dopamine receptor activation in both brain regions. Learn. Memory 12(3), 296–301 (2005)
18. Berridge, C.W., Waterhouse, B.D.: The locus coeruleus-noradrenergic system: modulation of behavioral state and state-dependent cognitive processes. Brain Res. Rev. 42(1), 33–84 (2003)
19. Rolls, E.T.: Précis of The brain and emotion. Behav. Brain Sci. 23(2), 177–191 (2000); discussion 192–233
20. Saddoris, M.P., Gallagher, M., Schoenbaum, G.: Rapid associative encoding in basolateral amygdala depends on connections with orbitofrontal cortex. Neuron 46(2), 321–331 (2005)

21. Mannella, F., Mirolli, M., Baldassarre, G.: The role of amygdala in devaluation: a model tested with a simulated robot. In: Berthouze, L., Prince, C.G., Littman, M., Kozima, H., Balkenius, C. (eds.) Proceedings of the Seventh International Conference on Epigenetic Robotics, Lund University Cognitive Studies, pp. 77–84 (2007)

Appendix: Mathematical Details of the Model

Throughout the Appendix, τ_x denotes the decay rate of a leaky quantity x, the sub-index $\cdot_p$ denotes the activation potential of the corresponding neuron, symbols $\mathbf{X}$, $\mathbf{x}$, and x are used respectively to denote matrices, vectors and scalars, the function φ is defined as $\varphi[x] = \max[0, x]$ and the function χ as $\chi[x] = 1\, if\, x \geq 0\, else\, \chi = 0$. The values of parameters are listed at the end of the Appendix.

LA-CeA: Functioning and Learning. INP (**inp**) processes the input signal from sensors $\mathbf{s} = (s_{li},\ s_{to},\ s_{fs},\ s_{ft})'$ with a leak function:

$$\tau_{\mathbf{inp}} \cdot \dot{\mathbf{inp}} = -\mathbf{inp} + \mathbf{s} . \tag{1}$$

LA is formed by four leaky neurons (**la**) activated as follows:

$$\tau_{\mathbf{la}} \cdot \dot{\mathbf{la}}_p = -\mathbf{la}_p + w_{\mathbf{inp,la}} \cdot \mathbf{inp} , \qquad \mathbf{la} = \varphi[\tanh[\mathbf{la}_p]] \tag{2}$$

where $w_{\mathbf{inp,la}}$ is the fixed weight of the connections from IMP to LA. The "double leak" processing of signals implemented by IMP and LA is used to smooth the derivative of LA (see Eq. (3)).

The trace of LA neurons (**la_tr**) is a leak function of the positive value of the derivative of their activation (**l̇a**):

$$\tau_{\mathbf{la\_tr}} \cdot \dot{\mathbf{la\_tr}}_p = -\mathbf{la\_tr}_p + b_{\mathbf{la\_tr}} \cdot \varphi[\dot{\mathbf{la}}] , \qquad \mathbf{la\_tr} = \varphi[\tanh[\mathbf{la\_tr}_p]] \tag{3}$$

where $b_{\mathbf{la\_tr}}$ is an amplification coefficient.

CeA is formed by two leaky neurons (**cea**) activated as follows:

$$\tau_{\mathbf{cea}} \cdot \dot{\mathbf{cea}}_p = -\mathbf{cea}_p + \mathbf{W}_{\mathbf{la,cea}} \cdot \mathbf{la} + \mathbf{W}_{\mathbf{bla,cea}} \cdot \mathbf{bla} \tag{4}$$
$$\mathbf{cea} = \varphi[\tanh[\mathbf{cea}_p]]$$

VTA is formed by a dopamine leaky neuron (da) which activates as follows:

$$\tau_{da} \cdot \dot{da}_p = -da_p + bl_{da} + \mathbf{w}_{\mathbf{cea},da} \cdot \mathbf{cea} , \qquad da = \varphi[\tanh[da_p]] \tag{5}$$

where bl_{da} is the dopamine baseline.

The weights of the LA-CeA connections ($\mathbf{W}_{\mathbf{la,cea}}$) are updated with a three-element Hebb rule involving CeA, LA's trace and dopamine:

$$\Delta\mathbf{W}_{\mathbf{la,cea}} = \eta_{\mathbf{la,cea}} \cdot (\chi[da - th_{da}] \cdot da) \cdot \mathbf{cea} \cdot \mathbf{la\_tr}' \cdot (1 - |\mathbf{W}_{\mathbf{la,cea}}|) \tag{6}$$

where $\eta_{\mathbf{la,cea}}$ is a learning rate, the term $(\chi[da - th_{da}] \cdot da)$ implies that learning takes place only when $da \geq th_{da}$, and the term $(1 - |\mathbf{W}_{\mathbf{bla}}|)$ keeps the weights in the range $[-1, 1]$.

BLA: Functioning and Learning. BLA is formed by four leaky neurons (**bla**) activated as follows:

$$\tau_{\mathbf{bla}} \cdot \dot{\mathbf{bla}}_p = -\mathbf{bla}_p + \mathbf{W_{bla}} \cdot \mathbf{bla} + (w_{\mathbf{la,bla}} \cdot \mathbf{la} + c_{\mathbf{bla}} \cdot \mathbf{la\_tr}) \tag{7}$$
$$\mathbf{bla} = \varphi[\tanh[\mathbf{bla}_p]]$$

where $c_{\mathbf{la\_tr}}$ is an amplification coefficient. According to this equation, with a transient constant input signal the activation of a BLA neuron presents a high initial peak (due to **la_tr**) followed by a lower constant value (due to **la**) and then by a smooth descent to 0 (due to the leak after the signal end): this activation has a derivative suitable for BLA learning (see below).

In order to train lateral connections of BLA, a trace of the derivative of the activation of BLA neurons **bla_tr** is computed as follows:

$$\tau_{\mathbf{bla\_tr}} \cdot \dot{\mathbf{bla\_tr}}_p = -\mathbf{bla\_tr}_p + \cdot \dot{\mathbf{bla}} . \tag{8}$$

Small values of this trace are ignored in the learning algorithm by considering the "cut trace" **bla_tr_cut** defined as: **bla_tr_cut** = **bla_tr** *if* |**bla_tr**| < $th_{\mathbf{bla\_tr}}$ *else* **bla_tr_cut** = 0. Given the activation dynamics of BLA (Eq. (7)), the corresponding derivative (and, with some delay, its trace) presents: (1) an initial peak at signal onset; (2) a negative peak at the end of the signal onset; (3) a negative peak at the signal offset. The key point of the learning algorithm of BLA is that a connection between two neurons is potentiated in coincidence of a negative peak of the presynaptic neuron and a positive peak of the postsynaptic neuron. These two events mark a pre-synaptic-onset/post-synaptic-onset sequence (or a pre-synaptic-offset/post-synaptic-onset one). The matrix **S**, reported below, captures these conditions for all couples of neurons:

$$\mathbf{S} = \chi[\mathbf{bla\_tr\_cut}] \cdot \chi[-\mathbf{bla\_tr\_cut}]' - \chi[-\mathbf{bla\_tr\_cut}] \cdot \chi[\mathbf{bla\_tr\_cut}]' . \tag{9}$$

Denoting with *pre* and *post* the presynaptic and postsynaptic neurons, **S** has an entry equal to 1 when $bla\_tr\_co_{pre} < 0$ and $bla\_tr\_co_{post} > 0$, equal to -1 when $bla\_tr\_co_{pre} < 0$ and $bla\_tr\_co_{post} > 0$, and equal to 0 otherwise. The learning rule of lateral connections is then:

$$\Delta \mathbf{W_{bla}} = \eta_{\mathbf{bla}} \cdot \chi[da - th_{da}]da \cdot (ltp_{\mathbf{bla}} \cdot \varphi[\mathbf{S}] + ltd_{\mathbf{bla}} \cdot \varphi[-\mathbf{S}])(1 - |\mathbf{W_{bla}}|) \tag{10}$$

where $\eta_{\mathbf{bla}}$ is a learning rate, $ltp_{\mathbf{bla}}$ is a long time potentiation coefficient, and $ltd_{\mathbf{bla}}$ is a short term depression coefficient.

Model's Parameters. The model's parameters were set as follows: $\tau_{\mathbf{inp}} = \tau_{\mathbf{la}} = \tau_{\mathbf{bla}} = 500$ ms, $\tau_{\mathbf{la\_tr}} = \tau_{\mathbf{bla\_tr}} = 5000$ ms, $\tau_{\mathbf{cea}} = 100$ ms, $\tau_{da} = 50$ ms, $w_{\mathbf{inp,la}} = 10$, $b_{\mathbf{la\_tr}} = 1000$, $w_{\mathbf{la,bla}} = 0.5$, $c_{\mathbf{bla}} = 60$, $bl_{da} = 0.3$, $th_{da} = 0.6$, $th_{\mathbf{la\_tr}} = 0.00001$, $\eta_{\mathbf{bla}} = 0.0005$, $\eta_{\mathbf{la,cea}} = 0.15$, $ltp_{\mathbf{bla}} = 1.0$, $ltd_{\mathbf{bla}} = 0.3$. Some connections, assumed to be innate or pre-learned, are clumped to 1 (*l*=learned): $w_{\mathbf{bla}\,fs,ft} = 1$, $\mathbf{w_{cea,da}} = (1, 0)$, $\mathbf{W_{la,cea}} = \begin{pmatrix} l & l & l & 1 \\ l & l & 1 & 1 \end{pmatrix}$, $\mathbf{W_{bla,cea}} = \begin{pmatrix} l & l & l & 1 \\ l & l & l & 1 \end{pmatrix}$. The model's equations were integrated with the Euler method with a 50 ms step.

Acquiring a Functionally Compositional System of Goal-Directed Actions of a Simulated Agent

Yuuya Sugita and Jun Tani

RIKEN Brain Science Institute, 2-1 Hirosawa, Wako-shi, Saitama, 3510198, Japan
{sugita,tani}@bdc.brain.riken.go.jp

Abstract. We propose a sub-symbolic connectionist model in which a functionally compositional system self-organizes by learning a provided set of goal-directed actions. This approach is compatible with an idea taken from usage-based accounts of the developmental learning of language, especially one theory of infants' acquisition process of symbols. The presented model potentially explains a possible continuous process underlying the transitions from rote knowledge to systematized knowledge by drawing an analogy to the formation process of a geometric regular arrangement of points. Based on the experimental results, the essential underlying process is discussed.

1 Introduction

In this study, we try to examine the mechanisms in our mind that are involved in the shift from unrelated rote knowledge acquired by learning examples of objects or events into a flexible conceptual system by which we can conceive something not experienced as a recombination of the examples. Tomasello reported in [1] that infants can appropriately use holophrases, which are indivisible sentences such as "lemme-see," in a communicative context before understanding reusable units, which include phrases and words such as "let," "me" and "see." In his usage-based accounts of language development, each transition of the performance is explained in terms of the acquisition of a new type of smaller and more abstract symbolic device.

It is, however, difficult to transfer the idea that the utilization of wholes precedes the emergence of parts from explanatory to computational models. One of the most substantial problems is how to implement the acquisition of a composition rule, which combines smaller units into a whole concept. Cognitive theories often neglect the composition rule, since the rule and the units are considered to be two sides of the same coin [2]. This belief is plausible only as far as symbolic manipulation is concerned. However, the realization of an embodied composition rule, which is the correspondence of the symbolic manipulation in reality, requires much more than the acquisition of a mere syntactic structure of the symbolic system. Let us consider the case in which an agent generates an action specified by a target object and an operation on it. It is quite easy to represent the action in a symbolic system; a pair of symbols representing the target and operation is enough. On the contrary, an embodied composition rule, which is required to generate an action relevant to the pair, is not so trivial. The problem is that the rule tends to be too abstract to be learned by examples, because the rule needs to capture

M. Asada et al. (Eds.): SAB 2008, LNAI 5040, pp. 331–341, 2008.
© Springer-Verlag Berlin Heidelberg 2008

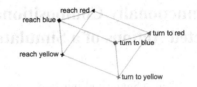

Fig. 1. Systematicity among concepts can be represented based on the geometric regularity

the anything residual that cannot be collected as symbols, which are usually grounded to something concrete. In fact, many computational models employ a pre-programmed composition mechanism, although their objectives are different from ours [3,4,5].

In order to avoid the difficulty concerning the abstractness of the explicit composition rule, this study investigates a novel embodied implementation of a functionally compositional system in the domain of goal-directed actions of a simulated agent. A functionally compositional system is one which does not keep any reusable units explicitly in the form of symbols but works like a conventional compositional system [6]. Instead of dealing with reusable parts explicitly, the functionally compositional system focuses on the systematic relationships among wholes. Each whole concept is embedded as points in a conceptual space implemented as an *n*-dimensional vector space. The geometric arrangement of these points represents the underlying combinatoriality among them. For example, a system of six actions specified by every possible combination of one of three objects and one of two operations is represented as a triangular prism, as shown in Fig. 1. Even if the positions of some actions are unknown, they can be inferred by utilizing the geometric regularity. Furthermore, this framework explains the transitions from rote knowledge to systematic knowledge in terms of a continuous internal process. The emergence of the regularity involved in the transition can be realized by the continuous motion of each point. It is also remarkable that each whole concept does not change through the transition. Only their relationships are altered, whereas the conventional implementation undergoes the replacement of a holistic symbol with a combination of elemental symbols.

In the following, we propose a computational model whereby the geometric regularity self-organizes through the learning of examples. Our experimental setting and connectionist architecture is explained in Section 2. In Section 3, the experimental results, which demonstrate that three different types of combinatorial generalizations are realized by the same model, are presented. An analysis of the result is shown in Section 4, and some related cognitive problems are discussed in Section 5.

2   Experimental Setting

In our experiment, a simulated mobile agent learns incomplete parts of a total of 36 different goal-directed actions; the actions are characterized by combinations of a target object, an operation on the target, and an optional verb modifier. The learning is conducted under the supervision of teaching programs written for this study. The agent is a model of the mobile robot depicted in Fig. 2(a). This robot has a color camera with a range of view of 120 degrees and two rotating motors driving each of its two wheels.

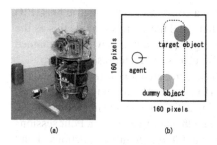

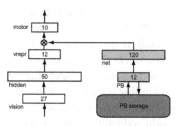

(a)   (b)

Fig. 2. The simulated agent. The agent is based on (a) a mobile robot that performs 36 types of goal-directed actions on a stage (b) where a target and an optional dummy colored object are placed randomly within a dashed square.

Fig. 3. The architecture of the learning network is presented. A rectangle represents a layer. The number of nodes contained in the layer is denoted by the number in the rectangle. The gray layers have a slower time constant than do the white layers. PB (parametric bias) storage is a working area.

In each experimental trial, the agent was required to perform either of two operations reach or turnto one of six colored objects (blue, cyan, green, yellow, orange, and magenta) in the environment shown in Fig. 2(b), where one or two objects, one of which is the target, are randomly placed. In actions involving reach, the agent is required to move toward the target and then to stop just before touching it. In the turnto actions, it has to pivot to the target. In our experiment, an operation turnto takes a verb modifier, which designates the offset angles (-30, -18, 0, +18, and +30) of the final position of the target in the visual field from the agent's center. A negative offset indicates the offset to the left, and zero is omitted. In the following, the action is denoted as a triplet consisting of the operation, target, and offset, for example, turnto-blue+18. It should be noted that this notation is used only for our convenience and the agent has no way to access it. In some situations, turnto with an offset is regarded as an operation, and, for example, is denoted as turnto+18.

As mentioned above, the actions are embedded in the concept vector space through the learning process. Unlike the conventional associative learning between an action and a vector, the vector is not provided *a priori*. Instead, the geometric arrangement of the vectors self-organizes the structure, reflecting the relationships among the actions, including unseen ones. The learning model which acquires this structure-preserving map is a connectionist network shown in Fig. 3. The network consists of two parts, each of which is tailored to its own functions. One part is a base-level network (base-net) which takes the visual information from the camera as input and outputs the angular velocities of the wheels (the left side of Fig. 3). The base-net is basically a conventional layered neural network except that it has second-order connections [7] between the vrepr and motor layers. This special mechanism enables the base-net to switch its function. Depending on an action to be performed, the base-net generates different motor values for identical vision input. The second-order connection is controlled by the meta-level network (meta-net) depicted on the right side of Fig. 3. The meta-net is also a conventional layered network. As input, it takes a vector encoding an action once at the beginning of the action, then, it outputs the weights of the second-order connections constantly until the action finishes. The input layer works just like a conventional parametric bias (PB) layer [8] which has an infinitely long time constant, and therefore we name this as the PB layer. PB storage is as the working area during the learning, as will be explained later in detail. It also keeps the self-organized PB vectors for a later test phase.

An experimental session consists of three phases: the creation of training data, the learning of the data, and the evaluation of the performance. The training data were created by sampling sensor-motor time series involving actions generated in an algorithmic manner. Then, the network learned a part of the data in an offline manner. Four sessions were conducted with supervised data consisting of a different number of actions: 4 and 21 out of 36 actions in the most sparse and dense cases, respectively. After the training error of the network decreased sufficiently, the performance was evaluated. A PB vector of an unseen action was computed by recognizing unused training data, as explained below. The agent, which was controlled by the network, was tested to determine whether it could make a previously unexperienced action in a novel environment with using the PB vector. In the remaining sections, each phase is explained in detail.

Phase 1: Generating Examples by Teaching Programs. For each of the 36 actions, 120 time series were recorded in different environments. In 20 out of the 120 cases, only a target object was placed in the stage, and in the remaining 100 cases, a dummy object was placed in addition to the target object. The dummy object was chosen from 5 objects, and therefore 20 time series were generated for each. Both a target and an optional dummy object were arranged at random positions within the area range shown by the dashed square in Fig. 2(b). Any arrangement where the target was occluded by the dummy at the home position of the agent was omitted.

Each exemplar time series consists of pairs of visual information and the corresponding motor value computed by a manually coded teaching program. This approach may seem inappropriate. However, if the agent learns the exemplars by rote, there would be no need to learn the action by using the network. The actual objective of the learning is, therefore, to establish the relationships among the provided exemplars in an unsupervised manner. Also, it should be mentioned again that a PB vector has no exemplar.

The teaching program calculates the desired rotation speed of the two wheels of the agent from the position of the specified target taken from a camera image at a constant time interval. At the same time, 27-dimensional visual and 10-dimensional motor information are recorded for later learning (see Fig. 3). The visual input vector does not have the position of the target explicitly. From the viewpoint of the network, the visual field is composed of nine vertically divided regions. Each region is represented by the fraction of the region covered by colored patches and the dominant hue of the patches in the region. The hue is encoded by the position $(\cos\theta, \sin\theta)$ in the color circle, where pure red, yellow, green, and blue are represented as $\theta = 0°, 90°, 180°$, and $270°$, respectively. The desired speed of the wheel takes a real value ranging from -0.2 to 1.0. A negative value indicates reverse rotation. The motor vector is composed of two five-dimensional real-valued vectors, each of which represents the speed of the wheel in the form of $[f(0), f(0.25), f(0.5), f(0.75), f(1.0)]$, where f is a Gaussian distribution with the mean of the desired speed and a sigma of 0.25. This increases the robustness against the re-generation error of the network.

Phase 2: Batch Learning. The network learns incomplete parts of the 36 actions in a batch manner by employing the data prepared in the previous phase as the supervising signal. The learning process is formulated as a conventional iterative, steepest descent optimization with respect to the error function E, defined in (1). The model has two types of parameters to be optimized: one is the vector W consisting of all the connection weight values of the network; and the other is the set PB consisting of PB vectors pb_i for all supervised actions $i \in \mathcal{A}$.

$$E(W, PB) = \sum_{i \in \mathcal{A}} E_i(W, pb_i) \tag{1}$$

$$E_i(W, pb_i) = \sum_{j=0}^{119} \sum_{t=0}^{l_{ij}} E_{ij}(t; W, pb_i) \tag{2}$$

$$E_{ij}(t; W, pb_i) = \|\hat{m}_{ij}(t) - m(v_{ij}(t); W, pb_i)\|^2 , \tag{3}$$

where l_{ij} is the length of the j-th training data of an action i, $\hat{m}_{ij}(t)$ is the desired motor vector corresponding to the visual vector $v_{ij}(t)$ at the time step t in the training data, and $m(v_{ij}(t); W, pb_i)$ is its actual value generated by the network under the condition that the connection weight is W, and the PB vector for the action is pb_i with the identical vision input. The parameters W and PB are updated simultaneously by learning all the provided data in a batch manner. The learning procedure is implemented by using the conventional back-propagation algorithm. At the beginning, all the connection weight values are randomized with a small value, and $pb_i, \forall i \in \mathcal{A}$ are set to the zero vector. All the PB vectors reside in the storage since the values of the PB nodes are switched so that the network can learn all the given actions at the same time. And then, the following procedure is conducted 30,000 times.

(1) Do the following for each actions i in $\mathcal{A}$:
    (1.1) Load the stored pb_i to the PB nodes.
    (1.2) For each of the 120 sensor-motor time series, calculate the delta errors of connection weights $\partial E_{ij}/\partial W (t; W(T), pb_i(T))$ and of PB vector $\partial E_{ij}/\partial pb_i (t; W(T), pb_i(T))$ by using the back-propagation algorithm.
    (1.3) Update pb_i by using the summation of all the delta errors of pb_i for all time steps t of all time-series j of the action i, and store the updated vector to the storage.
(2) Update W by using the summation of all the delta errors of W for all time steps t of all time-series j of all the provided actions i.

Thus, the connection weights capture the common characteristics among all the actions and play a background part while each PB vector is specialized to its corresponding vector. In the analysis of the experimental results, we observe the acquired geometric structure constructed by the PB vectors in the conceptual space.

Phase 3: Examining the Generalization Capability. Two aspects of the generalization capability of the agent, 1) transfer of the skill to a novel environment and 2) recombination of the supervised actions into an unexperienced action, were tested. For examining the transfer of skill to a novel environment, the agent was tested to determine if it could accomplish each known action in 280 novel environments where a target and dummy object were placed in a systematic manner. The PB vectors acquired through the second phase were employed. This test reveals the kind of information kept in the vectors. If the vector codes only specific trajectories of taught examples without generalization, it is impossible to generate a goal-directed action in a different environment. In order to investigate the recombination of supervised actions into an unexperienced action, the PB vector encoding a novel action $i' \notin \mathcal{A}$ should be examined. The vector can be computed by the *recognition* procedure. The algorithm is basically identical to the learning procedure except that W is not updated. By employing 30 out of 120 examples of the action i' produced in the first phase, $pb_{i'}$ is optimized with regard to the error function for the action i' defined in (2) by using W acquired in the second phase. Once $pb_{i'}$ is obtained for each unseen action, the generation test can be conducted in the same way as in the trained action cases.

3   Results

We next observe the changes of the generalization capability depending on the sparseness of the provided examples. The degrees of generalization are compared for four sessions of teaching data of different sparseness. In Fig. 4, a trained action is indicated by a black box. In all the experiments, all the trained actions were regenerated successfully; this means the agent could accomplish the goal in more than 80 percent of the test environments explained above. A gray box shows an action achieved by the combinatorial generalization without extra teaching. The criteria of success for the novel action are identical to that for the trained action. In the remaining sections, the results are discussed only from the viewpoint of the performance. We'll re-examine issues about the underlying mechanism in the next section.

Experiment 1: Learning by Rote. In this case, no combinatorial generalization was observed because of very sparse training data (Fig. 4(a)). This suggests that the agent regarded the provided actions as being holistic; namely, it could not find any re-usable parts such as an operation and a target.

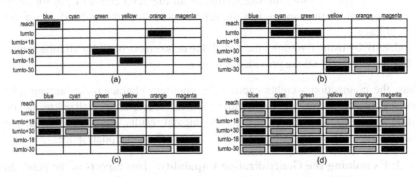

Fig. 4. The result of experiments 1-4 are shown in (a)-(d), respectively. A black box represents a trained action, and a gray box represents an action acquired as a recombination.

Experiment 2: A Local Compositional System. As training data increases, two novel actions `turnto-yellow-18` and `turnto-orange-30` were acquired without learning exemplars (Fig. 4(b)). This implies that the local compositional system is self-organized since one of the reusable operations `turnto-18` and `turnto-30` and one of the reusable targets `yellow`, `orange`, and `magenta` could be composed in any possible way, including unseen ones.

Experiment 3: Two Independent Compositional Systems / Categorization. Two separate local compositional systems emerge when further training data were added. One is system { `turnto, turnto+18, turnto+30` } × { `blue, cyan, green` }, and the other is system { `reach, turnto-18, turnto-30` } × { `yellow, orange, magenta` }. They are independent of each other since the targets of one system cannot be applied to the operations of the other system. The result can be interpreted as the categorization of targets based on operations applicable to these targets [9].

Experiment 4: Operation (Target, Offset). Finally, all the possible actions were acquired when the robot was trained with examples consisting of 21 out of the 36 actions. At least two incompatible interpretations of the results are possible. One is that all six operations have an equal relationship with one another. In this case, each operation is regarded as being a discrete symbol. The other possibility is that the similarity based on the offset values is understood. If so, the operations concerning `turnto` could have the structure `turnto` × { `-30, -18, 0, +18, +30` } (= `OFFSET`), and the `reach` operation exists apart of them. The result of an additional experiment proved that the latter interpretation is correct: the similarity based on the offset values was understood. The agent could re-generate some actions which have intermediate offsets such as `turnto-blue-24` by recognizing newly created examples of the actions. Last but not least, all six targets form `TARGET` class, since they can be applied to all operations equally. Thus, the roles of a target, an operation, and an offset emerge to organize the argument structure `TARGET` × ((`turnto` × `OFFSET`) + `reach`).

4   Analysis

For the analysis of the acquired PB structure, we discuss the underlying mechanism of the combinatorial generalization proposed in Section 1: the geometric regularity self-organized in the conceptual space. A main objective of the discussion is to bridge the gap between the symbolic behavior of the system and its sub-symbolic implementation.

Figure 5(a) shows a concept structure underlying the local compositional system observed in the second experiment. PB vectors for six actions included in the system are displayed (see also Fig. 4(b)). The displayed vectors are obtained through the learning process for a trained action and through the recognition process for an untrained one. The original 12-dimensional vectors are projected onto a 2-dimensional plane computed by applying the conventional principal component analysis (PCA) method to the six vectors. The accumulated contribution rate up to the second principal component (PC) is 0.79. A regular structure similar to the prism shown in Fig. 1 is observed in the figure, although the third and subsequent PCs show irregularity. Thus, an unexperienced

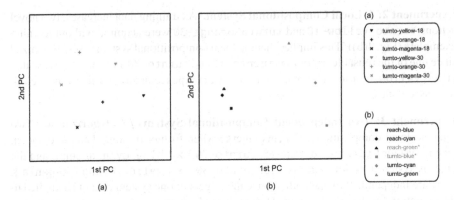

Fig. 5. Two parts of the PB space self-organized in experiment 2 are presented. Actions depicted in (a) constitute a geometric regularity as the underlying structure of the compositional system, whereas the actions depicted in (b) do not, as the agent failed to generate actions marked by *.

action has the "correct" position in the concept space. Two congruent triangles representing the two elements `turnto-18` and `turnto-30` are regarded as being equivalent in terms of their relation to the three elements `yellow`, `orange`, and `magenta`. This is a possible underlying representation of role-governed categories [10] of elements: these two groups have the role of an operation and a target, respectively. It should be noted that both roles emerge at the same time since both are defined in a circular manner. This is well represented geometrically in that the congruency of two triangles always accompanies the congruency of three lines corresponding to each of the targets. Meanwhile, no regularity is found in the plot of actions { `reach`, `turnto` } × { `blue`, `cyan`, `green` } (Fig. 5(b)). This is consistent with the performance that no combinatorial generalization was realized with regard to the abovementioned actions. A similar picture is found in the first experiment, where no generalization was realized.

In the third experiment, we can find two separate regular structures in accordance with the observed performance. They exist on different sub-spaces, although these sub-spaces are not orthogonal to each other. This explains the incompatibility of elements between the systems.

A new facet is discovered in the concept structure self-organized in the last experiment. Not only a structure representing the relationships among elements of different roles but also one representing the similarity among elements within each role are observed clearly. The former is the congruency of sub-units, which is similar to the structure found in the second and third experiments. The latter, with regard to each role of a target and an operation, are shown in Figs. 6(a) and (b), respectively.

The projection plane of Fig. 6(a) is chosen by applying PCA to representative vectors of the targets obtained by averaging the PB vectors for all operations for each of the targets. If a component of a target and an operation in the PB vectors are independent of each other, this method averages away the operation information. This assumption is shown to be true later in this paper. The accumulated contribution rate up to the second PC is more than 0.98, and so almost all the information is displayed in the plot. Six clusters corresponding to each of the targets are observed in the figure. This

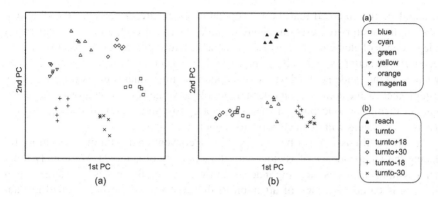

Fig. 6. The concept space self-organized in experiment 4. The space is projected to the planes in which the difference of 36 actions with respect to targets (a) and operations (b) are maximized.

implies that each target has its own representation in the subspace regardless of its surrounding context; namely, an operation takes the target as its argument. This can also be stated as follows: a subspace holding information of a specific role, a target in this case, emerges. Furthermore, the clusters are arranged in a circle like the continuum of color by hue. This arrangement suggests that the agent understands the similarity of color of the target. This is indirectly proven by the tendency to choose a target of a closely related but incorrect color. The more similar a dummy object is to a target with respect to color, the more easily the agent mistakes the dummy for the target. And so, the generalization of color is realized. When there is no specified target in the environment, the agent chooses an object that has a color similar to the target as a substitute.

In Fig. 6(b), the projection plane to see the difference among the operations is chosen by averaging the target information instead of the operation information. The accumulated contribution rate up to the second PC is more than 0.86. Here, both continuous and discrete sub-structures exist at the same time. In the first PC (x-axis of the figure), the continuum of operations turntos by offset emerges. Apart from that, the cluster of the operation reach is positioned. This implies that the second PC (y-axis of the figure) carries the distinction between reach and turnto. In addition, another continuum of the turnto operations by the absolute value of the offsets is found in the third PC, of which the contribution rate is approximately 0.12. Thus, the subspace of operations consists of three orthogonal components. In addition, the subspace of both targets and operations are also orthogonal to each other.

5   Discussion and Conclusion

We discover at last an underlying analog mechanism of the phenomenologic system of symbols inferred in Section 3 by considering the following correspondences:

(1) The analog correspondence of an elemental symbol is the center of gravity of a cluster of actions containing the element as a part (see Figs. 6(a) and (b)).
(2) The composition of symbols is realized by summing up their corresponding vectors.

It should be remembered that the conceptual elements are *dependent* on each other, although the vectors representing them are *independent* of each other. The superficial independence of elements strongly relies on the structure-preserving map between a PB vector and an action, which self-organizes in the connections of the network. The map provides all the fundamental devices to maintain the functional compositionality of the system, such as the composition rule and the role-governed categorization of elements. Without them, the conceptual elements cannot constitute an embodied whole action, just as chess pieces without a chessboard cannot constitute a game.

The above discussion can be transferred to a conventional symbolic system by replacing the independence of vectors with the atomicity of symbols. We usually think that a symbol carries its meaning independently; however, the discussion suggests that this idea is based on a lack of attention to the existence of the background mechanism. Keeping the connection between a symbol and its referent in the real world is not enough to maintain the coherency between an internal compositional system and the reality outside. The symbol grounding problem proposed by [11] should be reconsidered from the broader perspective of a system in which both symbols and their background are essentially inter-related.

To conclude, a sub-symbolic implementation of the recombination of goal-directed actions is presented. Three different types of functionally compositional systems emerge depending on the sparseness of the provided examples by using an identical learning model. Thus, our model provides a possible dynamical interpretation of conventional usage-based models. In future work, the transition process of the internal structure involved in the transition from a holistic system to a compositional one will be presented. Also, associative learning between goal-directed actions and sentences will be investigated by employing the technique proposed in [12].

References

1. Tomasello, M.: Constructing a Language: A Usage-Based Theory of Language Acquisition. Harvard University Press (2003)
2. Goldberg, A.: Constructions: A Construction Grammar Approach to Argument Structure. University of Chicago Press (1995)
3. Iwahashi, N.: Robots That Learn Language: Developmental Approach to Human-Machine Conversations. In: Vogt, P., Sugita, Y., Tuci, E., Nehaniv, C. (eds.) Symbol Grounding and Beyond, pp. 143–167. Springer, Heidelberg (2006)
4. Roy, D.: Learning visually grounded words and syntax for a scene description task. Computer Speech and Language 16, 353–385 (2002)
5. Cangelosi, A., Riga, T.: An embodied model for sensorimotor grounding and grounding transfer: Experiments with epigenetic robots. Cognitive Science 30(4), 673–689 (2006)
6. van Gelder, T.: Compositionality: A connectionist variation on a classical theme. Cognitive Science 14, 335–364 (1990)
7. Pollack, J.: The induction of dynamical recognizers. Machine Learning 7, 227–252 (1991)
8. Tani, J., Ito, M., Sugita, Y.: Self-organization of distributedly represented multiple behavior schemata in a mirror system: reviews of robot experiments using RNNPB. Neural Networks 17, 1273–1289 (2004)

9. Takamuku, S., Takahashi, Y., Asada, M.: Lexicon acquisition based on object-oriented behavior learning. Advanced Robotics 20(10), 1127–1145 (2006)
10. Markman, A., Stilwell, C.: Role-governed categories. Journal of Experimental and Theoretical Artificial Intelligence 13(4), 329–358 (2001)
11. Harnad, S.: The symbol grounding problem. Physica D 42, 335–346 (1990)
12. Sugita, Y., Tani, J.: Learning semantic combinatoriality from the interaction between linguistic and behavioral processes. Adaptive Behavior 13(1), 33–52 (2005)

Learning to Generalize through Predictive Representations: A Computational Model of Mediated Conditioning

Elliot A. Ludvig and Anna Koop

Department of Computing Science
University of Alberta
Edmonton, AB T6G 2E8
{elliot,anna}@cs.ualberta.ca

Abstract. Learning when and how to generalize knowledge from past experience to novel circumstances is a challenging problem many agents face. In animals, this generalization can be caused by mediated conditioning—when two stimuli gain a relationship through the mediation of a third stimulus. For example, in sensory preconditioning, if a light is always followed by a tone, and that tone is later paired with a shock, the light will come to elicit a fear reaction, even though the light was never directly paired with shock. In this paper, we present a computational model of mediated conditioning based on reinforcement learning with predictive representations. In the model, animals learn to predict future observations through the temporal-difference algorithm. These predictions are generated using both current observations and other predictions. The model was successfully applied to a range of animal learning phenomena, including sensory preconditioning, acquired equivalence, and mediated aversion. We suggest that animals and humans are fruitfully understood as representing their world as a set of chained predictions and propose that generalization in artificial agents may benefit from a similar approach.

The texture of our experience is often dotted by aversions and affinities that are only indirectly related to rewarding or punishing outcomes. For example, if I have a near-death experience in an ambulance on the way to the hospital, I am likely to shudder next time I hear a siren go by, even if the ambulance was not playing its siren during my traumatic episode. Or if I get really sick at a restaurant before ordering, I will certainly think twice about eating their food in the near future. In these two examples, stimuli that were never directly experienced in the offending situations still gain some of the residual response that memory of the initial situation provokes. In the animal learning literature, this indirect learning has been termed *mediated conditioning* and repeatedly reproduced in the laboratory with notable instances including sensory preconditioning [1,2], as in the ambulance example, acquired equivalence [3,4], and mediated aversion [5,6], as in the restaurant example. Generalization between stimuli based on their experienced history seems *prima facie* like a valuable asset to an animal or human (or animat) facing novel stimuli or situations and may even form part of the basis for categorization and conceptual knowledge [7].

M. Asada et al. (Eds.): SAB 2008, LNAI 5040, pp. 342–351, 2008.

This learned generalization hints at a larger puzzle that has troubled researchers in both machine learning and cognitive science: When should what you learn in one situation generalize to what you do in another? Generalization is a cornerstone of adaptive behavior that allows agents to take advantage of previous experience beyond the particulars of the original learning context. In the psychological literature, most studies of generalization have focussed on how responding generalizes amongst physically similar stimuli (e.g., tones of different frequencies). Mediated conditioning, however, presents an instance whereby long-term equivalences can be established between physically distinct stimuli, merely because of the animal's experience with the consequences, antecedents, and associates of those stimuli [4]. In this paper, we propose that the computational formalism of *predictive representations* [8,9,10] from reinforcement learning provides an efficient and effective mechanism for the mediated conditioning exhibited by many animals and humans. In this predictive representation (PR) approach to adaptive learning, stimuli are represented as the constellation of predicted future observations, rather than as composites of their physical properties. We leverage this idea to develop a real-time PR model and show how this reinfrocement-learning model explains the learned generalization observed in mediated conditioning experiments.

1   Predictive Representation Model

The key insight behind our model is that stimuli are represented as a collection of chained predictions about future observations [9]. This predictive representation for a stimulus implies that generalization will occur readily between stimuli that share similar predictions about the future—in a strong parallel to the manner that generalization occurs most readily between stimuli that share physical properties. These PRs play a similar role to the images or associatively activated representations in other theories of animal conditioning [4,6,11,12].

Figure 1 presents a schematic of the PR model, illustrating how these representations fit into the full learning scheme. Prediction generation in the model is a two-step process: On a given time step, the observations (stimuli) are first used to generate interim predictions for every potential stimulus. These interim predictions are then combined with the same initial observations to generate a new set of final predictions for that time step. These final predictions determine behavior, so, for example, in a simple learning task where a light is followed by a tone and then by food, the light would lead to a prediction of the tone which would lead to a prediction of the food. As a result, after learning, the light would also (indirectly) lead to a (weaker) prediction of the food and thereby elicit some of the associated conditioned responding.

We approach these tasks as a reinforcement-learning prediction problem, except that we calculate a separate value function for every stimulus—not only rewards. More formally, on every time step t, a value function V_t is computed for every potential observation as a semi-linear function of the vector $\mathbf{x}_t$ of the observation/prediction values x_t^i and the vector $\mathbf{w}_t$ of the learned weights w_t^i:

$$V_t = \sigma(\mathbf{w}_t^T \mathbf{x}_t) = \sigma(\sum_{i=1}^{n} w_t^i x_t^i) \tag{1}$$

where the squashing function, $\sigma(x) = \frac{1}{1+e^{-x}}$, is used to keep the value of V between 0 and 1. Half the components of the vector $\mathbf{x}_t$ are binary, indicating whether a particular stimulus was present on that time step (1) or not (0). Such a simplification is not strictly necessary, and real-valued noisy observations are surely possible, but not considered here for ease of exposition. The other half of the components are real-valued elements that correspond to predictions (see Fig. 1). The key to our model is that this computation is performed twice on each time step. The first iteration uses only the binary observations (with all predictions set to 0) to calculate an interim prediction. On the second iteration, this interim prediction becomes part of the stimulus representation and is used to generate the final prediction that is compared to future experience.

Learning in the PR model occurs on the ensuing time step when new observations are encountered, through the temporal-difference (TD) learning algorithm [10,13]. With this learning rule, an error δ_t is formed for each potential outcome, which is the difference between the current prediction and the sum of the new observations and resultant new predictions (as discounted by γ):

$$\delta_t = x_{t+1} + \gamma \tilde{V}_{t+1} - V_t . \tag{2}$$

Note that $\tilde{V}_{t+1}$ is the prediction as calculated using the vector of the new observations and predictions, $\mathbf{x}_{t+1}$, and the weight vector before being updated, $\mathbf{w}_t$, through the same two-step process described above. The discount factor, γ, determines the temporal horizon of the prediction. A low γ makes the model short-sighted, focusing the

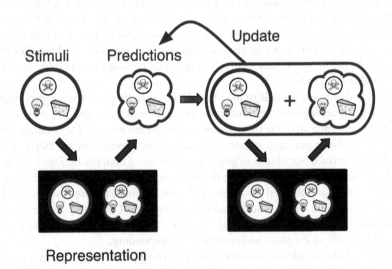

Fig. 1. Schematic of the PR Model. Observations (*blue circles*) at one time step are first used to generate interim predictions (*boxed green clouds*) for every stimulus. The same observations are then combined with these interim predictions to generate the final predictions (*larger green clouds*). Finally, on the next time step, the new observations and resultant new predictions are combined to update (*red arrow*) the weights based on the discrepancy between the predictions from the previous time step and these observed outcomes.

prediction on the near future; a higher γ extends the temporal window which the model is trying to predict.

The prediction calculation is updated on every time step by changing each weight, w^i, according to the TD error for that potential outcome:

$$w^i_{t+1} = w^i_t + \alpha \delta_t x^i_t \tag{3}$$

where α is a step-size parameter that influences the learning rate. In the model, events that have high salience, such as shocks, have a large step size, whereas less salient events, such as tones and lights, have a lower step size (and thus learning rate). For simplicity, we chose to only update the weights from the second iteration and force the weights in the first iteration to be identical to the corresponding weights in the second iteration. Other versions of this PR model with deeper (multi-layer or recurrent) predictions or with multiple cascading or independent learning updates are certainly possible and may even capture further empirical phenomena not considered here.

An important element of the PR model is that all experimental situations are modeled as real-time. In previous models of mediated conditioning and related phenomena [12,15], the flow of experience was often divided into discrete trials and punctate events—a structure which is not immediately apparent in the real world. The PR model allows stimuli to exist for multiple time steps, thereby predicting the continuation of themselves, a feature that is vital in explaining sensory preconditioning and mediated aversion (see Tables 2 and 3).

2 Results

We demonstrate successful performance of the PR model on three animal learning tasks that seem to involve mediated conditioning: acquired equivalence [3], sensory preconditioning [2], and mediated aversion [5]. For each of these experiments, we simulated the PR model with 100 trials in the first stage, 3 trials in the second stage, and a single test trial in the final stage. Sensory stimuli all lasted for 15 time steps, while food reward, shock, and illness lasted 3 time steps; an inter-trial interval of 60 time steps separated trials. In all simulations, the discount factor γ was .98, and the step size α was .4 for shock and illness, .3 for food reward, and .05 for other stimuli. Weights were initialized to 0 and capped at 3.

2.1 Application: Acquired Equivalence

When two stimuli are repeatedly followed by the same outcome, they often come to be treated more similarly in the future; that is, these stimuli acquire an equivalence relation [4,16]. For example, Honey and Hall [3] presented rats with three different stimuli (A, B, and C): A and B were always followed by food reward (f) while C was never rewarded (see upper part of Table 1). Rats then received parings of stimulus A with an electric shock (sh). When subsequently tested with stimuli B and C, rats showed significantly greater conditioned fear to stimulus B, which shared a common history with

Table 1. Experimental details and model interpretation of the acquired equivalence experiment from Honey and Hall [3]. Each column is a different stage of the experiment. A, B, and C are different sounds; f = food; sh = shock; pr = prediction.

Stage 1	Stage 2	Test	Result
Experiment: Honey and Hall [3]			
A→f	A→sh	B	B > C
B→f		C	
C			
PR Model Explanation:			
A→pr(f)	A→pr(f)→sh	B→pr(f)	B→pr(sh)
B→pr(f)	Therefore:	& pr(f)→pr(sh)	
	A→pr(sh)		
	pr(f)→pr(sh)		

shocked stimulus A. This transfer of conditioned fear to the stimulus that shared a training history with the shocked sound is the hallmark of an acquired equivalence relationship. This acquired equivalence by common consequences has also been demonstrated in pigeons [14] and humans [7,17].

Figure 2 presents simulation results from the PR model on this acquired equivalence task. As with animals, the model produced a greater prediction of shock (equivalent to more conditioned fear) with the stimulus (B) that shared a training history with the shocked stimulus (A). The lower part of Table 1 gives an intuitive account of how our model yields these results. After the first stage, both A and B produce a prediction of food. In the second stage of training, A produces a prediction of food, which is followed by shock. Thus, A produces a prediction of shock, and, here is the key point, the prediction of food also produces a prediction of shock. Finally, in the third stage, B still produces a prediction of food, which, in turn, produces a prediction of shock. This indirect prediction of shock is the basis of the acquired equivalence effect (and other forms of mediated conditioning) in our PR model.

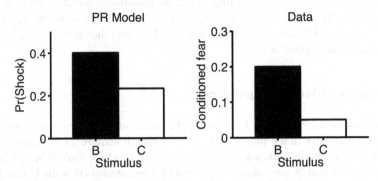

Fig. 2. PR model simulation results (*left*) and empirical data (*right*) from an acquired equivalence experiment. Data are re-plotted from Honey and Hall [3]. Stimulus B shared a training history with the shocked stimulus, while stimulus C did not.

2.2   Application: Sensory Preconditioning

Sensory preconditioning is another experimental situation wherein one stimulus gains an association with reward through the mediation of a second stimulus [1,2]. The upper portion of Table 2 displays the experimental design for a typical sensory preconditioning procedure. Animals are first trained with pairings of two previously neutral stimuli (A, B) as well as a third, unpaired stimulus (C). In the second stage, one of the paired stimuli (B) is then followed by a mild shock (sh). Finally, in the test stage, the other two stimuli are presented alone, and animals display greater conditioned fear to the paired stimulus (A) than the unpaired stimulus (C). The link established between B and A by their training history (in Stage 1) results in greater subsequent generalization between the two stimuli. This effect can be further augmented by presenting the stimuli simultaneously rather than sequentially in Stage 1 (bracketed conditions in Table 2).

Figure 3 displays the empirical data (right) and corresponding simulation results (left) from the real-time PR model in a sensory preconditioning procedure. As with real animals, in the model, sensory preconditioning results in greater generalization to the paired stimulus (A) from the first stage, most markedly for the simultaneous training case [2]. The lower portion of Table 2 schematizes how the PR model explains this generalized responding to stimulus A in the test stage. After the first stage of training, stimulus A produces a prediction of stimulus B. Because stimuli last for multiple time steps in the real-time PR model, all stimuli also learn to produce self-predictions. In the second stage, stimulus B produces a prediction of itself, so both the stimulus and its prediction are followed by the shock. As a result, both B and the prediction of B lead to predictions of shock. In the final, test stage, stimulus A leads to a prediction of B, which leads to a prediction of shock and the associated conditioned response. The simultaneous case shows greater sensory preconditioning than the sequential version in the PR model because, in the first stage, in the simultaneous case, the model additionally learns that stimulus B predicts stimulus A (bracketed value in Table 2). This additional

Table 2. Experimental details and model interpretation for a sensory preconditioning experiment. Each column is a different stage of the experiment. For clarity, only predictions directly pertinent to the explanation of the primary effect are included. Bracketed items refer to the simultaneous version of the task. A, B, and C are different stimuli; sh = shock; pr = prediction.

Stage 1	Stage 2	Test	Result
Experiment: Rescorla [2]			
A→B [AB]	B→sh	A	[A] > A > C
C		C	
PR Model Explanation:			
A→pr(A),pr(B)	B→pr(B)→sh	A→pr(A),pr(B)	A→pr(sh)
B→**pr(B)**,[pr(A)]	[B→pr(A)→sh]	& pr(B)→pr(sh)	[A→pr(sh)]
	Therefore:	[& pr(A)→pr(sh)]	
	B→pr(sh)		
	pr(B)→pr(sh)		
	[pr(A)→pr(sh)]		

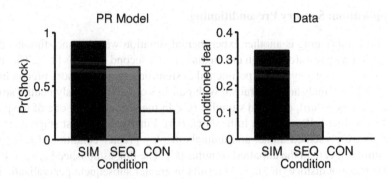

Fig. 3. PR model simulation results (*left*) and empirical data (*right*) from a sensory preconditioning experiment. Data are re-plotted from Figure 4 in Rescorla [2] as degree of response suppression. SIM = simultaneous; SEQ = sequential; CON = control.

prediction results in the prediction of A directly preceding shock in Stage 2, leading to greater generalization in Stage 3 because of the self-prediction of stimulus A.

2.3   Application: Mediated Aversion

A final set of empirical phenomena that nicely illuminate properties of this PR model are the series of mediated aversion experiments [5,6]. Table 3 shows the design for a typical experiment: Animals are first trained with 2 neutral stimuli (A, B) each paired with one of 2 different foods/flavours (f1, f2). In the second stage, animals are presented one of the two stimuli followed by injection with lithium chloride (LiCl), an illness-inducing agent. On the final, test stage, animals are presented with the 2 foods/flavours and will typically preferentially eat from the food whose associate was not paired with illness in the second stage.

Table 3. Experimental details and PR model interpretation for a mediated aversion experiment. Each column is a different stage of the experiment. For clarity, only predictions directly pertinent to the explanation of the primary effect are included. A, B = stimuli; f1, f2 = foods/flavours; G1, G2 = 2 groups of animals; pr = prediction; LiCl = Lithium Chloride, an illness-inducing agent.

Stage 1	Stage 2	Test	Result
Experiment: Holland [5]			
A→f1	G1: A→LiCl	f1	G1>G2
B→f2	& G2: B→LiCl		
PR Model Explanation:			
A→pr(A),pr(f1) A→pr(f1)→LiCl		**f1→pr(f1)**	f1→pr(LiCl)
f1→pr(f1)	Therefore:	& pr(f1)→pr(LiCl)	
	A→pr(LiCl)		
	pr(f1)→pr(LiCl)		

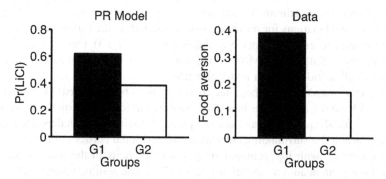

Fig. 4. PR model simulation results (*left*) and empirical data (*right*) from a mediated aversion experiment. Data are re-plotted from Figure 1 in Holland [5] as percent decrease in consumption of food f1 for the groups that had the same food (G1) or a different food indirectly devalued (G2).

Figure 4 shows how the PR model effectively captures the key result: In both the data and simulations, the group that had food f1 indirectly devalued showed a much greater prediction of illness than the second group. As sketched out in the lower part of Table 3, the PR model learns that stimulus A leads to a prediction of food f1 in the first stage. As a result, in the second stage, both stimulus A and the prediction of food f1 precede (and learn to produce predictions of) illness (LiCl administration). Finally, in the final test phase, food f1 leads to a prediction of itself which leads to a prediction of illness and the observed food aversion. Once again, self-prediction is an important component of the explanation, but this time in a different guise than with sensory preconditioning. In sensory preconditioning, self-prediction is the important feature in the second phase when stimulus B's self-prediction leads to the prediction of B producing a prediction of food (cf. Table 2). In the mediated aversion experiment, the crucial self-prediction occurs in the final phase when food predicts itself, leading to a prediction of illness (cf. Table 3).

3   Conclusions

In this paper, we have shown how mediated conditioning can be effectively modeled with our real-time PR network model. The PR model conceives of humans and animals as generating a network of chained predictions of future observations, which, in some ways, cashes out the "image" or "representation" of earlier theories of conditioning [4,6,11]. The selected empirical examples—acquired equivalence, sensory preconditioning, and mediated aversion—each illustrate additional properties of the model in explaining this form of learned generalization.

In their neural network model, Gluck and Myers [12] also address many of the same empirical phenomena. They suggest that redundancy compression and predictive differentiation are the two functions largely responsible for the increased generalization observed in mediated conditioning experiments. Here, we propose an alternate computational account, based on the notion that stimuli are represented as the chained predictions of all future observations. Similarities in this predictive space produce learned generalization between stimuli. In addition, the real-time dynamics of our PR model proffers novel explanations for more phenomena, including the difference between simultaneous and successive sensory preconditioning (see Fig. 3). Our model also bears some similarity to Sutton's TD Models [10], which allow artificial agents to incrementally learn a full-world model for better planning.

Where in the brain might all these iterative predictions be computed? One possibility is suggested from the few studies that have examined lesion effects on these tasks. We know that acquired equivalence and sensory preconditioning are both dependent on the hippocampus and the surrounding entorhinal and perirhinal cortices [12,18,19,20,21]. Moreover, humans with hippocampal atrophy show deficits in the transfer (generalization) stage of an acquired equivalence task [17]. These results, taken together, hint that the medial temporal areas might be responsible for creating new predictive representations for use by reinforcement learning systems elsewhere in the brain (e.g., basal ganglia; see [22]). These predictive representations could also provide a unifying framework for knowledge creation [23], including spatial learning and object memory, two of the more common processes attributed to the hippocampus and perirhinal cortex, respectively.

In conditioning, animals clearly learn more than a simple association between a neutral cue and a rewarding stimulus. They learn a panoply of interrelations among all the different stimuli in their environment—relationships that can be exposed through clever experimental manipulations, as in the generalization tests central to acquired equivalence, sensory preconditioning, and mediated aversion. These three examples of mediated conditioning or learned generalization demonstrate the value of trying to model animal learning as a network of chained predictions. This predictive promiscuity, as captured by our PR model, helps animals learn and adapt more quickly when confronted with novel situations and stimuli. No doubt the empirical story in each of these cases is more nuanced than this brief exposition has allowed (for more details, see [4,6]), but our model captures the core effects and offers a framework for thinking about how the empirical exceptions might constrain future modeling attempts.

Acknowledgments. The authors would like to thank Rich Sutton for inspiration without constraint, the Alberta Ingenuity Fund and iCore for support, and Karen Skinazi for editing help.

References

1. Brogden, W.J.: Sensory pre-conditioning. Journal of Experimental Psychology 25, 323–332 (1939)
2. Rescorla, R.A.: Simultaneous and successive associations in sensory preconditioning. Journal of Experimental Psychology: Animal Behavior Processes 6, 207–216 (1980)
3. Honey, R.C., Hall, G.: Acquired equivalence and distinctiveness of cues. Journal of Experimental Psychology: Animal Behavior Processes 15, 338–346 (1989)
4. Hall, G.: Learning about associatively activated stimulus representations: Implications for acquired equivalence and perceptual learning. Animal Learning & Behavior 24, 233–255 (1996)
5. Holland, P.C.: Acquisition of representation-mediated conditioned food aversions. Learning & Motivation 12, 1–18 (1981)

6. Holland, P.C.: Event representation in pavlovian conditioning: Image and action. Cognition 37, 105–131 (1990)
7. Hall, G., Mitchell, C., Graham, S., Lavis, Y.: Acquired equivalence and distinctiveness in human discrimination learning: Evidence for associative mediation. Journal of Experimental Psychology: General 132, 266–276 (2003)
8. Littman, M.L., Sutton, R.S., Singh, S.: Predictive representations of state. In: Advances in Neural Information Processing Systems, vol. 14, pp. 1555–1561 (2002)
9. Rafols, E.J.R., Ring, M.B., Sutton, R.S., Tanner, B.: Using predictive representations to improve generalization in reinforcement learning. In: Proceedings of the International Joint Conference on Artificial Intelligence, pp. 835–840 (2005)
10. Sutton, R.S.: TD models: Modeling the world at a mixture of time scales. In: Proceedings of the 12th International Conference on Machine Learning, pp. 531–539 (1995)
11. Wagner, A.R.: SOP: A model of automatic memory processing in animal behavior. In: Spear, N.R., Miller, R.R. (eds.) Information processing in animals: Memory mechanisms, pp. 5–47. Erlbaum, Hillsdale (1981)
12. Gluck, M., Myers, C.: Hippocampal mediation of stimulus representation: A computational theory. Hippocampus 3, 491–516 (1993)
13. Sutton, R.S.: Learning to predict by the methods of temporal differences. Machine Learning 3, 9–44 (1988)
14. Bonardi, C., Rey, V., Richmond, M., Hall, G.: Acquired equivalence of cues in pigeon autoshaping: Effects of training with common consequences and with common antecedents. Animal Leaning & Behavior 21, 369–376 (1993)
15. Honey, R.C., Ward-Robinson, J.: Acquired equivalence and distinctiveness of cues: I. Exploring a neural network approach. Journal of Experimental Psychology: Animal Behavior Processes 28, 378–387 (2002)
16. Lawrence, D.H.: Acquired distinctiveness of cues: I. Transfer between discriminations on the basis of familiarity with the stimulus. Journal of Experimental Psychology 39, 770–784 (1949)
17. Myers, C.E., Shohamy, D., Gluck, M.A., Grossman, S., Kluger, A., Ferris, S., Golomb, J., Schnirman, G., Schwartz, R.: Dissociating hippocampal versus basal ganglia contributions to learning and transfer. Journal of Cognitive Neuroscience 15, 185–193 (2003)
18. Coutureau, E., Killcross, A.S., Good, M., Marshall, V.J., Ward-Robinson, J., Honey, R.C.: Acquired equivalence and distinctiveness of cues: II. Neural manipulations and their implications. Journal of Experimental Psychology: Animal Behavior Processes 28, 388–396 (2002)
19. Lazzaro, S.C., Gournani, K., Ludvig, E.A., Gluck, M.A.: Lesions of the entorhinal cortex abolish sensory preconditioning in rats. In: Society for Neuroscience Abstracts, 997.12 (2005)
20. Nicholson, D., Freeman, J.: Lesions of the perirhinal cortex impair sensory preconditioning in rats. Behavioral Brain Research 112, 69–75 (2000)
21. Port, R.L., Patterson, M.M.: Fimbrial lesions and sensory preconditioning. Behavioral Neuroscience 98, 584–589 (1984)
22. Schultz, W., Dayan, P., Montague, P.R.: A neural substrate of prediction and reward. Science 275, 1593–1599 (1997)
23. Koop, A.: Understanding experience: Temporal coherence and empirical knowledge representation. Master's thesis, University of Alberta (2007)

Detection of Weak Signals by Emotion-Derived Stochastic Resonance

Shogo Yonekura[1], Yasuo Kuniyoshi[2], and Yoichiro Kawaguchi[1]

[1] Interfaculty Initiative in Information Studies
{yonekura,yoichiro}@iii.u-tokyo.ac.jp
[2] Graduate School of Information Science and Technology,
The University of Tokyo, 7-3-1, Hongo, Bunkyo-ku, Tokyo, Japan, 113-0033
kuniyoshi@isi.imi.i.u-tokyo.ac.jp

Abstract. This paper reports a new finding on functionalities of trembling, the bodily manifestation of fear and joy. We consider trembling of a physically-simulated agent consisting of a vision system and a neural system. It is demonstrated that the noise to visual streams generated by trembling enhances signal to noise ratio of the neural system.

Keywords: Fear, Joy, Emotions, Embodiment, Stochastic Resonance.

1 Introduction

Most organisms are capable to behave rationally in the real-world. They can infer a potential threat from subtle sensory signals [6]. They can select the most rational action from a vast amount of options[3]. They can recognize the mental state by observing a subtle changes of a facial expression[7][20]. One of the surprising scientific findings in recent decades is that these rational abilities are fundamentally based on emotion. That is, impairment of emotions, in particular of 'emotional states' such as drastic changes of blood pressure or skin conductance, leads immediately to impairment of learning, decision-making, and sociality [3].

Though functionalities of emotional 'states' have been well-addressed, emotional 'movements' such as laughter and trembling have been largely neglected and been regarded simply as movements representing "overflowed neural and mental states" [1][5]. In contrast, this paper argues that emotional movements, trembling in particular, provide something more than representations, and that in order to fully understand emotions it is crucial to investigate the effects of emotional movements on the flow of sensory information.

The main goal of this paper is to propose functions of trembling for fear and joy from the perspective of embodiment and physics. Later in this paper, we demonstrate immense benefits of trembling; trembling provides additive noise to sensory streams, and then, response of the nonlinear neural system is enhanced due to stochastic resonance. In the next section, we sketch prospective roles of emotional movements in contrast with the roles of emotional states, from the perspective of embodied artificial intelligence.

M. Asada et al. (Eds.): SAB 2008, LNAI 5040, pp. 352–361, 2008.

2   Prospective Functions of Emotional Movements

2.1   Beyond Representational and Biasing Roles of Emotions

Emotion manifests itself in diverse ways, some manifestations are in blood pressure, bodily temperature, neural state, or hormone density, and some are in facial expression, vocalization, bodily movement, or behavior. The former 'internal' manifestations are addressed as emotional states, roles of which are assumed to be fundamental for both survival and social activities. For example, modulated bodily states of anger or fear, are for the preparation of fight or flight [9]. Modulated neurochemical distributions due to emotions such as increased or decreased acetylcholine, serotonin, dopamine, are known to be serving effective control of behavioral tendencies via short-term as well as long-term modulation of neural states, cognitive modes, and mental states [26][27][29][30].

One of the most reliable accounts for the role of emotional states is that emotional states provide "markers" which represent values of events for an organism, and the markers in turn, bias attention, motivation, action-selection, and other cognitive activities [3][8][18][31]. On the contrary to the representational and relatively indirect biasing roles of the internal manifestations, we expect that external manifestations of emotions actively and directly deform, constraint, or even create sensory streams by actions and bodily movements.

2.2   Emotional Movements and Active Sensing

Beneath our expectation for dynamical influences of emotions, the theory provided by J.-P.Sartre is underlying, that emotion is a special kind of action to reduce frustration among one and the world [4]. For example, fainting, which tends to be regarded as a 'marker' of exhausted mental as well as biological states, is a means to "escape from frustrations". That is, emotion (e.g., anxiety or sadness) is addressed as escape for the sake of self-defense, and bodily manifestation, fainting, is addressed as a concrete means of the escape.

As far as bodily manifestation is a means of emotion, bodily manifestation is inseparable from the purpose of emotion. And, since bodily movements inevitably influence sensory data, it is reasonably expected that we can find benefits of emotional movements within the sensory stream. In particular, we expect that emotional movements provide 'extractor' of hidden information, or 'amplifier/canceler' of weak/over-gained sensory signals. The perspective of embodied artificial intelligence and active sensing support this expectation. For instance, the length and the shape of a rod are perceived by "shaking" [21]. Saliency-driven attention gives statistically well-tuned sensory information which is easy to be learned [28]. To further support, we note erection of the ears of mammals. Erection of ears is not to represent their arousal states, but to ascertain the source and nature of the danger [1].

In next sections, we first focus on trembling, a typical manifestation of fear and joy, and organize prospective situations when we fear and rejoice, and purposes why we fear and rejoice. And, we build a simulated creature to identify and quantify functions of emotional trembling.

3   Model of Emotion-Derived Stochastic Resonance

3.1   When and Why We Fear and Rejoice

Usually we fear when a predictable "deterministic" world starts ruining, and when indeterministic world appears where we can predict neither incoming sensory signals nor consequences of our actions. This apprehension can be adopted to several developmental studies. For example, fear of infants to strangers[22], which emerges later in the development, will be because strangers are not predictable to infants, compared to caregivers. Furthermore, fear of an infant sitting in the edge of a cliff, will be due to the inability to suppress the instability of the posture, and in fact, it is clearly shown that an infant fears because the infant cannot control the bodily-posture in the edge of a cliff[19]. Likewise, it is possible to claim that joy appears when we experience unpredictable events, and when indeterministic world appears, as joy occurs when we anticipated a reward, not when we actually get it[1][4], and as pleasant tickle sensation occurs when infants cannot predict incoming sensory signals [23]. Recent findings on brain science support this claim. Response of dopamine neurons, which is strongly correlated to joy, depends on the unpredictability of rewards [24][25].

As far as fear and joy appear in response to indeterminacy, the purpose of fear and joy is reasonably expected to be in the reduction of indeterminacy. Next, we propose our hypothesis on how trembling reduces indeterminacy, after introducing a physical phenomenon, stochastic resonance.

3.2   Stochastic Resonance Derived from Fear and Joy

Stochastic resonance is a phenomenon where the ability of a nonlinear system to detect sub-threshold weak signals is enhanced due to the presence of noise [10]. It is known that a lot of non-linear systems exhibit stochastic resonance not only under white noise [16], but also under temporally or spatially colored noise [12][13][14]. It is also known that in order to maximize the effect, the variance of noise should be reasonably large; small noise does not effect the system, whereas too large noise hides the signals [11][15][32].

The scenario of the emotion-derived stochastic resonance is simple. Let's say an agent is walking in a dark forest and hears something moving. The agent can anticipate the existence of a potential threat or reward, but cannot identify because it is too dark to see. Thus, the world around the agent becomes indeterministic, because the agent has no model to behave adaptively without the information of the potential threat or reward. And immediately, the agent will fear, or rejoice, will possibly start trembling, and the trembling body, head, and eyes, will result in a noisy visual stream. If the variance of the noise sequence is adequate, the neural system of the agent will be able to detect weak signals by exploiting stochastic resonance. And by this, the agent will successfully identify potential threat or reward, and reduce the unpredictability of the environment. If only the agent identifies the source of potential threat or reward, indeterminacy will disappear, and thus, the agent will be able to behave adequately.

4 Implementation

4.1 Design of Overall Architecture

To test our hypothesis, we implemented a physically-simulated active vision system. Implemented system consists of vision system, active joint, and neural system. Within the environment, we prepared a "windmill" in order to add periodic visual stimuli to the system. The schematic model of the active vision system and design of the physical agent are depicted in Fig.1. Overall simulation is integrated within Ageia PhysX as Fig.2 with time-step 0.005.

4.2 Active Vision System

The video camera samples gray-scaled images I_G^t at every simulation step at the resolution of 320×240 pixels, temporally subtracts I_G^{t-1} from I_G^t, and thus, gets ΔI_G^t. The sensory input, I^t, is calculated by down-sampling $|\Delta I_G^t|$ to 16×12 pixels. Note I_i^t, ith bin of I^t, consists of the information of movements of objects in the environment, s_i^t, and the camera itself, n_i^t.

The one motor of the simulated active vision system is controlled by the angle command $a_m(t)$, generated randomly by "trembling source" at frequency $f_t = 40$ Hz unless specified otherwise. In this paper, we adopt noise of uniform distribution for a_m, whose range is $[-\sigma, \sigma]$. For example, in case $\sigma = 0$, then

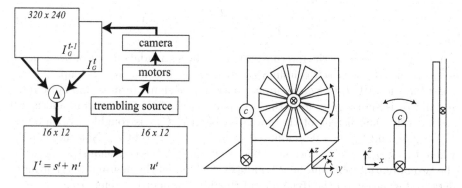

Fig. 1. Left: schematic model of the active vision system. I_G^t means gray-scale image sampled by the camera mounted on the top of the agent. I^t means sensory stimuli added to neural system u^t, where s^t is signal term and n^t is noise term induced by trembling. Angle of a motor is controlled by random position command generated by trembling source. Middle and right: schematic model of the physical agent and its surrounding environment. The character 'c' denote the camera, and the circles with a cross denote the joint controlled by motors. Pitch of the camera is tilt around y axis.

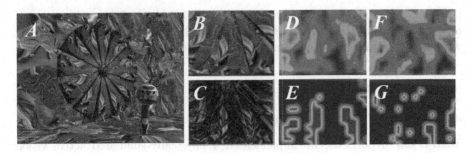

Fig. 2. Integration of the whole simulation. (A): physically-simulated environment and agent within Ageia PhysX. (B): original camera image. (C): $|\Delta I_G^t|$, image obtained by temporal subtraction of the gray-scaled images. (D) and (F): examples of I^t, 16×12 sensory input to the bistable neurons during trembling at 2.5 sec and 2.7 sec. (E) and (G): examples of 16×12 output of the bistable neurons at 2.5 sec and 2.7 sec.

the system keeps $a_m = 0$. And in case $\sigma = 20.0$, then, a_m takes random value between -20 and 20 [deg]. Note frequency of the bodily trembling depends on the function of f_t and characteristics of the body such as torque limit or stiffness.

4.3   Ensemble of Nonlinear Bistable Neurons

As a neural system, we consider an ensemble of 16×12 bistable nonlinear neurons, each of which has two potential well, that is two stable states $U_s^\pm$. The model of ith neuron is described as:

$$\tau \frac{du_i}{dt} = -u_i + J_{ii} f(u_i) + g_s I_i, \tag{1}$$

$$f(u_i) = \frac{2}{1 + \exp\left(-5(u_i - \theta_i)\right)} - 1, \tag{2}$$

where u_i and $f(u_i)$ are the internal state and the sigmoidal output of ith neuron, respectively. J_{ii} is a self-coupling coefficient, and is 2.5 in this paper. Time constant τ is 0.01. I_i is sensory input consisting of signal term s_i and noise term n_i, and g_s is signal gain. θ_i is the bias parameter of the ith neuron, and is 1.0 consistently. The height of the potential barrier, $U_0^\pm$, is roughly estimated as $U_0^+ \approx 2.66$ and $U_0^- \approx -0.66$ (shifted from $U_0^\pm \approx \pm 1.66$ by θ). Note input $g_s I_i$, where $U_0^- < g_s I_i < U_0^+$, does not effect switching of the neuronal state between the two potential well, and hence, induces 'locking' of the neural state. For more detail information on the dynamics of this kind of neurons, refer [16].

4.4   Calibration of Mean Noise $E(n)$

Sensory input to the ith neuron, I_i^t, takes the form $I_i^t = s_i^t + n_i^t$, where s_i^t is signal and n_i^t is noise to the ith bin. Since in order to confirm stochastic resonance, mean value of noise $E(n_i)$ should be zero, and since it is impossible to separate n_i from s_i, therefore, we implement time-window, and within the time-window, estimate mean noise of the ith bin, $\tilde{E}(n_i)$, as mean-value of I_i^t. Here, length of a time-window is 0.2 [sec]. Based on the estimated mean noise, we compute input to the ith neuron, $I_i^t = I_i^t - \tilde{E}(n_i)^{T^j}$.

5   Experiments and Results

5.1   Experimental Setup

Stochastic resonance is observed as the enhancement of the signal-to-noise ratio (SNR) by the presence of additive noise [15][16]. Therefore, in order to confirm stochastic resonance, following two conditions are considered: (1) input sensory-signal s_i should be far below U_0^+ and should be periodic; (2) mean value of noise should be zero or should be smaller than zero, that is, $E(n_i) \leq 0$. Note that $s_i + E(n_i)$ should be smaller than U_0^+, but $|s_i + n_i|$ must be large enough, otherwise, neurons can never switch between the two potential well.

To prepare periodic sensory input, we instantiate an environment where a windmill is "slowly" rotating, at the angular velocity, aV [rad/sec]. Here, if an environment is not textured, or textures are "flat", sensory-input goes to zero, because we are temporally differentiating camera images. Therefore, we select "complex" textures to avoid zero sensory-input (see Fig.2). Note movements of the windmill give approximate sinusoidal signals around 1.52 [Hz]. Here, since sensory signal s_i is far below the height of the potential barrier U_0^+, and since $U_0^- \approx -0.66$, the neuron without additive noise stays at U_s^- (see Fig.3).

And, for histogram analysis of embodied noise, we prepare an environment where the agent is trembling at $f_t = 20$, 40, and 200 Hz, whereas the windmill is not rotating. The distribution shown in Fig.3 indicates that embodiment distorted the uniform distribution of position command a_m.

5.2   Detection of Subthreshold-Signal by Trembling

We first demonstrate that trembling enables the agent to detect weak signals. We rotated the windmill at $aV = 0.8$ [rad/sec], generated trembling with $\sigma = 20$ deg., arbitrarily selected one neuron, and observed the neural output. The

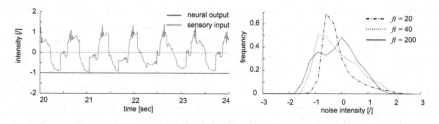

Fig. 3. Left: typical sensory input I_i to the ith neuron and the neural output without trembling. Note that the signal is far below the potential barrier U_0^+. Right: histogram of embodied noise, under $\sigma = 20$ degree and $f_t = 20$, 40, and 200 Hz.

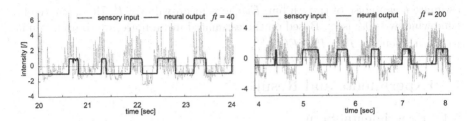

Fig. 4. Typical time-series of sensory input and neural output under $f_t = 40$ (left) and $f_t = 200$ (right). In both left and right figures, $\sigma = 20$ consistently. It is indicated that trembling successfully provided detection of the weak signals.

result shown in Fig.4 indicates that neuron successfully detects the input sensory signals both in $f_t =$40 and 200 Hz, which is too weak to detect without trembling.

5.3 Enhancement of Signal to Noise Ratio by Trembling

Signal to noise ratio (SNR) is estimated by the following equations [11];

$$SNR = 10\log\frac{S}{N} = 10\log\frac{2|Y(k_0)|^2}{\sum_{k=0}^{L-1}|Y(k)|^2 - 2|Y(k_0)|^2}. \tag{3}$$

Signal power is represented with the magnitude of the spectrum $Y(\omega)$ of neural output at the input frequency ω_0. The background noise spectrum $N(\omega_0)$ is estimated as the summation of the output power spectrum $Y(\omega)$, for $\omega \neq \omega_0$. Here, $Y(k)$ denotes output of kth bin of L-point DFT of neural output, and k_0 denotes the bin corresponding to the peak frequency ω_0.

Using this method, we observed SNR of output of arbitrary selected one neuron, while changing angular velocity of the windmill, and changing standard deviation σ of a_m. The result shown in Fig.5 indicates significant enhancement of SNR around $10 < \sigma < 20$ and decrement of SNR for larger σ, independently from trembling frequency and angular velocity.

6 Discussion and Conclusion

6.1 Biological Plausibility Implied by the Experimented Results

In order to demonstrate stochastic resonance by emotional trembling, we dared implement so simple agent that it could lose plausibility as a biological system. Despite of the simplicity, however, the experimented results strongly support our hypothesis that biological systems are exploiting stochastic resonance by emotional trembling. Fig.5 indicates that the remarkable effect exists within limited parameter space (i.e., within $10 < \sigma < 20$), regardless of the frequency of the visual stimuli and of trembling. Moreover, Fig.5 indicates that trembling at 20 Hz, which biological systems definitely can reproduce, is enough to generate stochastic resonance. Though fluctuation and randomness are required for trembling to induce stochastic resonance, it will not be a difficult task for any kind of embodied agent to learn how to exploit stochastic resonance by trembling.

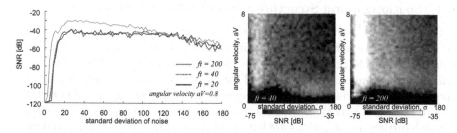

Fig. 5. Observed SNR profiles. Left: SNR profile under the arbitrary angular velocity, 0.8, and f_t =20,40, and 200. Right: SNR map under $f_t = 40$ and 200. Horizontal axis is standard deviation σ, and vertical axis is angular velocity of the windmill.

6.2   Emotional Trembling as a Novel Kind of Active Sensing

Most of perceptions of natural and artificial embodied agents are achieved by 'active sensing'. For example, "seeing" object surface from several different angles enables effective object discrimination [17]. The important factor to realize active sensing, which is derived by studies of embodied artificial intelligence, is that motor commands should be adequately controlled by and coupled with sensory data, otherwise, no movement provides any function [2]. In this context therefore, trembling does not seem to serve any information processing, because trembling is not much more than movements decoupled from sensory data. Nevertheless, it provides the role of fundamental importance to detect and to perceive weak-signals. Given that fear and joy serve reduction of indeterminacy, one possible interpretation of the result will be that emotional trembling is a special kind of active sensing exploiting embodied noise.

6.3   Ambivalence of Emotions

According to J.-P. Sartre, emotions are rational and intended. Whereas, our common-sensed apprehension is opposite; emotions are irrational and unintended. This contradiction is explained by the ambivalence of emotions, and by the char-acteristics of our consciousness and subconsciousness.

As demonstrated, in an environment where only weak signals exist, trembling might be the only one method to identify the source of a potential threat or a potential reward. In this context, we have no choice other than to generate trem-bling. Thus, ambivalence emerges here; we intentionally generate trembling by our rational thinking, while on the other suffer the pains of fear or feel awkward at the pleasure of joy, being exposed to the flood of sensory signals as a result of trembling. The reason why we intuitively tend to assume emotions as unin-tended products is because of self-deceit and because of lack of the monitoring process of the causal relationship between our subconscious intentions and the consequent painful or pleasant flood of sensory signals [4].

6.4 Spatiotemporal Structure within Embodied Noise

One might think that internal noise (e.g., neural noise) also can induce stochastic resonance. We agree that this claim is quite reasonable. However, emotional trembling seems to provide further functions in addition to stochastic resonance, which internal noise cannot provide. That is, embodied noise essentially involves spatial correlation due to the environmental structure and temporal correlation due to the agent's motion, whereas internal noise cannot have such structures. Though further investigations are left as future works, embodied noise and internal noise should have different roles, and we hope understanding the difference among roles of embodied and internal noise let us understand emotions further.

7 Conclusion

In this paper, we started from pointing out the effects of emotional movements on the sensory stream of an embodied agent, and then, we hypothesized that trembling for fear and joy, is active sensing to detect subthreshold weak signals. Using physically-simulated active vision system consisting of a bistable nonlinear neural system, we demonstrated stochastic resonance by trembling. We hope this research sheds light on the embodied basis of emotions.

Acknowledgments

This work is supported by JST CREST. The authors are grateful to Sejuti Rahman (the Univ. of Tokyo) for her valuable comments.

References

1. Darwin, C.: The Expression of the Emotions in Man and Animals, 3rd edn. Oxford University Press, Oxford (1996)
2. Pfeifer, R., Scheier, C.: Understanding Intelligence. MIT Press, Cambridge (1999)
3. Damasio, A.R.: Descartes' Error: Emotion, Reason, and the Human Brain. Grosset/Putnam (1994)
4. Sartre, J.-P.: Sketch for a Theory of the Emotions. Taylor and Francis, Abington (1971)
5. James, W.: What is an emotion? Mind 9(34), 188–205 (1884)
6. LeDoux, J.: The Emotional Brain: the Mysterious Underpinnings of Emotional Life. Simon and Schuster (1996)
7. Ekman, P., Friesen, W.V.: Unmasking the Face. Prentice Hall, Englewood Cliffs (1957)
8. Ehlers, A., Margraf, J., Rotch, W.T., Taylor, B., Birbaumer, N.: Anxiety induced by false heart rate feedback in patients with panic disorder. Behaviour research and therapy 26(1), 1–11 (1988)
9. Cannon, W.B.: Bodily Changes in Pain, Hunger, Fear and Rage: An Account of Recent Researches into the Function of Emotional Excitement. Appleton, New York (1915)
10. Wiesenfeld, K., Moss, F.: Stochastic resonance and the benefits of noise: from ice ages to crayfish and SQUIDs. Nature 373(5), 33–36 (1995)

11. Mitaim, S., Kosko, B.: Adaptive stochastic resonance. Proceedings of the IEEE 86(11), 2152–2183 (1998)
12. Wang, S., Liu, F., Wang, W.: Impact of spatially correlated noise on neural firing. Physical Review E 69, 011909 (2004)
13. Hamaguchi, K., Okada, M., Kubota, S., Aihara, K.: Stochastic resonance of localized activity driven by common noise. Biological Cybernetics 92, 438–444 (2005)
14. Nozaki, D., Collins, J.J., Yamamoto, Y.: Mechanism of stochastic resonance enhancement in neuronal models driven by 1/f noise. Physical Review E 60(4), 4637–4644 (1999)
15. Kosko, B., Mitaim, S.: Stochastic resonance in noisy threshold neurons. Neural Networks 16, 755–761 (2003)
16. Inchiosa, M.E., Bulsara, A.R.: Nonlinear dynamic elements with noisy sinusoidal forcing: Enhancing response via nonlinear coupling. Physical Review E 52(1), 327–339 (1995)
17. Beer, R.D.: Dynamical approaches to cognitive science. Trends in Cognitive Science 4(3), 91–99 (2000)
18. Cañamero, L.: Emotion understanding from the perspective of autonomous robots research. Neural Networks 18, 445–455 (2005)
19. Adolph, K.: Specificity of learning: Why infants fall over a veritable cliff. Psychological Science 11, 290–295 (2000)
20. Adolphs, R., Gosselin, F., Buchanan, T.W., Tranel, D., Schyns, P., Damasio, A.R.: A mechanism for impaired fear recognition after amygdala damage. Nature 433(6), 68–72 (2005)
21. Burton, G., Turvey, M.T.: Perceiving the lengths of rods that are held but not wielded. Ecological Psychology 2(4), 295–324 (1990)
22. Joseph, R.: Environmental influences on neural platicity, the limbic system, emotional developmental and attachment: a review. Child Psychiatry Hum. Dev. 29, 189–208 (1999)
23. Blackmore, S.-J., Wolpert, D.M., Frith, C.D.: Central cancellation of self-produced tickle sensation. Nature Neuroscience 1(7), 635–640 (1998)
24. Schultz, W., Dayan, P., Montague, P.R.: A neural substrate of prediction and reward. Science 275, 1593–1599 (1997)
25. Schultz, W.: Predictive reward signal of dopamine neurons. Journal of Neurophysiology 80, 1–27 (1998)
26. Yu, J.A., Dayan, P.: Acetylcholine in cortical inference. Neural Networks 15, 719–730 (2002)
27. Goldman, M.S., Golowasch, J., Marder, E., Abbott, L.F.: Global structure, robustness, and modulation of neuronal models. The Journal of Neuroscience 21(14), 5229–5238 (2001)
28. Lungarella, M., Pegors, T., Bulwinkle, D., Sporns, O.: Methods for quantifying the information structure of sensory and motor data. Neuroinformatics 3, 243–262 (2005)
29. Doya, K.: Metalearning and neuromodulation. Neural Networks 15, 495–506 (2002)
30. Doya, K., Uchibe, E.: The cyber rodent project: Exploration of adaptive mechanisms for self-preservation and self-reproduction. Adaptive Behavior 13(2), 149–160 (2005)
31. Levenson, R.W.: Autonomic Nervous System Differences among Emotions. Psychological Science 3, 23–27 (1992)
32. Gammaitoni, L., Hänggi, P., Jung, P., Marchesoni, F.: Stochastic Resonance. Reviews of Modern Physics 70(1), 223–287 (1998)

The Influence of Asynchronous Dynamics in the Spatial Prisoner's Dilemma Game

Carlos Grilo[1,2] and Luís Correia[2]

[1] Dep. Eng. Informática, Escola Superior de Tecnologia e Gestão
Instituto Politécnico de Leiria
grilo@estg.ipleiria.pt
[2] LabMag, Dep. Informática, Faculdade de Ciências da Universidade de Lisboa
Luis.Correia@di.fc.ul.pt

Abstract. We examine the influence of asynchronism in the Spatial Prisoner's Dilemma game. Previous studies reported that less cooperation is achieved with the asynchronous version of the game than with the synchronous one. Here, we show that, in general, the opposite is the most common outcome. This conclusion is only possible because a larger number of scenarios was tested, namely, different interaction topologies, a transition rule that can be tuned to emulate different levels of determinism in the choice of the next strategy to be adopted and different rates of asynchronism. The influence of stochastic and deterministic periodic updating in the outcome of the system is also compared. We found that these two update disciplines lead basically to the same result. This is an important issue in the simulation of social and biological behavior.

1 Introduction

Spatial evolutionary games are used in the area of evolutionary game theory as models to study, for example, how could cooperation ever emerge in nature and human societies [11]. They are also used as models to study how can cooperation be promoted and sustained in artificial societies [9]. In these models, a structured population of agents interacts during several time steps through a given game which is used as a metaphor for the type of interaction that is being studied. The population is structured in the sense that each agent can only interact with its neighbors. After each interaction session, some or all the agents, depending on the update method used, have the possibility to change their strategies. This is done using a so called transition rule that models the fact that agents tend to adapt their behavior to the context in which they live by imitating the most successful agents they know. It can also be interpreted as the selection step of an evolutionary process in which the least successful strategies tend to be replaced by the most successful ones.

The final outcome of these models, that is, the proportion of cooperating agents eventually achieved, can be influenced by, for example, the game that is being used, the interaction topology, the transition rule or the update method. The most used game in this area is the Prisoner's Dilemma game (see section 2.1).

M. Asada et al. (Eds.): SAB 2008, LNAI 5040, pp. 362–371, 2008.

There are some works where the influence of some of those aspects on the spatial version of this game is studied. For example, in [10] the influence of the interaction topology is analyzed. Also, in [7] the influence of the interaction topology, the transition rule and the update method are studied.

In this area, the discussion about using synchronous or asynchronous update methods started with a paper by Huberman and Glance [5]. Synchronous updating means that, at each time step, the revision of strategies happens for all agents simultaneously, while this is not the case for asynchronous updating. In that paper the authors contested the results achieved in [8] by Nowak and May who showed that cooperation can be maintained when the game is played in a regular 2-dimensional grid by agents which do not remember their neighbors' past actions. Huberman and Glance criticized the fact that the model used in [8] was a synchronous one, which is an artificial feature. They also presented the results of simulations where cooperation was no longer sustainable when an asynchronous updating was used. After this work, in [7] Nowak *et al* tested their model under several conditions, including synchronous and asynchronous updating and showed that cooperation can be maintained for many different conditions, including asynchronism. However, the results are presented through system snapshot images, which render difficult to measure the way they are affected by the modification from synchronous to asynchronous updating. Recently, in [6], a similar scenario was studied using various asynchronous update methods besides synchronous updating. The authors found that *"The most notable difference between the synchronous and asynchronous schemes, is that the synchronous updating scheme supports more cooperators than the other updating schemes"*.

In this paper, we show that, in general, more cooperation is achieved with the asynchronous version of the Spatial Prisoner's Dilemma game than with the synchronous one. The conclusions derived in [5] and [6] result from the fact that a limited number of conditions were tested, namely, the utilization of only the best-neighbor transition rule, according to which an agent always imitates the strategy of its most successful neighbor. Here, we use the transition rule used in [7] (see Section 2.3). This rule can be tuned to cover the spectrum between proportional updating, with which agents can imitate strategies other than the one used by their most successful neighbor, and the best neighbor rule. Different types of interaction topologies were also used so that the conclusions derived can have a more general character.

In what concerns to the update methods, we first used an asynchronous stochastic update method [2] that allows us to cover the spectrum from synchronous to sequential updating. Usually, asynchronous updating is understood as sequential updating, which means that, at each time step, only one agent updates its strategy after interacting with its neighbors. But, reality seems to lie somewhere between these two extremes and, so, sequential updating can be considered as artificial as synchronous updating. In a population of interacting agents, many interaction and decision processes can be occurring at the same time but not necessarily involving all the agents. If both were instantaneous phenomena we could model the dynamics of the system as if they occurred one after another but that is not usually the case. These processes can take some time, which means that their output is not available to other ongoing decision processes. Even if we consider them as being instantaneous, the time that information takes to be transmitted and perceived implies that their consequences are not immediately available to other agents.

We also compare the results achieved using the stochastic update method with the ones achieved with a deterministic periodic one. There are numerous examples of periodic behaviors where not all the population is necessarily synchronized [3][6]. This comparison can help us understand to what extent the results depend on this type of updating.

The paper is structured as follows: in Sec. 2 we describe the model we used in our simulations and in Sec. 3 we present and discuss the results. Finally, in Sec. 4 some conclusions are drawn and future work is advanced.

2   The Model

2.1   The Prisoner's Dilemma Game

In the Prisoner's Dilemma game (PD), players can cooperate (C) or defect (D). The payoffs are the following: R to each player if they both play C; P to each if they both play D; T and S if one plays D and the other C, respectively. These values must obey $T > R > P > S$ and $2R > T + S$. It follows that there is a strong temptation to play D. But, if both play D, which is the rational choice or the Nash equilibrium of the game, both get a smaller payoff than if they both play C, hence the dilemma. For practical reasons, the payoffs are usually defined as $R = 1$, $T = b > 1$ and $S = P = 0$, where b represents the advantage of D players over C ones when they play the game with each other. This has the advantage that the game can be described by only one parameter without loosing its essence [7].

2.2   Population Topologies

We used two types of topologies: *small-world networks* (SWNs) [13] and *scale-free networks* (SFNs) [1]. We build SWNs as in [12]: first, a toroidal regular 2-dimensional grid is built so that each node is linked to its 8 surrounding neighbors by undirected links; then, with probability ϕ, each link is replaced by another one linking two randomly selected nodes. Parameter ϕ is called the *rewiring probability*. In some works [7] self-links are allowed because it is considered that each node can represent not a single agent but a set of similar agents that may interact with each other. Here, we do not allow self-interaction since we are interested in modeling nodes as individual agents. Repeated links and disconnected graphs are also avoided. The rewiring process may create long range links connecting distant agents. For simplicity, we will call neighbors to all interconnected agents, even if they are not located at adjacent nodes. By varying ϕ from 0 to 1 we are able to build from completely regular networks to random ones. SWNs have the property that, even for very small values of the rewiring probability, the mean path length between any two nodes is much smaller than in a regular network, maintaining however a high clustering coefficient observed in many real systems including social ones.

Although frequently used to model real networks, SWNs do not have a power law degree distribution that is frequently observed on such networks [10]. SFNs are network models that have this property since their degrees follow the power law distribution $P(k) \sim k^{-\gamma}$. We build SFNs in the following way: the network is initialized with m_0 fully connected nodes. Then we add nodes, one at a time, until the network has the desired size. Each added node is linked to m_0 already existing nodes so that the probability of creating a link with some existing node i is equal to $\frac{k_i}{\sum_j k_j}$, where k_i is the degree of i, that is, the number of nodes to which it is connected. This method of link creation is called *preferential attachment*, since the more links a node has, the greater is the probability of creating links to it. This has the effect that a little proportion of nodes has a big connectivity while the most part has a very low connectivity.

2.3   Interaction and Strategy Update

On each time step, agents first play a one round PD game with all their neighbors. Agents can only play C or D and the only way they can change their strategy is by way of the application of the transition rule, after the interaction process. This rule is used to model the fact that agents tend to imitate the most successful agents they know. In order to be able to model intermediate levels of asynchronism in the strategy update process, we use an update method called *asynchronous stochastic dynamics* (ASD)[2]. When ASD is used, at each time step, each agent has a given probability $0 < \alpha \leq 1$ of applying the transition rule in order to decide which strategy to use next. The α parameter is called the *synchrony rate* and is the same for all agents. When $\alpha = 1$ we have synchronous updating and as $\alpha \to \frac{1}{n}$, where n is the population size, ASD approaches sequential updating.

To model the strategy update process, we used a generalization of the *proportional* transition rule [7]. Let G_i be the average payoff earned by agent i in the second stage, N_i be the set of neighbors of agent i, s_i be equal to 1 if i's strategy is C and 0 otherwise, and d a positive number. According to this rule, the probability that an agent i adopts C as its next strategy is

$$p_C(i) = \frac{\sum_{l \in N_i \cup i} s_l(G_l)^d}{\sum_{l \in N_i \cup i}(G_l)^d}. \tag{1}$$

The d parameter acts as a weight that favors the most successful neighbor's strategy B in the update process: the bigger d, the larger is the probability that i adopts B. When $d \to +\infty$ we have a deterministic best-neighbor rule such that i always adopts B as its next strategy. When $d = 1$ we have the proportional update rule. It can be viewed, as well, as the *deterministic degree* of the transition rule. We use average payoffs instead of total payoffs because agents may have a different number of neighbors.

3 Simulations and Results

All the simulations were done with populations of $50 \times 50 = 2500$ agents, randomly initialized with 50% of Cs and 50% of Ds. When the system is running synchronously, i.e., when $\alpha = 1$, we let it run during a transient period of 900 iterations. After this, we let the system run during 100 more iterations and, at the end, we take as output the average proportion of cooperators during this period, which is called the *sampling period*. When $\alpha \neq 1$ the number of selected agents at each time step may not be equal to the size of the population and it may vary between two consecutive time steps. In order to guarantee that these runs are equivalent to the synchronous ones in what concerns to the total number of individual updates, we let the system first run until 900×2500 individual updates have been done. After this, we sample the proportion of cooperators during 100×2500 individual updates and we average it by the number of time steps needed to do these updates. Each simulation is a combination of the ϕ/m_0, α, b and d parameters, and all the possible combinations of the values shown in Table 1 were tested. For each combination, 30 runs were made and the average of these runs is taken as the output.

Table 1. Parameter values used in the simulations

Parameter	Values
ϕ (SWNs)	0 (regular), 0.01, 0.05, 0.1, 1 (random)
m_0 (SFNs)	4, 8, 12
α (ASD)	0.1, 0.2, 0.3, 0.4, 0.5, 0.6, 0.7, 0.8, 0.9, 1 (synchronous)
b (PD game)	1, 1.1, 1.2, 1.3, 1.4, 1.5, 1.6, 1.7, 1.8, 1.9, 2
d (transition rule)	$+\infty$ (best neighbor), 100, 10, 8, 6, 4, 2, 1 (proportional)

In [4] we presented results of simulations done with the ASD update method where only SWNs were used. In the next section, we present an analysis of the results achieved based also on simulations done with SFNs. Most of the conclusions apply to both types of topologies but we will point the differences where they exist. In order to help the understanding of the conclusions, in Fig. 1 we show four typical charts produced by the simulations. In Sec. 3.2 we then compare the results achieved using ASD with the ones achieved using a periodic deterministic update method.

3.1 ASD Results

We can conclude that, for most conditions, the system is not very sensitive to small changes in the α value (we consider that the system is sensitive when its outcome changes by more than 0.1 when the α value is changed by 0.1). There are, however, some situations of big sensitivity (Ex: Fig. 1d). Also, the results show that, in general, the system responds monotonically to changes in the α value. The few situations of non-monotonicity happen almost only for SWNs

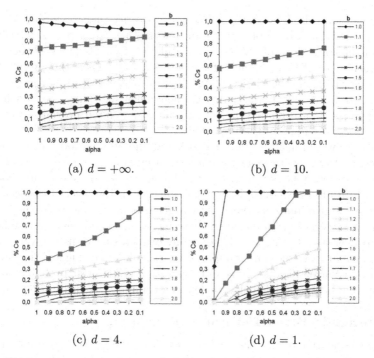

(a) $d = +\infty$.   (b) $d = 10$.

(c) $d = 4$.   (d) $d = 1$.

Fig. 1. % of cooperators for $m_0 = 4$ (SFNs) and different combinations of d, b and α

and mainly for large values of d, that is, when the probability that an agent imitates its most successful neighbor is high. As to the influence of the interaction topology on these two aspects, the results show that the system becomes more robust and monotonous as ϕ and m_0 are increased. We hypothesize that the real reason for this coincidence of results can be due to the fact that the mean path length between the nodes decreases as we increase ϕ and m_0, respectively, on SWNs and SFNs. Another possibility would be the clustering degree of the network. However, while the clustering degree decreases on SWNs as we increase the ϕ value, it can not decrease on SFNs as we increase the value of m_0. So, this result can not be ascribed to this property. More work must be done, however, in order to confirm this.

There is a somewhat unexpected result that can be phrased like this: the lower the value of the d parameter, the more is cooperation favored when we decrease the value of α. That is, as we decrease the value of d, the slope of the curves increases. This also means that the system becomes more sensitive to α changes as the d parameter is decreased. We call this effect the *small determinism degree and small synchrony rate* effect.

Possibly, the most relevant result is that, as we decrease the α value, the proportion of cooperators increases for the big majority of the simulations. That is, in general, asynchronism supports more cooperators than synchronism. This conclusion can only be derived because several scenarios were tested, namely, different interaction topologies and different values for the determinism degree of the transition rule. Indeed, this allowed us to verify that the situations where this conclusion does not apply are the ones where big values of the determinism degree of the transition rule, d, are used. Putting it another way, asynchronism is detrimental to the emergence of cooperation only when the probability of committing errors in the choice of the next strategy to use is very low. But reality is far from such perfection and phenomena like noise, be it deliberate or not, perception errors, or simply bad evaluations, often prevent agents from choosing what would seem the best choice.

Finally, we would like to stress that using an update method able to cover all the spectrum from synchronous to near sequential updating, as the ASD method is, allows a deeper analysis of the system being studied. It allows, for example, the identification of existing phase transitions: often the level of cooperation is 0 for $\alpha = 1$ and it remains there until a given $\alpha = c$ value is reached. Then, suddenly, the level of cooperation starts to increase, sometimes in a significative way, as α decreases from c to 0.1. For many of such situations c is very near to 1 or even equal to 1 (see, for example, Fig. 1d). This may suggest that in these cases the existence of some degree of cooperation is the most probable outcome in the system being modeled since it exists for almost the entire α domain.

3.2 Stochastic *Versus* Deterministic Updating

In [3] a classification of *random boolean networks* (RBNs) is given based mainly on the update methods used. The authors found that the behavior of asynchronous deterministic RBNs is much closer to the behavior of synchronous RBNs than to the one shown by asynchronous stochastic RBNs. In spite that our intention in this paper is not to make a comparison between the model studied here and RBNs, a question arises: is this also the case for spatial evolutionary games? An affirmative answer could help in the formulation of a general explanation for the influence of asynchronism on dynamical systems, if there is any. A negative answer is also important because, not only it takes this hypothesis off the path, but also because it can help us understanding the influence of deterministic updating on the modeled systems.

In order to model asynchronous deterministic dynamics, the authors of [3] used a sequential update method and another one (DGARBN)[1] that allows more than one agent to be updated at the same time. DGARBN is also a periodic update method: each agent has two parameters, p and q that, respectively, determine the period of the updating, and the phase (the translation of the update). Agents updated at the same time step are updated synchronously. The problem with this method is that it doesn't allow us to control the synchrony rate so that a direct comparison with the results achieved with ASD can be made. Besides, it doesn't provide the same number of updates for all agents when time grows. This means that agents can not be considered homogeneous anymore, which is a significant modification generating a different problem. Therefore, we changed the method so that the p parameter is the same for all agents. We call *asynchronous deterministic dynamics* (ADD) to this update method. As in the original method,

---

[1] DGARBN: Deterministic Generalized Asynchronous Random Boolean Networks.

$q < p$ and is defined randomly for each agent. This way, when $p = 2$, $p = 5$, or $p = 10$, respectively, 50%, 20% and 10% of the population is updated each time step, on average. These are the values we used in our simulations. They allow us to compare the results achieved with the ASD method for $\alpha = 0.5$, $\alpha = 0.2$ and $\alpha = 0.1$. It would be desirable that we could compare results for other α values, namely for $0.5 < \alpha < 1.0$, but we don't know about any deterministic and periodic update method that allows us to cover that interval.

Let us consider Res_{Sync}, Res_{ASD} and Res_{ADD} as the results achieved for a given combination of parameters with synchronous updating, ASD and ADD, respectively (each point in the charts of Fig. 1 corresponds to a given combination of parameters). In order to see which one of the two methods is closer to synchronous updating and how different they are, we first compute the quantities $dif_{ASD} = |Res_{ASD} - Res_{Sync}|$ and $dif_{ADD} = |Res_{ADD} - Res_{Sync}|$ for each combination of parameters. dif_{ASD} and dif_{ADD} are not considered in the following when $Res_{Sync} = Res_{ASD} = Res_{ADD}$ since in these cases it makes no sense to talk about which method is closer to synchronous updating. After this, we separately compute the averages of dif_{ASD} and dif_{ADD} values. Let us call these two quantities Avg_{ASD} and Avg_{ADD}, respectively. Finally, we compute the quantities $s = Avg_{ASD} - Avg_{ADD}$ and $r = x/y$, where $x = min(Avg_{ADD}, Avg_{ASD})$ and $y = max(Avg_{ADD}, Avg_{ASD})$. The s value tells us two things: which method is closer to synchronous updating and how different they are from each other in absolute terms. If $s > 0$, ADD is closer to synchronous updating than ASD and vice-versa. Also, as $s \to 0$ the two methods become closer. The r meaning is the following: if F and C are, respectively, the method that is farther and closer to synchronous updating, then $r < 0.5$ means that C is closer to synchronous updating than to F. If $r > 0.5$, C is closer to F than to synchronous updating.

The results are ($s = 0.011$, $r = 0.928$) for SWNs and ($s = 0.010$, $r = 0.922$) for SFNs. This means that ADD is closer to synchronous updating than ASD. However, we are talking about s values very close to 0. Furthermore, the r values are very close to 1, which means that the two methods are much closer to each other than to synchronous updating. Given that these values result from all the combinations simulated for each type of interaction topology, we later separated the results along the different values of each parameter in order to enquire if there are certain types of combinations for which there is a clear difference between the two methods. We made separations for all the parameters and, as Table 2 shows, the same conclusions derived above can be applied no matter how we separate the results. This means that, although ADD is almost always closer to synchronous updating than ASD, it is much closer to ASD than to synchronous updating. This result indicates that the conclusions derived for ASD also apply to ADD and that possible different levels of cooperation observed in the system being modeled are not due to the deterministic or stochastic nature of the updating scheme used by the agents. Also, this means that if we are willing to explain the influence of update methods on the level of cooperation we must look somewhere else away from their deterministic *versus* stochastic nature.

Table 2. s and r values for the different values of each parameter

SWNs						SFNs					
$\phi = 0.0$		$\phi = 0.1$		$\phi = 1.0$		$m_0 = 4$		$m_0 = 8$		$m_0 = 12$	
s	r	s	r	s	r	s	r	s	r	s	r
0.020	0.877	0.006	0.952	0.011	0.940	0.013	0.916	0.008	0.928	0.011	0.900

$m = +\infty$		$m = 100$		$m = 2$		$m = 1$	
s	r	s	r	s	r	s	r
0.009	0.876	0.007	0.919	0.012	0.939	0.017	0.932

$\alpha = 0.5 \equiv p = 2$		$\alpha = 0.2 \equiv p = 5$		$\alpha = 0.1 \equiv p = 10$	
s	r	s	r	s	r
0.025	0.792	0.011	0.933	−0.002	0.985

4   Conclusion and Future Work

In this work we examined the influence of asynchronism on the evolution of co-operation in models where agents try to adapt their behavior to the context in which they live. We showed that, in general, asynchronism supports more cooperators in the Spatial Prisoner's Dilemma game than synchronism. This conclusion contradicts previous ones but it results from a more general analysis, based on a bigger number of tested conditions, namely, different types of topologies, various values for the determinism degree of the transition rule and different levels of the synchrony rate. Besides this conclusion, the asynchronous update method used allowed us to derive some conclusions concerning, for example, the sensitivity and monotonicity of the model to changes in the synchrony rate. Finally, we found that the outcome of the studied model is approximately the same whether a stochastic or a deterministic asynchronous updating is used.

Future extensions to this work will explore the ASD update method with other games in order to verify if the results achieved with the PD game as, for example, the *small determinism degree and small synchrony rate* are also present. The results achieved in [12] with the Snowdrift game, where the best-neighbor $(+\infty)$ and the proportional $(m = 1)$ transition rules, as well as synchronous and sequential updating were used, seem to indicate that this is the case. However, only by exploring intermediate levels of asynchronism and intermediate levels of determinism of the transition rule we can confirm this. We will also try to answer the question "Is there an explanation for the influence of asynchronism on the emergence of cooperation?" and, if it turns out that there is a positive answer "Is this explanation extensible to other dynamical systems?".

Acknowledgments

We thank the GruVA members Pedro Santana, Carlos Cândido, Vasco Santos and Pedro Mariano for useful discussions and comments. This work was partially supported by FCT/MCTES grant No. SFRH/BD/37650/2007.

References

1. Barabasi, A.-L., Albert, R.: Emergence of scaling in random networks. Science 286, 509 (1999)
2. Fatès, N., Morvan, M.: An experimental study of robustness to asynchronism for elementary cellular automata. Complex Systems 16(1), 1–27 (2005)
3. Gershenson, C.: Classification of random boolean networks. In: Proceedings of the Eight International Conference on Artificial Life, pp. 1–8. MIT Press, Cambridge (2002)
4. Grilo, C., Correia, L.: Asynchronous stochastic dynamics and the spatial prisoner's dilemma game. In: Neves, J., Santos, M.F., Machado, J.M. (eds.) EPIA 2007. LNCS (LNAI), vol. 4874, pp. 235–246. Springer, Heidelberg (2007)
5. Huberman, B., Glance, N.: Evolutionary games and computer simulations. Proceedings of the National Academy of Sciences 90, 7716–7718 (1993)
6. Newth, D., Cornforth, D.: Asynchronous spatial evolutionary games: spatial patterns, diversity and chaos. In: Proceeding of the 2007 IEEE Congress on Evolutionary Computation, pp. 2463–2470 (2007)
7. Nowak, M., Bonhoeffer, S., May, R.M.: More spatial games. International Journal of Bifurcation and Chaos 4(1), 33–56 (1994)
8. Nowak, M., May, R.M.: Evolutionary games and spatial chaos. Nature 359, 826–829 (1992)
9. Oh, J.C.: Cooperating search agents explore more than defecting search agents in the internet information access. In: Proceedings of the 2001 Congress on Evolutionary Computation, CEC2001, pp. 1261–1268. IEEE Press, Los Alamitos (2001)
10. Pacheco, J.M., Santos, F.C.: Network dependence of the dilemmas of cooperation. In: Science of Complex Networks: From Biology to the Internet and WWW, CNET 2004, vol. 776, pp. 90–100 (2005)
11. Smith, J.M.: Evolution and the Theory of Games. Cambridge University Press, Cambridge (1982)
12. Tomassini, M., Luthi, L., Giacobini, M.: Hawks and doves on small-world networks. Physical Review E 73(1), 016132 (2006)
13. Watts, D., Strogatz, S.H.: Collective dynamics of small-world networks. Nature 393, 440–442 (1998)

A Study of Off-Line Uses of Anticipation

Giovanni Pezzulo

Institute of Cognitive Sciences and Technologies - CNR
Via S. Martino della Battaglia, 44 - 00185 Rome, Italy
giovanni.pezzulo@istc.cnr.it

Abstract. In a simulated guards-and-thieves scenario we study how
the behavioral system of an autonomous agent, which consists of multiple perceptual and motor schemas endowed with anticipatory mechanisms, self-organizes for satisfying its drives. Furthermore, we study how
schemas acquired for navigation can be re-used off-line, 'in simulation',
for forecasting future dangers, and planning trajectories leading to goal
locations. We argue that off-line simulations permit not only to coordinate with the present, but with the future, too, and to act goal-directed.

1 Introduction

In a previous study [15] we have shown that a schema-based agent architecture
endowed with anticipatory mechanisms (internal forward models) running on-
line with action can adaptively satisfy multiple motivational pressures (drives)
in a complex environment better than a purely reactive system, despite the
costs of prediction. Here we extend the analysis by investigating how the same
anticipatory mechanisms can be exploited off-line to produce internal, 'mental'
simulation of behavior for the sake of preventing dangers and planning goal-
directed action.

Our model includes elements of three theories. The first is Hesslow's *simulation hypothesis* [6], arguing that the brain reenacts sensorimotor structures
used for actual interaction for the sake of exploring the effects of one's own actions 'in simulation'; see also [4]. Several artificial systems have been proposed
[7,20,22,23,25] that generate long-term predictions by chaining short-term predictions, and use this capability for simulative planning (a virtual exploration of
multiple possible plans before—or instead of—attempting them in practice) or
for evaluating the outcome of their actions in advance. While those architectures
are based on a single neural network and have a single task, in our model multiple schemas run and simulate concurrently, and the agent deals with multiple
motivations.

The second is Baars' *global workspace theory* (GWT) [2], which introduces the
idea of broadcast of sensory content from the sensory cortex (either activated
externally, or self-activated, like in imagery) to a wide set of 'expert networks'
in the brain that compete for processing it (see [19] for a recent implementation
of the GWT). Here we implement planning via a broadcast mechanism from
a 'goal location neural unit' to the schemas repertoire: broadcast then triggers

M. Asada et al. (Eds.): SAB 2008, LNAI 5040, pp. 372–382, 2008.

competition among multiple schemas that simulate alternative paths to the goal location.

The third is Damasio's *somatic marker hypothesis* [3] in which 'as if' predictive loops serve for forecasting potential dangers and stopping planning or action if negatively-marked sensory states are predicted that were previously stored.

2   The Agent Architecture

Schema-based agent architectures [1] are inspired by ethological models of behavior, in which several behavioral units (e.g., for grasping or reaching) are encoded in separated modules called schemas, which cooperate and compete for realizing complex behavior which exceeds the capabilities of isolated schemas.

The agent architecture used in our experiments, implemented on the top of the AKIRA Schema Library (AKSL) [16], is shown in fig. 1 (left). It includes 16 schemas for navigation, 4 schemas for recognizing and catching the treasure, 6 schemas for detecting and escaping from guards, and 2 schemas for avoiding obstacles. The architecture also includes two *drives* (hunger and fear) and a *goal* (goal location unit), that play the role of simple and complex motivational units respectively. It also includes a *broadcast mechanism*, a lookup table of *somatic markers*, a *plan unit*, a *pool of limited resources*, and *sensors and actuators* (camera and wheels controllers) that receive asynchronous commands by schemas with different firing rates (see the two diamond-shaped edges, where *Fs* are firing rates of schemas). Arrows represent learned Hebbian links (see later).

2.1   Schemas Repertoire

In our architecture, like in several related systems (e.g., [1]), there are two kinds of schemas: *perceptual schemas* and *motor schemas*. Perceptual schemas (e.g., *detect treasure* and *detect guard*) control the vision of the agent by moving a camera

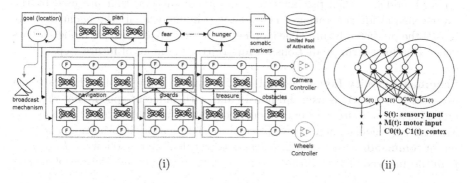

(i)   (ii)

Fig. 1. (i) The agent architecture; see explanation in the text. (ii) Each schema is implemented as a Jordan-type RNNs [10] that realizes inverse and forward modeling and includes a context loop. See explanation in the text. It includes 3 sensory inputs $S(t)$, 3 sensory outputs, 3 motor inputs $M(t)$, 3 motor outputs, 2 context inputs $C0(t)$ and $C1(t)$, 2 context outputs, 10 hidden nodes.

(the only sensor). Motor schemas (e.g., *catch treasure* and *escape guard*) receive sensry input from related perceptual schemas (e.g., *escape guard* from *detect guard*) and command the movements of the agent by moving its wheels. Moreover, each perceptual schema send sensory information to one motor schemas.

Each schema is a complete sensorimotor unit: it includes specific (neural) circuits for processing stimuli and generating motor commands (inverse models), plus a predictive component for generating sensory predictions (forward models) [15,24]. Both components are integrated in the same neural network, a Jordan-type RNN [10] illustrated in fig. 1 (right), which includes a sensorimotor loop and a context loop and permits both to control action (at time t) and to predict its sensory effects (at time $t+1$). The sensory input node represents three nodes (encoding positions in the three axis, $< x(t), y(t), z(t) >$) and the motor output node represents three nodes (encoding $< vx(t), vy(t), vz(t) >$, i.e., the motor command for reaching $< x(t+1), y(t+1), z(t+1) >$).

Generation vs. Simulation. Schemas run in 2 modes: generation or simulation.

Generation mode is the default. After preprocessing (features extraction) operated by hand-coded routines (see [15]) the RNN receives the estimated position $< x(t), y(t), z(t) >$ of the feature it is specialized to deal with (e.g., a red shape) with from the camera (in the case of perceptual schemas) or a perceptual schema (in the case of motor schemas), produces a motor command $< vx(t), vy(t), vz(t) >$ and sends it to the camera (or wheels) controller. The RNN also receives an efference copy of the final motor command executed by the camera (or wheels) controller and generates the sensory prediction $< x_p(t+1), y_p(t+1), z_p(t+1) >$.

Any schema can run off-line in *simulation mode*, too, for predicting the long-term sensory consequences of its motor commands. In this case its motor commands are inhibited (not sent to the actuators), but fed as sensory inputs to the forward model. It then produces new sensory predictions that are used by the inverse model for generating a new motor command 'as if' the agent actually sensed the predicted future. The loop between forward and inverse models allows generating long-term predictions for an arbitrary number of future steps.

Schemas Activity Level. Several schemas can be active at once. Each schema executes its operations asynchronously and with different speed, that depends on its current *activity level act* (calculated anew at the beginning of each cycle). More active schemas then receive more up-to-date sensory information, and send motor commands with higher firing rate. *act* is calculated as follows: $rel + links + plan$ (normalized in $[0,1]$) if this sum is larger than *pool*, otherwise to *pool*, where:

- *rel* indicates how much the schemas is expected to be successful in the current context; in our model (like in [24]) this depends on its *prediction accuracy*: position sensed and predicted (by the frward model) are compared, and *rel* is set to $(1 - || < x(t), y(t), z(t) > - < x_p(t), y_p(t), z_p(t) > ||)$.
- *links* is activation received by drives or other schemas due to Hebbian links;

- *plan* is activation received by plan units due to Hebbian links;
- *pool* is the amount of resources currently available in the limited pool.

Each schema has a threshold *thr* (set to 0.3 in our simulations). If $act < thr$ the schema functions normally but its motor commands to the actuator are inhibited. See [16] for a more detailed specification of the parameters.

2.2   Other Components of the Agent Architecture

Drives. The architecture includes two internal, motivational states, *fear* and *hunger*, modeled as simple homeostatic variables whose satisfaction is a primary source of behavior [9]. *Hunger* is raised (of 0.1 every 10 cycles) by a 'biological clock', and is set to 0 when the treasure is grabbed. *Fear* is set to 1 when a guard is detected, and decreases (of 0.1 every 5 cycles) otherwise. In the learning phase (see later) *fear* and *hunger* develop associative links with schemas. Their main roles are steering behavior (by activating schemas for escaping guards and reaching the treasure respectively when the context is appropriate), and creating positive feedback and persistence (aka hysteresis). A Fuzzy Cognitive Map (FCM) [13], a hybrid neural network - fuzzy system, is used for calculating drives activity level (in [0,1]): *fear* and *hunger* are modeled as two nodes in the FCM and have mutually inhibitory links set to -0.6.

Goal Location Units. Goal location units are self-exciting units (with weight 1) that encode a specific position in the map $< x_g, y_g, z_g >$ and *broadcast* them to location schemas for triggering goal-directed action. Moreover, they are able to recognize when the agent reaches the location $< x_g, y_g, z_g >$. In our simulations the (normalized) activity level of goal location units is set to 0.7. Goal location units represent complex motivations that 'compete' with drives for determining the agent's behavior: the former introduces top-down pressures, while the latter allows remaining opportunistic and responsive to dangers.

Broadcast mechanism. The broadcast mechanism, implemented as a black-board (see [16]), permits to feed sensory states from the goal location unit to all navigation schemas, or from a navigation schema to all others. As it will become clearer in the tasks description, when a schema is successful in 'simulation mode' it is allowed to broadcast its simulated sensory state; schemas that receive a broadcast start simulating, too, and if successful they can broadcast. Goal location unit broadcast for triggering goal-directed navigation, too.

Plan Units. While schemas are specialized for fine-grained movements, plan units permit to realize more complex ones by storing and then reenacting sequences of schemas. Plans are first 'imagined' then executed: the plan in fact stores sequences of schemas whose simulations were effective. Plan units implementation follows the *competitive queuing* model in [8]: they run an internal context signal (a 'clock') and create on the fly Hebbian links between states of the clock and schemas which broadcast (i.e., those effective in simulation). When the clock is replayed, the same schemas are triggered in sequence.

Somatic Markers. Somatic markers are implemented as a simple lookup table that stores sensory states in which high (> 0.9) fear was experienced during learning. When a sensory state included in the lookup table is perceived or simulated, the value of *fear* is raised to 1. If a schema has simulated a dangerous state, it is *inhibited*: its activity level set to 0.

Limited Pool of Resources. The total amount of resources shared by the schemas is limited. This represents *negative feedback* among them, that together with *positive feedback* provided by the associative links is an essential element of emergent, self-organizing phenomena [11]. The total amount of resources, *total_pool*, is set to 5.0 in our simulations.

Sensors and Actuators. The two actuators (camera and wheel motors) function in the same way. They receive asynchronously motor commands from schemas, with different firing rate (F in fig. 1), and fuse them. There is then no actual selection of one single schema for execution: the agent's behavior depends on the graded contribute of all active schemas, whose influence is proportional to their activation (this action selection scheme is called 'emergent' in [21]).

3   Experimental Set-Up

The experimental set-up is intended to test the off-line re-enactment of anticipatory mechanisms originally developed to be used on-line. For this reason, several agent's mechanisms are kept simple and learning serves only for bootstrapping a reliable architecture for supporting off-line simulation and its broadcast.

The agent plays the role of a thief in a guards-and-thieves scenario located in a simulated 'house' composed of 19 rooms whose positions vary between -10000 and +10000 in the three axes. In the design we used the 3-D engine Irrlicht (irrlicht.sourceforge.net), having realistic physics based on ODE (www.ode.org). During the learning phase, the agent first learns its four sets of schemas and then integrates them in a coherent schema repertoire. The agent is then tested in two tasks, consisting in catching the treasure (whose position can be known or unknown) and, at the same time, avoiding being captured by guards.

Schemas learning. Perceptual and motor schemas for dealing with guards, treasure, and obstacles are learned in a supervised way. During learning, treasure and obstacles appear in fixed locations and guards follow predictable trajectories. Circular and oval trajectories having different amplitudes were used. The RNN of each schema is trained with the BPTT algorithm [18]. One example (of guard, treasure, or obstacle) was sampled every twelve, with a total of thirty-six; learning stopped when the error of at least one forward model (the Euclidean distance between the actual and predicted position in 3D, $0.1 * 10^{-6}$) was less than 0,0000001.

Perceptual and motor schemas for navigation are instead learned in two phases by navigating in the house map (without treasure and guards). In the *first phase*

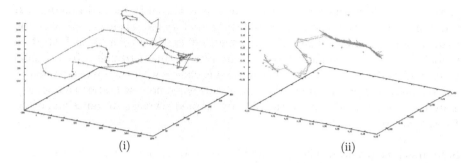

(i)   (ii)

Fig. 2. (i) Landmarks (crosses) placed in the first phase in the 3D house map. (ii) Movements executed while learning in the second phase (L_start, right, and L_goal, left, are in two different rooms). Each segment is a unitary movement, crosses are landmarks.

the agent navigates all the map at constant velocity with a wall-following pre-defined routine (encoded by hand) and a coupled perceptual-motor schema for obstacle avoidance, and places landmarks in the environment. For each 10 cycles of the wall-following routine a landmark is randomly placed in the neighbors of the agent. The landmarks are shown in fig. 2, left (note that in proximity of obstacles or turns the agent is slower and then it places more landmarks).

In the *second phase* the agent performs motor babbling for learning to navigate among landmarks, and its movements are selected or discarded on the basis of the simulated annealing (SA) algorithm [12]. The agent is placed in a random landmark location (L_start). Another landmark (L_goal) is selected randomly having a predefined distance (300 $\pm v$ units, where v is a learned parameter). The agent has to learn a schema for navigating from L_start to L_goal which allows accurate control and sensory prediction (this is why a schema's 'size' matters). The agent then begins its motor babbling. The next movement is selected by using the method described in [14] that conducts the agent in one of the adjacent locations by performing a unitary movement in one of the two axes x or y. The simulated annealing algorithm is used for evaluating if that movement gets the agent closer to L_goal. If this is the case, the agent actually moves there. Otherwise, it executes that movement with a probability of $e^{\delta E/T}$ where $\delta E/T$ is calculated as *current goal location minus expected goal location*, and T is a function of the *temperature* parameter of simulated annealing (which decreases for each attempted movement). If no movement is executed, the next execution cycle of simulated annealing starts. When the agent reaches L_goal, the whole 'history' of the movements is stored (a quadruple: position of the agent, desired position, movement executed, position reached). Fig. 2 (right) shows the movements executed while learning to navigate in a portion of the environment.

The history is then used as the training set for the navigation schemas (RNNs) with the BPTT algorithm [18] (with the same method used for the other schemas). If a schema does not reach the desired level of accuracy in pre-diction it is discarded, the v parameter is lowered (of 20 units) and the second phase restarts from the same L_start, but a new L_goal is selected randomly. If learning succeeds, the schema is stored (note that stored schemas generalize the data set and permit navigation in a space that is wider than 'from L_start to L_goal'). Now a new random landmark in a fixed range (10 units) from L_goal is selected as the new L_start, a new L_goal is choose with the same criterion as before, and the process restarts. Learning ends when no new L_start can be generated that lies outside any of the already learned schemas. In our simulations 16 schemas were learned.

Schemas integration. The agent architecture now includes four sets of schemas learned independently: the challenge is integrating them. For this purpose the agent (that now includes all the schemas) navigates again in the environment in which two guards dwell in fixed locations. All schemas are active but their motor commands are inhibited. The agent is instead controlled by the wall following routine used in the first learning phase. Schemas do not learn their internal models any more. Instead, energetic links among all the schemas and drives are learned with the following Hebb rule [5]: $\Delta W_{ji} = \eta a_i a_j$, where η is the learning rate (set to 0.2) and a_i, a_j represent the activations of two schemas, or a schema and a drive (normalization is applied at each cycle). This means that schemas and drives which are active in the same span of time develop associative links (e.g., fear with schemas for avoiding guards). The associative topology of the network implicitly encode possible trajectories or location/drive associations. Somatic markers are learned in this phase, too, by simply recording in the lookup table the sensory context in which guards are detected.

Task 1: finding the treasure. After learning, the agent is placed in a random location in the house and its *hunger* is set to 0.5. Two guards are in the same positions experienced during learning, and two more appear in random locations. The task consists in satisfying hunger by finding the treasure, which appears in random locations; it ends either with the treasure grabbed (success) or the agent captured by guards (failure). This tasks serves to test the trade-off between simulation/imagination and action: simulations allow discovering future dangers, but come at the risk of 'distracting' from immediate dangers (schemas can not 'generate' when they are simulating). In this task we compared two agent architectures.

Agent 1 (AG1) can only run schemas in 'generation' mode. Thus, although it exploits (short-range) anticipatory mechanisms, its behavior is driven by the environment's immediate affordances (guards and treasure) only.

Agent 2 (AG2) selects actions depending on its long-term predictions, too. It operates alternatively in 'generation' and 'simulation' modes. In 'generation' mode AG2 navigates and all its schemas compete for recognizing its current context/position by generating sensory predictions; the more accurate ones gain activity and priority over the effectors. When the predictions of one schema are accurate enough over a certain span of time (set to 3 consecutive good predictions), AG2 is ready for generating reliable simulations (in [20] this state is referred as *situated*). It thus switches to 'simulation' mode, produces long-term sensory predictions, and then broadcasts them to all navigation schemas. Those able to match (for 3 times) the simulated sensory stimuli they receive via broadcast are allowed to simulate and broadcast, and so on. This mechanism, that chains several schemas' predictions and (in principle) can 'image' (or 'plan') paths in the whole house, is used by AG2 for preventing future dangers. Every sensory prediction generated by schemas is compared with the 'dangerous' sensory states stored in the *somatic markers* lookup table. If a schema simulates a dangerous state, it is inhibited (its activity level is set to 0), and this prevents that path to be selected; see [3]. Moreover, *fear* is set to 1: this causes activation of avoidance schemas in anticipation of possible dangers. Otherwise, the path is considered 'safe' and the agent's actions planned by schemas in simulation are executed. No memory of the sequence of simulating schemas is maintained, but the first simulating schema runs in generation mode (by restoring the last context it had in generation mode). This is likely to produce the appropriate sensory context for the second simulating schema, and so on. Note that, due to dynamics in the environment and errors in prediction, there is no guarantee that the 'imagined' sequence of schemas will be actually executed.

We have run 100 experiments, and compared the average number of times AG1 and AG2 successfully reach the treasure, and the average time of success. Our results (see tab. 1, left) show that AG2 performs significantly better than AG1. The costs of running simulations (time spent in imagination rather than action) are by far overwhelmed by the advantages of predicting possible dangers in dynamic and dangerous environments.

Task 2: reaching known goal locations. Task 2 consists in reaching the treasure, which appears in a random location (but 'known' by the agent and set as its goal location unit) without being captured by the guards, which appear randomly. The task ends when the agent grabs the treasure or is captured by a guard. Differently from task 1, here the treasure location is known, and thus simulations can be used for planning and not only for avoiding dangers. In [17] we have argued that the simplest form of adaptivity for an agent consists in exploiting self-organization of its behavior repertoire for *coordinating with its present needs and the most immediate affordances*, but in order to realize purposive, goal-directed action it has to *coordinate with the future*, too. Simulative planning permits to realize such coordination encompassing present and future states without losing situatedness.

Table 1. Results of the tasks 1 and 2. (a) AG2 vs. AG1 is significant with ANOVA both for success % (p<,001) and Time (p<,001). (b) AG4 vs. AG3 is significant with ANOVA both for success % (p<,01) and Time (p<,001).

Agent	Success %	Time (in secs.)	Agent	Success %	Time (in secs.)
AG1	.505	157.8	A3	.872	71.2
AG2	.803	81.3	A4	.927	56.4
(a) Results of Task 1			(b) Results of Task 2		

In this task we compared two agent architectures. Agent 3 (AG3) is the same as AG2 but includes a goal location unit that *broadcasts* the goal (treasure) location (the sensory input $< x_g, y_g, z_g >$) to all navigation schemas, that then run in simulation mode (since no actual context information can be used, a sample of their possible sensory inputs is used instead). As for AG2, those which predict dangerous sensory states are inhibited. Those which successfully predict achievement of the goal location can either run in generation mode, if their sensory context is appropriate (i.e., if they are 'one step' from the goal location), or broadcast their context ($< x_g(t-1), y_g(t-1), z_g(t-1) >$) as the new goal location, and so on. This continues until a schema has the appropriate context for running in generation mode and can send a valid motor command. This in turn is likely to produce the right context for one or more other schemas, and so on, until the goal location is reached. The functioning of AG3 then differs from AG2 since the former 'plans' by using all its navigation schemas as a backward chaining mechanism (from goal to current location), and the latter as a forward chaining mechanism (from current to possible locations). Again, due to dynamics in the environment, there is no guarantee that the 'planned' sequence of schemas will be actually executed.

Agent 4 (AG4) is the same as AG3 but includes a plan unit, too. For each experiment, it stores the sequence of successfully simulating schemas, from goal location to start location. When the goal location is reached in simulation, the (reversed) context signal is replayed for activating the right sequence of schemas (this is done by setting their activity level to 1). However, all the other schemas remain active, too (although with a lower activity level). This means that AG4's behavior depends both on 'top-down' commands received by the plan (that activate schemas in the pre-planned sequence) and on 'bottom-up' influences of all the other schemas that do not belong to that sequence (that are still able to temporarily gain priority over the effectors if their predictions are accurate). This permit to follow the plan and, at the same time, to remain opportunistic and responsive to novel events such as guards.

We have run 100 experiments, and compared the average number of times AG3 and AG4 successfully reach the treasure, and the average time of success. Both agents had a high success rate and were able to 'simulate' a valid path to the goal location in all experiments: simulative planning can thus be considered an effective strategy for coordinating with future, desired goal states. Our results (see tab. 1, right) show that AG4 performs significantly better than AG3: once a good plan has been built, it is better to 'store' and use it to control action top-down, even with the risk of being less opportunistic. Qualitative observation indicates that AG3 and AG4 are often diverted from their simulated trajectories (e.g., by guards), but plan units in AG4 provide stronger *persistence* toward the goal location.

4   Conclusions

There are currently several limitations in our agent architecture that need to be addressed in future work. Learning is mainly supervised, the implementation of several mechanisms (e.g., somatic markers, broadcast) is very simple, there is no replanning, and some parameters are tuned by hand. However, our design choices are motivated by the necessity of realizing the complex architectural set-up necessary for a study of simulative capabilities (the aim of our paper): anticipatory schemas that can be reliable used on-line and off-line are difficult to obtain [25].

Consistently with recent simulative theories of cognition [3,4,6], the results of our experiments indicate that *mental simulation* is an effective strategy for avoiding dangers and planning in dynamic environments despite the fact that 'imagination' can in principle make an agent less efficacious in its current sensorimotor interaction. As argued in [4,17], we believe that off-line, mental simulation is a suitable, embodied alternative to 'reasoning by symbol-crunching' of traditional AI systems, since it permits internal manipulation of (anticipatory) representations without losing grounding and situatedness. Although further investigation is necessary for studying the trade-offs of engaging with 'the present' or 'the future', this study contributes to shed light on how mental simulation enabled increasingly complex cognitive capabilities and the role it played in the passage from present-directed to goal-directed, purposive action.

References

1. Arbib, M.A.: Schema theory. In: Shapiro, S. (ed.) Encyclopedia of Artificial Intelligence. 2nd edn., vol. 2, pp. 1427–1443. Wiley, Chichester (1992)
2. Baars, B.J.: A Cognitive Theory of Consciousness. Camb. Univ. Press, NY (1988)
3. Damasio, A.R.: Descartes' Error: Emotion, Reason and the Human Brain. Grosset/Putnam, New York (1994)
4. Grush, R.: The emulation theory of representation: motor control, imagery, and perception. Behavioral and Brain Sciences 27, 377–396 (2004)
5. Hebb, D.O.: The organisation of behaviour. Wiley, New York (1949)
6. Hesslow, G.: Conscious thought as simulation of behaviour and perception. Trends in Cognitive Sciences 6, 242–247 (2002)
7. Hoffmann, H.: Perception through visuomotor anticipation in a mobile robot. Neural Networks 20, 22–33 (2007)
8. Houghton, G.: The problem of serial order: A neural network model of sequence learning and recall. In: Dale, R., Mellish, C., Zock, M. (eds.) Current Research in Natural Language Generation, pp. 287–319. Academic Press, London (1990)
9. Hull, C.L.: Principles of Behaviour. Appleton-Century-Crofts, New York (1943)
10. Jordan, M.I., Rumelhart, D.: Forward models: Supervised learning with a distal teacher. Cognitive Science 16, 307–354 (1992)
11. Kelso, J.A.S.: Dynamic patterns: the self-organization of brain and behavior. MIT Press, Cambridge (1995)
12. Kirkpatrick, S., Gelatt, C.D., Vecchi, M.P.: Optimization by simulated annealing. Science 4598, 671–680 (1983)
13. Kosko, B.: Neural Networks and Fuzzy Systems. Prentice Hall Int., Singapore (1992)
14. Marsaglia, G.: Choosing a point from the surface of a sphere. The Annals of Mathematical Statistics 43, 645–646 (1972)

15. Pezzulo, G., Calvi, G.: A schema based model of the praying mantis. In: Nolfi, S., Baldassarre, G., Calabretta, R., Hallam, J.C.T., Marocco, D., Meyer, J.-A., Miglino, O., Parisi, D. (eds.) SAB 2006. LNCS (LNAI), vol. 4095, pp. 211–223. Springer, Heidelberg (2006)
16. Pezzulo, G., Calvi, G.: Schema-based design and the akira schema language: An overview. In: Butz, M.V., et al. (eds.) ABiALS 2006. LNCS (LNAI), vol. 4520, pp. 128–152. Springer, Heidelberg (2007)
17. Pezzulo, G., Castelfranchi, C.: The symbol detachment problem. Cognitive Processing 8(2), 115–131 (2007)
18. Rumelhart, D.E., Hinton, G.E., Williams, R.J.: Learning internal representations by error propagation. In: Parallel distributed processing: Explorations in the microstructure of cognition, vol. 1. MIT Press, Cambridge (1986)
19. Shanahan, M.P.: A cognitive architecture that combines internal simulation with a global workspace. Consciousness and Cognition 15, 433–449 (2006)
20. Tani, J.: Model-based learning for mobile robot navigation from the dynamical systems perspective. IEEE Trans. on Systems, Man, and Cybernetics 26, 421–436 (1996)
21. Tyrrell, T.: Computational Mechanisms for Action Selection. PhD thesis, University of Edinburgh (1993)
22. van Dartel, M.F.: Situated Representation. PhD thesis, Maastricht University (2005)
23. Vaughan, R.T., Zuluaga, M.: Use your illusion: Sensorimotor self-simulation allows complex agents to plan with incomplete self-knowledge. In: Nolfi, S., Baldassarre, G., Calabretta, R., Hallam, J.C.T., Marocco, D., Meyer, J.-A., Miglino, O., Parisi, D. (eds.) SAB 2006. LNCS (LNAI), vol. 4095, pp. 298–309. Springer, Heidelberg (2006)
24. Wolpert, D.M., Kawato, M.: Multiple paired forward and inverse models for motor control. Neural Networks 11(7-8), 1317–1329 (1998)
25. Ziemke, T., Jirenhed, D.-A., Hesslow, G.: Internal simulation of perception: a minimal neuro-robotic model. Neurocomputing 68, 85–104 (2005)

An Individual-Based Model of Task Selection in Honeybees

Thomas Schmickl and Karl Crailsheim

Department for Zoology, Karl-Franzens-University Graz, Universitaetsplatz 2,
A-8010 Graz, Austria
thomas.schmickl@uni-graz.at,
karl.crailsheim@uni-graz.at

Abstract. Adaptive division of labour is one key characteristic of eusocial insect colonies and of high relevance in biology, ethology, swarm intelligence and robotics. We constructed an individual based model of division of labour in a honeybee colony. Our model incorporates distinct worker cohorts (foragers, storers, nurses), unemployed bees and larvae. Our goal was a model as accurate as possible, thus we implemented a heterogeneous environment, agents' physiology and the flow of nutrients within the colony. In our model, the bees decide which task to choose, depending on the intensity of stimuli and on individual thresholds, which are modulated in response to task performance. We describe the main aspects of this model and demonstrate the stability of the emerging division of labour. The model predicts the energetic costs of sudden perturbations (removing/adding cohorts of workers of one task), as well as the resulting shifts in task cohort sizes.

1 Introduction

In social insects thousands of workers generate a characteristic heterogeneous environment inside of their colony's housing. The workers cooperate and share the colony's workload (nursing the brood, foraging and processing for food, ...). Eusocial insect species developed the ability of workers to specialize in distinct sets of tasks. In some species this specialisation is permanent (e.g., due to specialized morphology), in other species such specialisation is temporary. In honeybees, division of labour is highly related with workers' age under normal conditions, but is also quite flexible and adaptive after perturbations.

Studies showed, that specialised worker bees tend to locate in specific areas in the hive, which correlate to the performed task: nurse bees are most prominently found in the brood nest. Foragers are found at food sources and near the hive entrance, and food storer bees, which are often found near the entrance area and in the honey storage area.

Several models have been discussed about how the observed age-correlated but still adaptive division of labour can be explained. While the ultimate colony-level effects are well known, the underlying proximate mechanisms are still discussed: Early (and some recent) studies suggest a rather fixed age-related scheme [10] [4] [5], based on age-related physiological and morphological differences of workers.

M. Asada et al. (Eds.): SAB 2008, LNAI 5040, pp. 383–392, 2008.

Other studies suggest that workers motion and search for work determines the chosen task [2]. In contrast to that, other authors suggest that division of labour is driven by an adaptive process inside of workers, which adapts thresholds according to the work history of the worker [3].

Since the studies of Nobel laureate K. Lorenz [6] the importance of behavioural thresholds has been known. These thresholds determine whether or not a stimulus of a given intensity triggers a corresponding behaviour in an animal. The exposure to stimuli as well as the performance of a behaviour can modulate these thresholds. The processes of *sensitisation* and *facilitation* are of high relevance for the mechanisms depicted in our model presented here: If a stimulus once triggers a certain behaviour, the animal is more likely to repeat this behaviour again, even at lower stimulus levels. Thus, the performance of behaviours can lower the corresponding thresholds. After some time without such a stimulus, these effects disappear, thus the thresholds rise again, in the absence of the corresponding behaviour. Such adaptations were suggested by recent studies [12] [3] to explain task specialisation in social insects, an approach that we follow in our model.

Several of the model approaches mentioned above have been investigated in mathematical terms, and in several simulation studies. These studies showed that the suggested models suffice to explain observed aspects of division of labour. However, these models, and also the published simulation studies, are very abstract. They often treat the movement of the agents and the heterogeneous environment in very simple ways, like chess figures that jump from field to field on a chess board. Significant aspects like the agents' energetic expenditures, which are related to the performed task, are neglected as well as other physiological parameters. Our model depicts the regulation of division of labour in a honeybee colony. In contrast to other models our model incorporates many honeybee specific constraints and the most important aspects of the stimuli that can trigger the behavioural tasks. We describe the model's implementation, its basic ideas and we investigate the stability and the efficiency of the predicted emerging division of labour in the simulated colony.

2   The Model

We constructed a multi-agent model of a honeybee colony (*TaskSelSim*) consisting of 700 adult bees and 100 larvae. The model is implemented in NetLogo 3.1.4 [13]. The details of the model's implementation (equations, all parameters) are described in [9]. The paper at hand focuses on using this model for simulation experiments that investigate the stability and efficiency of the predicted division of labour. The model depicts the environment in the hive as a discrete set of patches organized as a grid of 31 x 52. See figure 1 for details on the spatial organisation of the modelled colony. These patches can contain quantities of larvae (one per patch) or nectar (40 units per patch). In this model, we use the terms 'nectar' and 'honey' synonymously. All other honeybee nutrients (pollen, jelly)

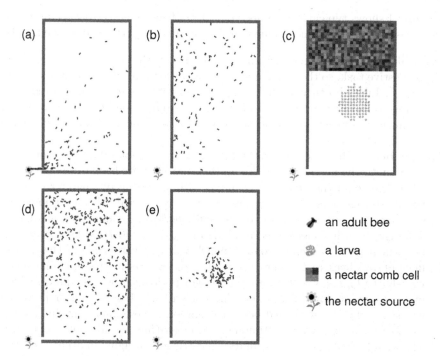

Fig. 1. Spatial organization of the model. In all sub-figures the border of the hive is shown as a rectangle with the hive's entrance/exit as small hole in the lower left corner. All figures show exemplary a typical situation at time $t = 3000 steps$. (a) Typical distribution of forager bees. (b) Typical distribution of storer bees. (c) In the upper part of the hive the nectar storage area is shown (honeycomb). In the center of the hive, the brood is located. (d) Typical distribution of unemployed bees. (e) Typical distribution of nurse bees. All figures: The nectar source is shown as a symbolized flower left of the hive. To simulate plausible foraging distances, the motion speed of foragers is slowed down as soon as they leave the hive.

were neglected in our model. The basic unit of fluid quantities (nectar/honey) is one crop load of an average adult bee. The discrete patches also hold the stimuli that can be emitted by agents. These stimuli are: vibration signals emitted by bee dances, touch signals performed by returning foragers and chemical signals emitted by larvae to signal their hunger status. Although the environment is discreet, the agents can move in continuous space. In each time step, the model executes the following procedures (once per time step):

1. Process the status of each patch (decay of chemical signals).
2. All agents emit the appropriate stimuli for that step, chemical stimuli are diffused.
3. Process the agents' physiological processes (consumption of nectar).
4. Perform one behavioural step per agent (e.g., navigation or nursing).

Figure 1 shows typical spatial distributions of bees assigned to working tasks. These tasks determine which behavioural programs are performed by the agents:

Unemployed bees: These bees move randomly in the hive. They perform no exhaustive task, so they consume their nectar load in their crop at a low rate (cr_{low}). Like all other adult bees, they can refill their nectar crop load at storage cells in the upper part of the hive. In our model, bees had to switch to this unemployed state at least for one time step before they could engage in a different task.

Forager bees: These bees leave the hive with a low (but sufficient) crop load. They fly to the nectar source, fill up their crop and fly back to the entrance. There they emit the *unloading* stimulus within a radius of 1 patch to attract nearby storage bees which can take over the nectar load. Then they move for some time t_{rest} in the hive and occasionally perform a communication dance: A forager bee that has waited too long ($T_{search} >= 50steps$) for a receiving bee will perform a tremble dance, which will motivate some of the nearby unemployed bees to become storer bees. In contrast, if the queuing delay was short ($T_{search} <= 20steps$) the forager performs a waggle dance, which motivates unemployed bees to become forager bees. This regulation principle was shown by [11] with real bees. To represent these dances in our model, the forager agents emit a stimulus locally (within a radius of 3 patches). The waggle dance acts as a stimulus that triggers foraging in other bees, the tremble dance lowers the threshold associated with the task 'storing'. These dance stimuli are only emitted for several time steps and disappear as soon as the emitter stopped to produce them. After the resting time is over, they leave the hive again towards the nectar source.

Storer bees: These bees wait near the entrance for returning foragers. They take the crop load of returning foragers and head towards the storage area. They drop their nectar load there and head back towards the entrance.

Nurse bees: These bees navigate (uphill) in the chemical stimulus emitted by hungry larvae. If they are located on a patch containing a hungry larva, they start to feed this larva until it is saturated or the nurse is almost empty. These feedings last for several time steps.

Larvae: The brood resides in cells (patches) in the central broodnest area (see figure 1c). Larvae cannot move. If they have low nectar reserves, they emit a chemical hunger signal, which is stored in the patch they are located in. A diffusion process allows the signal to spread and a decay process removes the stimulus from the system over time. In contrast to the dance stimuli, the chemical stimulus stays a longer time present in the local area after the emitter stopped the emission.

2.1   Simulated Physiology

All adult bees basically consume their nectar loads in their crops at low rates ($cr_{low} = 0.0004units/step$). Flying forager bees are the only exception: They consume nectar at higher rates ($cr_{high} = 0.001units/step$). Larvae consume

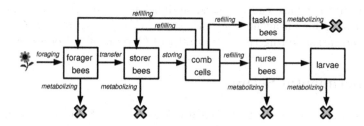

Fig. 2. Schematic representation of the flow of nectar in our model. Boxes represent cohorts of worker bees, larvae or combs. Arrows indicate nectar flows. The flower symbolises the outside nectar source and the cross-like symbols represent sinks.

nectar at the rate $cr_{larva} = 0.0004 units/step$. If an agent (bee or larva) runs out of nectar, it dies and is removed from the system. Figure 2 shows the resulting flow and consumption of nectar among and within the cohorts of bees. For a detailed reasoning of these values, please see [9].

2.2 The Key Aspects of Our Model of Division of Labour

The most important aspect in our model is the implementation of the task selection mechanism. We followed the approaches of [3] and implemented a threshold based system. Each patch can hold a set of stimuli of varying intensities, depending on the type of each stimulus and depending on the emission of stimuli of that type by nearby agents. Each type of stimulus can motivate an employed adult bee agent (task = 'no-task') to join one of the tasks $m \in \{$ 'foraging', 'storing', 'nursing' $\}$. Whenever one of these stimuli exceeds an individual threshold of an agent i located on that patch x, the agent engages in the associated task m. Each of these thresholds is modelled in a non-linear manner, as is shown by equation 1. $p_{i,m}$ models the probability to engage in task m in one time step. $s_{x,m}$ is the local intensity of the task-associated stimulus. $\Theta_{i,m}$ is used to shift the threshold individually up and down, n is used to express the degree of non-linearity in these behavioural decisions.

$$p_{i,m} = \frac{s_{x,m}^n}{s_{x,m}^n + \Theta_{i,m}^n} \tag{1}$$

Employed bees switch back to the unemployed state with probabilities of $\lambda'_{nursing'} = \lambda'_{storing'} = 0.005/step$ and $\lambda'_{foraging'} = 0.001/step$. To allow specialisation within this systems, the levels of the thresholds are adapted individually during run time. In the case that an unemployed agent engages in task m', the $\Theta_{i,m}$ is reduced by ξ_m, making it more likely that the agent will engage in this task in future. Whenever an umployed agent does not engage in a task, the corresponding threshold is increased by φ_m, making it more unlikely that these behaviours will be triggered later on. In our simulations, all values of ξ were set to $\xi = 0.1$ and all values of φ were set to $\varphi = 0.001$. It was shown in [9] that these parameter values lead to plausible division of labour. In our simulations we used $n = 2$ for all agents and all agents initially started with Θ values of 0.001 for all tasks. During run time, values of Θ are confined between 0 and 1.

3   Experimental Setup

We investigated how well the modelled colony can handle sudden disturbances in the division of labour. Additionally, we investigated the impact of such disturbances on the colony's nectar accumulation. We started our simulations with 700 adult bees randomly distributed in the hive (685 unemployed bees, 15 foragers). 100 larvae with randomized nectar reserves were distributed randomly (normal distribution) around the hive's center and nectar storage cells were randomly filled. Using these initial conditions, we simulated 10000 time steps to allow the colony to converge to a stable equilibrium of task cohorts. The whole simulation structure was saved on hard disk (at $t = 10000$). This structure was then reloaded multiple times, every time a perturbation was performed after this 'reloading'. Then the model was evaluated for additional 10000 time steps. Each experiment was repeated 6 times, our figures show mean values of these 6 repetitions. All runs are compared to the mean values of 6 repetitions of a control run (no perturbation was performed). During all performed simulation runs, no adult bees died during the runtime of the experiments. Up to 6 larvae disappeared in consequence of the induced fluctuation (removal of worker bees).

4   Results and Discussion

4.1   Experiment 1: Removal of Forager Bees

Figure 3 shows the predicted colony-level reactions to a sudden loss of forager bees. Compared to the control experiment ('no change'), the colony's cohort of foragers was affected for approx. 1500 time steps. After that, the worker cohorts reached a stable equilibrium again. For the period with the lowered amount of foragers, also the cohort of food storer bees was decreased, which is in good agreement with empiric results gained with real bees. The cohort of nurse bees showed almost no reaction to the induced disturbance. Both cohorts (foragers, storers) reached a lowered equilibrium in the second phase of the experiment, which can be explained by the fact that a fraction of the adult worker bees was removed, thus the total population size decreased.

4.2   Experiment 2: Removal of Storer Bees

Figure 4 shows the predicted colony level effects to a sudden removal of storer bees. Compared to the control experiment, the colony's cohort of foragers was also affected by a decrease in the storer cohort. The decreased cohort of storer bees could unload less foragers per time step, thus queueing delays of foragers increased. This lowered the number of waggle dances (less foragers were recruited)

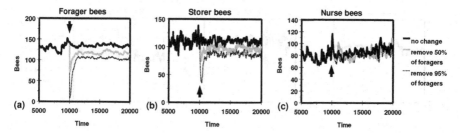

Fig. 3. Colony-level reaction to removal of several fractions of the foraging cohort. Mean values of 6 repetitions, arrows indicate the timing of the removal.

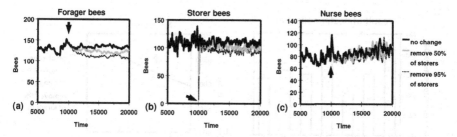

Fig. 4. Colony-level reaction to removal of several fractions of the storer cohort. Mean values of 6 repetitions, arrows indicate the timing of the removal.

and increased the number of storer bees quickly to the new (lowered) equilibrium. Again, the cohort of nurse bees was almost unaffected. The adaptation of the cohort of storer bees worked rather quickly compared to compensation in the cohort of foragers in experiment 1, demonstrating that the group of storer bees is regulated stronger under the regime of tremble dances than is the cohort of foragers by the performance of waggle dances.

4.3 Experiment 3: Removal of Nurse Bees

Figure 5 shows the predicted colony level effects to a sudden removal of nurse bees. Like in the previous experiment, the cohort of foragers showed no sudden response (no peak), but the perturbation decreased the colony's population, leading to a lowered stable equilibrium. Also the cohort of storer bees was not strongly affected. As shown in figure 5c, the cohort of nurse bees was strongly affected by the perturbation, but the cohort size quickly increased again after the perturbation. This demonstrates the strong regulation of nursing by the chemical signals emitted by hungry brood. Please note that in consequence to our manipulation, the cohort of nurse bees reached a higher equilibrium than before. This 'over-compensation' is typical for homeostatic systems and can be explained by the delayed effects of the larvae's simulated physiology: The removal of nurse bees resulted in a less feeding of larvae, resulting in more hungry larvae

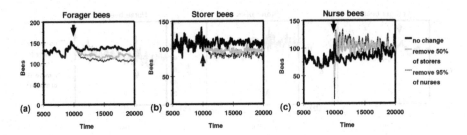

Fig. 5. Colony-level reaction to removal of several fractions of the nurse bee cohort. Mean values of 6 repetitions, arrows indicate the timing of the removal.

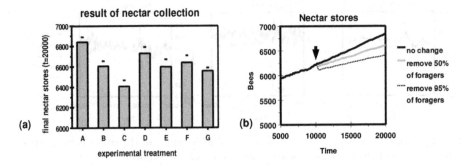

Fig. 6. (a) Final nectar stores after perturbations: A=no perturbation, B=50% foragers removed, C=95% foragers removed, D=50% storers removed, E=95% storers removed, F=50% nurses removed, G=95% nurses removed. Bars indicate mean values of 6 repetitions each, markers indicate standard deviation within repetitions. (b) Dynamics of the size of nectar stores resulting from a sudden removal of forager bees.

that emit more hunger signal. Empiric studies showed, that after a removal of nurse bees, forager and storer bees are recruited to the job of nursing [7], as we observed it in our simulations.

4.4   Analysis of Efficiency

The costs of adaptation are very important for a species: Each disturbance of a well-balanced system will induce costs in term of energy losses or population losses. The better (more precise) and the faster the adaption works, the higher is the expected fitness of such a species. This fitness is important for a species' chances in natural selection during biological evolution [1]. To investigate this aspect with our model, we analyzed the dynamics of the nectar stores, as the accumulation of nectar is one of the main indicators for assessing the fitness of a colony. As figure 6 shows, the perturbations we induced affected the efficiency of the nectar collection. All perturbations reduced the final nectar stores compared to the control run (A). The more bees were removed due to our perturbations, the less nectar was accumulated, although each bee is also a consumer. This demonstrates that our model predicts a net gain of nectar per bee, what is important to ensure plausibility of the models predictions. Figure 6 shows that the removal of foragers (B-C) had the most detrimental effect on the nectar stores, compared to the other perturbations (D-G). Figure 6b shows the typical dynamics of nectar stores in response to forager removal: The removal act resulted in a sudden drop of nectar reserves. The more foragers were removed, the more prominent was this effect. But also after the adaptation period the nectar dynamics differed significantly from the pre-perturbation period. The new equilibrium in division of labour was shifted to a lower number of foragers, thus the slope in nectar gain was lowered. These dynamics are in good agreement with other studies, which we achieved with a different model that focused on foraging target decisions in honeybees [8].

5 Conclusion

Our analyses showed that our model is able to predict the adaptive division of labour in a honeybee colony. The predicted results are in good agreement to known empiric studies. We could identify how 'deeply' the induced perturbation disturbed the complex system of division of labour and how (fast) the system responded by recruiting additional bees to the perturbed task cohort. Our model is the first one that incorporates different types of stimuli in a near-nature way, what was important to interpret the dynamics of the observed adaptation phases. Also these dynamics correspond well to empiric findings. In addition, our model allowed us to predict also the energetic costs of perturbations. Without implementing a simulated physiology and the correct flow of nectar among the different groups of agents, such insights are impossible. This implementation of agents' physiology showed that the observed division of labour allowed all agents to maintain (in most cases) a sufficient energy supply. Also the minimal losses of adult bees showed that the colony itself could maintain a high level of homeostasis. To the best of our knowledge, our model is currently the only one incorporating these important factors. Other models can predict changes in task allocation, but without observing the survival of bees and brood in those analysis and without observing the energetic gains and costs, the efficiency of suggested adaptation mechanisms is hard to assess. In future we will also incorporate comb building and introduce age-related differences among bees, to further investigate the honeybee system.

Acknowledgements

T. Schmickl is supported by : EU-IST FET project I-Swarm, no. 507006; FWF-grant Temperature-induced aggregation of young honeybees, no. P19478-B16.

References

1. Darwin, C.: On the Origin of Species By Means of Natural Selection. John Murray, London (1859)
2. Franks, N.R., Tofts, C.: Foraging for work: How tasks allocate workers. Animal Behaviour 48, 470–472 (1994)
3. Gautrais, J., Theraulaz, G., Deneubourg, J.-L., Anderson, C.: Emergent polyethism as a consequence of increased colony size in insect societies. Journal of Theoretical Biology 215, 367–373 (2002)
4. Huang, Z.-Y., Robinson, G.E.: Regulation of honey bee division of labor by colony age demography. Behavioural Ecology and Sociobiology 39, 147–158 (1996)
5. Johnson, B.R.: Limited flexibility in the temporal caste system of the honey bee. Behavioral Ecology and Sociobiology 58, 219–226 (2005)
6. Lorenz, K.: Vergleichende Verhaltensforschung. Springer-Verlag Wien, NewYork (1978)
7. Robinson, G.E., Page, R.E., Strambi, C., Strambi, A.: Colony integration in honey bees: Mechanisms of behavioral reversion. Ethology 90, 336–348 (1992)
8. Schmickl, T., Crailsheim, K.: Costs of environmental fluctuations and benefits of dynamic decentralized foraging decisions in honey bees. Adaptive Behaviour 12(3-4), 263–277 (2004)
9. Schmickl, T., Crailsheim, K.: Taskselsim: A model of the self-organisation of the division of labour in honeybees. Mathematical and Computer Modelling of Dynamical Systems 14(2), 101–125 (2008)
10. Seeley, T.D.: Adaptive significance of the age polyethism schedule in honeybee colonies. Behavioural Ecology and Sociobiology 11, 287–293 (1982)
11. Seeley, T.D.: The tremble dance of the honey bee: message and meanings. Behavioral Ecology and Sociobiology 31, 375–383 (1992)
12. Theraulaz, G., Bonabeau, E., Deneubourg, J.-L.: Response threshold reinforcement and division of labour in insect societies. Proceedings of the Royal Society of London B 265, 327–332 (1998)
13. Wilensky, U.: NetLogo. Center for Connected Learning and Computer-Based Modeling. Northwestern University, Evanston, IL (1999),
http://ccl.northwestern.edu/netlogo/

Distributed Adaptation in Multi-robot Search Using Particle Swarm Optimization

Jim Pugh and Alcherio Martinoli*

Swarm-Intelligent Systems Group
École Polytechnique Fédérale de Lausanne
1015 Lausanne, Switzerland
{jim.pugh,alcherio.martinoli}@epfl.ch

Abstract. We present an adaptive strategy for a group of robots engaged in the localization of multiple targets. The robotic search algorithm is inspired by chemotaxis behavior in bacteria, and the algorithmic parameters are updated using a distributed implementation of the Particle Swarm Optimization technique. We explore the efficacy of the adaptation, the impact of using local fitness measurements to improve global fitness, and the effect of different particle neighborhood sizes on performance. The robustness of the approach in non-static environments is tested in a time-varying scenario.

1   Introduction

Designing even simple behaviors for robots that are efficient and robust can be very difficult for humans; it is often not hard to implement a rudimentary controller that accomplishes the task, but achieving near-optimal performance can be very challenging. Unsupervised robotic learning allows for automated design of efficient, robust controllers, which saves much design time and effort. Learning is also essential for allowing robots to adapt to situations where the task/environment is unknown beforehand or is constantly changing.

Particle Swarm Optimization (PSO) is a promising new optimization technique which models a set of potential problem solutions as a swarm of particles moving about in a virtual search space. PSO achieves optimization using three primary principles: evaluation, where quantitative fitness can be determined for some particle location; comparison, where the best performing location for some particle can be selected out of multiple possibilities; and imitation, where the qualities of better particles are mimicked by others. The algorithm can be used to evolve parameters for robotic controllers.

In the field of robotics, locating targets within an unknown environment is a task well-suited to mobile robots. Robots can be equipped with sensors to detect targets and programmed to explore the area in search of their goals. The automated nature of this approach may save much time and effort as compared

---

* Jim Pugh and Alcherio Martinoli are currently sponsored by a Swiss NSF grant (contract Nr. PP002-116913).

M. Asada et al. (Eds.): SAB 2008, LNAI 5040, pp. 393–402, 2008.

to other search methods. Performance may be further improved by using multiple robots in parallel, which will decrease the time needed to complete the search task and increase robustness to failures of individual robots. Robotic search is especially preferable when the area is hazardous or inaccessible to humans (for example, finding victims in a disaster area [10]).

In the PSO algorithm, groups of virtual agents interact in order to achieve optimization. In collective robotics, groups of robots interact to accomplish their goals. It is therefore possible to make a one-to-one parallel between particles and robots and to implement PSO in a distributed fashion for learning in multi-robot systems. Each robot is responsible for a single particle, which it evaluates at each iteration. After each evaluation, the robots communicate to share the fitness information needed to progress to the next iteration of the algorithm. By running the algorithms in this fashion, we need no external supervisor to oversee the learning process, and the speed of learning is significantly improved, as many robots evaluating in parallel decreases the number of required controller evaluations and therefore decreases the total learning time.

2 Background

The original PSO method was developed by James Kennedy and Russell Eberhart [12]. Every particle in the population begins with a randomized position $(x_{i,j})$ and randomized velocity $(v_{i,j})$ in the n-dimensional search space, where i represents the particle index and j represents the dimension in the search space. Candidate solutions are optimized by flying the particles through the virtual space, with attraction to positions in the space that yielded the best results. Each particle remembers the position at which it achieved its highest performance $(x_{i,j}^*)$. Each particle is also a member of some neighborhood of particles, and remembers which particle achieved the best overall position in that neighborhood (given by the index i'). This neighborhood can either be a subset of the particles (local neighborhood), or all the particles (global neighborhood). For local neighborhoods, the standard method is to set neighbors in a pre-defined way (such as using particles with the closest array indices as neighbors modulo the size of the population, henceforth known as a "ring topology") regardless of the particles' positions in the search space. The equations executed by PSO at each step of the algorithm are

$$v_{i,j} = w \cdot v_{i,j} + pw \cdot rand() \cdot (x_{i,j}^* - x_{i,j}) + nw \cdot rand() \cdot (x_{i',j}^* - x_{i,j})$$
$$x_{i,j} = x_{i,j} + v_{i,j}$$

where w is the inertia coefficient which slows velocity over time, pw is the weight given to the attraction to the previous best location of the current particle and nw is the weight given to the attraction to the previous best location of the particle neighborhood. $rand()$ is a uniformly-distributed random number in $[0, 1)$.

Unsupervised robotic learning has been studied extensively in the past, including some focus on multi-robot learning. Several multi-robot learning methods were shown to work in a wide variety of scenarios [1] [21]. Techniques for increasing individual learning speed via multi-robot learning were studied in [11] and [14]. Recently, PSO has been used for unsupervised learning of robotic behaviors by evolving weights for Artificial Neural Networks (ANNs), both in the case of single-robot learning [18] and distributed multi-robot learning [19].

Some exploration has been done in the past on multi-robot search and similar tasks. This work has been relatively disjoint thus far, with most studies focusing on a particular scenario which is not explicitly connected to other related work. The cost of using additional robots in a search task was explored and tested with simulation [7]. Detailed analysis has been done for swarms following a gradient [16]. In 2001, a contest at the International Joint Conference on Artificial Intelligence on collective robotic urban search and rescue [17] prompted some research on the topic [10]. Other publications explore multi-robot search strategies in simulation [6], for infrared tracking with simulation and real robots [7], and for odor source localization with real robots [8].

In [5], PSO was used to tune the parameters of a PSO-inspired multi-robot search strategy. Besides this, we are not aware of any previous publications on synthesis of multi-robot search behavior, and no previous studies have considered multi-robot search adaptation in changing environments.

3   Bio-inspired Multi-robot Search

The algorithm we use on our robot group for localizing targets in an unknown environment is inspired by the chemotaxis behavior of some types of bacteria, such as *E. coli*. By changing the rotation direction of their flagella, these bacteria can either swim in a straight line or to tumble in place. When moving, if the bacterium observes that the chemical gradient is positive, it is likely to continue to movement in the same direction. If it observes that the chemical gradient is negative, it is more likely to tumble and therefore assume a new random direction. This behavior results in overall movement in a positive gradient direction [2]. This type of chemotaxis behavior has inspired several effective robotic search strategies in the past [4] [9] [13]. However, none of these strategies used collaboration between robots in the search process.

In our algorithm, robots begin at some random locations within a bounded environment containing one or more target. We assume that robots are capable of perceiving the intensity of some emission from targets which fades non-linearly with increasing distance from the target. The robot will measure the perceived emission intensity and make a forward movement for a fixed distance. The robot will then measure the new perceived emission intensity. If the intensity is higher, the robot will maintain the same direction and make another step. If the intensity is lower, the robot will assume a new bearing and make a step in that direction. The process is then repeated. If a robot encounters any obstacles while moving (i.e. walls or other robots), it will turn to avoid the obstacle using a reactive obstacle avoidance algorithm.

The collaboration aspect of our algorithm arises when a robot is choosing a new bearing. Robots are assumed to be capable of relative localization and communication with other nearby robots within a certain range (this could be accomplished using an on-board module such as the one described in [20]). In this scenario, relative localization is not restricted to line-of-sight, though this assumption may not hold on a real robotic platform. Using relative localization, robots continually share their current position and most recent perceived emission intensity. When choosing a new angle, if a robot detects at least one other robot in range with higher perceived intensity than its own, it will choose a bearing directly towards the robot with highest perceived intensity. If it detects no other robots with higher perceived intensity, it will uniformly select a random bearing within some arc in the approximately opposite direction that it currently faces.

This algorithm has four free parameters which can be adjusted to optimize the behavior for different environments: STEP_SIZE, RL_RANGE, CW_LIMIT, and CCW_LIMIT. STEP_SIZE is the distance robots move forward at each step of the algorithm. RL_RANGE is the maximum range of the relative localization and communication system (only robots with distance less than or equal to RL_RANGE will be perceivable by another robot). CW_LIMIT is the maximum angular offset from 180 degrees which the robot will consider when choosing a random bearing in the clockwise direction, and CCW_LIMIT is the maximum angular offset from 180 degrees which the robot will consider when choosing a random bearing in the counter-clockwise direction. The bearing is therefore uniformly randomly selected from an arc of size CW_LIMIT+CCW_LIMIT.

Algorithm parameters can be seen graphically in Fig. 1a.

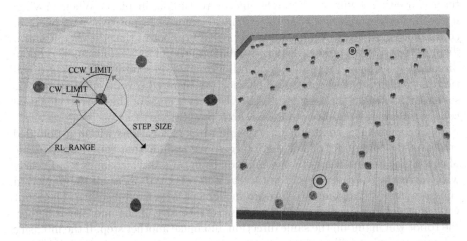

Fig. 1. (a) Graphical depiction of multi-robot search algorithm parameters with simulated e-puck robots. (b) Robot arena with e-puck robots and targets. Targets have been circled.

4 Distributed PSO for Parameter Adaptation in Multi-robot Search

We now apply distributed Particle Swarm Optimization to adapt the free parameters of our multi-robot search algorithm in a realistically simulated environment.

4.1 Experimental Setup

For our adaptation technique, we use the noise-resistant PSO algorithm from [19]. At each iteration, the algorithm reevaluates the previous best locations, averaging the new fitness value with previous ones to get a more accurate measure of the actual fitness. Although this requires twice as many fitness evaluations at each iteration as their standard counterparts, this technique prevents noisy fitness evaluations from severely disrupting the learning process and often gives better results given the same number of evaluations of candidate solutions.

Initial particle elements are randomly generated in the range $[0.0, 1.0]$. The elements are allowed to change to any value during evolution, but are bounded to this range and scaled appropriately during evaluation (i.e. any negative value is considered as 0.0 and any value greater than 1.0 is considered as 1.0 and then scaled during evaluation). Velocity is randomly initialized in the range $[-0.5, 0.5]$ and prevented from ever going outside this range. We use a swarm size of 50, with $pw = nw = 2.0$ and $w = 0.6$.

To explore the impact of particle neighborhood size in this scenario, we consider three different particle neighborhoods in experimentation: an *lbest* local neighborhood in a ring topology with one neighbor on each side (2 neighbors total), a *gbest* global neighborhood where all particles are neighbors (49 neighbors total), and an intermediate neighborhood in a ring topology with five neighbors on each side (10 neighbors total) which we denote *ibest*.

We use Webots, a realistic simulator, for our robotic simulations [15], using 50 e-puck[1] robots [3]. Every robot is responsible for the evaluation of a single, unique particle from the PSO swarm. The robot(s) operate in a 4.0 m x 4.0 m square arena with no additional obstacles (see Fig. 1b). Several targets are randomly placed in the environment. If a robot comes within range $r = 0.10$ m of a target, the target is considered to be "found", and it is randomly moved to a new location in the arena. The emission intensity perceived by a robot i is equal to:

$$I_i = \eta(.) + \sum_j \frac{P_j}{d_{ij}^2}$$

where P_j is the power of target j, d_{ij} is the distance between robot i and target j (in meters), and $\eta(.)$ is random background noise, given by the absolute value of a sample taken from a Gaussian probability density function with mean 0 and standard deviation $\sigma = 10$.

---

[1] http://www.e-puck.org

For the robot controller, we use the bio-inspired multi-robot search algorithm described previously, with parameters STEP_SIZE, RL_RANGE, CW_LIMIT, and CCW_LIMIT determined by the distributed PSO algorithm; these parameters are linearly scaled from the $[0.0, 1.0]$ range given by PSO to the ranges $[0.1, 1.0]$, $[0.0, 3.9]$, $[0, \pi]$, and $[0, \pi]$, respectively. To evaluate a candidate solution, we run the multi-robot search algorithm with the specified parameters for a span of 120 seconds. During that time, robots average their perceived emission intensity, sampling every 64 milliseconds, and use this value as their fitness (normalized to a maximum of 1.0, assuming an upper intensity limit of 255). This fitness function rewards robots who spend more time in close proximity to a target (where emission intensity is higher) over those who remain farther away.

For our initial experiments, we use three targets with emission power $P_j = 10$ for all targets. Each of the 50 robots is responsible for a single member of the PSO swarm (i.e. they must evaluate that member at each iteration and communicate the resulting fitness measure). We assume all robots are synchronized in their evaluation here, something which might not necessarily be possible on real multi-robot systems. With 100 iterations of the PSO algorithm, the adaptation requires a total simulated learning time of approximately 6 hours 40 minutes (100 iterations comprised of 2 evaluations each lasting 2 minutes).

4.2   Results

The average progress of individual robot fitness throughout adaptation for the three different particle neighborhoods can be seen in Fig. 2a. Clear improvements in fitness for all neighborhood types can be observed over the course of the algorithm, although the overall change in fitness is not dramatic (approximately a 30% increase from the initial average fitness value). There do not appear to be major differences in performance for different particle neighborhood types; using an *lbest* neighborhood seems to cause slower initial improvements, but final performances are similar for all neighborhood types, with *gbest* doing only slightly better than *ibest*, which does slightly better than *lbest*.

While individual robot fitness is useful for adaptation, it does not necessarily provide us with a good indicator of the performance of the robot swarm as a whole. To measure this, we use the number of targets "found" at every evaluation of multi-robot search as a group fitness value. The average progress of group fitness throughout adaptation can be seen in Fig. 2b. We again see clear improvements in fitness for all neighborhood types, though with a significantly larger improvement (more than a 100% increase from the initial average fitness value); this shows that our individual and group fitness measures are very well-aligned. We also notice more pronounced differences between results for different particle neighborhoods: *lbest* improves quite slowly throughout adaptation; *gbest* improves more quickly initially, but plateaus early in the adaptation process; *ibest* offers a compromise between the two, improving more slowly than *gbest* but continuing throughout adaptation to eventually achieve a higher fitness than the other two neighborhood types. This is similar to what was observed in [19], where an intermediate neighborhood size gave optimal performance.

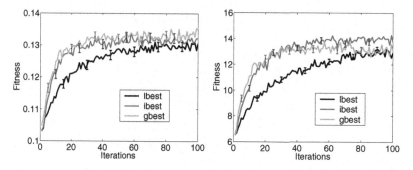

Fig. 2. Average individual fitness of particles throughout adaptation averaged over 250 runs for *lbest*, *ibest*, and *gbest* particle neighborhoods for (a) individual fitness, and (b) group fitness. Error bars represent standard error.

When using PSO for optimization, the standard approach is to run the algorithm for repeated iterations until a termination criterion is reached (e.g., a certain number of iterations have occurred) and then select the best found solution. While this is straight-forward for numerical optimization, it can be much more difficult in robotic learning. Determining which solution is actually best requires repeated evaluations to obtain low-noise fitness measures, which may require significantly more running time for robots. In addition, once a final solution is selected by all robots, the robotic swarm can no longer adapt to changing environmental parameters which may occur in its surroundings. For these reasons, there are clear advantages to running distributed PSO optimization without termination, where robots continue to adapt their parameters indefinitely. The possible drawback to this method is that the performance of robot swarm may be significantly lower than if it were to use the best found solution.

In order to assess whether continuing adaptation incurs a performance penalty, we compare the final average group fitness and the best found solution group fitness after 100 iterations of adaptation. The final average group fitness was taken as the average group fitness over the last five adaptation iterations. The best found controller was selected by evaluating the final personal best solutions for all particles five times and selecting the one with the highest average individual fitness; this solution was run on all robots to determine its group fitness. The average performance over 250 runs for different neighborhood types can be seen in Fig. 3. The final average group fitness is as high as the best found solution group fitness for all neighborhood types. This indicates that continuing adaptation indefinitely will not result in a significant performance decrease. Oddly, for *gbest*, the average group fitness is actually higher than the best found controller fitness. One possible explanation is that the *gbest* neighborhood may overfit on individual fitness, causing a slight decrease in the group fitness when an over-optimized solution is selected. We can also confirm here that the *ibest* neighborhood significantly outperforms *lbest* and *gbest* using an ANOVA test, both for the best found solution (P-value of 0.0084 for *lbest* and 0.0005 for *gbest*) and average group fitness (P-value of $1.6 \cdot 10^{-8}$ for *lbest* and 0.0014 for *gbest*).

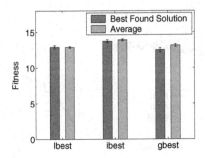

Fig. 3. Final average group fitness and best found solution group fitness averaged over 250 runs for *lbest*, *ibest*, and *gbest* particle neighborhoods. Error bars represent standard error.

5   Adaptation in Non-static Environments

A major potential benefit to unending robotic adaptation is that robots could automatically adjust to changes that might arise in the environment. To test this possibility, we rerun our experiments in a non-static environment. We use the same simulated scenario as in the previous experiments. At the start of adaptation, we use three targets with emission power $P_j = 3$ for all targets. After 50 iterations, we switch to using ten targets with emission power $P_j = 10$ for all targets. We again run 100 iterations total using 50 robots.

The average of individual robot fitness throughout adaptation in the non-static environment can be seen in Fig. 4a (fitness in the last 50 iterations is significantly higher due to the increased power and number of targets and is shown at a reduced scale in the plot, with the scaling factor chosen to best align the data). We observe the same trend as in the previous experiment for the first 50 iterations. At this point, there is a fitness shift for all neighborhood types as the simulations switches over to the new environmental parameters and fitness scaling. In the last 50 iterations, we see a slightly larger increase in individual fitness compared to what was observed in the static environment, with similar final fitness for all neighborhood types.

The group fitness throughout adaptation in the non-static environment can be observed in Fig. 4b (fitness in the last 50 iterations is significantly higher due to the increased power and number of targets and is shown at a reduced scale, with the scaling factor chosen to best align the plots). Progress in the first 50 iterations is the same as for the static environment. In the last 50 iterations, we see continuing fitness improvements for *lbest* and *ibest* neighborhoods, with larger gains than were observed in the last 50 iterations in the static environment, particularly for *lbest*. This indicates that robots successfully adapt from the initial scenario to the new environmental parameters. The fitness for the *gbest* neighborhood remains approximately constant. This could be caused by premature convergence, which would lead to lower particle diversity and prevent further adaptation. Higher diversity in the other neighborhoods allow them to

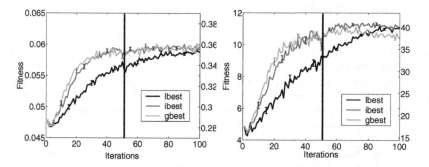

Fig. 4. Average fitness of particles throughout adaptation in a non-static environment averaged over 250 runs for *lbest*, *ibest*, and *gbest* particle neighborhoods for (a) individual fitness, and (b) group fitness. Fitness in the later iterations is scaled to match fitness in the initial iterations. Error bars represent standard error.

continue to adapt in non-static environments, particularly in the case of *lbest*, which achieves final performance as high as *ibest* in this scenario.

6   Conclusion and Outlook

We have shown that distributed Particle Swarm Optimization can be used for adaptation in multi-robot systems, illustrated with the case study of multi-robot search. Best performance is obtained by using an intermediate particle neighborhood between *lbest* and *gbest*, which offers fast optimization without premature particle convergence in this scenario. Adaptation can be continued indefinitely without a significant performance penalty, making it possible for the robot swarm to automatically adapt in non-static environments.

In this paper, we have devoted our focus to the multi-robot adaptation process using distributed PSO and spent little time studying the multi-robot search algorithm itself. Observing the final solutions found by the adaptation process could give us insight into the impact of each of the different algorithmic parameters and how they influenced the robots' performance. The bio-inspired multi-robot search technique should also be compared to other common search strategies for similar scenarios (such as standard "optimal" search techniques or PSO-inspired search) to evaluate its overall potential.

References

1. Balch, T.: Behavioral diversity in learning robot teams. PhD Thesis, College of Computing, Georgia Institute of Technology (1998)
2. Berg, H.C.: E. coli in motion. Springer, NY (2003)
3. Cianci, C., Raemy, X., Pugh, J., Martinoli, A.: Communication in a swarm of miniature robots: The e-puck as an educational tool for swarm robotics. In: Şahin, E., Spears, W.M., Winfield, A.F.T. (eds.) Swarm Rob. Workshop. LNCS, vol. 4433, pp. 103–115. Springer, Heidelberg (2007)

4. Dhariwal, A., Sukhatme, G.S., Requicha, A.A.G.: Bacterium-inspired robots for environmental monitoring. In: Proc. of the IEEE Intl. Conf. on Robotics and Automation, New Orleans, LA, USA, April 26 - May 1, 2004, pp. 1436–1443 (2004)

5. Doctor, S., Venayagamoorthy, G.K., Gudise, V.G.: Optimal PSO for Collective Robotic Search Applications. In: Proc. of the IEEE Congress on Evolutionary Computation, Portland, OR, USA, June 19-23, 2004, pp. 1390–1395 (2004)

6. Goldsmith, S.Y., Robinett, R.: Collective search by mobile robots using alpha-beta coordination. In: Drogoul, A., Tambe, M., Fukuda, T. (eds.) Collective Robotics, pp. 136–146. Springer, Berlin (1998)

7. Hayes, A.T.: How Many Robots? Group Size and Efficiency in Collective Search Tasks. In: Proc. of the 6th Intl. Symp. on Distributed Autonomous Robotic Systems DARS 2002, pp. 289–298. Fukuoka, Japan (2002)

8. Hayes, A.T., Martinoli, A., Goodman, R.M.: Swarm Robotic Odor Localization: Off-Line Optimization and Validation with Real Robots. In: McFarland, D. (ed.) Special Issue on Biological Robots. Robotica, vol. 21, pp. 427–441 (2003)

9. Holland, O., Melhuish, C.: Some adaptive movements of animats with single symmetrical sensors. In: 4th Intl. Conf. on Simulation of Adaptive Behaviour. MIT Press, Cambridge (1996)

10. Kantor, G., Singh, S., Peterson, R., Rus, D., Das, A., Kumar, V., Pereira, G., Spletzer, J.: Distributed search and rescue with robot and sensor teams. In: Proc. of the 4th Intl. Conf. on Field and Service Robotics, Japan (2003)

11. Kelly, I.D., Keating, D.A.: Faster learning of control parameters through sharing experiences of autonomous mobile robots. Intl. J. of System Science 29(7), 783–793 (1998)

12. Kennedy, J., Eberhart, R.: Particle swarm optimization, Neural Networks. In: Proceedings, IEEE Intl. Conf., vol. 4, pp. 1942–1948 (1995)

13. Marques, L., Nunes, U., de Almeida, A.T.: Olfaction-based mobile robot navigation. Thin Solid Films 418, 51–58 (2002)

14. Mataric, M.J.: Learning in behavior-based multi-robot systems: Policies, models, and other agents. In: Sun, R. (ed.) Special Issue on Multi-disciplinary studies of multi-agent learning. Cognitive Systems Research, vol. 2(1), pp. 81–93 (2001)

15. Michel, O.: Webots: Professional Mobile Robot Simulation. Int. J. of Advanced Robotic Systems 1, 39–42 (2004)

16. Ögren, P., Fiorelli, E., Leonard, N.E.: Cooperative Control of Mobile Sensor Networks: Adaptive Gradient Climbing in a Distributed Environment. IEEE Transactions on Automatic Control 49(8), 1292–1302 (2004)

17. Osuka, K., Murphy, R., Schultz, A.C.: USAR Competitions for Physically Situated Robots. IEEE Robotics and Automation Magazine, 26–33 (September 2002)

18. Pugh, J., Zhang, Y., Martinoli, A.: Particle swarm optimization for unsupervised robotic learning. In: Swarm Intelligence Symp., Pasadena, CA, pp. 92–99 (2005)

19. Pugh, J., Martinoli, A.: Multi-Robot Learning with Particle Swarm Optimization. In: Intl. Conf. on Autonomous Agents and Multiagent Systems, Hakodate, Japan, May 8-12, 2006, pp. 441–448 (2006)

20. Pugh, J., Martinoli, A.: Relative Localization and Communication Module for Small-Scale Multi-Robot Systems. In: Proc. of the IEEE Intl. Conf. on Robotics and Automation, Miami, Florida, USA, May 15-19, 2006, pp. 188–193 (2006)

21. Stone, P.: Layered Learning in Multi-Agent Systems. PhD Thesis, School of Computer Science, Carnegie Mellon University (1998)

Homeotaxis: Coordination with Persistent Time-Loops

Mikhail Prokopenko, Astrid Zeman, and Rongxin Li

CSIRO Information and Communication Technology (ICT) Centre
Locked bag 17, North Ryde, NSW 1670, Australia
mikhail.prokopenko@csiro.au

Abstract. We present a novel approach to self-organisation of coordinated behaviour among multiple resource-sharing agents. We consider a hierarchical
multi-agent system comprising multiple energy-dependent agents split into local
neighbourhoods, each with a dedicated controller, and a centralised coordinator
dealing only with the controllers. The coordinated behaviour is required in order
to achieve a balance between the overall resource consumption by the multi-agent
collective and the stress on the community. Minimising the resource consumption
increases the stress, while reducing the stress may lead to unrestricted and highly
unpredictable demand, harming the individual agents in the long-run. We identify underlying forces in the system's dynamics, suggest a number of quantitative measures used to contrast different strategies, and introduce a novel strategy
based on persistent sensorimotor time-loops: *homeotaxis*. Homeotaxis subsumes
the homeokinetic principle, extending it both in terms of scope (multi-agent self-
organisation) and the state-space, and allows to select the best adaptive strategy
for the considered system.

1 Introduction

In general, the ability to coordinate (e.g., synchronise) multiple individual actions
within large multi-agent groups is an adaptive response observed in many biological
systems. As noted by Trianni and Nolfi [17], "synchrony can increase the efficiency of
a group by maximising the global outcome or by minimising the interference among
individuals". An investigation of Baldassarre *et al.* [3] characterised coordinated motion in a swarm collective as a self-organised activity of the constituent independently-
controlled modules, and measured the increasing organisation of the group on the basis
of Boltzmann entropy. The emergent *common direction* of motion, with the chassis orientations of the robots spatially aligned, was related to high synchrony and coordination
within the group.

While synchronisation has been extensively studied in a variety of applications, ranging from swarm robotics [3,17], to coordinating sensors in wireless networks [18], to
models of fireflies flashing in unison in biology [11], the inverse problem of desynchronisation has received less attention, as noted by Patel *et al.* [13]. Patel *et al.* consider
desynchronisation as the task of spreading a given set of identical oscillators throughout a time period, resulting in a round-robin schedule, and argue that this can be useful
in several applications. For example, in wireless sensor networks sensor nodes can (i)
desynchronise their sampling times to distribute the energy cost, while still providing
efficient coverage, and (ii) desynchronise their transmission times to avoid collisions

M. Asada et al. (Eds.): SAB 2008, LNAI 5040, pp. 403–414, 2008.

and message loss. The study of Patel *et al.* is motivated by biology: "cells, acting as oscillators, control animal gaits and regulate heart valves through desynchronisation" [13], leading to biologically-inspired algorithms for achieving desynchronisation.

Thus, both synchronisation and desynchronisation may be desirable at different stages in a multi-agent system. We believe, synchronisation enables better exploitation, while desynchronisation allows the system to explore alternatives. In general, a system that is well-balanced in terms of synchronisation and desynchronisation (i.e., in terms of exploitation and exploration) achieves an adequate coordination of its components through space and time, and across multiple activities. In other words, synchronisation and desynchronisation are interleaved sub-tasks of a more generic task: multi-agent spatiotemporal coordination.

A coordinated system may involve a degree of centralisation. For example, a bio-inspired control architecture for artificial muscle materials is proposed by Odhner *et al.* [12]. The active material is broken up into many small cells, coordinated to produce a combined force or displacement. A single central controller uses only one input and one output: it measures only the summed displacement of all of the cells, producing a feedback signal that is broadcast to the cells. Each cell controls its displacement with a stochastic automaton, that is "a small local control automaton containing a pseudo-random number generator, so that it contracts and relaxes stochastically, with a probability distribution dictated by the input from the central feedback controller" [12]. This closed-loop system results in a smooth and predictable motion (e.g., tracking a desired position), and is scalable to many cells.

This paper solves the task of coordination in a model setting (distributed power load management system) using a novel domain-invariant and bio-inspired principle: *homeotaxis*, that incorporates homeokinesis and self-organisation of perception-action loops. The coordinated behaviour in our model system is required in order to balance the overall energy consumption by a multi-agent collective and the stress on the community. Minimising the energy consumption strains the system, while removing the stress typically leads to unrestricted and highly unpredictable demand, harming the individual agents in the long-run.

Section 2 introduces our model problem: resource-sharing agents connected to a power load management system, and presents the motivation for the study. Domain-invariant approaches to multi-agent coordination are reviewed in Section 3. The following sections describe the proposed approach, and present simulation results, followed by conclusions.

2   A Distributed Power Load Management System

Electricity distribution is a complex system, consisting of loads (appliances), generators, transformer stations and distribution networks, and influenced by a market. During most times of the year, the market price is low (e.g., less than AUD\$20 per megawatt hour), but it can also be very high (near or at the maximum price, e.g., AUD\$10,000 per megawatt hour) during a peak demand, when energy supply is under extreme pressure.

Energy demand management is a new technology proposed to cope with the unpredictability of the energy market and provide a rapid response when supply is strained by

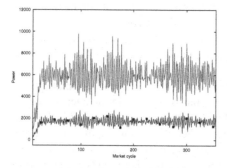

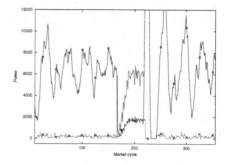

Fig. 1. Strategy $q_t = l_t$. Circles on solid line: x_t; crosses on dashed line: l_t; dotted line: u_t.

Fig. 2. Strategy $q_t = u_t$. Circles on solid line: x_t; crosses on dashed line: l_t; dotted line: u_t.

demand. Energy demand management essentially equips appliances with simple agents, and enables these agents to defer their electricity consumption when price suddenly skyrockets. We consider a hierarchical multi-agent system comprising multiple appliance-level agents split into local neighbourhoods, each with a dedicated neighbourhood's controller (middle-level), and a high-level centralised coordinator dealing only with the middle-level controllers. Appliances within a neighbourhood can be switched on or off by their middle-level controller, within the constraints defined by customer preferences. The neighbourhood's controller agent in turn receives an energy quota from the high-level coordinator — this energy quota is the limit on the total energy consumption for the neighbourhood.

In this paper, we focus on the behaviour of the high-level coordinator, i.e., on how to choose the optimal strategy setting the real-time quota to a middle-level controller. The high-level coordinator needs to incorporate information from the market (e.g., the current local price of energy p_t), as well as inputs from middle-level controller. The coordinator is required to balance the risk of exposure to volatile prices, and reduce strain in the load network.

The local energy demand limits for a neighbourhood are obtained by each middle-level controller as two values per interval: 1) the minimal consumption l_t of the local neighbourhood, and 2) the unconstrained (default) consumption u_t of the neighbourhood. The first value is defined as the total minimal consumption over an entire future market cycle (e.g., five minute interval). The minimal load requirements of appliances depend on the hardware and operating limits of individual appliances. The unconstrained consumption of the neighbourhood refers to its total consumption when the individual agents operate under normal conditions without any external influence. The limits l_t and u_t are computed by each middle-level controller in the beginning of a market cycle, using an optimisation procedure described elsewhere [9].

The coordinator sets the quota q_t for the next market cycle, as a value inclusively between the values of the minimal and unconstrained demand: $l_t \le q_t \le u_t$. In response to this quota the middle-level controller selects the best control plan for each load in its neighbourhood [9], resulting in actual total consumption x_t in the neighbourhood. These interactions essentially create a control loop where multiple individual agents plan their consumption for the next interval t, producing the aggregate limits l_t and u_t, while the coordinator reacts to these inputs (as well as the dispatch price p_t) by setting the quota q_t. This results in the actual consumption x_t, affecting the agents' plans for the next interval $(t + 1)$ and the future limits l_{t+1} and u_{t+1}, and so on. Thus, the limits depend on the quotas set in the past.

One may assume that an optimal quota-setting strategy would be the minimal achievable consumption for a neighbourhood: $q_t = l_t$, that is, restricting the agents' energy consumption levels to the lowest possible value that they can cope with. This does reduce the total cost $S(T) = \sum_t^T p_t x_t$. However, when the minimal quota is imposed, some of the energy which is not consumed in one period will need to be consumed at a later time. The longer a minimal quota is imposed, the more loads are pushed to a stressful limit. The results of such undesirable synchronisation are shown in Fig. 1: the minimal quota increases the minimal achievable consumption l_t after some initial period during which the agents can still defer their demand. The agents become eventually synchronised when none of them can defer the demand any longer, stressing the neighbourhood. Such stress would result in a peak demand if a quota is released, and can cause negative effects such as shortage of energy or a peak price, threatening system stability. In short, when $q_t = l_t$, the minimal consumption l_t itself grows over time, and while the cost S is low, the stress (the total uncompromisable "incompressible" demand), defined as $R(T) = \sum_t^T l_t$, is very high.

Similarly, setting the quota to another possible extreme $q_t = u_t$ (except during the period around cycle 200 when an extremely high price p_t forces the coordinator to use the default $q_t = l_t$) results in an unbalanced outcome (Fig. 2). In this case, the stress $R(T)$ is minimal, but the cost $S(T)$ is high. The stability of the actual consumption x_t, measured as the variance $\sigma_x^2(T)$, is also quite significant.

Hence the coordinator needs to define an optimal quota which not only reduces energy costs, but also avoids instability and stress within the system. More precisely, optimality of the quota over a time period T depends on (i) the total cost $S(T)$; (ii) stability $\sigma_x^2(T)$ of the actual consumption x_t; and (iii) the stress $R(T)$. In order to derive an optimal quota-setting strategy we turn to domain-invariant principles of multi-agent coordination.

3   Domain-Invariant Principles of Adaptive Behaviour

A few approaches were recently proposed in order to characterise and achieve spatiotemporal coordination in a general way. For example, a modular robotic system modelling a multi-segment snake-like organism, with actuators ("muscles") attached to individual segments ("vertebrae") was evolved according to generic information-theoretic measure (excess entropy or predictive information, defined in Shannon sense) [16]. In general, one may argue that *information-driven self-organisation* is one of the main evolutionary forces that may be used in both design (e.g., information-driven evolutionary design [15]) and biological evolution [14]. An example of such information-driven dynamics is the acquisition of information from the environment: there is evidence that pushing the information flow to the information-theoretic limit (i.e., maximisation of information transfer through the system's perception-action loop) can give rise to intricate behaviour, induce a necessary structure in the system, and ultimately adaptively reshape the system [8].

Information-driven self-organisation that relies on dynamics of predictive information [15,10] is related to the search of domain-invariant principles carried out by Der *et al.* [4,7,6,5]. While traditionally an objective function, measuring the distance between the current and a desired behaviour, is provided explicitly by the designers, Der *et al.* consider self-referential adaptive systems: the systems for which the objective function is derived from the dynamics of the system itself, i.e. "adaptive, embodied systems where the objective of adaptation is a function of the robots sensor values alone" [5]. In doing so they follow the principle of homeokinesis — the dynamical analogue of homeostasis [1]. The principle of homeokinesis was developed by Der *et al.* [4,7] as a general domain-invariant principle for self-organisation in robot behaviour. The principle is based on the assumptions that the robot is able to (i) learn an internal representation (self-model) of its current behaviour, and (ii) adapt its behaviour by minimising the difference between the self-model and true behaviour in the real world. Importantly, the principle does not lead to stabilisation of stationary states (that would result in a "do-nothing" behaviour), but rather suggests a smooth and predictable kinetic regime.

In developing the principle of homeokinesis, Der and Liebscher [7] have introduced a reversal of time in the modelling process, capturing a time-loop error. The time-loop error is formalised in Section 4 — at this stage we note, following [7], that it is driven by two opposite forces. On the one hand, the time-loop error is small if the current behaviour is well represented by the internal model, producing behaviours that correspond to smooth and predictable sensor values. The second tendency is that the time reversal in the modelling process inverts a stable behaviour into unstable, and vice versa. Specifically, if the behaviour is stable in the forward time direction, it is unstable if the time is reversed, and therefore, the time-loop error is minimised if the kinetic behaviour of the robot is unstable in the forward time direction. This feature eliminates trivial behaviours (e.g., a "do-nothing" behaviour). Nevertheless, the instability cannot continue unrestricted, as the first tendency (the faithful self-model) demands smooth trajectories in the sensor space. This closed-loop interplay between a smooth and predictable sensor space exploration and an unstable kinetic behaviour balances exploration and exploitation aspects of the behaviour.

Importantly, both the information-driven self-organisation and the principle of homeokinesis emphasise the role of behaviour's predictability as well as non-stationary sensorimotor dynamics in achieving the desired balance between exploitation and exploration. A possible unification is described by Ay *et al.* [2] and Der *et al.* [5].

4   Homeotaxis as Coordination with Persistent Time-Loops

In deriving the optimal quota-setting strategy we extend the homeokinetic principle to the task of multi-agent coordination, by considering the consumption x_t within a multi-agent system to be among sensory inputs of the high-level coordinator, and the coordinator's output q_t to be its actuation. This extends the principle to *homeokinetic coordination*. Homeotaxis is achieved by enhancing the time-loop error with persistence error, and is motivated by *taxis*: "an innate behavioural response by an organism to a directional stimulus" [19][1].

4.1 Example

Let us begin by exemplifying the original use of the time-loop error, following a simple homeokinesis model of Der and Liebscher [7]. Consider the linear system with dynamics

$$x_{t+1} = c_t x_t + \xi_t \tag{1}$$

where c_t is the controlling variable, and ξ_t is the part not handled by the model. The system can be modelled forward in time:

$$\hat{x}_{t+1} = c_t x_t , \tag{2}$$

as well as backward in time:

$$\breve{x}_t = x_{t+1}/c_t = (c_t x_t + \xi_t)/c_t = x_t + \xi_t/c_t \tag{3}$$

Here $\hat{x}_{t+1}$ is the predicted value given the current observation x_t, while $\breve{x}_t$ is the reconstructed value, given the latest observation x_{t+1}. Traditionally, having observed x_{t+1}, one uses $E = (x_{t+1} - \hat{x}_{t+1})^2$ as the error to minimise, i.e. as a feedback signal. The time-loop error is defined as

$$W = (\breve{x}_t - x_t)^2 , \tag{4}$$

that is, it is obtained by going forward in time from x_t to x_{t+1}, followed by a step (3) from x_{t+1} to the reconstructed state $\breve{x}_t$ backward in time. The full sequence $x_t \rightarrow x_{t+1} \rightarrow \breve{x}_t$ is called the time loop. Using gradient descent to minimise the time-loop error (4) $W = \xi_t^2/c_t^2$, with respect to the controlling variable c_t, yields the update rule (for a small number $\epsilon > 0$ determining the rate of descent)

$$c_{t+1} = c_t + \epsilon \, \xi_t^2/c_t^3 \tag{5}$$

As mentioned in the previous section, the time-loop error W is small if the current behaviour is well represented by the internal model (2), producing behaviours that obtain smooth and predictable ("exploitable") sensor values. On the other hand, the time-loop error is minimised if the kinetic behaviour is unstable in the forward time direction, eliminating trivial stationary behaviours, and encouraging exploration.

4.2 General Case

In general, one considers a system with an adaptive controller, defined by the controller's parameter vector $\mathbf{c}_t$, and output that depends on the sensor values $\mathbf{x}_t$ observed at time t. The adaptive model M aims to predict the true sensor values $\mathbf{x}_{t+1}$ at $(t+1)$:

$$\hat{\mathbf{x}}_{t+1} = \mathbf{x}_t + M(\mathbf{x}_t, y_t; m) , \tag{6}$$

---

[1] Taxis differs from *kinesis*, "a non-directional change in activity in response to a stimulus that results in the illusion of directed motion due to different rates of activity depending on stimulus intensity" [19].

where m is the predictor's parameter. Abbreviating $M(\mathbf{x}_t, y_t; m)$ as $M(\mathbf{x}_t)$, and minimising the prediction error

$$E = (\mathbf{x}_{t+1} - \hat{\mathbf{x}}_{t+1})^2 = (\mathbf{x}_{t+1} - \mathbf{x}_t - M(\mathbf{x}_t))^2 = (\triangle \mathbf{x}_t - M(\mathbf{x}_t))^2 \qquad (7)$$

by gradient descent yields an update rule for the controller and the predictor, respectively (the latter does not have to be updated in real-time), where $\epsilon > 0, \eta > 0$:

$$\mathbf{c}_{t+1} = \mathbf{c}_t - \epsilon \frac{\partial E}{\partial \mathbf{c}_t} \qquad m_{t+1} = m_t - \eta \frac{\partial E}{\partial m} \qquad (8)$$

Assuming the dynamics

$$\mathbf{x}_{t+1} = \mathbf{x}_t + M(\mathbf{x}_t) + \xi_t , \qquad (9)$$

where ξ_t is the vector of perturbations not covered by the model M, the model can be written backward in time [7] as

$$\check{\mathbf{x}}_t = \mathbf{x}_{t+1} + M^{(-)}(\mathbf{x}_{t+1}) , \qquad (10)$$

where we define the reverse model $M^{(-)}$ as follows:

$$\mathbf{x}_t = \hat{\mathbf{x}}_{t+1} + M^{(-)}(\hat{\mathbf{x}}_{t+1}) \qquad (11)$$

The definition (11) is symmetric to the expression (6). This definition corrects the one given in [7] that specifies $\mathbf{x}_t = M^{(-)}(\mathbf{x}_t + M(\mathbf{x}_t))$, which is equivalent to $\mathbf{x}_t = M^{(-)}(\hat{\mathbf{x}}_{t+1})$. The time-loop error then is given by

$$W = (\check{\mathbf{x}}_t - \mathbf{x}_t)^2 = (\xi_t + M^{(-)}(\mathbf{x}_{t+1}) - M^{(-)}(\hat{\mathbf{x}}_{t+1}))^2 \qquad (12)$$

yielding the update rules for the controller and the predictor, respectively:

$$\mathbf{c}_{t+1} = \mathbf{c}_t - \epsilon \frac{\partial W}{\partial \mathbf{c}_t} \qquad m_{t+1} = m_t - \eta \frac{\partial W}{\partial m} \qquad (13)$$

4.3   Extension

In developing the principle of homeotaxis and applying it to the distributed energy management, we consider a system with an adaptive controller and output

$$q_{t+1} = \begin{cases} l_{t+1} & \text{if } c_{t+1}x_t < l_{t+1} \\ u_{t+1} & \text{if } c_{t+1}x_t > u_{t+1} \\ c_{t+1}x_t & \text{otherwise} \end{cases} \qquad (14)$$

during a transition from t to $(t+1)$. Specifically, x_t is the last observed actual consumption (the sensor value) achieved in response to the quota q_t; the current limits l_{t+1} and u_{t+1} are known as well. The quota-controlling parameter c_{t+1} is to be updated before setting the quota (the output) q_{t+1}.

The adaptive model aims to predict the true sensor value

$$x_{t+1} = c_{t+1}x_t + \xi_t \qquad (15)$$

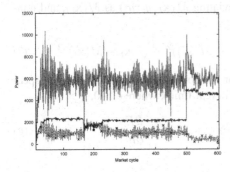

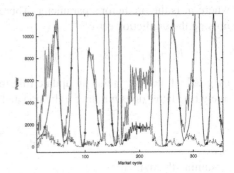

Fig. 3. Minimising error Q. Circles on solid line: x_t; crosses on dashed line: l_t; dotted line: u_t.

Fig. 4. Basic homeokinesis. Circles on solid line: x_t; crosses on dashed line: l_t; dotted line: u_t.

that will result when the quota is set:

$$\hat{x}_{t+1} = q_{t+1} \tag{16}$$

The quota dynamics may also be described recursively:

$$q_{t+1} = q_t + \zeta_t \tag{17}$$

for some perturbation ζ_t that needs to be minimised to maintain persistent quotas. The persistence error can be formulated, by applying relationships (17) and (14), as

$$Q = (q_t - q_{t+1})^2 \tag{18}$$

In our case, $Q = (q_t - c_{t+1}x_t)^2$ is minimised by

$$\hat{c}_{t+1} = q_t/x_t \tag{19}$$

The control strategy using this simplistic update rule balances cost, stability and stress reasonably well, but is not responsive to sudden changes in the underlying demand. Fig. 3 shows the dynamics for a scenario where there is a spike in demand at market cycle 500 — it is evident that the optimal balance is lost after the spike. However, it can be complemented by a signal produced by the time-loop error:

$$W_Q = (\breve{x}_{t;Q} - x_t)^2 = (x_{t+1}/\hat{c}_{t+1} - x_t)^2 = ((\hat{c}_{t+1}x_t + \xi_t)/\hat{c}_{t+1} - x_t)^2 = \xi_t^2/\hat{c}_{t+1}^2 \tag{20}$$

The error W_Q exemplifies *persistent time-loop error*:

$$W_Q = (\breve{x}_{t;Q} - x_t)^2 = (\xi_t + M^{(-)}(x_{t+1}; \hat{c}_{t+1}) - M^{(-)}(\hat{x}_{t+1}))^2 \tag{21}$$

where $\hat{c}$ is determined by minimisation of the persistence error Q (18). The controller update rule that minimises both the persistence error Q and the usual time-loop error is given by

$$c_{t+1} = \hat{c}_{t+1} - \epsilon \frac{\partial W_Q}{\partial \hat{c}_{t+1}} = \hat{c}_{t+1} + \epsilon' \frac{\xi_t^2}{\hat{c}_{t+1}^3} \tag{22}$$

In summary, the *persistent time-loop error* W_Q is obtained by (i) minimising the persistence error (18), and (ii) using the corrected controlling variable in the time-loop reconstruction. Essentially, it explicitly demands smoothness and predictability of trajectories in both the sensor-space and action-space. It is interesting to note, at this stage, that a basic homeokinetic strategy where update rule is given by (5) does not work in this case: as shown in Fig. 4, the actual consumption is too unstable, while being reasonably smooth and predictable on average.

The next step is to estimate the perturbation ξ_t. We used the following approximations:

$$\xi_t^{(1)} = |\breve{x}_t - x_t| \qquad \xi_t^{(2)} = |x_{t+1} - x_t| \qquad \xi_t^{(3)} = |\breve{x}_t - q_t| \qquad (23)$$

Perfect control is indicated by $c = 1$. According to (20), $\xi_t^{(1)}$ yields $\hat{c} = 1$, while $\xi_t^{(2)}$ yields $c = 1$ according to (15), and $\xi_t^{(3)}$ yields $\hat{c} = 1$ according to (20) and (19) considered together.

5   Results

In this section we present experimental results of three different homeotaxic strategies that minimise persistent time-loop error with approximations $\xi_t^{(1)}$ (Fig. 5), $\xi_t^{(2)}$ (Fig. 6), and $\xi_t^{(3)}$ (Fig. 7). Each experiment involved 100 individual agents in the neighbourhood coordinated over a 5000 cycle run, including a short period around cycle 200 during which the coordinator used the default $q_t = l_t$, and an instantaneous spike in demand to $min(5000; u_t)$ at cycle 500. The performance was measured in terms of the cost S, stability σ_x^2, stress R over $T = 5000$ cycles, and their linear combination

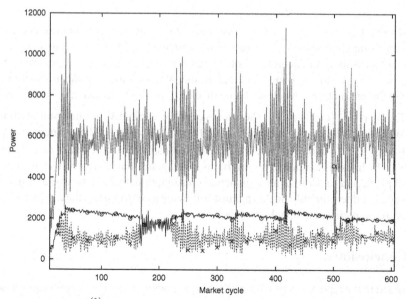

Fig. 5. Minimising $\xi_t^{(1)}$. Circles on solid line: x_t; crosses on dashed line: l_t; dotted line: u_t.

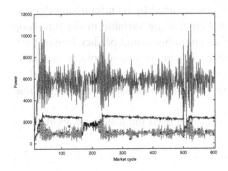
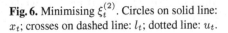
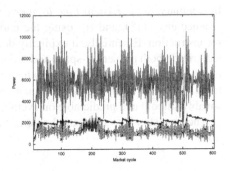

Fig. 6. Minimising $\xi_t^{(2)}$. Circles on solid line: x_t; crosses on dashed line: l_t; dotted line: u_t.

Fig. 7. Minimising $\xi_t^{(3)}$. Circles on solid line: x_t; crosses on dashed line: l_t; dotted line: u_t.

Table 1. Comparative results (bold: best in each category)

Strategy	$S(T)$	$\sigma_x^2(T)$	$R(T)$	Loss	RMSE$_c$
homeotaxis $\xi_t^{(1)}$	376,477	**38,806**	5,405,235	95,581	0.04371
homeotaxis $\xi_t^{(2)}$	385,528	41,967	5,101,962	**93,769**	0.04440
homeotaxis $\xi_t^{(3)}$	373,629	47,850	5,551,528	97,663	0.04388
persistent quota	547,045	394,167	2,941,011	123,531	**0.03922**
$q_t = l_t$	**334,945**	109,560	8,498,646	129,437	c_t not used
$q_t = u_t$	833,075	3,454,456	**973,996**	438,493	c_t not used
basic homeokinesis	723,124	20,900,555	1,751,512	2,179,883	0.52995

(the "loss"): $0.1(S + \sigma_x^2) + 0.01R$ (assuming, e.g., that cost and stability contribute ten times as much as stress). All homeotaxic strategies achieve good results (Table 1), easily outperforming the strategies considered above (minimisation of persistence error; $q_t = l_t$; $q_t = u_t$; and basic homeokinesis). We used the root mean square error (RMSE$_c$) in estimating how close the controller c_t is to the perfect case ($c_t = 1$), i.e. RMSE$_c = \sqrt{\sum_1^T (c_t - 1)^2/T}$. The success can be explained by the combination of homeokinetic exploration and exploitation of persistent actions. The optimal strategies based on $\xi_t^{(1)}$, $\xi_t^{(2)}$ and $\xi_t^{(3)}$ have also shown robustness to selfishness of individual agents (selfishness is defined here as a fraction of agents that always refuse to follow a control plan), and to additive noise in actual consumption, that can be due to a fraction of agents that intermittently decide against following a control plan (these experimental results are omitted due to a lack of space).

6   Conclusions

We argued that adaptive coordination of multiple resource-sharing agents requires both synchronisation and desynchronisation, balancing exploitation and exploration stages.

This balance can be derived (learned) from dynamics of the multi-agent system itself, if there is a feedback measuring the distance between current and a desired behaviour. A model hierarchical system comprised multiple energy-consuming agents split into local neighbourhoods, each with a dedicated controller, and a centralised coordinator dealing only with the controllers. The overall consumption within a neighbourhood was considered as a sensory input of the high-level coordinator, while the coordinator's output formed its actuation. This interpretation allowed us to develop homeotaxis: a generic domain-invariant approach, based on the homeokinetic principle, extended in terms of scope (multi-agent self-organisation) and the state-space (predictable perception and action spaces). A number of homeotaxic strategies, based on persistent sensorimotor time-loops, were introduced and experimentally verified, achieving a balance between resource consumption and stress within the multi-agent community.

Acknowledgments. The authors are grateful to Oliver Obst for suggesting the term *homeotaxis*, and to Peter Corke for pointing out the work of Odhner *et al.* [12].

References

1. Ashby, W.R.: Design for a Brain. Chapman and Hill, London (1954)
2. Ay, N., Bertschinger, N., Der, R., Güttler, F., Olbrich, E.: Predictive information and explorative behavior of autonomous robots. European Journal of Physics: Complex Systems (submitted, 2007)
3. Baldassarre, G.: Self-organization as phase transition in decentralized groups of robots: A study based on boltzmann entropy. In: Prokopenko, M. (ed.) Advances in Applied Self-organizing Systems, pp. 127–146. Springer, Heidelberg (2007)
4. Der, R.: Self-organized robot behavior from the principle of homeokinesis. In: Proceedings of the SOAVE Workshop, Ilmenau (2000)
5. Der, R., Güttler, F., Ay, N.: Predictive information and emergent cooperativity in a chain of mobile robots. Santa Fe Institute Working Paper 08-02-006 (2008)
6. Der, R., Hesse, F., Martius, G.: Rocking stamper and jumping snake from a dynamical system approach to artificial life. J. Adaptive Behavior 14, 105–116 (2005)
7. Der, R., Liebscher, R.: True autonomy from self-organized adaptivity. In: Proceedings of The Workshop on Biologically Inspired Robotics, The Legacy of W. Grey Walter, August 14-16, 2002, HP Bristol Labs (2002)
8. Klyubin, A.S., Polani, D., Nehaniv, C.L.: All else being equal be empowered. In: Cap-carrère, M.S., Freitas, A.A., Bentley, P.J., Johnson, C.G., Timmis, J. (eds.) ECAL 2005. LNCS (LNAI), vol. 3630, pp. 744–753. Springer, Heidelberg (2005)
9. Li, R., Poulton, G.: Optimal management of multiple resource consuming devices during periods of supply restriction. Technical report, CSIRO ICT Centre, Macquarie University (May 2007)
10. Lizier, J.T., Prokopenko, M., Zomaya, A.Y.: Local information transfer as a spatiotemporal filter for complex systems. Phys. Rev. E 77, 026110 (2008)
11. Mirollo, R., Strogatz, S.: Synchronization of pulse-coupled biological oscillators. SIAM Journal of Applied Math 50, 1645–1662 (1990)
12. Odhner, L., Ueda, J., Asada, H.: Broadcast feedback of stochastic cellular actuators. In: Proc. of The 10th International Symposium on Experimental Robotics (ISER), Rio de Janeiro (2006)

13. Patel, A., Degesys, J., Nagpal, R.: Desynchronization: The theory of self-organizing algorithms for round-robin scheduling. In: Proceedings of The First IEEE Conference on Self-Adaptive and Self-Organizing Systems (SASO), pp. 87–96 (2007)
14. Piraveenan, M., Polani, D., Prokopenko, M.: Emergence of genetic coding: an information-theoretic model. In: eCosta, F.A., Rocha, L.M., Costa, E., Harvey, I., Coutinho, A. (eds.) ECAL 2007. LNCS (LNAI), vol. 4648, pp. 42–52. Springer, Heidelberg (2007)
15. Prokopenko, M.: Design vs self-organization. In: Prokopenko, M. (ed.) Advances in Applied Self-organizing Systems, pp. 3–17. Springer, Heidelberg (2007)
16. Prokopenko, M., Gerasimov, V., Tanev, I.: Evolving spatiotemporal coordination in a modular robotic system. In: Nolfi, S., Baldassarre, G., Calabretta, R., Hallam, J.C.T., Marocco, D., Meyer, J.-A., Miglino, O., Parisi, D. (eds.) SAB 2006. LNCS (LNAI), vol. 4095, pp. 558–569. Springer, Heidelberg (2006)
17. Trianni, V., Nolfi, S.: Minimal communication strategies for self-organising synchronisation behaviours. In: Proceedings of The First IEEE Symposium on Artificial Life, pp. 199–206 (2007)
18. Werner-Allen, G., Tewari, G., Patel, A., Nagpal, R., Welsh, M.: Firefly-inspired sensor network synchronicity with realistic radio effects. In: Proceedings of The Third ACM Conference on Embedded Networked Sensor Systems (SenSys), San Diego, pp. 142–153 (2005)
19. Wikipedia,: Taxis — wikipedia, the free encyclopedia (2007) [Online; accessed January 10, 2008]

Noise-Induced Adaptive Decision-Making in Ant-Foraging

Bernd Meyer[1], Madeleine Beekman[2,3], and Audrey Dussutour[3]

[1] FIT Centre for Research in Intelligent Systems, Monash University, Melbourne
[2] Behaviour and Genetics of Social Insects Lab and Centre for Mathematical Biology
[3] School of Biological Sciences, University of Sydney
bernd.meyer@acm.org, {mbeekman,adussutour}@usyd.edu.au

Abstract. Ant foraging is a paradigmatic example of self-organized behavior. We give new experimental evidence for previously unobserved
short-term adaptiveness in ant foraging and show that current mathematical foraging models cannot predict this behavior. As a true extension, we
develop Itô diffusion models that explain the newly discovered behavior
qualitatively and quantitatively. The theoretical analysis is supported by
individual-based simulations. Our work shows that randomness is a key
factor in allowing self-organizing systems to be adaptive. Implications for
technical applications of Swarm Intelligence are discussed.

1 Introduction

Groups of humans or animals often make decisions collectively without any central control or coordination. The paradigmatic example of self-organized groups
are colonies of social insects, such as ants and bees, whose strikingly organized
and seemingly purposeful behavior at the group level is organized without any
central "master plan" [4]. Their complex behavior at the colony level emerges
from simple interactions between myriads of individuals that only process local
information [19]. Such decentralized coordination exhibits a number of properties that are highly desirable in technical applications, specifically robustness,
adaptiveness and parallelism. Hence, social insect behavior has been used as an
inspiration for a wide range of engineering tasks [2,13].

Despite the central importance of self-organized decision-making for many
natural and technical systems, the theoretical understanding of its fundamental
properties is still in its infancy. A core problem is that there is very little insight
into how self-organized processes work in dynamic environments. Most research
addresses only static conditions. Yet, being able to *adapt in dynamic environments* on all timescales is crucial for almost all natural systems and for many
engineering artifacts using similar organizational principles.

Foraging in mass-recruiting ant species is arguably the best understood form
of self-organized behavior, and detailed knowledge about different forms of organization as well as quantitative and qualitative mathematical models are available [4]. However, until recently this research has focused almost entirely on static

M. Asada et al. (Eds.): SAB 2008, LNAI 5040, pp. 415–425, 2008.

environments where the available resources are constant. The conventional wisdom is that the ability of ants to quickly adapt to a changing environment is generally very limited, and this is in agreement with the predictions of existing mathematical models. Recently new experiments with dynamic environments have started to cast some doubt on the scope of these models. These new experiments clearly demonstrate that at least some mass recruiting ant species can flexibly and quickly adapt their foraging behavior to environmental changes. As the standard models cannot explain this adaptive decision-making, a more powerful way of modelling is required. The purpose of the present paper is to introduce such a type of model.

We will show that the conventional models have limited predictive power because they are mean field descriptions based on the general assumption that the net influence of noise is zero. As we will show, this is not the case in dynamic environments. Here, noisy communication has a crucial role in allowing ant colonies to behave adaptively. Our technical contributions are three-fold: Firstly, we give new experimental evidence for short-term adaptation in the foraging behavior of the mass recruiting ant species *Pheidole megacephala*. Secondly, we introduce continuous stochastic models based on Itô diffusions [12] and Fokker-Planck equations [18] as a new tool for theoretical research in ant behavior. Thirdly, we apply this method to our experiments obtaining a quantitative mathematical model that explains the adaptation and coincides with the conventional model in the noise-free limit. Our mathematical analysis is supported by individual-based simulations which confirm our experimental findings and validate the formal model.

2   Standard Models of Ant Foraging

Mass recruiting ant species, such as *Lasius niger* and *Pheidole megacephala*, coordinate their foraging activities mainly using pheromone communication [14]. When returning from a food source to the nest, the foragers deposit pheromones on the paths they use. New foragers that venture out from the nest in search of food probabilistically follow the pheromone gradients in the environment and pheromone gradually evaporates over time. In this way a self-limiting positive feedback loop is

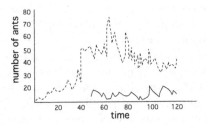

Fig. 1. Average Number of Foragers at Better Source [4]. Dashed: Both Sources Presented Simultaneously; Solid: Better Source Presented with Delay.

created and increasingly more pheromone is deposited on the paths to food sources. As the amount of pheromone deposit is generally (directly) modulated by food source quality or (indirectly) by the distance of the food source to the nest, the majority of the traffic tends to converge on the paths to more desirable food sources [4].

A classical binary decision experiment to investigate this behavior is the "Y-Bridge" [4]. In this experiment the colony is presented with two food sources at the two ends of a forking path from the nest (Fig. 3, top). When the quality of the food sources or the path lengths to the sources are different the ants typically make the correct collective choice: the vast majority of foragers converge on the shorter path or on the richer source, respectively. However, this is generally only true if both choices are present right from the beginning. If at first only the inferior choice is presented (longer path or worse quality) and the superior one is only added after the traffic has stabilized, the colony will generally *not* adaptively adjust and will stick with the inferior solution instead [4] (Fig. 1).

A widely used class of models that explains these and related experiments in *static environments* is based on dynamic systems using ordinary differential equations [8,17]. Let the amount of pheromone on the two branches be denoted by c_1, c_2. In the standard model [8], the probabilities of an individual ant to choose either branch when leaving the nest are $p_i = \frac{(k+c_i)^\alpha}{\sum_{j=1}^{2}(k+c_j)^\alpha}$ where k, α are non-negative constants fitted to experimental data. Each individual forager deposits an amount of pheromone q_i upon its return to the nest. While the total number of foragers leaving the nest per time unit in reality depends on the amount of trail pheromone present, it has been shown that this flux may be assumed to be constant without changing the characteristics of the model [17]. Assuming a constant total flux of Φ foragers, the number of foragers on Branch i is $n_i = p_i \Phi$, and the development of the two pheromone levels is $\frac{dc_i}{dt} = p_i Q_i - \rho c_i$ where ρ is the rate constant for pheromone evaporation, $Q_i = \Phi q_i$.

Let Branch 1 be the superior path. There are two reasons why it may attract more traffic than Branch 2: Either the food sources are of different quality and pheromone deposit is actively modulated by food quality ($q_1 > q_2$) or Branch 1 is shorter and thus receives pheromone deposits by returning foragers earlier, i.e. it gets a head start in the competition. Pheromone deposit on the shorter path may also be (indirectly) modulated, for example due to a lower trail fidelity or a higher rate of U-turns on the longer branch [4] or due to home range marking [9].

It is well known that this model exhibits three fixpoints [4,17]. The first fixpoint corresponds to a proportional usage of paths (i.e. pheromone on both paths and in the case of equal deposits $q = q_1 = q_2$ both sources will be equally exploited). The other two fixpoints correspond to the situation where predominantly one source is exploited and the other resource is exploited far less, even for $q_1 = q_2$. The

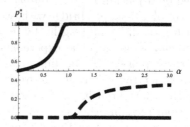

Fig. 2. Standard Model Fixpoints

exact proportion of exploitation depends on the parameters k, α, Φ, ρ and the proportion q_1/q_2. The first fixpoint is only stable in a limited parameter range. The fixpoint diagram for p_1 in Fig. 2 summarizes the situation for $k = 0$ [15]. For

$\alpha < 1$ only the first fixpoint $p_1^* = 1/\left(1 + \left(\frac{q_2}{q_1}\right)^{\frac{\alpha}{1-\alpha}}\right)$ is stable, but it exchanges its stability with the other two fixpoints at $\alpha = 1$. Interestingly, for $\alpha > 1$ two fixpoints are stable: the one that corresponds to exclusive exploitation of the superior source as well as the one that corresponds to only exploiting the inferior source. The model has experimentally been matched to the behavior of real ant colonies, specifically for *L. niger* with $\alpha = 2, k = 6$ [4,1]. When $k \gtrsim 0$ the situation remains in principle the same, but there will always be a residual amount of exploitation of the less used resource. This amount depends on k.

Note that the model so far does not fully predict which of the two stable fixpoints will be observed in any given experiment. The decision depends on the differences between food source qualities and path lengths and on the times at which the two sources are first discovered. However, it has been observed that the decision is successfully made for the superior choice in most cases if the differences (and thus deposit ratio q_1/q_2) and the flux Φ are large enough [4]. Once a stable fixpoint has been reached the behavior is locked regardless of changes in the reward ratio. It is thus in agreement with the model that the colony will not adapt if the better source is presented with significant delay.

3   Experiments with Dynamic Environments

To investigate the adaptiveness of foraging in the mass recruiting "big headed ant" *P. megacephala* we used a slightly modified version of the classical Y-bridge experiment. In each experiment, a colony starved for 5 days was given access to two equal food sources ($3ml$ of $1M$ sucrose solution) placed on a platform ($70 \times 70mm$) at the ends of a Y-shaped bridge with two branches of different length: $180mm$ and $60mm$ (Fig. 3). The experiment had three phases. In the first phase

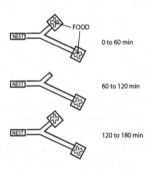

Fig. 3. Experimental Set-up

(top) we let the colony forage freely for one hour at both sources. After this time the short branch was blocked at the level of the food source preventing ants from reaching this source in Phase 2 (middle). One hour later the short branch was reopened for Phase 3 (bottom). A control experiment was performed with 3 hours of duration in which no changes were applied to the set-up. We replicated the experiment with 21 colonies of approximately 2,000 workers each with brood and queen kept at room temperature ($25 \pm 1°C$) with a $12 : 12$ L/D photoperiod. Fig. 4 shows the average proportion of the number of foragers on the branches. The behavior of *P. megacephala* is surprisingly different from that of *L. niger*: This species does indeed manage to adapt to the change in food sources (or their accessibility). It is obvious that the ants initially commit to the more desirable short branch and switch to the longer branch when the shorter one becomes unavailable in Phase 2. In Phase 3 the colonies generally switch back to the more desirable shorter branch.

We can attempt to model our experiment using the standard model and changing the deposit constants q_i when switching from one phase to another. There is nothing surprising about the switch from Phase 1 to Phase 2. As the food source on the short branch becomes unavailable, this branch does not receive any further pheromone in Phase 2 and the pheromone left from Phase 1 simply evaporates until nothing is left. At the same time the longer branch still receives new recruitment pheromone. The ants therefore must eventually switch to the longer branch. The model predicts this. However, the switch from Phase 2 to Phase 3 is not as expected from the predictions of the standard model. At the end of Phase 2 sufficient time has elapsed for the colony to focus predominantly on the longer branch, and we know from control experiments that by this time the pheromone on the blocked shorter branch has evaporated sufficiently to not influence the path choice anymore. From Section 2 we can be certain the ants should not be able to refocus on the shorter branch when it is re-opened, because the traffic has reached a stable fixpoint.

It is crucial to realize that the standard model cannot be coaxed into any other behavior by tuning its parameters. Specifically, the only way to tune the model such that it switches the predominant exploitation from one resource to the other when the deposit ratio q_1/q_2 changes after one branch has reached saturation is to set $\alpha \leq 1$. However, for $\alpha \leq 1$ the model has only one fixpoint and we should *always* observe this fixpoint in real experiments. What we do see, however, is that every trial has two possible outcomes. The standard model thus fails to predict Phase 3 correctly. The necessity to modify it is clear.

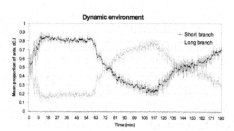

Fig. 4. Average Proportion of Foragers

4   Noise-Induced Adaptiveness

The question is, which crucial aspect of *P. megacephala*'s decision making does the standard model ignore. We will show that no new mechanism needs to be invoked. It is simply and somewhat counter-intuitively the noisiness of the decision making behavior which enables the colonies to behave adaptively. Despite the fact that the role of individual error in foraging has long been recognized [7,16], current foraging models are generally mean-field descriptions, and errors are simply treated as average error rates. The fundamental problem with this is that mean-field models, strictly speaking, only give us a correct description if the behaviors of all individuals are identical (conforming to the average of their true behaviors). This simplification is not always justified. In the presence of non-linear responses, behavioral deviations of only a few individuals from the mean can trigger qualitative changes in the colony behavior. To model these fluctuations, we must use stochastic models. Leading researchers have called for analytic stochastic models of ant foraging in the past already [17]. This call is still echoed in the most recent surveys [13], but it has gone unanswered as yet. In this section we will derive a stochastic model of mass recruitment based on the standard model and show how it explains the newly observed effects.

We rewrite the two-dimensional model into a single dimension, exploiting $p_1 = 1 - p_2$. It is convenient to introduce a "re-normalized deposit" $z = Q_2 c_1 + Q_1 c_2$ for which $\frac{dz}{dt} = Q_1 Q_2 (p_1 + p_2) - \rho (Q_2 c_1 + Q_1 c_2) = Q_1 Q_2 - \rho z$ Observe that z develops completely deterministically with finite limit $lim_{t \to \infty} z(t) = \frac{Q_1 Q_2}{\rho}$. Note that z equilibrates quickly. We run each phase of the experiment long enough so that the pheromone levels are saturated. Thus, to describe the switching behavior between the phases we can replace $z(t)$ with its limit and eliminate c_2 from the model. From the standard model with $\alpha = 2$ we obtain

$$\frac{dc_1}{dt} = Q_1 \frac{(k + c_1)^2}{(k + c_1)^2 + (k + \frac{\frac{Q_1 Q_2}{\rho} - Q_2 c_1}{Q_1})^2} - \rho c_1 \qquad (1)$$

c_2 is given implicitly by $c_2(t) = \frac{z(t) - Q_2 c_1(t)}{Q_1}$. We plot the general shape of $\frac{dc_1}{dt}$ for the relevant parameter region in Fig. 5. Clearly c_1 will eventually converge on one of the axis intersections. The first and last axis intersection correspond to stable fix points (exclusive use of a single source), whereas the middle point is unstable.

Let c_1 stand for the pheromone level on the *longer* branch. The stability of the highest fixpoint of c_1 in the mean field model (corresponding to most pheromone on the longer/inferior path and thus its almost exclusive use) is what makes it impossible for the model to switch to the exploitation of the superior food source when it is re-introduced at the commencement of Phase 3. However, if we take random fluctuations into account this is no longer true:

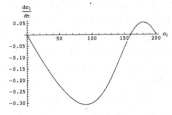

Fig. 5. Drift term

Any fluctuation that pushes the pheromone level on the longer branch momentarily to the left of the second axis intersection will be amplified and lead to a switch to the short branch. As we observe reasonably large random fluctuations of the forager counts on each branch in experiments, the assumption of such random fluctuations is justified. The first reason why the counts must be "noisy" is that each ant makes an individual binary decision with decision probability p_i (Bernoulli trial). Thus the distribution of the number of foragers selecting path i is given by the binomial distribution $B(N, p_i)$ with mean $N \cdot p_i$ and variance $\sigma^2 = N p_i (1 - p_i)$. If we want to be able to model possible developments of individual trials (instead of just averages), we have to replace p_i with $(p_i + \sigma W)$ where W is a (Gaussian) random variable with mean 0 and unit variance. In addition to this intrinsic noise, it is reasonable to assume random fluctuations in the ants' behavior due to individual variations, additional (random) inter-actions between colony members, physical limitations of their sensory systems, and environmental influences. The behavioral fluctuations enter into the model as fluctuations of the model parameters during the course of an experiment. We can determine the influence of parameter fluctuations on p_i for fixed c_i by the Delta-Method [6], which approximates the variance of a function $f(X)$ of a ran-dom variable X with mean μ_X and variance σ_X^2 as $\text{Var}(f(X)) = \left(\frac{d}{dx}f(\mu_X)\right)^2\sigma_X^2$. It is easy to verify that even small fluctuations in α have significant influence on p_i, especially when c_1 is close to the unstable fixpoint (the decision point). We could endeavor to model the noise precisely using the Delta-Method for all rate constants in Eq. 1. However, this would be a very treacherous form of precision given that all model parameters are only estimates, that the basic model itself is only an approximation and that many other environmental factors can cause fluctuations. A good and simple approximation of the experimental data is to set σ to an appropriately fitted constant. We substitute p_i with $(p_i + \sigma W)$ in the basic model equation and write the result formally as a diffusion in the form of an Itô stochastic differential equation where $\mu(x)$ describes the deterministic development and σdW describes the influence of noise [12].

$$dx = \mu(x)dt + \sigma dW \tag{2}$$

$$\mu(x) = Q_1\frac{(k+x)^2}{(k+x)^2 + (k + \frac{\frac{Q_1Q_2}{\rho} - Q_2x}{Q_1})^2} - \rho x \tag{3}$$

5   Fokker-Planck Analysis of Itô Diffusion Model

We have obtained the Itô diffusion Eq. (2,3) as our basic model for the pheromone level on the longer branch, where σ is a constant noise level. We are mainly interested in the steady-state pheromone levels. Instead of just a mean, the model allows us to obtain an explicit density function $\pi(c)$ for the probability to find a pheromone level c on the longer branch after the system has run for a long time. Let $f(t, y)$ be the probability density for the longer branch to carry y units of pheromone at time t. The time-development of $f(\cdot, \cdot)$ is described by the Fokker-Planck Equation (FPE [18]).

$$\partial_t f(t, y) = -\partial_y \left[\mu(y) f(t, y)\right] + \partial_{yy}\left[\frac{1}{2}\sigma^2(y) f(t, y)\right] \tag{4}$$

The steady state $\pi(c) = f(c, t)$ is time-independent, so Eq. 4 reduces to an ODE

$$0 = -\frac{d}{dy}\left[\mu(y)\,\pi(y)\right] + \frac{1}{2}\frac{d^2}{dy^2}\left[\sigma^2(y)\,\pi(y)\right] \tag{5}$$

Since the pheromone level on the longer branch is physically limited to the interval $(0, Q_1/\rho)$, we must assume reflecting boundaries at both interval ends which prevents x from becoming negative and implies zero net flow of probability across the interval boundaries. The solution of Eq. 5 then is

$$\psi(x) = e^{\int_a^x (2\mu(y)/\sigma^2(y))dy}$$
$$\pi(x) = C\frac{\psi(x)}{\sigma^2(x)}$$
(6)

where C is a suitable normalization constant [12]. We plot the solution of Eq. 6 for $q_1 < q_2$ in Fig. 6 (here, $q_1/q_2 = 1/3$). Obviously, the vast majority of the probability mass is assigned to the vicinity of zero. The interpretation of this is that in the long run all pheromone will almost certainly vanish from the longer branch, provided both branches are available. Regardless of the start state of the system, the long branch will always be disused and the exploitation will switch back to the shorter (better) branch. This is indeed what we see in the *P. megacephala* experiments. However, our model must be able to account for both types of behavior, where the switching takes place and where it does not occur.

By inspection of the drift term $\mu(x)$ in Fig. 5 we may suspect that something special is happening in the range of the upper limit of c_1. We magnify the range of the saturation level of c_1 in the inset of Fig. 6 and observe a (tiny) local maximum of probability. The key to why this local maximum explains both cases is that the stationary distribution is only reached after infinite time. To know how long it will take for the colony to disuse the longer branch

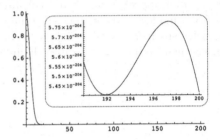

Fig. 6. Stationary Distribution π

after the shorter branch is reopened, we must find the expected time that it takes for the system to reach a very low pheromone level on the longer branch after starting it with the longer branch being (almost) completely saturated. This expected time is given by the so-called "First Passage Time" which can be obtained by solving the Kolmogorov-Backward Equation [12] for Eq. 2. For a reflecting boundary at the upper end b of the interval, the expected time $t(x)$ for the pheromone level on the longer branch to fall from x to a is

$$t(x) = \int_a^x \left(\frac{2}{\psi(y)} \int_y^b \frac{\psi(z)}{\sigma^2(z)} dz \right) dy$$
(7)

The numeric solution to Eq. 7 for $a = 100$ (in minutes) is plotted on the left in Fig. 7 for a variance of 0.04 which is approximately matched to our experiments for the 30 minutes before switching to Phase 3 ($k = 16$ for *P. megacephala*). We note that this escape time is short, so we expect the pheromone level on the longer branch (and thus the number of foragers) to fall of relatively quickly, exactly as observed in the experiments. In the experiments with *L. niger* we see significantly lower noise levels [11]. Solving Eq. 7 for such low levels, we find that the escape time is dramatically higher (Note the scale in Fig. 7, right). We

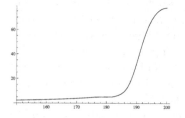 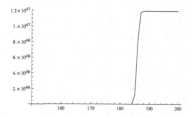

Fig. 7. Escape Time. left: $\sigma^2 = 0.04$, *P. megacephala*, right: $\sigma^2 = 0.002$ *L. niger*

would thus not expect to see such a switch happen in real experiments or only in very rare cases. This is consistent with all experiments and demonstrates that the same stochastic model explains the *P. megacephala* experiments where the switch does happen and the *L. niger* experiments, where it is not observed.

To verify our findings we have implemented an individual-based simulation of the *P. megacephala* experiments (Java code can be obtained by request from the first author). Our model represents each ant in the colony as an individual object. All actions are controlled with probabilistic decisions independently for each object, including resting, feeding, leaving the nest, moving and branch choice. The individual-based model also includes details from which the analytic model abstracts, specifically deposit reduction over time and stochastic total flux, to validate that the inclusion of these factors does not influence our results. Parametric studies are in agreement with the analytic model and confirm that the qualitative behavior of the model does not depend on these factors.

6 Discussion and Conclusions

We have given experimental evidence for truly adaptive behavior in ant foraging for *P. megacephala*. This behavior is qualitatively different from the behavior of other species, specifically *L. niger*, and to the best of our knowledge this is the first report of evidence for fully adaptive foraging in ants. We note that *L. niger* is non-invasive and relies on permanent food sources, while *P. megacephala* is an invasive species and opportunistic feeder exploiting ephemeral sources.

Surprisingly no new functional mechanisms needed to be invoked to explain the adaptive behavior. Randomness in the decision making turned out to be the crucial factor that allows the self-organized process to be adaptive. The standard models for ant foraging are unable to capture this behavior because they are mean field descriptions. We have presented a new type of analytic model based on stochastic differential equations that explains this behavior. Our findings are confirmed by individual-based simulations.

On a more general level, our main finding is that noise can enable self-organizing systems to adapt readily to changing environments. While self-limiting positive feedback loops are a core mechanism for self-organizing group activities, randomness is the driving factor that allows such systems to be adaptive. As always, there is a price to be paid: higher noise levels result in less efficient decision making in static environments. Cost and benefit of noise must thus be balanced. One could come to the bold conjecture that each ant species should have evolved to the optimum noise level for its typical environment. For example, when food patches differ little in quality, adaptiveness may not be required and less noise in the system would allow a quick and committed decision, i.e. generally more efficient foraging. On the other hand, when sources with large quality differences occur, it may be advantageous to use more noise in the system to be able to compare a new find to the currently used source. The slightly decreased general efficiency will be more than compensated for by the ability to adapt.

Our results are not only interesting in the context of natural systems but may have far reaching implications for applications of swarm intelligence [2]. Ant Colony Optimization algorithms, for example, are very directly modelled on the foraging behavior of real ants [3]. For some algorithms the resemblance is so close that the algorithms' function can fundamentally be understood as a simulation of ant foraging in an abstract phase space, and the fundamental model equations describing the algorithms' dynamic are virtually identical to those for real ant colonies. We may thus expect the behavior of these systems to be governed by the same rules. Similar considerations hold for swarm robots [2], self-organized routing algorithms [10], and other applications of swarm intelligence. Our results indicate that noise should be taken into account as a constructive component when engineering such systems. For example, we may wish to use controlled injection of noise to trigger adaptation when the environment changes.

Systems of coupled feedback equations comparable to those used in ant foraging models are very common across a large variety of models for self-organized behavior and decision making. For example, the development of market trends [20] or the dispersion of innovations [5] can be modelled in this form. In consequence it is highly likely that here, too, we will need stochastic models to fully understand the characteristics of these mechanisms in dynamic environments. For one case we have demonstrated how this can be achieved using stochastic differential equations and Fokker-Planck equations. We are confident that this class of models has wide applicability to other self-organizing systems.

References

1. Beckers, R., Deneubourg, J.L., Goss, S.: Modulation of trail laying in the ant Lasius niger and its role in the collective selection of a food source. Journal of Insect Behaviour 6, 751–759 (1993)
2. Bonabeau, E., Dorigo, M., Theraulaz, G.: Swarm Intelligence—From Natural to Artificial Systems. Oxford University Press, Oxford (1999)
3. Bonabeau, E., Dorigo, M., Theraulaz, G.: Inspiration for optimization from social insect behaviour. Nature 406, 39–42 (2000)
4. Camazine, S., et al.: Self-Organization in Biological Systems. Princeton University Press, Princeton (2001)
5. Capasso, V., Bakstein, D.: An Introduction to Continuous-Time Stochastic Processes. Birkhäuser, Basel (2005)
6. Casella, G., Berger, R.L.: Statistical Inference, 2nd edn. Duxbury (2002)

7. Deneubourg, J.L., Pasteels, J.M., Verhaeghe, J.C.: Probabilistic behavior in ants: A strategy of errors. Journal of Theoretical Biology 105, 259–271 (1983)
8. Deneubourg, J.L., et al.: The self-organizing exploratory pattern of the argentine ant. Journal of Insect Behaviour 3, 159–168 (1990)
9. Devigne, C., Detrain, C.: How does food distance influence foraging in the ant Lasius niger. Insectes Sociaux 53, 46–55 (2006)
10. Dressler, F.: Self-Organization in Sensor and Actor Networks. John Wiley, Chichester (2008)
11. Dussutour, A., Deneubourg, J.-L., Fourcassié, V.: Amplification of individual preferences in a social context: the case of wall-following in ants. Proceedings of the Royal Society London B 272, 705–714 (2005)
12. Gardiner, C.W.: Handbook of Stochastic Methods. Springer, Heidelberg (2004)
13. Garnier, S., Gautrais, J., Theraulaz, G.: The biological principles of swarm intelligence. Swarm Intelligence 1, 3–31 (2007)
14. Hölldobler, B., Wilson, E.O.: The Ants. Springer, Heidelberg (1990)
15. Meyer, B.: On the convergence behaviour of ant colony search. In: Asia-Pacific Conference on Complex Systems, Cairns (December 2004)
16. Nicolis, G., Prigogine, I.: Exploring Complexity. W.H.Freeman, New York (1989)
17. Nicolis, S.C., Deneubourg, J.L.: Emerging patterns and food recruitment in ants: an analytical study. Journal of Theoretical Biology 198, 575–592 (1999)
18. Risken, H.: The Fokker-Planck Equation. Springer, Heidelberg (1989)
19. Sumpter, D.J.T.: The principles of collective animal behaviour. Philosophical Transactions of the Royal Society B 361(1465), 5–22 (2003)
20. Weisbuch, G., Stauffer, D.: Hits and flops dynamics. Physica A 287, 563–576 (2000)

Division of Labour in Self-organised Groups

Roderich Groß[1], Shervin Nouyan[2], Michael Bonani[1],
Francesco Mondada[1], and Marco Dorigo[2]

[1] LSRO, Ecole Polytechnique Fédérale de Lausanne, Lausanne, Switzerland
roderich.gross@ieee.org,{michael.bonani,francesco.mondada}@epfl.ch
[2] IRIDIA, CoDE, Université Libre de Bruxelles, Brussels, Belgium
{snouyan,mdorigo}@ulb.ac.be

Abstract. In social insect colonies, many tasks are performed by higher-order entities, such as groups and teams whose task solving capacities transcend those of the individual participants. In this paper, we investigate the emergence of such higher-order entities using a colony of up to 12 physical robots. We report on an experimental study in which the robots engage in a range of different activities, including exploration, path formation, recruitment, self-assembly and group transport. Once the robots start interacting with each other and with their environment, they self-organise into teams in which distinct roles are performed concurrently. The system displays a dynamical hierarchy of teamwork, the cooperating elements of which comprise higher-order entities. The study shows that teamwork requires neither individual recognition nor inter-individual differences, and as such might contribute to the ongoing debate on the role of such characteristics for the division of labour in social insects.

Keywords: cooperation, division of labour, foraging, group transport, hierarchy, path formation, self-organisation, swarm robotics, teamwork.

1 Introduction

The field of swarm intelligence draws inspiration from decentralised and self-organising biological systems in general and from the collective behaviour of social insects in particular [1,2]. At present, little is known about the mechanisms that regulate such biological systems, and in particular, about how such mechanisms could enhance the design of swarm intelligence systems. Thus, it is not surprising that the complexity exhibited in current implementations of swarm intelligence systems does neither come close to the complexity of biological systems, nor does it come close to the complexity of systems men built following the more traditional "top-down" approach.

In this paper, we investigate the conditions under which complexity can "emerge" in swarm intelligence systems. One way of measuring complexity is to look at the structural organisation of individuals when performing a task. In an insect colony, various organisational levels can be observed. Both behaviours at the individual level as well as at the colony level have been extensively studied [3]. "However, between these two extremes, numerous functional adaptive

M. Asada et al. (Eds.): SAB 2008, LNAI 5040, pp. 426–436, 2008.
© Springer-Verlag Berlin Heidelberg 2008

units, or 'parts' exist" [4, page 291]. These *intermediate-level* parts comprise *groups* and *teams*.

Teamwork is widely observed in vertebrates. Here, *individual recognition* is believed to be an important factor [5]. Fewer examples of teamwork are known in invertebrates. Oster and Wilson ([6]; reviewed in [7]) argue that members of social insect colonies can not form teams as a consequence of their low grade of discrimination: social insects can discriminate "nest mates from aliens, [and] members of one caste as opposed to another" [6, page 151], however, "there is very little evidence that social insects can recognise each other as individuals (but see Tibbetts [8])" [7, page 6]. In contrast, in the recent literature [3,7,9], biologists suggest that teams are indeed formed in social insects, and do not require individual recognition. Another aspect that is subject of the ongoing debate is whether inter-individual differences (e.g., members of different castes) are fundamentally required in teamwork [3,10,11].

In the following, we investigate whether tasks that require a complex division of labour fundamentally require individual recognition or inter-individual differences. We illustrate the methods and results of a series of experimental works in which a set of "identical" robots is required to perform a complex, cooperative task. At the beginning of a trial, the robots are randomly scattered in a bounded arena that contains two objects—the *prey* and the (static) *nest*. The task is to retrieve the prey to the nest. The following constraints are given:

- the prey requires concurrent, physical handling by multiple robots to be moved,
- each robot's perceptual range is small when compared to the distance between the nest and the prey; moreover, perception is unreliable,
- no robot has any (explicit) knowledge about the environment beyond its perceptual range,
- communication among robots is unreliable and limited to a small set of simple signals that are locally broadcast.

In the following we use the terms groups and teams as defined by Anderson and Franks [11]. In particular, a *group* is a set of individuals that tackle a group task; a *team* is a set of individuals that tackle a team task. A *group task* is a task that "requires multiple individuals to perform the same activity concurrently"; a *team task* is a task that "requires different subtasks to be performed concurrently" (page 535). Furthermore, a *partitioned task* is "a task that is split into two or more subtasks that are organised sequentially (Jeanne [12]; reviewed in Ratnieks and Anderson [13]; Anderson and Ratnieks [14])" [7, page 4]. Anderson and Franks [7,11], and Anderson and McMillan [15] found that the definition of teamwork, developed primarily from studies of social insects, also applies more generally to societies of other animals, including humans, and robots.

Fig. 1 (left) summarises the division of labour present in our robotic colony. Overall, the robots accomplish a partitioned task comprising three subtasks that are organised sequentially: (i) *path formation* requires robots to explore the environment and form a path in between the nest and the prey that can be traversed in both directions; (ii) *recruitment* requires some robots to maintain

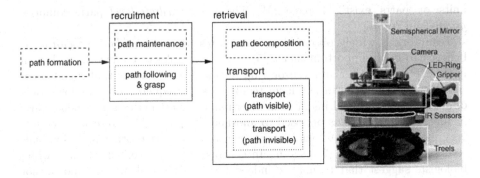

Fig. 1. Left: illustration of the division of labour in our robotic colony that accomplishes a group foraging task. The task is a partitioned task (see arcs). Individual tasks, group tasks and team tasks are framed respectively by dotted, dashed and solid lines. Right: side view of a colony member, the *s-bot* robot.

the path, while other robots follow the path from the nest to the prey and then grasp either the prey directly or other robots already gripped onto it; (iii) *retrieval* requires some robots to decompose the path, while other robots transport the prey along the path to the nest (until the prey, or a robot transporting it, is in physical contact with the nest). *Path formation* itself is a group task, because only a group of robots can establish a path. Similarly, *path maintenance* and *path decomposition* are group tasks. *Recruitment* is a team task, because it requires two different sub-tasks to be performed concurrently—*path maintenance* and *path following & grasp*, where the latter is an individual task. *Retrieval* is a team task as some robots have to engage in *transport*, while others, at the same time, have to reside in the path to guide the transport robots towards the nest. *Transport* can be considered a (nested) team task, as (i) multiple robots are required to transport the prey, and as (ii) the transporting robots, when unable to perceive the path, need to perform distinct actions to avoid that the group transport is ineffective.

The remainder of this paper is organised as follows. In Section 2, we detail the methods, that is, the robot's hardware, the controller and the experimental setup. In Section 3 we present the results. Finally, in Section 4, we discuss the results and conclude the paper.

2   Methods

2.1   Hardware

We use a robotic system called swarm-bot lying at the intersection between collective and reconfigurable robotics [16]. The system is composed of basic robotic units, called *s-bots*, which are fully autonomous and mobile, and capable of connecting to each other. Fig. 1 (right) shows the physical implementation of the *s-bot*. The robot has a total height of 19 cm and weighs approximately 700 g. An

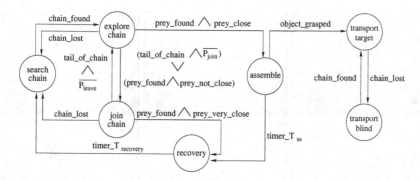

Fig. 2. State diagram of the finite state machine that controls each *s-bot*. Circles represent states (i.e., behaviours). Edge labels specify conditions that trigger transitions between the corresponding states. The initial state is **search chain**. Variable $\overline{P_{join}}$ (and $\overline{P_{leave}}$) is *True* with probability P_{join} (P_{leave}), and *False* otherwise.

s-bot can connect with another by grasping the connection ring with its gripper, and it can receive connections on more than two thirds of its perimeter. The chassis can be rotated in any horizontal direction. This allows *s-bots*, once assembled into a physical entity, to move in a common direction. A 2-D traction sensor, mounted between the *s-bot's* turret and the chassis, measures the mismatch between the direction in which the chassis is trying to move and the direction in which the connected group is trying to move. For the purpose of communication, the *s-bot* is equipped with an omni-directional camera, four microphones, eight RGB LEDs, and two loudspeakers. For a comprehensive description of the *s-bot's* hardware, see [16].

2.2   Controller

The controller consists of a collection of basic behaviours that are implemented using either the motor schema paradigm, neural networks, or simple hand written commands. A comprehensive description of the individual behaviours is available in [17,18,19,20]. Following the behaviour-based approach [21] we could comfortably merge all basic behaviours into a common framework, which is illustrated by a state diagram in Fig. 2.

Fig. 3 shows a sequence of images taken from the experiment. The *s-bots* are initially located at random positions. If an *s-bot* does neither perceive a (visually connected) chain of *s-bots* nor the nest, it performs a random walk (state **search chain**). An *s-bot* that finds a chain or the nest follows the perimeter of the encountered structure (state **explore chain**). The nest can be considered as the root of all chains. When the *s-bot* reaches the tail of a chain, it will join the chain with probability P_{join} per time step (state **join chain**). *S-bots* that are part of a chain do not leave it unless they are situated at the chain's tail, in which case they leave it with probability P_{leave} per time step. The process of

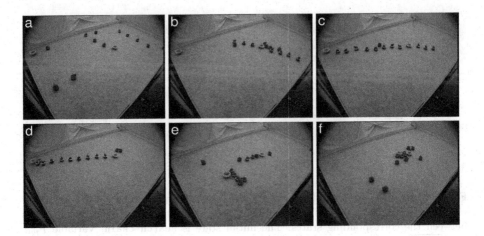

Fig. 3. Sequence of images taken for the trial with group size $N = 12$ *s-bots* and distance $D = 240$ cm between the nest (blue cylindrical object) and the prey (red cylindrical object). This trial last 15 minutes. For a detailed description, see Section 3.

probabilistically joining/leaving a chain is at the basis of the exploration of the environment as it allows the formation of new chains in unexplored areas.

If a chain member perceives the prey it does not leave the chain; thereby the chain becomes stable. If the prey is still far, other *s-bots* can still join to extend the chain in the direction of the prey; otherwise a path connecting the nest to the prey has been formed that can be traversed in both directions. Once a path is formed, it is maintained and in this way automatically recruits other *s-bots* to assemble to the prey (state `assemble`). *S-bots* that do not succeed self-assembling within a hard-coded time period, move back to the nest to rest for a while (state `recovery`). Once a sufficient number of *s-bots* has assembled to the prey, the transport effectively starts; the *s-bots* pull the prey towards the tail of the chain (state `transport target`). In the event that some *s-bots* can not perceive the path, they use their force sensors to estimate the direction of transport (state `transport blind`). When the prey reaches the tail of the chain, the corresponding *s-bot* leaves the chain and moves back to the nest to rest for a while (state `recovery`). In this way the transporting *s-bots* are guided from node to node of the dissolving chain to eventually reach the nest. An *s-bot* leaving the chain to rest at the nest emits a sound signal for a period of 30 s. Transporting *s-bots* respond to this signal by temporarily suspending the transport. This gives the chain *s-bot* sufficient time to move away.

2.3   Experimental Setup

The experiments take place in a bounded arena of size 500 cm × 300 cm. The nest is positioned in the centre of the arena. The prey is put at distance D away from the nest towards one of the four corners. N *s-bots* are positioned on a grid

Table 1. Left: number of *s-bots* required to accomplish sub-tasks *path formation* (N_p), *recruitment* (N_r) and *retrieval* (N_t) for different initial distances (D in cm) between the nest and the prey. Right: overall level of success achieved for setups (N, D): no success (0), sub-task *path formation* accomplished (1), sub-task *recruitment* accomplished (2), and sub-task *retrieval* accomplished (3). Entries in parentheses denote setups that were not tested as the number of *s-bots* N is clearly not sufficient to solve the task. Grey levels of cells represent the best achievable level of success: white denotes no success, light grey denotes success level 1, and dark grey denotes success level 3.

D	60	90	120	150	180	210	240
N_p	1	2	3	4	6	7	8
N_r	3	4	5	6	8	9	10
N_t	3	4	5	6	8	9	10

D / N	1	2	3	4	5	6	7	8	10	12
60	1	1	3	3	3	3	3	3	3	3
90	0	1	1	3	3	3	3	3	3	3
120	0	0	1	1	3	3	2	3	3	2
150	0	0	0	1	1	1	3	2	3	3
180	(0)	(0)	0	0	1	1	1	3	3	3
210	(0)	(0)	(0)	0	0	1	1	0	2	2
240	(0)	(0)	(0)	(0)	(0)	0	0	0	3	3

composed of 60 points uniformly distributed in the arena. The initial position of each *s-bot* is assigned randomly by uniformly sampling without replacement. An *s-bot*'s initial orientation is chosen randomly from a set of 12 possible directions.

We study a wide range of experimental setups, with group sizes $N = 1, 2, 3, 4, 5, 6, 7, 8, 10$ and 12, and distances (in cm) $D = 60, 90, 120, 150, 180, 210$ and 240. For each of these 70 setups we conduct a single trial.

The number of *s-bots* required to form a path connecting the prey with the nest depends on the initial distance between the two objects. To calculate lower bounds for the number of *s-bots*, we assume the *s-bots* to be organised in a single chain that is perfectly linear and directed towards the prey. Then, the lower bound values are computed based on the programmed (and measured) distances between adjacent *s-bots* (27 cm) and between the first chain member and the nest (30.5 cm), as well as the programmed (and measured) maximum distance of the last chain member from the prey (38.5 cm). For the accomplishment of the overall task, two additional *s-bots* are required (at the same time) to engage in transport. The lower bound values so computed are shown in Table 1 (left).

3 Results

Table 1 (right) gives an overview of the results. In 46 out of the 70 setups sub-task path formation can in principle be accomplished (see light grey cells). In 44 out of the corresponding 46 trials the *s-bots* succeeded in forming a path. For setups (N, D) = (5, 180) and (6, 210) a path was formed even though the number of *s-bots* was thought to be insufficient. A path of five (six) *s-bots* has a maximum predicted length of 177 cm (204 cm), which is 3 cm (6 cm) less than the distance that needs to be covered, and therefore still within the range of perceptual error of the *s-bots*' cameras. In 33 out of the 46 setups, also sub-tasks *recruitment* and *retrieval* can in principle be accomplished (see dark grey cells in Table 1 (right)). In 27 out of these 33 setups, the *s-bot* group was able to do so, thereby the entire task was completed.

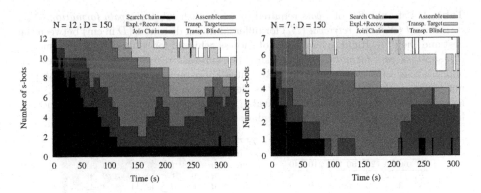

Fig. 4. State dynamics observed in trials for two setups (N, D). The respective grey levels indicate the number of *s-bots* in states `search chain`, `explore chain` and `recovery`, `join chain`, `assemble`, `transport target` and `transport blind`.

Fig. 4 shows the state dynamics for two trials that are discussed below.

- $(N, D) = (12, 150)$: all *s-bots* start in state `search chain`. Once the nest has been found, they aggregate into chains. At $t \approx 80\,s$, a path to the prey consisting of five chain members is established. Even though a path to the prey is formed, other *s-bots* that find the nest self-organise into an additional chain. The formation of the path is not explicitly communicated among the *s-bots* of the group. However, as the *s-bots* in the newly formed chain leave this chain with a constant positive probability, after some time only the chain forming the path remains. At time $t \approx 130\,s$ a first *s-bot* is recruited and grasps the prey, joined by a second *s-bot* about 15 s later. While the prey is transported towards the nest, the chain gradually dissolves. During the transport, additional *s-bots* try to assemble with the pulling structure. Two of them succeed, whereas others fail because the pulling structure is in motion. By looking at the state diagram in Fig. 4 (left), one can see that some of the *s-bots* engaged in transport are not capable of perceiving the path (see white area). Thus, we have an example where the *s-bots* exhibit a hierarchy of teamwork: the group of *s-bots* that cannot perceive the path needs to interact with the group of *s-bots* that can perceive the path; thereby these groups form a team. This team, which is composed of all transport *s-bots*, can be considered a higher-order entity. It forms part of another team which includes another higher-order entity—the group of *s-bots* maintaining or decomposing the path. This nested structure is illustrated in Fig. 1 (left).
- $(N, D) = (7, 150)$: at time $t \approx 30\,s$ a path between nest and prey is already established. At time $t \approx 100\,s$, two *s-bots* have been recruited and are assembled with the prey. The five remaining *s-bots* are aggregated in the chain forming the path. During the transport, chain members disaggregate once in the immediate vicinity of the prey, and follow the path back

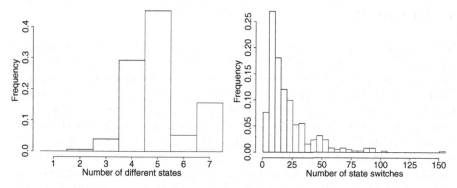

Fig. 5. Left: number of distinct behavioural roles an *s-bot* performed during a trial. Data from all *s-bots* and all trials. Right: number of times an *s-bot* changed its behavioural role during a trial. Data from all *s-bots* and all trials.

to the nest to rest. After some time, the very same *s-bots* resume activity, follow the path, and eventually two of them assemble with the pulling structure and participate in transport. This is an example of how the composition of teams can adapt to changes in the workload of the underlying sub-tasks.

Fig. 3a–f show a sequence of images taken during the trial with group size $N = 12$ and distance $D = 240$ (in cm). During the path formation phase, two chains are formed concurrently (b), and it takes several rearrangements of the chains until a path is formed. This path consists of a chain of eight *s-bots* (c). Shortly thereafter, two *s-bots* get recruited and assemble with the prey (d). During retrieval, most of the *s-bots* of the pulling structure loose sight of the path, which is gradually dissolving, and the prey is moved in the wrong direction (e). However, the path gets re-established by a new *s-bot* extending the chain in the direction to the prey. As a consequence, the transport resumes and can be completed (f). This is an example of a situation in which teamwork among higher-order entities (such as teams or groups) requires a participating entity to adapt its configuration to unexpected environmental circumstances.

Fig. 5 (left) shows the number of distinct behavioural roles (i.e., states) individual *s-bots* performed during the experiment. In 75% of the cases, an *s-bot* performed either four or five of the seven roles. This suggests that the *s-bots* are indeed inter-changeable. Only in 4% of the cases, an *s-bot* performed less than four behaviours during the trial. In 15.7% of the cases, an *s-bot* performed all seven behaviours.

Fig. 5 (right) shows the number of times an *s-bot* changed its behavioural role during the trials of our experiments. The most frequently observed number of changes in behaviour belongs to the six to ten changes range. Note, however, that both mean and median number of changes are higher than this range of values (20.9 and 14.5, respectively).

4   Discussion

In this paper, we have presented an experimental study in which a colony of autonomous robots has to solve a complex foraging task. The task requires a range of sub-tasks to be performed including (i) exploration of the environment, (ii) formation of a path between a prey and a nest, (iii) recruitment of nest mates to the prey, (iv) self-assembly into pulling structures, and (v) group transport of the prey back to the nest. Due to the limited abilities of the robots, the accomplishment of the task requires the concurrent activity of at least i robots ($i \in \{3, 4, 5, 6, 8, 9, 10\}$), where i depends on the experimental setting. Moreover, the accomplishment of the task requires division of labour, in other words, the robots need to perform different sub-tasks concurrently. Such constraints are typically not considered in other studies of group foraging systems (e.g., [22,23,24,25,26,20]), which often make use of some form of global perception or communication, and which often do not require a complex division of labour. In general, we believe that the investigated problem provides a framework that captures the essence of a variety of problems that are addressed at the collective level in social insect colonies.

Inspired by the behaviour of the natural counterparts, we developed a relatively simple, decentralised control algorithm. Although most of our primitive behaviours as well as the overall framework, a finite state machine, were manually designed, we believe that similar types of rules can result from natural or artificial evolutionary processes. In this respect, such systems could be adaptive to changes in the environment.

A series of experimental results from systematic trials with up to twelve physical robots confirm the efficacy of the system. In almost all of the trials where the group size is sufficient to accomplish the overall task, the group succeeded in retrieving the prey to the nest. Video recordings from the experiments are available at http://iridia.ulb.ac.be/supp/IridiaSupp2008-008.

The colony displayed a self-organised and dynamically changing hierarchy of teamwork in which collaboration took also place among high-order entities including groups and teams. The higher-order entities (including the entire system) proved surprisingly robust with respect to the inaccurate and sometimes malfunctioning behaviour of their component modules—parts of a robot such as the tracks, entire robots, and even groups of robots broke down or exhibited unexpected behaviour.

We believe that these experiments are among the most sophisticated examples of self-organisation in robotics to date. The study confirms in a new way that complex forms of division of labour can indeed result from the interactions of individuals that follow relatively simple and local rules. The study also demonstrates that teamwork requires neither individual recognition (the robots we use are inter-changeable) nor inter-individual differences (the robots we use are homogeneous in terms of "morphology" and "brain"), and as such might contribute to the ongoing debate on the role of such characteristics for the division of labour in social insects.

Acknowledgement

This work was supported by the Sixth Framework Programme of the European Community in the form of the IST FET project "SWARM-BOTS" (grant no. IST-2000-31010) and of a Marie Curie Intra-European Fellowship (contract no. MEIF-CT-2006-040312), and by the Scientific Research Directorate of the French Community of Belgium in the form of the "ANTS" project, an "Action de Recherche Concertée". It reflects only the authors' views. The European Community is not liable for any use that may be made of the information. Marco Dorigo acknowledges support from the Belgian F.R.S.–FNRS, of which he is a research director. The authors thank Nigel R. Franks for stimulating discussions that helped in the preparation of the manuscript.

References

1. Bonabeau, E., Dorigo, M., Theraulaz, G.: Swarm Intelligence: From Natural to Artificial Systems. Oxford Univ. Press, New York (1999)
2. Garnier, S., Gautrais, J., Theraulaz, G.: The biological principles of swarm intelligence. Swarm Intelligence 1, 3–31 (2007)
3. Hölldobler, B., Wilson, E.O.: The Ants. Harvard Univ. Press, Cambridge (1990)
4. Anderson, C., McShea, D.W.: Intermediate-level parts in insect societies: Adaptive structures that ants build away from the nest. Insectes Soc. 48, 291–301 (2001)
5. Wilson, E.O.: Sociobiology. Harvard Univ. Press, Cambridge (1975)
6. Oster, G.F., Wilson, E.O.: Caste and ecology in the social insects. Princeton Univ. Press, Princeton (1978)
7. Anderson, C., Franks, N.R.: Teamwork in ants, robots and humans. Adv. Stud. Behav. 33, 1–48 (2004)
8. Tibbetts, E.A.: Visual signals of individual identity in the wasp *Polistes fuscatus*. Proc. R. Soc. Lond. B 269, 1423–1428 (2002)
9. Franks, N.R.: Teams in social insects: Group retrieval of prey by army ants (*Eciton burchelli*, Hymenoptera: Formicidae). Behav. Ecol. Sociobiol. 18, 425–429 (1986)
10. Beshers, S.N., Fewell, J.H.: Models of division of labor in social insects. Annu. Rev. Entomol. 46, 413–440 (2001)
11. Anderson, C., Franks, N.R.: Teams in animal societies. Behav. Ecol. 12, 534–540 (2001)
12. Jeanne, R.L.: The evolution of the organization of work in social insects. Monit. Zool. Ital. 20, 119–133 (1986)
13. Ratnieks, F.L.W., Anderson, C.: Task partitioning in insect societies. Insectes Soc. 46, 95–108 (1999)
14. Anderson, C., Ratnieks, F.L.W.: Task partitioning in insect societies: Novel situations. Insectes Soc. 47, 198–199 (2000)
15. Anderson, C., McMillan, E.: Of ants and men: Self-organized teams in human and insect organizations. Emergence 5, 29–41 (2003)
16. Mondada, F., Gambardella, L.M., Floreano, D., Nolfi, S., Deneubourg, J.-L., Dorigo, M.: The cooperation of swarm-bots: Physical interactions in collective robotics. IEEE Robot. Autom. Mag. 12, 21–28 (2005)
17. Groß, R., Bonani, M., Mondada, F., Dorigo, M.: Autonomous self-assembly in swarm-bots. IEEE Trans. Robot. 22, 1115–1130 (2006)

18. Groß, R., Mondada, F., Dorigo, M.: Transport of an object by six pre-attached robots interacting via physical links. In: Proc. 2006 IEEE Int. Conf. Robot. Autom., pp. 1317–1323. IEEE Comp. Soc. Press, Los Alamitos (2006)
19. Nouyan, S., Campo, A., Dorigo, M.: Path formation in a robot swarm: Self-organized strategies to find your way home. Swarm Intelligence 2 (2008)
20. Tuci, E., Groß, R., Trianni, V., Bonani, M., Mondada, F., Dorigo, M.: Cooperation through self-assembling in multi-robot systems. ACM Trans. on Autonomous and Adaptive Systems 1, 115–150 (2006)
21. Arkin, R.: Behavior-Based Robotics. MIT Press, Cambridge (1998)
22. Kube, C.R., Zhang, H.: Collective robotics: From social insects to robots. Adapt. Behav. 2, 189–218 (1993)
23. Werger, B., Matarić, M.: Robotic food chains: Externalization of state and program for minimal-agent foraging. In: Maes, P., Matarić, M.J., Meyer, J.A., Pollack, J., Wilson, S.W. (eds.) From Animals to Animats 4, Proc. of the 4th Int. Conf. on Simulation of Adaptive Behavior, pp. 625–634. MIT Press, Cambridge (1996)
24. Krieger, M.J.B., Billeter, J.B., Keller, L.: Ant-like task allocation and recruitment in cooperative robots. Nature 406, 992–995 (2000)
25. Kube, C.R., Bonabeau, E.: Cooperative transport by ants and robots. Robot. Auton. Syst. 30, 85–101 (2000)
26. Yamada, S., Saito, J.: Adaptive action selection without explicit communication for multirobot box-pushing. IEEE Trans. Syst., Man, Cybern. C 31, 398–404 (2001)

Social Control of Herd Animals by Integration of Artificially Controlled Congeners

Nikolaus Correll, Mac Schwager, and Daniela Rus

Computer Science and Artificial Intelligence Laboratory
Massachusetts Institute of Technology
Cambridge, MA 02139

Abstract. We study *social* control of a cow herd in which some of the animals are controlled by a sensing and actuation device mounted on the cow. The control is social in that it aims at exploiting the existing gregarious behavior of the animals, rather than controlling each individual directly. As a case study we consider the open-loop control of the herd's position using location-dependent stimuli. We propose a hybrid dynamical model for capturing the dynamics of the animals during periods of grazing and periods of stress. We assume that stress can either be induced by the sensing and actuation device or by social amplification due to observing/overhearing nearby stressed congeners. The dynamics of the grazing part of the proposed model have been calibrated using experimental data from 10 free-ranging cows, and various assumptions on the animal behavior under stress are investigated by a parameter sweep on the hybrid model. Results show that the gregarious behavior of the animals must be increased during stress for control by undirected stimuli to be successful. We also show that the presence of social amplification of stress allows for robust, low-stress control by controlling only a fraction of the herd.

1 Introduction

We wish to study the potential of low-stress managing a cow herd by exploiting the cows's innate gregarious behavior using a small number of controlled animals. The idea of controlling herd animals using robotic agents [1] or by devices mounted on the animals ("smart collars") [2,3,4,5] bears great potential for revolutionizing animal husbandry. Cows seem to be particularly well suited for this endeavor as keeping them in their natural environment is labor intensive and costly.

Recently, the idea of controlling groups of animals by integrating artificially controlled agents into the animal society and leveraging the natural effects of gregarious behavior has been brought forward in [6]. In [6], cockroaches were presented with a dark and a bright shelter in a circular arena, and usually aggregate under the dark shelter, which they prefer. Miniature robots [7] were then impregnated with cockroach pheromones and integrated into the cockroach swarm. Unlike the cockroaches, the robots were programmed with a preference for the bright shelter. As shelter selection was shown to be a social decision, the robots could thus bias the choice of the cockroaches.

M. Asada et al. (Eds.): SAB 2008, LNAI 5040, pp. 437–446, 2008.

Similarly for husbandry, we wish to identify social behavior in farm animals that can be exploited by integration of a small number of controlled animals into the herd, instead of controlling each individual separately. In this paper, we approach this goal for a cow herd by using a hybrid model of the herd dynamics. The model has been developed based on physical principles and calibrated using a system identification process based on data collected from 10 cows on the United States Department of Agriculture, Agricultural Research Services (USDA-ARS) experimental ranch in Las Cruces, New Mexico, USA [8].

We hypothesize that two social effects will enable the control of the herd. The first social effect is that when an animal is stressed, its gregarious tendencies are increased, thus drawing it to the center of the herd. The second social effect (which is backed-up by observations in [9]) is that a stressed animal may pass on its stressed state to its neighbors. Stress induced by a device mounted on the animal, e.g. due to an aural stimulus, does not only alter the behavior of the animal itself, but also affects nearby animals, which see or hear a stressed congener. This mechanism bears the potential for amplifying the effect of external stimuli provided to only a few cows. In this paper we extend the model from [8] to include these two social effects. The increased agreggation tendency of stressed animals is parameterized by a constant factor, and the propagation of stress from one animal to another is modeled by a radius of detection and a time constant allowing for temporal stress decay. The resulting dynamical model is a hybrid system with two modes: stressed and grazing.

Using extensive simulations of the hybrid dynamical model, we study the potential for social herding by employing a mobile virtual fence [9], which induces stress in those animals that are outside the fenced area and wear sensing and actuation devices. The mechanism that we rely on to control the herd is quite different from those described in previous virtual fencing studies, however. We do not give the animal any directional information from the cue itself. That is to say, when an animal is cued in our model, it does not know where it is "supposed" to go. It simply reacts in a stressed manner, which causes it to draw closer to the center of the herd while propagating its stressed state to its neighbors. In contrast, previous approaches have used graduated cuing intensity or stereo cuing to give directional information to animals [4]. Directional cuing may be useful, especially for animals that have been trained to interpret the directional cues. However, it has been noted in human subjects that it is difficult to detect meaningful directional information from such directional cuing [9]. The results of our simulation study imply that a herd can still be managed effectively with a control system that gives no directional information by leveraging the animals' natural gregarious instincts. Our results show both the potential and the conditions for social control of a cow herd. Furthermore, we provide an agenda for further field experiments, which are required to identify the parameters of the extended model proposed in this paper and validate our approach.

2   Data-Collection Experiments

Cows were equipped with a small light-weight box for data collection during field experiments [8], see Figure 1. The box contains electronics for recording the GPS location of the animal as well as other sensor data (e.g. position of the head, body orientation)

Fig. 1. The sensing and actuation box [8] is roughly 21.5cm×12.0cm×5.5cm and weighs around 1kg. It is equipped with a GPS receiver, wireless networking features, and a suit of sensing and actuation capabilities. The Lithium-Ion batteries and solar panel allow for indefinite operation under normal conditions. The sensor box is mounted to the head of a cow with a leather strap designed to use the cow's ears to keep the box in position. Pictures courtesy of I. Vasilescu.

and environmental data (e.g. temperature). The box also contains electronics for networking the herd. The sensing box was used for collecting trajectory data from 10 free-ranging mature cows (Hereford and Hereford × Brangus genetics) in a 466ha area on the USDA-ARS's Jornada Experimental Range (JER). This site has an undulating topography of predominantly sandy soil, populated with grasses and shrubs. Trajectory data were collected over 3 days at a rate of 1Hz. Parts from these data were then used for calibrating the parameters for the model used in this paper. The calibration process as well as the animal experiments are described in more detail in [8]. For future experiments, the box provides a two-tier animal control system consisting of a set of speakers for applying arbitrary, differential sound stimuli and a set of electrodes that enable the application of differential electric shock. The animal control system was not used during the collection of the data used in this paper.

3   Model

In [8] we developed a linear-in-parameter model that provides good qualitative and quantitative agreement with various individual and collective metrics when used for simulating grazing cows. In this paper, we introduce a hybrid model, whose single discrete state variable corresponds to the behavioral modes grazing and stressed. The dynamics in each behavioral mode only differ by an increased motivation to aggregate and increased speed of the animal. The dynamics in both grazing and stressed behavioral modes are governed by two naturally distinct mechanisms. First, each agent is given internal dynamics to enforce the constrains of Newton's laws. Second, a force is applied to each agent from its interaction with each of the other agents in the group. All remaining effects are modeled as a white noise process.

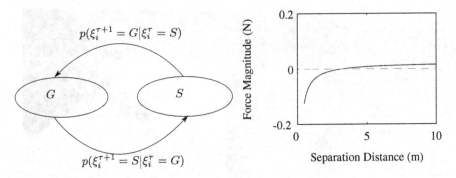

Fig. 2. *Left*: Probabilistic Finite State Machine modeling an agent's behavioral mode and its transitions. Each behavioral mode has its own linear-in-parameter dynamics. *Right*: The magnitude of the agent-to-agent interaction force for $\theta_1 = .0225$, $\theta_2 = .0732$ and $\alpha = 1$, which corresponds to the average value measured between 10 cows.

3.1 Individual Agent Dynamics

Given a group of m agents, every individual is modeled by a hybrid model consisting of a continuous dynamical part, describing position and velocity of an individual, and a discrete part modeling the individual's behavioral mode. This model is illustrated by the Finite State Machine depicted in Figure 2, left.

We distinguish between two different behavioral modes, which determine the dynamics of the agent. Agent i's behavioral mode at any time τ is given by $\xi_i^\tau \in \{G, S\}$, which corresponds to grazing ($\xi_i^\tau = G$) or stressed behavior ($\xi_i^\tau = S$). We define the behavior mode transition rules as

$$p(\xi_i^{\tau+1} = S | \xi_i^\tau = G) = \begin{cases} 1 & \text{if } \mathcal{N}_{S,i}^\tau(R) > \mathcal{N}_{G,i}^\tau(R) \vee u_i^\tau = 1, \\ 0 & \text{otherwise} \end{cases} \tag{1}$$

where $\mathcal{N}_{S,i}^\tau(R)$ and $\mathcal{N}_{G,i}^\tau(R)$ are the number of agents in behavioral mode $\xi_i^\tau = S$ or $\xi_i^\tau = G$, respectively, within a radius of R around agent i. Also, $u_i^\tau \in \{0, 1\}$ is the binary control input to the cow, so that $u_i^\tau = 0$ corresponds to no control stimuli at time τ, while $u_i^\tau = 1$ when the animal is receiving a stimuli at time τ. Similarly, we define

$$p(\xi_i^{\tau+1} = G | \xi_i^\tau = S) = \frac{1}{T} \tag{2}$$

where T is a time constant modeling stress decay.

The dynamics of agent $i \in \{1, \ldots, m\}$ can be written in state-space, difference equation form as

$$x_i^{\tau+1} = \begin{bmatrix} 1 & 0 & \Delta t & 0 \\ 0 & 1 & 0 & \Delta t \\ 0 & 0 & a_i & 0 \\ 0 & 0 & 0 & a_i \end{bmatrix} x_i^\tau + \begin{bmatrix} 0 & 0 \\ 0 & 0 \\ 1 & 0 \\ 0 & 1 \end{bmatrix} \left(\frac{1}{m} \sum_{j=1, j \neq i}^m f_{ij}(p_i^\tau, p_j^\tau, \xi_i^\tau) + w_i^\tau(\xi_i^\tau) \right) \tag{3}$$

The state of agent i is given by the vector x_i^τ, consisting of its East position, North position, Eastern component of velocity, Northern component of velocity. An agent's position is thus given by, $p_i^\tau = [e_i^\tau \; n_i^\tau]^T$. The time step Δt is given by $t^{\tau+1} - t^\tau$, and we assume it is constant for all τ. The term a_i represents damping, $a_i = 1$ for zero damping, and $|a_i| < 1$ for stable systems. The function $f_{ij}(p_i^\tau, p_j^\tau, \xi_i^\tau)$ determines the coupling force applied by agent j to agent i as a function of the agent's behavioral mode ξ_i^τ. Finally, $w_i^\tau(\xi_i^\tau)$ is one of two zero-mean, stationary, Gaussian white noise processes used to model the unpredictable decision-motive processes of agent i. Which of the two processes that is applied to the model at a given time is determined by the behavioral mode, ξ_i^τ. The two white noise signals are distinguished by different covariance matrices, and each is uncorrelated with $p_j \; \forall j$. Nonholonomic constraints of the cows are neglected in this treatment, though they could be incorporated with an increase in the complexity of the model structure. Note that the force terms are only applied to affect changes in velocity in accordance with Newton's second law.

3.2   Agent-to-Agent Interaction Force

Dropping the τ superscripts for clarity, the form of the agent coupling force $f_{ij}(p_i, p_j, \xi_i)$ is given by

$$f_{ij}(p_i, p_j, \xi_i) = \left(\alpha(\xi_i)\theta_1 - \frac{\theta_2}{\|p_j - p_i\|} \right) n_{ij}, \qquad (4)$$

where $n_{ij} = (p_j - p_i)/\|p_j - p_i\|$ is the unit vector along the line from p_i to p_j (henceforth, $\| \cdot \|$ will denote the ℓ^2 norm).

The parameter $\alpha(\xi_i)$ is used to model the animal's tendency to aggregate as a function of its behavioral mode. During grazing $\alpha(\xi_i = G) = 1$, which corresponds to nominal behavior. When the animal is stressed, $\alpha(\xi_i = S) > 1$, corresponds to an increased attraction to neighbors. In this work we investigate, among other things, the effects of different values of $\alpha(\xi_i = S)$. Whereas in [8] we calibrated parameters θ_1 and θ_2 specific to each cow pair, we use values for θ_1 and θ_2 averaged over all cow pairs in this paper, which allows us to extrapolate the models to larger herds. For illustration, (4) given by $\|f_{ij}\| = \alpha(\xi_i)\theta_1 - \theta_2/\|p_j - p_i\|$ is shown in the right of Figure 2.

4   Experimental Setup

We are interested in the impact of the number of controlled animals on the performance of an open-loop control algorithm that moves the herd from an initial position to a defined final goal position by using the concept of a virtual fence [5,9]. In all our experiments, the fence is modeled by a circular disc of 25m diameter. The fence coordinates and their time evolution are assumed to be known to the sensor box. If the cow leaves the fenced area and is wearing an actuation device, it will be given a stimulus.

At its initial position the center is at $(0m, -50m)$. The center of the fence then moves with speed $v_f = 20\frac{m}{h}$ northwards. After $5h$ simulated time the experiment is stopped. We then measure the *average* number of individuals within the fence over the whole experiment as well as the final number of individuals within the fence at time $5h$.

Table 1. Parameters used for simulating the system defined by Equations 3 and 4

Δt	$= 1s$	a_i	$= .9294$	$w_i(\xi_i = G) =$	$\begin{pmatrix} 1.22e-2 & -9e-4 \\ -9e-4 & 1.46e-2 \end{pmatrix}$
θ_1	$= .0225$	θ_2	$= .0732$	$w_i(\xi_i = S) =$	$\begin{pmatrix} 2.44e-2 & -18e-4 \\ -18e-4 & 2.92e-2 \end{pmatrix}$

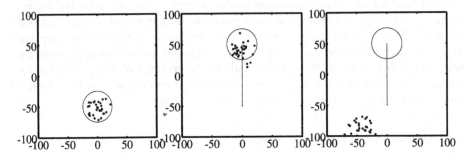

Fig. 3. *Left*: Random initial deployment of 30 cows and virtual fence configuration. *Middle*: Example of a final configuration after successful control (all animals are controlled). The herd oscillates around the center of the fenced area; red dots (cows outside the fence) denote cows being stressed. *Right*: Example of unsuccessful control (no animals were controlled). The herd does a random walk around the center of its initial deployment area.

Figure 3 illustrates the experimental setup, showing a random initial distribution (left), a successful final configuration (middle), and an unsuccessful final configuration (right). All simulations have been conducted with 30 cows. The covariance of the driving noise, w_i^T, during stress was set to be twice as high as during grazing, which roughly corresponds to observations from [9]. The time constant for stress decay has been set to $T = 30s$. All parameter values being used are summarized in Table 1.

Simulations are performed using the MATLAB Distributed Computing Toolbox on the CSAIL computational cluster. The ratio between simulated and computational time is roughly 15 on a modern 64bit workstation.

5   Results

As we are unaware of the effective amplification of the aggregation force α as well as the existence and specific values of the radius within which cows are socially affected by the behavioral mode of neighboring cows, we perform a parameter sweep over $\alpha(\xi_i = S) = \{1, 2, 4\}$ and $R = \{0, 5, 10\}$ for 1–10, 15, 20, 25 and 30 controlled cows.

Figure 4, left, shows the average number of individuals within the fence over the whole experiment for $\alpha = 2$, i.e. the attraction force between cows is twice as high during stress when compared with grazing, and different values of R for 50 simulations per data point (2250 simulations in total). Figure 4, right, shows the ratio of experiments where more than 15 cows where within the fence after 5h. These results confirm that

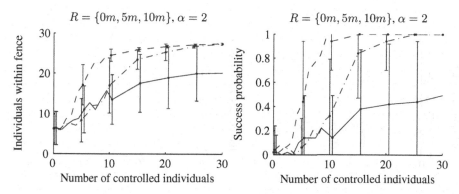

Fig. 4. *Left*: Average number of individuals within the fenced area during $5h$ of simulated time for $R = 0, 5, 10$ (– .,– –,— / blue, green, red) with $\alpha(\xi_i = S) = 2$ vs. the number of controlled individuals. Without assuming propagation of stress ($R = 0$) we observe a significant increase in performance for 10 controlled individuals. *Right*: Ratio of successful to unsuccessful simulation runs (success is defined as $> 50\%$ of the animals within the fence at the final fence location). 50 experiments per data-point, error bars are standard deviation.

manipulating 10 of 30 cows is sufficient for moving to a desired location using a virtual fence, assuming that the aggregation forces are twice as large during stress as during aggregation and that social amplification exists for a radius of $R = 5m$. For $R = 10m$, however, the controller performance decreases. This can be explained by the fact that, for a large radius, the whole herd tends to be in the same mode, inhibiting the controlling action of the virtual fence. For example, if most animals are unstressed, a large number of unstressed neighbors are always within the radius of any given cow, thereby outnumbering the stressed neighbors and inhibiting the propagation of stress. Conversely, if there is a large number of stressed animals, it is likely that the whole herd becomes stressed and the virtual fence is rendered ineffective as crossing the fenceline will make no difference as the animals are already stressed. We refer to this effect as a "stampede" since it bears an obvious resemblance to that phenomenon in natural herds.

We were then interested in testing the influence of the presumed increase in gregarious behavior for stressed animals. We tested two extreme cases: $\alpha(\xi_i = S) = 1$, so that stressed animals experience no greater attraction to their neighbors than grazing animals, and $\alpha(\xi_i = S) = 4$, so that stressed animals experience four times the attraction to their neighbors. Results for various values of R are shown in Fig.5 for both of these cases. We clearly see that $\alpha(\xi_i = S) > 1$ is a necessary condition for *social* herding, and is independent of the number of artificially modified individuals, as well as potential social amplification. We also observe that social amplification seems to become less important for high values of $\alpha(\xi_i = S)$.

We are also interested in the impact of the time constant of stress T and the size of the driving noise covariance matrix. We therefore ran simulations for $\alpha(\xi_i = S) = 2$ and $R = 5$ for $T = \{1s, 30s, 60s\}$ and multiplied the covariance matrix of the noise applied to the agents by 1, 2 or 4. Results for various T are shown in Fig. 6, left. We observe that stress indeed needs to be maintained for some time after the stimulus as

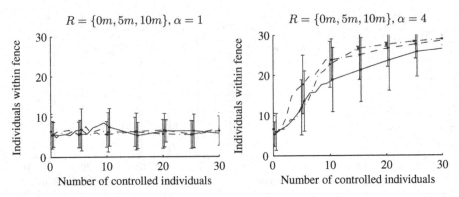

Fig. 5. Average number of individuals within the fenced area for the same experimental conditions as in Figure 4 but with $\alpha(\xi_i = S) = 1$ and $\alpha(\xi_i = S) = 4$ (left and right plot). For $\alpha(\xi_i = S) = 1$, herd control fails (*left*). 50 experiments per data-point, error bars are standard deviation.

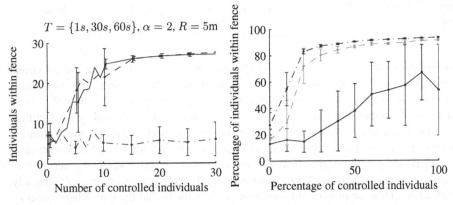

Fig. 6. *Left:* Average number of individuals within the fenced area for the same experimental conditions as in Figure 4, but for $T = \{1, 30, 60\}$ (– .,– –,— / blue, green, red). *Right:* Percentage of individuals within fence vs. percentage of controlled individuals for team sizes of 10, 30, and 100 animals (—,– –,-. / red, green, blue). 10 simulations per data point.

$T = 1$ yields poor performance. Although we did not explicitly test larger values than 60s for T, we conjecture that high values of T will eventually lead to all animals being in the stressed mode, which will cause the fence to be ineffectual. With respect to the driving noise covariance, we do not observe any significant difference in performance for the values that we tried.

Finally, we are interested how social control scales and experimented with the ratio of manipulated animals in team sizes of 10, 30 and 100 animals (Figure 6, right). Results show that modifying only a part of the herd becomes increasinlgy efficient for larger herds, and that controlling as little as 20% of the animals might be sufficient for the fence geometry and speed being chosen. Whereas performance for 30 and 100 animals is similar, control of herds with only 10 animals seems to be generally more difficult.

6 Discussion

Our results show the potential for social control of a cow herd given that a) the gregarious behavior increases under stress, and b) stress propagates through the herd by mutual observation. While further field experiments will allow us to better understand the differences in the proposed behavioral modes, we argue that the gregarious behavior might be stimulated to the necessary extent by providing *directional* impulses to the cows, e.g. using the sound system available on our sensor box. In this case, results from Fig. 5 suggest that high levels of gregarious behavior, be it innate to the species or artificially stimulated, can achieve the same level of performance (in terms of the number of modified individuals required) as a system with weak gregarious behavior but social amplification of stress.

In this paper, our model assumes that attraction and repulsion forces are the same for every neighbor. Results from [8] and observations on feral cattle [10] suggest, however, the existence of social preferences among the cows. Depending on the strength of such preferences, modifying a specific subset of the herd might increase controller performance. While kinship is usually well known in a domestic herd, further studies might reveal a relation between phenomenological properties and social leadership. It would then seem beneficial to apply stimuli to known social leaders. In [9] also spatio-temporal preferences within the herd (e.g. shadow seeking behavior or specific habits) are observed. Although such preferences might jeopardize an open-loop control approach, they might also be exploited for more effective control by planning trajectories, which are easily followed by the herd.

7 Conclusion

We extended the dynamical model for a cow herd presented in [8] by a hybrid structure, which differentiates between a grazing and a stressed behavioral mode. Using this model, we show that open-loop control of the cow herd using virtual fences is possible if the gregarious behavior is sufficiently stronger during stress than during grazing. Control of the herd exploits the natural gregarious behavior of the animal and does not require any learning of stimulus/action patterns. Moreover, we showed that when stress is propagated to neighbors (social amplification), robust control can be achieved by endowing only a fraction of animals with sensing and actuation.

Although we did not calibrate parameters for the animal behavior during stress, systematic simulations of our model suggest that driving noise covariance and stress decay-time are of little importance for the performance of social control, whereas an increase in gregarious behavior and stress propagation within the herd seem to be of utmost importance. In the future, we plan to artificially induce stress using aural signals of varying strength. We would then like to quantitatively validate the existence of distinct behavioral modes (grazing and stressed) and quantify its parameters. Specifically, we are interested in the increase in gregarious behavior and the mechanisms of social amplification. Given such an enhanced model for this particular species, we would then like to generalize the methodology to other gregarious animals and develop closed-loop control schemes for robust, low-stress control of farm animals.

Acknowledgments

The authors would like to thank Dean Anderson, Carrick Detweiler, and Iuliu Vasilescu for providing the animal data. N. Correll is sponsored by a grant from the Swiss National Science Foundation under contract number PBEL2–118737. Also, support for this research has been provided in part by NSF EFRI 0710252, SEI 0513755, ITR 0426838, and ARL SWARMS MURI. We are grateful for it.

References

1. Vaughan, R., Sumpter, N., Henderson, J., Frost, A., Cameron, S.: Robot control of animal flocks. In: Proceedings of the 1998 IEEE International Symposium on Intelligent Control (ISIC), pp. 277–282. IEEE Press, Piscataway (1998)
2. Quigley, T., Sanderson, H., Tiedemann, A., McInnis, M.: Livestock control with electrical and audio stimulation, June 1990. Rangelands (1990)
3. Tiedemann, A., Quiqley, T., White, L., Lauritzen, W., Thomas, J., McInnis, M.: Electronic (fenceless) control of livestock, United States Department of Agriculture, Forest Service, Tech. Rep. PNW-RP-510 (January 1999)
4. Anderson, D., Hale, C.: Animal control system using global positioning and instrumental animal conditioning, US Deparment of Agriculture, Tech. Rep. US Patent 6,232,880 (May 2001)
5. Anderson, D.: Virtual fencing - past, present and future. The Rangeland Journal 29, 65–78 (2007)
6. Halloy, J., Amé, J.-M., Detrain, G.S.C., Caprari, G., Asadpour, M., Correll, N., Martinoli, A., Mondada, F., Siegwart, R., Deneubourg, J.-L.: Social integration of robots in groups of cockroaches to control self-organized choice. Science 318(5853), 1155–1158 (2007)
7. Caprari, G., Colot, A., Siegwart, R., Halloy, J., Deneubourg, J.-L.: Building mixed societies of animals and robots. IEEE Robotics & Automation Magazine 12(2), 58–65 (2005)
8. Schwager, M., Detweiler, C., Vasilescu, I., Anderson, D., Rus, D.: Data-driven identification of group dynamics for motion prediction and control. Journal of Field Robotics (to appear, 2008)
9. Butler, Z., Corke, P., Peterson, R., Rus, D.: From robots to animals: Virtual fences for controlling cattle. I. J. Robotic Res. 25(5-6), 485–508 (2006)
10. Lazo, A.: Social segregation and the maintenance of social stability in a feral cattle population. Animal Behavior 48, 1133–1141 (1994)

Aggregating Robots Compute: An Adaptive Heuristic for the Euclidean Steiner Tree Problem

Heiko Hamann and Heinz Wörn

Institute for Process Control and Robotics,
Universität Karlsruhe (TH), 76131 Karlsruhe, Germany

Abstract. It is becoming state-of-the-art to form large-scale multi-agent systems or artificial swarms showing adaptive behavior by constructing high numbers of cooperating, embodied, mobile agents (robots). For the sake of space- and cost-efficiency such robots are typically miniaturized and equipped with only few sensors and actuators resulting in rather simple devices. In order to overcome these constraints, bio-inspired concepts of self-organization and emergent properties are applied. Thus, accuracy is usually not a trait of such systems, but robustness and fault tolerance are. It turns out that they are applicable to even hard problems and reliably deliver approximated solutions. Based on these principles we present a heuristic for the Euclidean Steiner tree problem which is NP-hard. Basically, it is the problem of connecting objects in a plane efficiently. The proposed system is investigated from two different viewpoints: computationally and behaviorally. While the performance is, as expected, clearly suboptimal but still reasonably well, the system is adaptive and robust.

1 Introduction

With the increase of interdisciplinary research new concepts were developed in the last decades. For example, swarm intelligence [2] applied to computational problems leads to powerful meta-heuristics [4] and applied to robotics it results in large-scale distributed robotic systems [16]. In this paper we try to pursue these approaches and to combine swarm intelligence with robotics and heuristics.

The scientific approach to computation was significantly governed by the computational devices used in the past. Thus, it began with sequential devices, later parallel machines with shared memory were studied, and even later the focus was on distributed but fully connected systems. All these approaches have determinism and explicit communication in common.

A new philosophy is introduced by applying concepts of swarm intelligence, e.g. simple local rules, indirect communication (stigmergy), and cooperation [2]. We can consider swarm intelligence as the final step of the process of getting away from centralized and deterministic systems towards fully distributed and probabilistic systems. Swarm intelligence was applied to computational problems by using software agents. We provide our computational devices with actuators making them mobile. Hence, they become real, embodied agents in the form of

M. Asada et al. (Eds.): SAB 2008, LNAI 5040, pp. 447–456, 2008.

robots. The idea of using a group of autonomous agents as processing elements, that are embedded in the environment, that sense and compute based only on local information was published by Payton et al. [14] and propagated as 'world-embedded computation'. This is related to an old question, whether an ant colony's struggle of survival might be viewed as computation or not [9]. Thus, we note that problem solving by adaptive and cooperative behaviors might be considered computation.

Combining the local and randomized approach of swarm intelligence with an emphasis on positional information, results in an interesting computing paradigm or also in a method of generating emergent behavior. The position of a robot in the physical world becomes the building block of collective information processing [8]. In a recent work Litus et al. [13] give a good summary of this idea:

> The key insight that underlies our methods is that the physical locations of the robots themselves could be considered as an approximate solution to the entire problem. An individual robot can move itself, thus refining the current solution approximation. No representation of the problem, or the current solution, needs to be held by any robot: they manifest the solution by their physical configuration.

Although these approaches suffer in principle from the same problems of intractability as classical approaches concerning hard problems, they might lead, nevertheless, to more efficient implementations. Such systems might be cheaper than classical devices and easier to maintain due to less complexity.

We focus on the Euclidean Steiner tree (EST) problem which is basically the problem of connecting objects in a plane efficiently. The agent-based heuristic, that we investigated here, was shortly introduced in [8]. The objects in the plane are connected by placing mobile relay stations, that we call robots in the following, instead of using wires. Starting with a uniformly distributed population of robots they aggregate in a way similar to diffusion-limited aggregation (DLA) [17]. Unlike DLA all robots are always moving and turn to avoid collisions between two moving robots. The objects to be connected serve as seeds from which trees of aggregated robots 'grow'. The use of such 'random trees' for planning in robotics was introduced in [12]. In contrast to our approach these random trees are only virtual and are centrally controlled. As we are growing several trees or clusters at the same time this approach is also connected to diffusion-limited cluster-cluster aggregation [11]. However, the clusters in the present work are static.

In the following section we define the EST problem and give a short survey of the related work. In section 3 we present a robot control algorithm generating a collective problem solving system for the EST problem. Thereafter, some results and validations are given in section 4 which are discussed in section 5.

2   The Euclidean Steiner Tree Problem

The EST problem is named after the swiss mathematician Jakob Steiner and is defined as follows: A given set Z of N points or terminals in a plane has to be connected by lines of minimal length and, in contrast to the related minimal spanning tree problem, it is allowed to add a set of extra points S, called *Steiner points* (for an example see Fig. 2(d)). The resulting network is a graph $G = (V, E)$ with nodes $V = Z \cup S$ and edges E accordingly defined. The probably better known instance of the Steiner tree problem class is defined on graphs where Steiner points can be picked from a finite set of points instead of placing them anywhere in a plane.

There are many applications to this problem in circuit design, mining, network design, and routing in ad hoc networks. Computing an optimal EST is NP-hard, i.e. no efficient algorithm is known and is unlikely to be found, and the discretized variant is NP-complete [5]. It is even hard to find an approximation within 95/94 of the optimum [3]. A lot of work has been done to find both better exact algorithms as well as polynomial time heuristics [10,15]. The best known heuristic was presented in [19] usually yielding solutions close to the optimum within a few seconds at least for $N < 1000$. A software to compute exact EST is the GeoSteiner package [18].

An amusing anecdote reported by Aaronson [1] gives a reason to suspect that anyway not too much intelligence might be of need for approximations of lower quality: "Yet a well-known piece of computer science folklore maintains that, if two glass plates with pegs between them are dipped into soapy water, then the soap bubbles will rapidly form a Steiner tree connecting the pegs, this being the minimum-energy configuration." Aaronson even experimented with real soap bubbles. He observed correct solutions but also cycles, incomplete trees, entirely different trees for the same configuration, and a relaxation process of several seconds leading to better soap bubble configurations. He follows that soap bubbles do not solve NP-complete problems in polynomial time. However, we note that (very) alternative approaches might be quite productive, too.

3 Growing Random Trees

3.1 Preconditions

We restrict ourselves to a two-dimensional setting but the proposed algorithm would work in the same way in three dimensions. We assume objects called *seeds* being placed in a bounded plane. Each seed represents a terminal out of Z of a considered Steiner tree problem. Furthermore we need some kind of robots that can move in the given environment. Whether these robots drive, crawl, fly, swim, or submerge is not relevant as long as they are able to move and to remain with sufficient accuracy at one spot. They have to be equipped with sensors allowing them to perceive other robots and seeds within a very short range compared to the dimensions of the bounded plane. Furthermore, they should be able to communicate to and measure an approximate bearing of other robots in their neighborhood. Their control mechanism suffice to be reactive. A large group of such mobile robots is positioned uniformly distributed over the whole plane. This could be done by the swarm autonomously in a previous phase.

3.2   Algorithm

In the following we explain the control algorithm of the robots. See Fig. 1 for a schematic overview. In general the robots move forward, try to find seeds, and listen for pings by robots being already connected to a seed. When a robot finds a seed (not shown in the schematic) and does not receive any pings it stops next to it. The robot generates a tree identification number (tree ID), that is with high probability unique, for example, by using a big random number. Then it starts to ping. When another robot receives this signal it checks back with the sender, if there is still an open slot (only 3 connections per robot are allowed). If so it maximizes the angles between itself, the sender, and the sender's neighbors. In case of 1 neighbor it forms a straight line. In case of 2 neighbors they try to reach a configuration with 120°, but only if this requires the movement of no more than 2 robots (local optimization). This can be achieved by some communication overhead and relying on relative angles only (see [7] for details). This restriction of 3 connections per robot and the consequent angle of 120° is not arbitrarily chosen, as the optimal solution of a Steiner tree problem always consists of Steiner points of degree 3 and angles of 120° only. Additionally, we want to cover as much space in the plane as possible with a minimal number of stopped robots. The optimal solution to this tiling problem is provably the hexagon as it is found in honey-combs [6]. However, the robots form only partial hexagons because the result should be a tree, i.e. a cycle-free graph.

Using this control algorithm the robots perform a process similar to diffusion-limited aggregation [17]. Provided with a sufficient number of robots and time a tree will grow at each seed. At some time, 2 trees will be connected. This is the case if a robot approaches 2 aggregated robots of different trees virtually at the same time. By communicating their tree ID they ensure not to form a loop and agree upon a new tree ID, which is propagated through the new tree. Later, more trees will join. When only one huge tree is left, cf. Fig. 2(a), or at an assigned time (reduction condition), a new process is started: All robots being connected to only one other robot, i.e. they are leafs, will cut this connection and leave. In a chain reaction all unnecessary robots cut their connections and a tree consisting of a relatively small number of robots is left, cf. Fig. 2(b). After this reduction robots being connected to 3 other robots represent Steiner points. By straightening the connections between the seeds and the Steiner points, i.e. releasing surplus robots in between (for details see [7]), we get a first approxima-tion to the optimal Steiner tree, cf. Fig. 2(c). Note, the positions of the Steiner points were determined dynamically during the tree growth process. Interpreting this in the sense of swarm intelligence: the Steiner point set S is the result of a collective decision that emerges from the numerous agent-agent interactions. An additional improvement is achieved by locally rearranging the Steiner points to-wards their optimal position (an approximation to the Fermat-Torricelli point). This is achieved by moving the robots at the Steiner point towards the direc-tion of the smallest angle. This might result in the optimal solution as shown in Fig. 2(d) but in general the optimal configuration of the Steiner points is only achievable by global optimization.

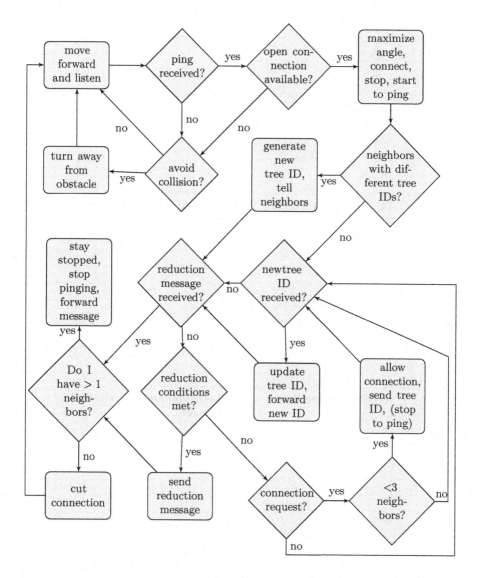

Fig. 1. Schema of the robot controller for the random tree growth algorithm

Keeping a reasonable amount of redundant robots in the lines this heuristic is robust to breakdowns of single robots although it might seem very inefficient. In addition, it is scalable because of its totally local approach. Whether this method can be a fast way of approximating a Steiner tree is a question of the reaction times and speed of the robots. At the time mass production of such devices will become possible, this scenario might actually be feasible [16].

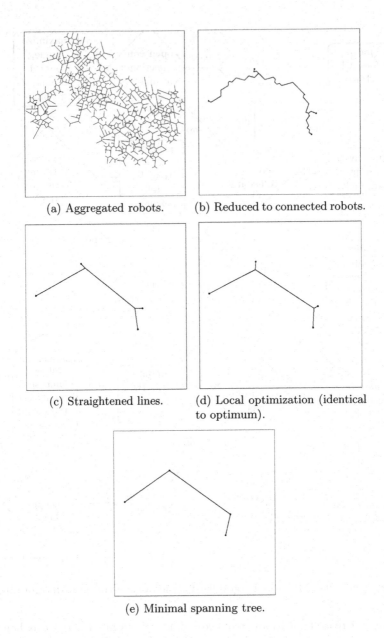

(a) Aggregated robots.   (b) Reduced to connected robots.

(c) Straightened lines.   (d) Local optimization (identical to optimum).

(e) Minimal spanning tree.

Fig. 2. Phases of the heuristic and minimal spanning tree

4   Results

The following results were obtained using a first-order geometric simulation with continuous space. Our emphasis is on the general behavior of the agent system

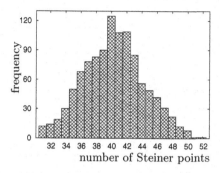

Fig. 3. Histogram of Steiner points generated in 550 samples by the heuristic for a problem instance of 100 terminals showing a mean of about 40.2, optimal are 38 Steiner points, and maximally possible are 98.

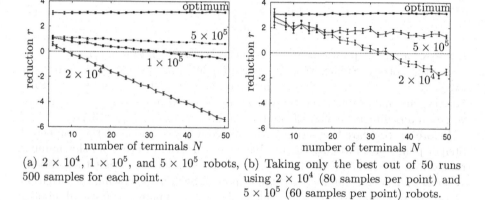

(a) 2×10^4, 1×10^5, and 5×10^5 robots, 500 samples for each point.

(b) Taking only the best out of 50 runs using 2×10^4 (80 samples per point) and 5×10^5 (60 samples per point) robots.

Fig. 4. Comparing the reduction of the heuristic for different robot numbers and varied problem size N to the optimal solution; error bars are 95% confidence intervals

which we claim to be covered by this kind of simulation. A more complex simulation would have multiplied the computational complexity. However, already this abstract simulation kept a personal computer busy for days computing thousands of runs using 10^5 and more robots. As we do not yet have such quantities of robots or other computational devices available we had to simulate our massively parallel heuristic serially.

We compare the results of our heuristic to the optimal solution and the minimal spanning tree—the typical benchmark problem for Steiner tree heuristics. For this purpose we compare the reduction in percent r of the (suboptimal) Steiner tree length L_steiner to the minimal spanning tree length L_spanning

$$r = \frac{L_\text{spanning} - L_\text{steiner}}{L_\text{spanning}} \cdot 100\% \tag{1}$$

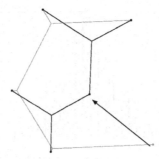

Fig. 5. Superimposed optimal solutions for the beginning set of terminals (gray) and the final set of the adaptivity test scenario. The arrow indicates the rearranged terminal.

with

$$L = \sum_{(\mu,\nu)\in E} \|\mu - \nu\|_2,$$ (2)

$\mu, \nu \in \mathcal{R}^2$ are node positions and $\|\cdot\|_2$ denotes the Euclidean norm.

For the optimal Steiner tree this value ranges from instance to instance between $r_{\min} = 0\%$ (minimal Steiner and spanning tree identical) and $r_{\max} > 10\%$. For the minimal Steiner tree the reduction averaged over many instances converges almost independently of the terminal number N to $\bar{r}_{\mathrm{opt}} \approx 3.1\%$.

Before we compare the actual performance we have a look at the number of Steiner points generated by the random tree heuristic. In principal this number ranges from 0 (spanning tree) to $N - 2$. However, it turns out that the heuristic generates reasonable numbers of Steiner points. See Fig. 3 showing the situation of a single problem instance with a mean that overestimates the optimal number of Steiner points by 5.8%.

Now we compare our heuristic to the exact solution focusing on the governing parameter, the robot number. We omit a time analysis since the time consumption can be kept constant with increasing robot number (due to strictly local actions of each robot). Only the serialized simulation of the heuristic suffered from the complexity of high robot numbers. The solutions become better the more robots are used as shown in Fig. 4(a). The reduction decreases linearly in problem size. Due to the probabilistic characteristic of the proposed heuristic the average performance is improved by repeated runs. In Fig. 4(b) we show the performance achieved by selecting the best solution out of 50 runs. Especially for bigger instances $N > 50$ the performance could be furthermore improved by increasing the number of robots.

While our heuristic is inferior to the state-of-the-art heuristics in the performance we identify its advantage in its adaptivity and due to the decentralized approach also in its robustness. We test the adaptivity by replacing a terminal after 40 time steps. The terminal configuration shown in Fig. 5 and 10^4 robots were used. Only 8.2% were irregular approximations (not all terminals connected) evaluating 500 samples. The average reduction dropped by 42% compared to the heuristic started with the final terminal configuration.

5   Discussion and Conclusion

The heuristic proposed in this paper is definitely not superior to the state-of-the-art heuristics concerning quality and computing time (the best known heuristic typically delivers approximations within about 4% from optimum [19]). However, it shows a reasonable degree of adaptivity. Furthermore, it is supposed to be quite robust as there is no single point of failure. Comparing the fault tolerance of our method to the classical approach corresponds to answering the question: How wrong can wrong be? One might argue that this comparison is unfair because the difference is only due to different output methods: explicit and physical (motion, positions, angles) compared to symbolic (numbers, calculations). However, this actually is the fundamental difference in the method of information processing between these approaches. The classical computer processes abstract symbols while the agent system processes physical positions. Therefore, we consider a comparison to be fair and the difference in the wrongness is big. A single error might cause almost infinitely high deviations using a symbol-based approach as there are virtually no limitations for what could happen to a symbol in the CPU or the memory, e.g. a single bit shift might lead to negative distances. This is in contrast to our agent system, where a single error might break two subtrees but arbitrary deviations are impossible. This is one advantage of the strictly bounded operating range of the robots, limiting not only their possibilities but also the consequences of errors. An analysis of the proposed algorithm would obviously be very hard due to its properties that might be considered 'emergent'. For example, the growth of a random tree is dependent on its relative position to other trees. Any model describing the tree growth independently would have little explanatory power. Including all trees into the model would increase its complexity significantly.

Acknowledgments

Hamann is supported by the German Research Foundation (DFG) within the Research Training Group GRK 1194 Self-organizing Sensor-Actuator Networks.

References

1. Aaronson, S.: NP-complete problems and physical reality. ACM SIGACT News 36(1), 30–52 (2005)
2. Bonabeau, E., Dorigo, M., Theraulaz, G.: Swarm Intelligence: From Natural to Artificial Systems. Oxford Univ. Press, Oxford (1999)
3. Chlebík, M., Chlebíková, J.: Approximation hardness of the Steiner Tree problem on graphs. In: Penttonen, M., Schmidt, E.M. (eds.) SWAT 2002. LNCS, vol. 2368, pp. 170–179. Springer, Heidelberg (2002)

4. Dorigo, M., Caro, G.D.: Ant Colony Optimization: A new meta-heuristic. In: Angeline, P.J., Michalewicz, Z., Schoenauer, M., Yao, X., Zalzala, A. (eds.) Proceedings of the 1999 Congress on Evolutionary Computation (CEC 1999), Piscataway, NJ, pp. 1470–1477. IEEE Press, Los Alamitos (1999)
5. Garey, M.R., Graham, R.L., Johnson, D.S.: Some NP-complete geometric problems. In: Annual ACM Symp. on Theory of Computing, pp. 10–22 (1976)
6. Hales, T.C.: The honeycomb conjecture. Discrete and Computational Geometry 25(1), 1–22 (2001)
7. Hamann, H.: Modeling and investigation of robot swarms. Master's thesis, University of Stuttgart, Germany (2006)
8. Hamann, H., Wörn, H.: Embodied computation. Parallel Processing Letters 17(3), 287–298 (2007)
9. Hofstadter, D.R.: Gödel, Escher, Bach. Basic Books (1979)
10. Hwang, F.K., Richards, D.S., Winter, P.: The Steiner Tree Problem. North-Holland, Amsterdam (1992)
11. Kolb, M., Herrmann, H.J.: The sol-gel transition modelled by irreversible aggregation of clusters. J. Physics A 18(8), L435–L441 (1985)
12. LaValle, S.M., Kuffner, J.J.: Rapidly-exploring random trees: Progress and prospects. In: Donald, B.R., Lynch, K.M., Rus, D. (eds.) Algorithmic and Computational Robotics, Wellesley, MA, USA, pp. 293–308. A. K. Peters (2001)
13. Litus, Y., Zebrowski, P., Vaughan, R.: Energy-efficient multi-robot rendezvous: Parallel solutions by embodied approximation. In: Workshop on Algorithmic equivalencies between biological and robotic swarms, Atlanta, USA (June 2007)
14. Payton, D., Daily, M., Estowski, R., Howard, M., Lee, C.: Pheromone robotics. Autonomous Robots 11(3), 319–324 (2001)
15. Robins, G., Zelikovsky, A.: Improved Steiner Tree approximation in graphs. In: 11th ACM-SIAM Symposium on Discrete Algorithms, pp. 770–779 (2000)
16. Seyfried, J., Szymanski, M., Bender, N., Estaña, R., Thiel, M., Wörn, H.: The I-SWARM project. In: Şahin, E., Spears, W.M. (eds.) Swarm Robotics Workshop, pp. 70–83. Springer, Heidelberg (2005)
17. Tom, J., Witten, A., Sander, L.M.: Diffusion-limited aggregation, a kinetic critical phenomenon. Phys. Rev. Lett. 19, 1400–1403 (1981)
18. Warme, D., Winter, P., Zachariasen, M.: Geosteiner homepage, http://www.diku.dk/geosteiner/
19. Zachariasen, M., Winter, P.: Concatenation-based greedy heuristics for the Steiner tree problem in the Euclidean plane. Algorithmica 25, 418–437 (1999)

Emergence of Interaction among Adaptive Agents

Georg Martius[1,2,3], Stefano Nolfi[4], and J. Michael Herrmann[1,2,5]

[1] Bernstein Center for Computational Neuroscience Göttingen
Bunsenstraße 10, 37073 Göttingen, Germany
[2] Institute for Nonlinear Dynamics, University of Göttingen
Bunsenstraße 10, 37073 Göttingen, Germany
[3] Max Planck Institute for Dynamics and Self-Organization
Bunsenstraße 10, 37073 Göttingen, Germany
[4] Institute of Cognitive Sciences and Technologies,
LARRAL Via S. Martino della Battaglia 44, 00185, Roma, Italy
[5] School of Informatics, Institute for Perception, Action and Behavior
Edinburgh University, 11 Crichton Street, Edinburgh EH8, Scotland, U.K.
georg@nld.ds.mpg.de, stefano.nolfi@istc.cnr.it, mherrman@inf.ed.ac.uk

Abstract. Robotic agents can self-organize their interaction with the environment by an adaptive "homeokinetic" controller that simultaneously maximizes sensitivity of the behavior and predictability of sensory inputs. Based on previous work with single robots, we study the interaction of two homeokinetic agents. We show that this paradigm also produces quasi-social interactions among artificial agents. The results suggest that homeokinetic learning generates social behavior only in the the context of an actual encounter of the interaction partner while this does not happen for an identical stimulus pattern that is only replayed. This is in agreement with earlier experiments with human subjects.

1 Introduction

The concept of self-organization describes the formation of specific structures in the presence of unspecific driving forces. It is considered to be relevant for the ontogenesis of living beings, the generation of behavior in autonomous robots, and the emergence of higher functions in agents. We have shown [1,2], that interesting behaviors can be generated in robotic agents [3] by a self-organizing controller that follows the cybernetic principle of *homeokinesis*[1] [4]. It combines the maximization of sensitivity with respect to external stimuli and the avoidance of unpredictable behaviors. Driven by this principle, an agent becomes engaged in a vivid interaction with its environment, it starts to move autonomously and escapes from blockage as well as from unpredictable i.e. quasi-random situations. In this way the agent shows a preference for states where the control actions are effective. If the environment also contains other agents then the adaptation capabilities of the agents may play a role in the emergence of communication [5].

---

[1] Greek: homoios (equally, likewise) and kinesis (movement), meaning the adaptive control of a kinetic quantity and thereby self-regulate internal parameters.

M. Asada et al. (Eds.): SAB 2008, LNAI 5040, pp. 457–466, 2008.

Whether or not the adaptivity is indeed sufficient to enable the communication will be the main question of this contribution. Although the interaction is contingent, i.e. may be negotiated to any level from no interaction to strongly correlated behavior, it turns out that the homeokinetic principle induces a bias toward intense interaction which is counterbalanced only by the self-generated instability of the agents. We will explore the characteristics of the behavior exhibited by two homeokinetic agents that interact in a shared environment, where direct internal representations of other agents are not possible. Moreover, internal states in the agents will turn out to be unnecessary. The experiments reported here are based on a variation of the perceptual crossing scenarios introduced and studied on human subject by Auvray et al. [6] in which two human subjects can move an object left or right along a single dimension and can perceive through a tactile sensor an object corresponding to the other agent or to a "shadow" image of the other agent. The interesting aspect of this scenario is that human subjects display a good ability to discriminate between sensing and non-sensing objects (corresponding to the other agent or to its shadow) which move exactly in the same way, without the need of a specific training [6]. This can be explained by considering the way in which the agents react to the perception or to the lack of perception of the other agent. The same type of ability is demonstrated by the ability of young children to spontaneously discriminate between a video showing the behavior of their mother interacting live with the child with respect to a video showing a pre-recorded interaction [7]. This discrimination skill has been reproduced in evolutionary robotics experiments where agents have been selected for the ability to discriminate between real agents and insensitive "shadows" [8] or between interacting agents and their exact behavioral replay [9]. From these experiments it has been concluded that the detection of social contingency does not require complex cognitive skills [8].

We will demonstrate how an ability to discriminate social contingency arises spontaneously in self-organizing agents which have been rewarded for paying attention to objects independently on whether objects correspond to agents or to their shadowed images. Indeed, as soon as agents develop a preference for any tactile perception, they tend to display a preference for situations which lead to a bi-directional sensation. This tendency is due to the characteristics of the homeokinetic paradigm which maximize the sensitivity with respect to external stimuli while minimizing unpredictability. These results suggest that agency detection might result indirectly from the entrainment of the behavior of the two agents which emerge spontaneously (i.e. even in absence of a direct reward) in self-organizing homeokinetic agents.

The next section introduces homeokinetic learning. In Sections 3 and 4 we describe result from our experiments which are discussed in Section 5.

2   Self-organized Control

Based on the concept of homeokinesis [4] and the concept of self-organization, we developed in previous studies a controller [4,1,2] that establishes interesting sensorimotor couplings in a closed loop setup. This is achieved by the simultaneous maximization of the sensitivity with respect to sensory stimuli and the maximization of the predictability of future inputs. Both predictability and sensitivity are defined with respect to an adaptive internal representation of the sensorimotor loop, cf. Fig. 1. From the difference of the sensor readings and the model estimates an energy function is obtained that is used to modify the parameters of the controller by gradient descent. The internal model is simultaneously improved. Details can be found in the appendix.

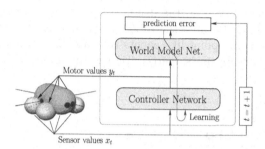

Fig. 1. Schematic view of the self-organizing controller attached to a wheeled robot

The capabilities of homeokinetic control become obvious by considering some examples that were realized on various robotic platforms [3]. The "rocking stamper" [2] consists of an inverse pendulum mounted on a bowl-like trunk. It exhibits different rocking modes, preferably at the eigenfrequencies of the system.

A more complex example for the self-organization of *natural* behaviors is provided by a spherical robot [10] which is actuated by three internal massive weights that can be moved along orthogonal axes. After an initial phase, the system prefers to keep one mass fixed at one axis while performing a coordinated movement of the other two such that the robot rotates around the first axis. Thus the robot moves forward like a wheel or sometimes turns on the spot. The behavior is changed every few tens of revolutions by an internal reorganization that occurs even in the absence of external stimuli. Furthermore, high-dimensional systems such as snake- or chain-like robots, quadrupeds, and wheeled robots [2] have been successfully controlled, where it is of particular interest that the control algorithm induces a preference for movements with a high degree of coordination among the various degrees of freedom. All the robotic implementations demonstrate the emergence of play-like behavior. However, it is possible to shape the development of behaviors with reinforcement [11].

3   Interaction among Homeokinetic Agents

We use a set-up similar to Refs. [6,12], where two agents are moving along a one-dimensional track with cyclic boundary conditions. Each agent has a copy (shadow) that moves along at a fixed distance, cf. Fig. 2 (left). The agents in our experiments have an independent parameter dynamics and are initialized randomly in order to avoid any implicit knowledge about the partner.

The controllers remain adaptive during the entire experiment, i.e. there is no training phase. The behaviors are acquired on the fly rather than being acquired during a previous evolutionary process. Each of the agents are equipped with a tactile sensor, see Fig. 2 (right), that is activated in the same way when being close to either the other agent or its copy. In contrast to Refs. [6,12], we use a sensor with a continuous characteristic, which fits better into the paradigm of dynamical systems and does not require internal states in the agents. The sensor values depends only on the distance between the centers of the robots and objects, cf. Fig. 2. Because the robots move in a one-dimensional world, it is reasonable to assume point-like agents that are, however, assumed to obey realistic physical laws for mass, inertia, friction, and motoric forces.

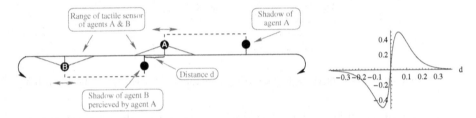

Fig. 2. (*left*) Schematic setup with two agents and the respective shadows.(*right*) Activation function of the tactile sensor with a sensor range of $R = 0.3$.

The motor forces affect the agent by the following dynamical equations

$$a_{t+1} = (y_t - \mu v_t)/m \qquad a\text{: acceleration, } \mu\text{: friction, } m\text{: mass} \qquad (1)$$
$$v_{t+1} = a_{t+1}\Delta t + v_t \qquad v\text{: velocity, } \Delta t\text{: time step} \qquad (2)$$
$$p_{t+1} = \text{Env}(v_{t+1}\Delta t + p_t) \qquad p\text{: position} \qquad (3)$$

where y is the force produced by the motor and $\text{Env}(\cdot)$ is a function that maps the position into the cyclic environment $[-1, 1)$. The shadow has the fixed offset position $s_t = \text{Env}(p_t + 0.6)$.

The motor value y is the controller output and hence depends on the sensory inputs obtained by a velocity sensor $x_1 = v_t$ and a tactile sensor x_2. The latter responses to the nearest object (either agent or shadow) at position o as

$$x_2 = \frac{e}{2}\gamma d\, e^{-|\gamma d|} \qquad d = \text{Env}(p_t - o), \ \gamma = 6/R \qquad (4)$$

All sensor values are subject to a weak noise. The term "tactile" is understood as in Ref. [12], practically it could be realized by a light sensor. The sensor should however be spatially localized or provide a good signal at ranges where the robot moves in a few time steps. We have tested tactile sensor characteristics different from (4) without obtaining qualitatively different results. The function that connects the sensory inputs to the motor output y is adaptive and is updated every simulation step, cf. Appendix. For all experiments the following parameters were used: $\epsilon = 0.1$ (Eqs. 10,11), $\Delta t = 0.01$, $m = 0.1$, $\mu = 1.5$, $R = 0.3$.

Above a certain distance from the other agent the agent experiences hardly any sensory errors and increases thus its sensitivity to sensory inputs, especially to the velocity sensor which leads to high speeds. This and the fact that the presence of the other agent causes large prediction errors make extended interactions with the other agent unlikely. In order to achieve reasonable search times, we applied a reinforcement scheme [11] that introduces a preference for both the other robot and its shadow in the same way. The agents receive a reward for states with an activated tactile sensor

$$r(t) = \begin{cases} 2 & |d| < R \\ 0 & \text{otherwise} \end{cases} \tag{5}$$

The objective function E (9) of the self-organizing controller is modulated by the reward,

$$E_r = (1 - \tanh(r(t)))E, \tag{6}$$

which may be interpreted as a learning rate modulation. Small learning rates correspond to high rewards and vice versa, such that the agent develops a tendency to stay in the rewarded areas, i.e. interact with objects in the range of the tactile sensor.

Fig. 3 displays the time that agent A received tactile input from agent B, agent B's shadow, and neither of both, respectively. The values are normalized with respect to their prior frequencies, which are given by the relative area where to encounter the events (agent and shadow have size 0.6 each, which leaves 0.8 for the free space).

In the case with reward the agent spends almost twice the time interacting with the other agent than with the other's shadow. This can be explained by the same

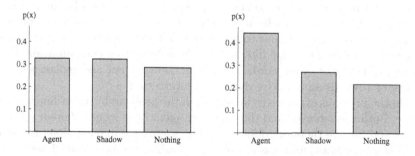

Fig. 3. Normalized interaction time of agent A with agent B, with the shadow of agent B, and with neither of the two, respectively for runs of 60 minutes simulated real time. (*left*) Self-organizing behavior according to the homeokinetic principle without rewards. (*right*) Homeokinetic learning with additional reinforcement of situations where tactile input was received.

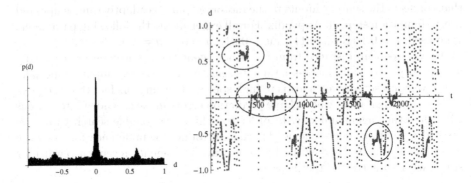

Fig. 4. Interaction of homeokinetic agents with reinforcement. (*left*) Distribution of distances between agent A and agent B. (*right*) Time course of the distance for four minutes (out of 60). (*a*) agent A senses shadow of agent B, (*b*) both agents oscillate around each other and (*c*) agent B senses shadow of agent A.

reasoning that applies to the experiments with human subjects. If both agents see each other then they explore less, i.e. the changes in behavior become smaller, possibly in order to appear more predictable. However, if agent A senses the shadow of agent B then agent B does not receive any tactile input and remain explorative. In Fig. 4 the distribution of distances between both agents and the time course of the distance for a short time is displayed. The histogram of distances shows clearly the preference of a direct interaction between the two agents. Interaction with the shadow results in the the small bumps at -0.6 and 0.6.

4   Agency Detection

In order to study the detection of agency, i.e. the distinction of the live interaction with a partner and a passive replay, we have performed a variant of the above experiment where one of the agents is replaced by an exact replay of an earlier experiment. This set-up has been already studied in an evolutionary approach [9] and is actually more similar to the double TV experiment [7,13]. In Ref. [9] the agents developed a stable interaction, which leads to a compensation of the influence of noise. In other words, the distinction of agent and replay is due to a tendency towards an adaptive dynamical interaction.

As in the previous experiment we approach the problem by an on-line learning scheme. Both the replayed and the exploring agent have again different parameter sets and behave therefore differently. The agents are moving along the same cyclic one-dimensional arena as above. Here no shadows are involved, see Fig. 5.

The agents are rewarded when sensing each other in the same way as above (5). In Fig. 6 the statistics of the behavior of the agents is plotted for both experiments. It is clearly visible that the agent spends more time interacting with the real agent than with the non-reactive replay. This implies that it is easier to synchronize if both agents are adapting.

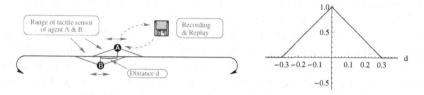

Fig. 5. (*left*) Schematic setup of the experiment with replay; (*right*) Sensor response in dependence of the distance between the agents

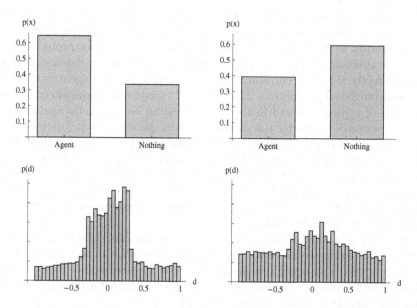

Fig. 6. Statistics of the behavior of the agents for life interaction (*left*) and for replay scenario (*right*). Top row depicts the relative time agent A spends on interacting with agent B and nothing, respectively. In the bottom row the distribution of distances between both agents is plotted.

5   Discussion

We have studied the social interaction of two agents that are capable of self-organizing their behavior. The experiments are inspired by psychological effects found in human subjects [6,7,13] and are intended to investigate the mechanisms of the experience of agency. Phenomenologically the observations are similar to those presented for the case of evolving robots [12,9], where the two agents were assumed to have an identical configuration, while we have used an on-line learning algorithm with random initial conditions. In a sense this is similar to a real life situation where the knowledge about the partners is limited. Due to the quasi-random occurrence of external inputs and the general destabilization of the parameter dynamics (Eq. 10,11) by the homeokinetic learning algorithm, the agent's parameters do not converge.

In the setup with perceptual crossings (Sect. 3) the agents showed a clear distinction of the contingently reacting partner and the non-contingent shadow. This is achieved even though the agents were rewarded for each tactile sensation regardless of whether it was caused by a real agent or its shadow. One reason for the successful distinction is the qualitatively different behavior of an agent interacting with another agent and the agent searching quickly through an environment. These results reproduce the findings in humans subjects [6] as well as in evolving robots [12].

The second experiment (Sect. 4) demonstrates that the self-organizing agents can perform an agency detection merely by self-sufficient behavioral adaptation. The interaction with the reactive agent turned out to be more probable than with the statistically identical replay. It suggests that agency detection might result indirectly from the entrainment of the behavior of the agent within the robot-environmental interaction. This behavior emerges spontaneously i.e. even in absence of a direct reward in the self-organizing homeokinetic agents. It is thus not necessary to specify the distinction as a specific goal, it follows rather from the adaptive explorativity level which requires predictable reactions from the environment, while in the replay case a similar behavior as in the quasi-random case is generated. Overall, this result confirms and extends the evidence summarized above which indicates that detection of social contingency can be properly characterized as a property of coupled dynamical systems that are regulated by simple control rules.

In Ref. [14] we studied a different approach to the problem of self-organization of interaction among agents which was based on an explicit maximization of the learning progress. This scheme, however, relied on the accumulation of previous knowledge and a discrete representation of the environmental information and is thus not a minimal model as required in [9].

Studies of the present kind are relevant to an emerging formal theory of social interaction. Results from numerical approaches essentially show that certain assumptions about the agents complexity are not necessary. However, the precise formulation of the forces in a social interaction cannot be identified by computer simulations. Nevertheless, we can conclude from the current study that the emergence of interaction with contingent agents might by rooted in a fundamental requirement to the sensorimotor loop, namely the maximal integration of the environment into the sensorimotor flow.

Acknowledgments. G.M. thanks the LARAL for their hospitality during his stay in Rome in October 2007. This work was partially supported by a grant (01GQ0432) from the BMBF in the NNCS program and the ECAGENTS project funded by the Future and Emerging Technologies programme (IST-FET) of the European Community under EU R&D contract IST-1940.

References

1. Der, R., Hesse, F., Martius, G.: Learning to feel the physics of a body. In: Proc. CIMCA-IAWTIC 2006, Washington, DC, pp. 252–257. IEEE Computer Society, Los Alamitos (2005)
2. Der, R., Hesse, F., Martius, G.: Rocking stamper and jumping snake from a dynamical system approach to artificial life. Adaptive Behavior 14, 105–115 (2006)
3. Der, R., Martius, G., Hesse, F.: Research webpage (2008), http://robot.informatik.uni-leipzig.de/Videos
4. Der, R., Steinmetz, U., Pasemann, F.: Homeokinesis - a new principle to back up evolution with learning. In: Computational Intelligence for Modelling, Control, and Automation, vol. 55, pp. 43–47. IOS Press, Amsterdam (1999)
5. Nolfi, S.: Emergence of Communication in Embodied Agents: Co-Adapting Communicative and Non-Communicative Behaviours. Connection Science 17, 231–248 (2005)
6. Auvray, M., Lenay, C., Stewart, J.: The attribution of intentionality in a simulated environment: the case of minimalist devices. In: Tenth Meeting of the Association for the Scientific Study of Consciousness, Oxford (2006)
7. Trevarthen, C.: The self born in intersubjectivity: The psychology of an infant communicating. In: Neisser, U. (ed.) The perceived Self, pp. 121–173. Cambridge Univerity Press, Cambridge (1993)
8. Di Paolo, E.A., Rohde, M., Iizuka, H.: Sensitivity to social contingency or stability of interaction? modelling the dynamics of perceptual crossing. New Ideas in Psychology, Special Issue on Dynamics and Psychology (in press, 2008)
9. Iizuka, H., Di Paolo, E.A.: Minimal agency detection of embodied agents. In: Almeida e Costa, F., Rocha, L.M., Costa, E., Harvey, I., Coutinho, A. (eds.) ECAL 2007. LNCS (LNAI), vol. 4648, pp. 485–494. Springer, Heidelberg (2007)
10. Der, R., Martius, G., Hesse, F.: Let it roll – emerging sensorimotor coordination in a spherical robot. In: Rocha, L.M., et al. (eds.) Artificial Life X: 10th Int. Conf. on the Simulation and Synthesis of Living Systems, pp. 192–198. MIT Press, Cambridge (2006)
11. Martius, G., Herrmann, J.M., Der, R.: Guided self-organisation for autonomous robot development. In: Almeida e Costa, F., Rocha, L.M., Costa, E., Harvey, I., Coutinho, A. (eds.) ECAL 2007. LNCS (LNAI), vol. 4648, pp. 766–775. Springer, Heidelberg (2007)
12. Rohde, M., Di Paolo, E.: An evolutionary robotics simulation of human minimal social interaction. In: Nolfi, S., Baldassarre, G., Calabretta, R., Hallam, J.C.T., Marocco, D., Meyer, J.-A., Miglino, O., Parisi, D. (eds.) SAB 2006. LNCS (LNAI), vol. 4095, pp. 485–497. Springer, Heidelberg (2006)
13. Stormark, K.M., Braarud, H.C.: Infants' sensitivity to social contingency. Infant Behavior & Development 27, 195–203 (2004)
14. Herrmann, J.M.: Dynamical systems for predictive control of autonomous robots. Theory in Biosciences 120, 241–252 (2001)

A   Derivation of the Learning Rule

The agents receive a vector of sensor values $x_t \in \mathbb{R}^n$ at time steps $t = 0, 1, 2, \ldots$. The actions of the robot are determined by a controller described by a function K that maps sensor values $x \in \mathbb{R}^n$ onto motor values $y = (y_1, \ldots, y_m)^T \in \mathbb{R}^m$,

$$y_i = K_i(x) = \tanh\left(\sum_j C_{ij}x_j + h_i\right),$$

(7)

where in terms of neural networks C_{ij} denotes the weights and h_i the biases. The robot is further equipped with an adaptive model of its environment that is also realized by a neural network. It approximates a function F that predicts new sensor values based on earlier sensor and motor values via

$$x_t = F(x_{t-1}, y_{t-1}) + \xi_t.$$

(8)

The world model F is realized by a one-layer feed forward neural network. The "noise" term ξ_t in Eq. 8 is the modeling error which is assumed to be of finite variance. Inserting Eq. 7 into 8, we obtain a dynamical system

$$x_t = \psi(x_{t-1}) + \xi_t$$

representing the dynamics of the sensorimotor loop. The internal representation ψ is used to define an error function. Predictability is achieved by minimizing the prediction error ξ. Sensitivity can be expressed by the Jacobian matrix,

$$L_{ij}(x) = \frac{\partial}{\partial x_j}\psi_i(x),$$

which specifies the linear response of the map ψ to a perturbation. High sensitivity requires large values of all eigenvalues of L which can be achieved by the maximization of the smallest eigenvalue of L or equivalently by a minimization of the largest eigenvalue of L^{-1}. The error function is thus defined as

$$E = \left\|L^{-1}\xi\right\|^2 = \xi^T\left(LL^T\right)^{-1}\xi.$$

(9)

The parameters of the controller are updated by gradient descent on E, where

$$L_{ij} = \sum_k \left(\frac{\partial}{\partial y_k}F_i(x,y)\right)\tanh'\left(\sum_l C_{kl}x_l + h_k\right)C_{kj},$$

which leads to the following learning rule for the controller parameters.

$$\varepsilon^{-1}\Delta C_{ij} = \zeta_i v_j - 2\zeta_i \rho_i y_i x_j$$

(10)

$$\varepsilon^{-1}\Delta h_i = \qquad - 2\zeta_i \rho_i y_i$$

(11)

where $\rho = Cv$, $v = L^{-1}\xi$, $\zeta_i = g_i'\mu_i$, $\mu = (\frac{\partial}{\partial y}F(x,y))^T(LL^T)^\dagger\xi$ and $M^\dagger$ denotes the Moore-Penrose inverse of M. Note that the learning rate ε is chosen such that the parameters change at the same time scale as the behavior, sometimes even synchronously. The interplay between synaptic and state dynamics of the controller induces the potential for a high dynamical complexity of the sensorimotor loop.

Acquisition of Human-Robot Interaction Rules via Imitation and Response Observation

Takatsugu Kuriyama[1,2] and Yasuo Kuniyoshi[1,2]

[1] Department of Mechano-Informatics,
Graduate School of Information Science and Technology, The University of Tokyo
7-3-1, Hongo, Bunkyo-ku, Tokyo, 113-8656 Japan
{kuriyama,kuniyosh}@isi.imi.i.u-tokyo.ac.jp
http://www.isi.imi.i.u-tokyo.ac.jp/
[2] JST ERATO Asada Synergistic Intelligence Project

Abstract. We aim to realize human-robot social game interaction as a kind of communication. We proposed a hypothetical development of social game interaction between an infant and a care-giver from a mechanism-sided standpoint, based on developmental psychology. Social games have rules, specific relationship between action and response. Applying the hypothesis, we also propose a scheme to design a robot in which a partner can teach interaction rules through interaction. To investigate the proposed scheme, we built a dynamic model which realizes imitation and ruled interaction and switches them observing partner's response. In the experiment, the partner can teach and the robot can acquire a rule adaptively through interaction without explicit teaching and subsequently it is also achieved about another rule without reset.

Keywords: human-robot communication, interaction rule, imitation, response observation, contingency detection.

1   Introduction

In human-robot communication, it is important to design how a robot should interact with a partner. It is necessary that a robot has an ability to extend communicative behaviors adaptively, because a partner probably feels a sense of alienation and boredom otherwise. For natural communication, it is important for the robot to acquire new communicative behaviors through interaction, as opposed to inhuman one-way programming. To realize this acquisition, we utilize fundamental and general principles because we want to treat various problems with a uniform framework. Here we focus on the principles of *imitation* and *contingency detection*, which are key developmental mechanisms for infants [1].

In human-robot communication, it is believed that an important ability for a robot is to imitate humans [2,3]. Of course, imitation is not enough for communication. In multi-agent robots, it is suggested that not only imitation but also negotiation, which can maximize information transfer, contributes to communication for lexicon learning [4]. In human-robot communication, it is indicated

M. Asada et al. (Eds.): SAB 2008, LNAI 5040, pp. 467–476, 2008.

that responsiveness, which increases the rate of information transfer, improves the impression of communication [5]. A robot can realize *joint attention*, which is a significant communicative behavior, by finding the causality between a human face pattern and an appropriate response [6]. Detecting information transfer and causality can be realized by *contingency detection*.

As stated above, *imitation* and *contingency detection* are expected to play substantial role in robot's acquisition of communicative behavior. We can say that to realize acquisition of communicative behavior using these principles may help understanding development of infant communication.

We aim to realize a human-robot social game interaction as a kind of communication. Researchers have shown that social games have rules and structures [7,8]. We focus on how, through interaction, a robot can learn and a partner can teach an interaction rule which is a relationship between a specific action and the other's corresponding response in a social game. In this paper, we propose a scheme to have a robot acquire interaction rules based on the principles of *imitation* and *contingency detection*. We also build and investigate a system using the proposed scheme, in which a partner can teach interaction rules to a robot.

2   Development of Social Game Interaction

To realize a natural form of human-robot social game interaction, we refer to ideas about the development of infant's social game interaction. If a robotic system is equipped with this developmental mechanism, it should be able to interact with its partner in a natural way.

Development of infant's social game interaction such as "peek-a-boo" and "roll-the-ball" progresses through the four stages described below [7,9]. Infants gradually move from a passive role to an active one in the course of the game.

1. **passive observation** — The infant merely observes the care-giver passively.
2. **taking part in one of the game's elements** — The infant takes part in one of game's elements and eventually grows to initiate more of elements.
3. **sharing of the game's activities** — Each player takes a turn in a well-organized fashion based on the convention of the game.
4. **generating modifications** — The infant generates variations within the rules of the game. The infant has a sufficient understanding of the game's rule structure to be able to add new rules.

3   For Designing Social Game Robot

3.1   Development from Mechanism-Sided Standpoint

Considering *imitation* and *contingency detection*, which are general developmental mechanisms in infants (as we noted at section 1), and based on the four developmental stages of social game interaction (we have noted at section 2), we propose a hypothesis about developmental stages from a mechanism-sided standpoint (Fig. 1). Of course, these stages are continuous and the interaction may regress to a previous stage.

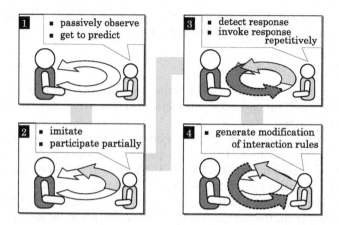

Fig. 1. A Hypothesis about Developmental Stages of Social Game Interaction from Mechanism-sided Standpoint

1. **passive observation** — The infant passively observe the care-giver and get to predict the care-giver's action and the game flow.
2. **taking part in one of the game's elements** — The infant just imitates the care-giver's actions. Thereby the infant participates the game flow partially as a result.
3. **sharing of the game's activities** — The infant detects the care-giver's response to own game-relevant action because the care-giver is expected to respond clearly to the infant's action if it is game-relevant. The infant understands and uses the rule which is the relationship between the infant's action and the care-giver's response. The infant is willing to invoke the response repetitively. Thereby the infant comes to play active role in the game.
4. **generating modifications** — The infant varies a game-relevant action and watches the care-giver's response to find another stable loop of the game flow. Thereby the infant generates modification of interaction rules.

3.2   A Scheme to Design Robot Behavior

We also propose a primary scheme to design a social game interaction robot. The scheme partially covers the hypothesis described above (stage 1 to 3). The scheme is described below.

A robot has functions of *imitation, contingency detection* and *adaptive action selection* of imitative or game-relevant action. At first, the robot imitates the partner's action by the *imitation* function. In a game, the partner is expected to respond to the robot's action clearly if the action is game-relevant. Thereby the robot finds the game rule via response observation by the *contingency detection* function. The rule is the relationship between the robot's game-relevant action and the partner's corresponding response. The robot tends to perform the game-relevant and game-ruled action while the contingency appears and tends to perform imitative actions if the contingency disappeared. This tendency is controlled by *adaptive action selection* function.

3.3 A Model in the Scheme

To investigate the proposed scheme described above, we build a dynamic model which realizes imitation and ruled interaction, and switches them depending on the partner's response.

To acquire interaction rules, a robot adaptively selects whether imitation or ruled interaction to perform. This model produces two interaction phases: imitation phase and ruled interaction phase. At the outset, the robot performs only imitation; this is called the imitation phase. Once the robot finds a rule where the partner clearly responds to the robot's game-relevant action, the robot starts to perform the action frequently. Then the interaction will be mainly governed by the rule; this is the ruled interaction phase. If the partner breaks the rule, the robot goes back to the imitation phase.

Taniguchi et al. have built a model in which a robot acquires interaction rules based on an idea of role reversal imitation [3]. But in this model, interaction type (imitation or ruled response) is explicit and completely separated. It is unlike a real interaction. In our model, interaction type is adaptively changed via observing partner's response.

We now summarize the mechanism of our model (Fig. 2). The model has three modules: an action observation module, a contingency detection module, and an action selection module.

For imitation, the action observation module estimates the partner's action tendency. Also, for each action a, the module calculates the occurrence frequency $w_{act}(a)$ as its score. If the robot selects its action just according to the score $w_{act}(a)$, the robot imitates the partner's action tendency. This module gradually forgets the tendency as well.

For ruled interaction, the contingency detection module estimates the co-occurrence frequency of pairs of robot's action and partner's response. This module also scores each action a by degree of causality from the robot's action to the partner's first action after it as its priority $w_{res}(a)$. If the robot selects its action purely according to the score $w_{res}(a)$, the robot performs ruled interaction. This module also gradually forgets the tendency.

To realize adaptive switching between imitation and ruled interaction, the action selection module scores each action a as the sum of the two scores ($w_{act}(a)$, $w_{res}(a)$) as its priority $w_{out}(a)$. This module selects the robot's next action. The switching is realized by the balance of the two scores.

3.4 Implementation

We now explain the implementation of each module. In the experiment, the posture of the robot and the partner is represented discretely. Time is also represented discretely. a is a posture transition, by the human or the robot, from a posture to another or the same. r is a first action of the partner after an action of the robot a. Time of r is defined as the next step of time of a.

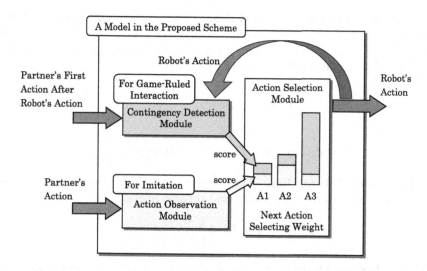

Fig. 2. A Model in the Proposed Scheme with Information Flow

Action Observation Module. The input is an observed action a of the partner. The output is $w_{act}(a)$ for each a. This module has occurrence counters $N(a)$ for each a to estimate the partner's action tendency. $N(a)$ is a continuous real number within the range from 0 to the parameter max_{act}. When an observed a is input, $N(a)$ is updated as (1). To implement gradual forgetting, $N(a)$ is updated at each time step as (2). δ_{act} is the parameter of the forgetting term. The score $w_{act}(a)$ is calculated as (3) as the estimated occurrence probability $P(a)$.

$$N(a) \leftarrow N(a) + 1 \quad for\ observed\ a\ . \tag{1}$$

$$N(a) \leftarrow N(a) - \delta_{act} \quad for\ each\ a\ . \tag{2}$$

$$w_{act}(a) = P(a) \approx \frac{N(a)}{\sum_i N(a_i)}\ . \tag{3}$$

Contingency Detection Module. The input is a performed action a of the robot and the first action r of the partner after the action of the robot. The output is $w_{res}(a)$ for each a. This module has occurrence counters $N(a,r)$ for each pair (a,r) to estimate the tendency of the relationship between a and r. $N(a,r)$ is a continuous real number within the range from 0 to the parameter max_{res}. When a given (a,r) is observed, $N(a,r)$ is updated as (4). To forget the tendency gradually, $N(a,r)$ is updated at each time step as (5). δ_{res} is the parameter of the forgetting term.

$$N(a,r) \leftarrow N(a,r) + 1 \quad for\ observed\ (a,r)\ . \tag{4}$$

$$N(a,r) \leftarrow N(a,r) - \delta_{res} \quad for\ each\ (a,r)\ . \tag{5}$$

—— **Acquisition of Interaction Rules.** $P(a,r)$ is the occurrence probability of the pair (a,r) as estimated in equation (6). $Ruled(a,r)$, the degree of confidence of a rule from a to r, is evaluated in (7) as relativeness between a and r with Mutual Information [10] (denoted I in (7,8)).

$$P(a,r) \approx \frac{N(a,r)}{\sum_{i,j} N(a_i, r_j)} . \tag{6}$$

$$Ruled(a,r) = I(a;r) = P(a,r) \, log_2 \frac{P(a,r)}{P(a)P(r)} . \tag{7}$$

—— **For Ruled Interaction.** The score $w_{res}(a)$ is calculated in (8) as the degree how a invokes ruled responses and is game-relevant.

$$w_{res}(a) = I(a;R) = \sum_r Ruled(a,r) . \tag{8}$$

Action Selection Module. For this module, the input is the scores w_{act} and w_{res} calculated by the action observation module and the contingency detection module. The output is a next robot's action a which is selected at the probability $P_{out}(a)$ calculated as (9). α is the parameter which biases the balance of imitation and ruled interaction. Larger α leads to more ruled interaction.

$$P_{out}(a) = w(a) = \frac{w_{act}(a) + \alpha \, w_{res}(a)}{\sum_i (w_{act}(a_i) + \alpha w_{res}(a_i))} = \frac{w_{act}(a) + \alpha \, w_{res}(a)}{1 + \alpha \sum_i w_{res}(a_i)} . \tag{9}$$

4   Simulation Experiment

4.1   Setup

The experiment is conducted in simulation. We suppose human and robot are sitting face-to-face (Fig. 3).

In Fig. 3, the partner is on the left. The partner's hand position is controlled by the examiner. On the right is the robot, which is controlled by the system described above. Arm posture is represented discretely so that the hand is located at one of four positions (Fig. 4). a and r is a set (p_0, p_1) which represents a posture transition from p_0 to p_1. The parameters are set as follows: $max_{act} = max_{res} = 10$. $\delta_{act} = \delta_{res} = 0.2$. $\alpha = 10$.

4.2   Task

We investigate whether the partner can teach and the robot can acquire novel interaction rules where the partner responds in a specific way to a specific action of the robot. The task is for the partner to teach the robot two rules (rule A(Fig. 5) and rule B(Fig. 6), noted later) in order. The aim of the task is to

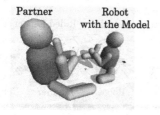

Partner   Robot with the Model

Fig. 3. Simulation Environment

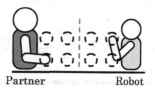

Partner   Robot

Fig. 4. Discrete Postures of Arm

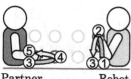

Partner   Robot

Fig. 5. Rule A

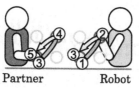

Partner   Robot

Fig. 6. Rule B

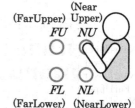

(Near
(FarUpper) Upper)
FU   *NU*

FL   *NL*
(FarLower) (NearLower)

Fig. 7. Names of Postures

confirm that the robot can find appearance and disappearance of a rule and find the correct rule intended by the partner within many rule candidates.

We now explain the two rules. As in Fig. 7, the label of a posture corresponds to the name of the circle where the hand is located. In rule A, the robot acts $[NL{\rightarrow}NU{\rightarrow}NL]$, and the partner should respond with $[NL{\rightarrow}FL{\rightarrow}NL]$. In rule B, the robot acts $[FL{\rightarrow}NU{\rightarrow}FL]$, and the partner should respond with $[NL{\rightarrow}FU{\rightarrow}NL]$. Each ruled interaction is not realized by only imitation.

In this task, the partner basically performs the following four policies described below in order (Fig. 8).

1. **Policy 1 — Demonstration A**
   — The partner demonstrates action $[NL{\rightarrow}NU{\rightarrow}NL]$ which the robot should perform in rule A and have the robot imitate the partner.
2. **Policy 2 — Ruled Response A**
   — The partner responds to the robot in rule A.
3. **Policy 3 — Demonstration B**
   — The partner demonstrates action $[FL{\rightarrow}NU{\rightarrow}FL]$ which the robot should perform in rule B and have the robot imitate the partner.
4. **Policy 4 — Ruled Response B**
   — The partner responds to the robot in rule B.

4.3   Result

The upper part of Fig. 9 represents transition of the partner's and the robot's posture through the experiment. The lower part of Fig. 9 also represents transition

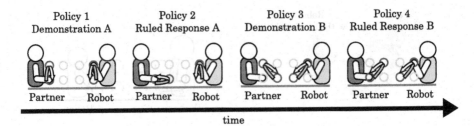

Fig. 8. Procedure of the Experiment

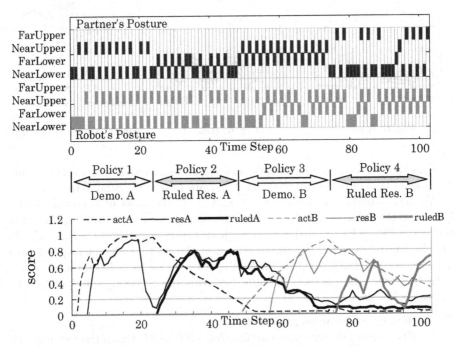

Fig. 9. Transition of Posture and Some of Internal States

of $\{w_{act}(NL \rightarrow NU) + w_{act}(NU \rightarrow NL)\}$ (named $actA$) which is the score for imitation of demonstration A, $\{w_{res}(NL \rightarrow NU) + w_{res}(NU \rightarrow NL))\}$ (named $resA$) which is degree of response contingency to the action, and $\{Ruled(NL \rightarrow NU, FL \rightarrow NL) + Ruled(NU \rightarrow NL, NL \rightarrow FL)\}$ (named $ruledA$) which is degree of confidence of rule A . $actB$, $resB$ and $ruledB$ are in the same manner.

5   Discussion

In the period of policy 1, the robot imitates the partner. $actA$ rises. This is considered the process in which the robot is recognizing the partner's demonstration

A. At the same time, *resA* is rising. This means that the robot started imitation and the partner still demonstrates and thereby, as a result, the robot recognizes as if the partner imitated the robot and the partner's action were contingent on the robot's action.

In the period of policy 2, the interaction is almost in rule A. *actA* is falling. This is considered the process in which the robot is recognizing that the partner is no longer in demonstration A. At the same time, *resA* fell and subsequently *resA* and *ruledA* rose. This means that the robot comes to recognize that the partner is no longer in the imitative response but responds in rule A.

These phenomena are alike in the period of policy 3 and 4. But the effects of the previous periods remain: we see that *resA* and *ruledA* continue to have certain values. So the phenomena are a little more complex than the previous periods. We see that the robot's posture is sometimes in *NL*. Thereby the partner has a little difficulty in teaching rule B. In the latter part of the period of policy 4, the interaction in rule B is stable and *resB* and *ruledB* rise.

The partner succeeded in adaptively teaching the robot different interaction rules without reset. It may be effective to break a previous contingency and the robot's recognition intentionally to change interaction to a new one.

6   Conclusion and Future Work

We proposed a hypothetical development of social game interaction between an infant and a care-giver from a mechanism-sided standpoint, based on developmental psychology. Applying the hypothesis, we also propose a scheme to design a robot in which a partner can teach interaction rules through interaction. To investigate the proposed scheme, we built a dynamic model which realizes imitation and ruled interaction and switches them observing partner's response. In the experiment, the partner can teach and the robot can acquire a rule adaptively through interaction without explicit teaching and subsequently it is also achieved about another rule without reset. We suggest that the proposed ideas contribute to improve social robot's adaptivity.

A problem is that the model breaks a previous rule when it acquires a new one. We aim to realize switching multiple game rules depending on the context of the game flow. Another problem is that the shown adaptation is realized by just changes of the internal state matrix. For developmental interaction, we think that action primitives or rules should be represented hierarchically. A crucial problem is that action primitives are represented discretely. Sugita et al. have built a model which acquires relationships between sentences and behavioral patterns [11]. We consider that RNNPB, the core component in that research, may contribute to extracting and switching primitives or rules from continuous interaction. Another future work is to avoid endless repetition of a specific interaction. Watson reported that an infant examines contingency between the infant's action and a toy's response and once the infant gets to know whether the contingency exists or not, the infant stops the examination [12]. We estimate that this knowledge may lead to an idea which realizes the fourth stage "generating modification" of the development and aim to investigate it.

This paper is concerned with gesture-based interaction, but it is pointed out that there is analogy between social game and language [8]. We think that humans utilize similar fundamental abilities for both processes, so this research is linked to explanation of language acquisition and its implementation in robots.

Acknowledgments. This research is supported by JST ERATO Asada Synergistic Intelligence Project.

References

1. Rochat, P.: Mechanisms of Infant Development. In: The Infant's World. Harvard University Press (2001)
2. Andry, P., Gaussier, P., Moga, S., Banquet, J.P., Nadel, J.: Learning and Communication via Imitation: An Autonomous Robot Perspective. IEEE Transactions on Systems, Man, and Cybernetics -Part A: Systems and Humans 31(5), 431–442 (2001)
3. Taniguchi, T., Iwahashi, N.: Computational model of role reversal imitation through continuous human-robot interaction. In: Proceedings of the 2007 workshop on Multimodal Interfaces in Semantic Interaction, pp. 25–31 (2007)
4. Fleischer, J., Shapiro, J.: Imitation is not enough for lexicon learning. In: From Animals To Animats 8: Proceedings of the 8th International Conference on the Simulation of Adaptive Behavior, pp. 477–486 (2004)
5. Nakata, T.: Expression with Informatical Factor in Human Robot Interaction. In: Proceedings of the ACM CHI 2003 Workshop on Subtle Expressivity of Characters and Robots, pp. 11–14 (2003)
6. Sumioka, H., Yoshikawa, Y., Asada, M.: Causality Detected by Transfer Entropy Leads Acquisition of Joint Attention. In: Proceedings of the 6th IEEE International Conference on Development and Learning, CD-ROM, No. 93 (2007)
7. Bruner, J.S., Sherwood, V.: Peekaboo and the Learning of Rule Structures. In: Bruner, J.S., Jolly, A., Sylva, K. (eds.) Play—Its Role in Development and Evolution, pp. 277–285 (1976)
8. Ratner, N., Bruner, J.: Games, social exchange and the acquisition of language. Journal of Child Language 5(3), 391–401 (1978)
9. Rome-Flanders, T., Cossette, L.: Comprehension of Rules and Structures in Mother-Infant Games: A Longitudinal Study of the Early Two Years of Life. International Journal of Behavioral Development 18(1), 83–103 (1995)
10. Shannon, C.E., Weaver, W.: The Mathematical Theory of Information. University of Illinois Press (1949)
11. Sugita, Y., Tani, J.: A Connectionist Approach to Learn Association between Sentences and Behavioral Patterns of a Robot. In: From Animals To Animats 8: Proceedings of the 8th International Conference on the Simulation of Adaptive Behavior, pp. 467–476 (2004)
12. Watson, J.S.: The perception of contingency as a determinant of social responsiveness, Origin of the Infant's Social Responsiveness. Halsted Press Division of Wiley, Chichester (1997)

On Modeling Proto-Imitation
in a Pre-associative Babel

Elpida Tzafestas

Institute of Communication and Computer Systems
National Technical University of Athens
Zographou Campus, 15773 Athens, Greece
brensham@softlab.ece.ntua.gr

Abstract. In this paper we present a model of generative proto-imitation that replicates external signals without associating with objects, as in higher-level imitation. A mixed population of adults, that have fixed associations objects-signals, and infants, that do not have associations but imitate unconditionally, endowed with a kinship and interaction structure, allows infants to develop signal affinity with their kin in a variety of conditions and within an initial random world, i.e. in a Babel. Our results indicate that the communicative value of imitation can be discovered after the basic apparatus is in place, rather than that communication is the end to which imitation is the means.

Keywords: Imitation, proto-imitation, kinship, Baldwin effect, development.

1 Introduction

Research in imitation spreads in various disciplines (for an overview see [1]) and generally centers around two major themes: the role of imitation in social interaction and communication and the mechanism by which imitated responses are produced. Functional studies related to social behavior and communication are common both in psychology and theoretical biology and rely on the implicit assumption that imitation is mainly a means to (learn to) communicate or interact socially [2][3][4]. Studies related to the neural mechanisms behind imitative response generation investigate mainly phenomena of neonatal or early infant imitation and are tackling questions such as the degree to which neonatal imitation is goal-directed, motivated and selective [5][6]. Other important issues we retained from the literature as specifications for modeling are:

- Imitation should start as a reactive or impulsive process and subsequently catalyze itself through the social interaction itself. This self-catalysis may be through direct training by adults; adult turn-taking in imitation is one such way of training [4].
- Imitation is a multi-level process found in many different forms of varying degrees of complexity in a vast number of animal species [7]. However, even animals that can imitate cannot learn to communicate at a

M. Asada et al. (Eds.): SAB 2008, LNAI 5040, pp. 477–487, 2004.

human-like level of complexity and this is an instance of the major question in cognitive science "what makes humans different from animals" [8].

- When used in a communicative context, human-level imitation manages to co-exist with diverse, arbitrary sign systems, which is apparently contradictory with the effect of imitation [9][10].
- Disturbed versions of imitative mechanisms may be found in a spectrum of developmental disorders [11] and this is an indication that imitation cannot be studied outside a development context, at least functionally [12].

We are therefore developing a model that allows proto-imitation outside explicit communication but may allow emergence of communication in the medium or long term ontogenetically. This model separates response imitation from response association to external meaning so as to make room both for species that can imitate but not associate meaning and for disabled humans that cannot associate well or even proto-imitate well. We work at the response imitation level to show that a wealth of phenomena may later emerge and especially those related to direct associations.

2   The Agent Model: Generative Imitation

The usual agent model found in the literature (see for example [3]), supposes the existence of M objects with one signal associated with each object. The term "object" can denote anything from an individual or an inanimate object to an action or an event, in short any external thing that can be referred to. The signals are supposed to transfer information about the objects and can take values in any physical medium that an agent is able to use; however, the typical case is to think of signals as vocalizations. An imitative process is one that allows one agent to learn to use another agent's signal to refer to the same object. As such, imitation allows to two or more agents to communicate by using the same signals for the same objects. One common formalization of this imitation process [9] in a population of agents is through the use of a MxL language matrix per agent where each entry denotes the probability for the agent to refer to object i (i=1,2,...,M) by using the signal j (j=1,2,...,L). This setup presupposes that associations between objects and signals should exist for imitation to take place and that the role of imitation is to make the associations of different agents converge to one common language. This assumption is partly due to the adoption of the language domain as the experimental field of imitation *par excellence*.

Our own model of imitation of adult agent responses does not assume any prior association to any external object. Instead we model the way an infant agent recognizes and reproduces an observed signal coming from an adult agent and standing for the response to a perceived object without having access to the object itself and thus without associating with it. Our model is a *functional* model of a neural structure that generates responses to match and replicate an external stimulus, i.e. the signal received (a similar generative model has been reported in [13]). We opted for a study with the aid of a functional model of the neural structure rather than the structure itself, because we did not want to constrain our results within the possibilities of a given structure. We are rather seeking the organizational properties that such a structure should have to allow imitation.

We begin therefore by adopting the view that the ontegenetic development at the neural level follows the same principles as Darwinian evolution at the population level ([14]) and that any novel responses should be generated internally and selected within the environment rather than be directly "taught" by it. By applying this view to the proto-imitation level of low-level reproduction of external signals, we get a model where continuous internal generation of responses combined with an environmental selection (i.e. reinforcement through actual response matching to external signal) allows various imitative phenomena within a population of agents. Association of signals to objects can appear later, if at all. We use real-valued signals as in [2] instead of the more common discrete or symbolic signals ([3][15] etc.), because we feel we should not ignore the continuous nature of real-world signals (especially vocalizations) and because we want to explicitly model the distance between agent responses.

An agent possesses a number of internal "frequencies" or eigenfrequencies (these could correspond to real vocal parameters or to neural patterns) that have varying degrees of affinity to a given signal: because frequencies as well as signals take real values in [0,1], affinities of frequencies will also take real values in the same interval.

```
// Frequencies: f[] - size K, f[i] in [0,1]
// External Signal: x in [0,1], Meaning/Object: n

// 1. Matching step: Compute affinity of each frequency
foreach frequency f[i] (i=1…K)
{         diff = |f[i]-x|
          if (diff <= T[i]) // T[i] = a threshold
             affinity[i] = (1-diff) (affinity in [0,1])
             else affinity[i] = 0;         }
// The (indirect through imitation) response of the
// agent to the external meaning/object n is:
language[n] = f[i] with max affinity[i]
Agent's total_affinity = avg(affinity[i],i=1…K); in [0,1]

// 2. Selection step:
//     Reproduce frequencies proportionally to affinity
foreach frequency f[i] (i=1…K)
{         // Pop = Population size of clones of f[i]
          pop = (affinity[i] * K / total_affinity);
          for j=1…pop
             //add new frequency in the range
             // [f[i]-T[i],f[i]+T[i]]:
             newf[j] = f[i]-T[i]+(random()*2*T[i])); }
// 3. Normalization step:
//     Inject new random frequencies
//     if less than K frequencies created (newf array)

// 4. Mutation step:
//     Replace randomly K1 of the frequencies:
          K1 = random()*ExplorationFactor*K
```

Fig. 1. The eigenfrequency-based functional model of generative imitation

The response to a signal is the eigenfrequency with the highest affinity. At each step, new eigenfrequencies are generated proportionally to the affinity of the previous ones. The highest matching frequencies reproduce massively, while the lowest ones vanish and are replaced by newly generated random eigenfrequencies. An exploration factor is also defined, which is the maximum percentage of random eigenfrequency replacement independently of affinity. The overall affinity of an agent to an external signal is the average affinity of all its frequencies, thus it is internally generated and not externally imposed/designed in any way. This measure expresses *how "well" an agent recognizes and can reproduce a signal* and may therefore serve as a basis for subsequent emergence of communication. This model is summarized in Fig. 1.

3   Experiments I: Adults, Infants and Kinship

We have simulated populations of agents that follow the previous imitation model. In the same spirit as [3][15] we have separated the learning phase from the adult mode of behavior. Because in our system we have no predefined associations between objects and vocalizations we use two classes of agents, the adult and the infant agents. The former are supposed to have stabilized and to always respond to a particular object with one particular vocalization, while the latter are in a learning stage and are imitating the agents they encounter. Social encounters are based on the obvious intuition that infants are nurtured by a limited set of "parents" or kin agents and therefore they encounter most often or even exclusively these agents. All such parental relations are initialized randomly within the overall population with the aid of the kin factor parameter, which is the number of parents or kin agents per infant.

In each simulation cycle, every infant encounters an adult selected randomly from its kin population, unless social noise is present. Each encounter involves the adult agent acting as a sender of one or more signals and the infant agent acting as a receiver that imitates the received signals. This process uses two parameters, the number M of actual external objects refered to by the adult agents and the imitation factor which is the maximum number of signals received and imitated on every encounter (imitation factor $\leq$ M). All adult vocalizations as well as all infant frequencies are initialized randomly in the range [0,1]. This combination of uniformed reactive imitation with a tight social interaction structure allows infants to develop high affinity with their kin, where *affinity with one or more agents is defined as the average affinity for all external objects*. We remind however that an infant cannot associate objects and vocalizations and its current signal (word) for a particular object is the last emitted one, as shown in Fig. 1. Figures 2 and 3 show how average infant affinities develop in the case of a single or two parents, when refering to one or more objects. In all cases, the affinity with kin is clearly higher than the affinity with adults in general and it develops slower as the number of refering objects increases. Final kin affinity is also lower as the number of kin agents increases. It is relevant to note that while average kin affinity rises, average adult affinity remains at best the same, thus the affinity with non-kin agents is on average negatively affected by the buildup of affinity with kin. Note also that in our simulations we have chosen K (the number of frequencies) to be high (50), to induce high initial affinity and learning speed.

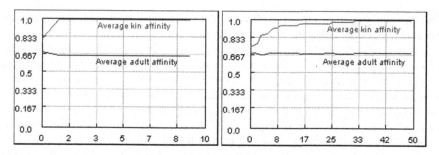

Fig. 2. *(x=time, y=affinity in [0,1])* 10 adults, 10 infants, random parental relations. (a) Kin factor = 1, M=1, imitation factor = 1 (single parent and object). (b) Kin factor = 1, M=3, imitation factor = 2 (single parent, many objects).

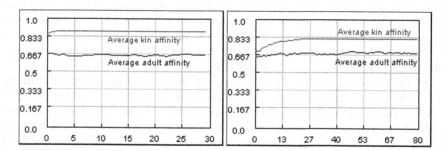

Fig. 3. *(x=time, y=affinity in [0,1])* 10 adults, 10 infants, random parental relations. (a) Kin factor = 2, M=1, imitation factor = 1 (many parents, single object). (b) Kin factor = 2, M=3, imitation factor = 2 (many parents, many objects).

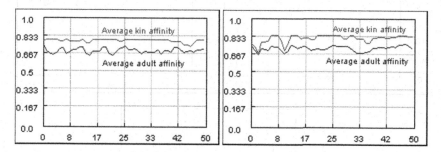

Fig. 4. *(x=time, y=affinity in [0,1])* 10 adults, 10 infants, random parental relations. Kin factor = 2, M=1, imitation factor = 1 (many parents). (a) Social exploration factor = 0.1, interaction factor = 5. (b) Signal noise probability: 0.1.

In figure 4 we are introducing noise in the system either as social noise or signal noise. Social noise is implemented with the aid of a social exploration factor which is the probability that an infant will encounter a non-kin agent belonging to a larger group of predefined agents (the interaction factor is the total number of agents that form an infant's social environment, kin and non-kin). Signal noise is implemented as a probability with which the received signal from another agent will be mutated to a random value (again in the range [0,1]). Even when noise is present, the infants are still able to develop higher affinities with kin than with other agents.

In figure 5, we replicate the experiment of [3] as can be implemented in our system: we examine role model or fitness-based matching and random matching as social interaction rules in place of our kin-based rule. Fitness-based matching means that an adult is selected for interaction by an infant with a probability proportional to its communicative fitness, that is in our case its average affinity with the agents it interacts with. Although the system is expected to be self-catalyzed by its own behavior (accidental communicative success will lead to higher probability to interact again in the future), it does not seem to allow consistent development of higher affinity with interacting agents. This is obvious in figure 5b, where the affinities for one particular infant are depicted: average kin affinity fluctuates above and below the average adult affinity levels without being able to settle (notice that in this case, where no kinship is defined, what we measure as kin affinity is actually the affinity with the interacting agents). The results are similar in the case of random matching. The lesson that can be drawn from this experiment is that informed matching is unnecessary for development of affinity with other agents; rather a blind observational imitation mechanism is sufficient, as has been also pointed out by [15].

Our final experiment, whose results are given in fig. 6, involves cultural learning from generation to generation. More specifically, we design a system where after 100 cycles of execution (recall that, as shown in previous figures, the populations stabilize in less than 30 cycles), the whole population is replaced by a new population where the adults are swept from the system, the infants become adults with no possibility to

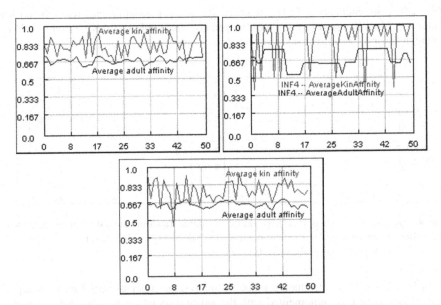

Fig. 5. *(x=time, y=affinity in [0,1])* 10 adults, 10 infants, M=1, imitation factor = 1, no noise. (a) Fitness-based matching, average population measurements, (b) One typical infant's measurements in the previous experiment, (c) Random-based matching, average population measurements.

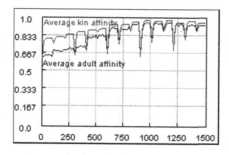

Fig. 6. *(x=generation, y=affinity in [0,1])* 10 adults, 10 infants, kin factor = 2, M=5, imitation factor = 3, no noise. 15 generations of 100 cycles each.

imitate anymore and newly born, randomly initialized infants enter the population in the place of the former infants. We run this system for a number of generations and we find, not too unsurprisingly, that subsequent generations achieve much higher average adult and kin affinity scores, approaching the theoretical maximum of 1, even for big language sets (M=5, imitation factor=3 in the example). This is an example of how the Baldwin effect [16] can arise in the cultural domain: cultural instead of genetic inheritance combined with generation overlapping can play the role of the acquired features transmission mechanism, thus in the long term leading to a system that has learnt adaptive or even "optimal" behaviors. It also parallels the results of [17] who also found that language becomes highly regular if it is transmitted from generation to generation. Finally, this result further supports our initial hypothesis that the communicative function of an individualistic sign system can evolve after the structural apparatus for proto-imitation is in place: it is not difficult to imagine that an initial Babel population of agents with fully diverse vocalization sets endowed with such a rudimentary imitation mechanism can develop and "discover" a few generations later that signs are shared and can use them in a communicative manner, although the mechanics of such an extension will be most probably fairly complex.

4 Experiments II: Social Structure and Individual Development

Having completed the experiments of the previous section, we wondered how the otherwise obvious infant-adult attachment relations could be dissected in detail so as to identify exactly how social structure, age or developmental differences can influence the emergence of affinity or dynamic kinship in agent populations. In figure 7, we give the results of a single-parent system where one infant is reinitialized with a new parent before stabilization. Correspondingly, figure 8 gives the results of a single-parent system where one infant is reinitialized with a new parent after it has stabilized. In both cases, if the agent is given enough time it can find the new stable position of maximum affinity with the new parent. However, re-initialization slows down the learning process because the infant frequency system at reinitialization is not completely random but more or less tuned to the former target, thus re-learning is slower than initial learning.

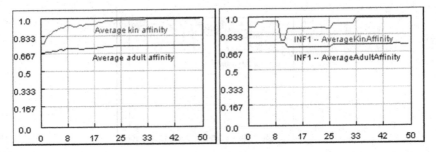

Fig. 7. *(x=time, y=affinity in [0,1])* 10 adults, 10 infants, kin factor = 1, M=3, imitation factor = 2, no noise. At t=10 reinitialization of an infant with a new parent. (a) Average population measurements, (b) The infant measurements.

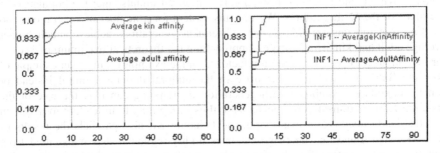

Fig. 8. *(x=time, y=affinity in [0,1])* 10 adults, 10 infants, kin factor = 1, M=3, imitation factor = 2, no noise. At t=30 reinitialization of an infant with a new parent. (a) Average population measurements, (b) The infant measurements.

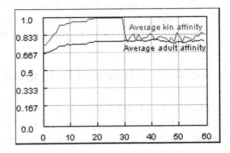

Fig. 9. *(x=time, y=affinity in [0,1])* 10 adults, 10 infants, kin factor = 1, M=3, imitation factor = 2, no noise. At t=30 allow infants to match randomly and imitate all adults. Average adult affinity does not change. Compare the second part of the chart (after t=30) with figure 5c. After the system has stabilized, random matching and imitation does not seem to harm.

A final experiment concerning the passage to adult age consists in reverting to random matching and imitation when the system has stabilized. In this case, counter-intuitively, the average adult affinity does not drop, but the average kin affinity drops and fluctuates a little above the adult level (fig. 9). This result combined with the

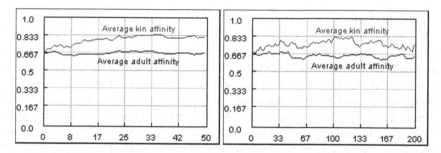

Fig. 10. *(x=time, y=affinity in [0,1])* 10 adults, 10 infants, M=3, imitation factor = 2, no noise. (a) Every infant has one parent and one friend infant. (b) Every infant has two parents and as many brethren as defined by shared parental links.

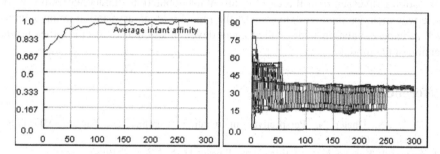

Fig. 11. No adults, 20 infants, M=3, imitation factor = 2, no noise, random matching model. (a) *(x=time, y=affinity in [0,1])* Average population affinity, (b) *(x=time, y=signals in [0,1] x 100)* All 20 agents vocalizations for one of the objects. If given sufficient time, they converge to one common vocalization for all agents. Vocalizations are amplified for visualization purposes.

previous ones means that the proto-imitative system is very vulnerable during stabilization from a random starting point but fairly robust when stabilized.

Because in reality, a young agent does not only interact with adults and it is not logical to assume that even then it only imitates adults, we extend the kinship relations by including either random other infants (friends) or the brothers of the infant. In figure 10, we give the results of a pair of such experiments. In both cases, and despite inferior convergence to high kin affinity values, there is clear separation of the trends for average adult and average kin affinity.

Although the distinction between infants and adults is defined only with respect to their ability to imitate, it makes sense to investigate what would happen in the extreme case of an all-infant population with random interactions. As shown in fig. 11, the system develops very high average affinity close to the theoretical maximum of 1. Interestingly, if given enough time such a system converges to a single shared language (and this is why final average affinity is so high). So, what can only evolve culturally if age-dependent or other imitation differences exist, can appear within one generation that is imitative and initially "tabula rasa". Once more this supports our hypothesis that proto-imitation precedes any notion of communication, it is purely reactive and does not need any association with meaning. Associations and the discovery of communicative value only come at a later stage.

5 Conclusion

We have presented a model of generative proto-imitation of signals without association with an external meaning. Our model purported to show that an initial non-functional "Babel" of diversely vocalizing agents, if endowed with a non-associative proto-imitation mechanism, allows communication to emerge at a subsequent, possibly associative, stage. We have shown how our model can yield a wealth of population phenomena that are generally taken with given explicit associations and especially how affinity can develop between initially tabula rasa imitative infants and adult agents to which infants are attached in a variety of ways. Other prominent results include the cultural development from generation to generation and convergence to one single "language", as well as an impressive one-generation convergence in the case an all-infant population. Immediate future work underway includes both an extension of the model with an association component as well as experimentation with explicit agent development that is absent in the current version of the model.

References

1. Hurley, S., Chater, N.: Perspectives on imitation in animals and humans. MIT Press, Cambridge (2005)
2. Suzuki, J., Kaneko, K.: Imitation games. Physica D 75, 328–342 (1994)
3. Nowak, M.A., Plotkin, J.B., Krakauer, D.C.: The evolutionary language game. J. Theor.Biol. 200, 147–162 (1999)
4. Eckerman, C.O., Stein, M.R.: How imitation begets imitation and toddlers' generation of games. Devel. Psychol. 26, 370–378 (1990)
5. Gergely, G., Bekkering, H., Király, I.: Rational imitation in preverbal infants. Nature 415, 755 (2002)
6. Bekkering, H., Wohlschläger, A., Gattis, M.: Imitation of gestures in children is goal-directed. Quarterly J. of Exp. Psychol. 53A, 153–164 (2000)
7. Caldwell, C.A., Whiten, A.: Evolutionary perspectives on imitation: Is a comparative psychology of social learning possible? Anim. Cogn. 5, 192–208 (2002)
8. Oliphant, M.: Rethinking the language bottleneck: Why don't animals learn to communicate? In: Presented at the 2nd Intern. Conf. on the Evolution of Language (1988)
9. Krakauer, D.C.: Selective imitation for a private sign system. J. Theor. Biol. 213, 145–157 (2001)
10. Grassly, N.C., von Haeseler, A., Krakauer, D.C.: Error, population structure and the origin of diverse sign systems. J. Theor. Biol. 206, 369–378 (2000)
11. Stieglitz Ham, H., Corley, M., Rajendran, G., Carletta, J., Swanson, S.: Imitation of meaningless gestures in individuals with Asperger syndrome and high-functioning autism. J. Autism Dev. Disord. 38, 569–573 (2008)
12. Karmiloff-Smith, A.: Development itself is the key to understanding developmental disorders. Trends in Cognitive Sciences 2, 389–398 (1998)
13. Tzafestas, E.S.: Compromising algorithmicity and plasticity in autonomous agent control systems: The autonomous cell. J. Intell. Syst. 9, 135–176 (1999)

14. Edelman, G.M.: Neural darwinism. Basic Books, New York (1987)
15. Oliphant, M.: The learning barrier: Moving from innate to learned systems of communication. Adaptive Behavior 7, 371–384 (1999)
16. Baldwin, J.M.: A new factor in evolution. Amer. Natur. 30, 441-451 &536–553 (1896)
17. Vogt, P.: Cumulative cultural evolution: Can we ever learn more? In: Proceedings Simulation of Adaptive Behavior (2006)

Evolution of General Driving Rules of a Driving Agent

Ivan Tanev[1], Hirotaka Yamazaki[2], Tomoyuki Hiroyasu[2], and Katsunori Shimohara[1]

[1] Department of Knowledge Engineering and Computer Sciences,
Doshisha University, Kyoto, Japan
hyamazaki@mikilab.doshisha.ac.jp, tomo@is.doshisha.ac.jp
[2] Department of Information Systems Design, Doshisha University, Kyoto, Japan
{itanev,kshimoha}@mail.doshisha.ac.jp

Abstract. We present an approach for automated design of the functionary of driving agent, able to operate a software model of fast running car. Our objective is to discover a single driving rule (if existent) that is general enough to be able to adequately control the car in all sections of predefined circuits. In order to evolve an agent with such capabilities, we propose an indirect, generative representation of the driving rules as algebraic functions of the features of the perceived surroundings of the car. These functions, when evaluated for the current surrounding of the car yield concrete values of the main attributes of the driving style (e.g., straight line velocity, turning velocity, etc.), applied by the agent in the currently negotiated section of the circuit. Experimental results verify both the very existence of the general driving rules and the ability of the employed genetic programming framework to automatically discover them. The evolved driving rules offer a favorable generality, in that a single rule can be successfully applied (i) not only for all the sections of a particular circuit, but also (ii) for the sections in several a priori defined circuits featuring different characteristics.

1 Introduction

The success of the computer playing sport games (e.g., chess [7], checkers, backgammon, tic-tac-toe [2], etc.) has long served as touchstone of the progress in the field of artificial intelligence (AI). The expanding scope of applicability of AI for the implementation of an agent with autonomous "learning" abilities includes soccer [9] and F1 racing [15] etc. [4], etc. Focusing in the domain of computerized car racing, in this work we consider the problem of the automated design of driving agent, able to operate a model of a fast running car. In order to drive a car fast around the circuit, the driver needs to define the best driving line and the way of approaching turns in the circuit. To realize the optimal line, the driver (agent) is also required to make a precise judgment about the state (i.e., position, orientation and velocity) of the car and its surrounding (distance to the apex of the turn, apex homing angle, etc.) and to react timely and precisely.

The *objective* of our work is to design the functionality of a driving agent, able to control a fast running software model of the remotely controlled scaled racing car in

M. Asada et al. (Eds.): SAB 2008, LNAI 5040, pp. 488–498, 2008.

various circuits with diverse characteristics. The agent should be able to learn the most general rules about how to control the fast running car not only on a particular circuit, but also on several a priori defined circuits. The possibility to discover such general driving rules (if existent) implies that the agent, when trained on a set of circuits with diverse characteristics, would be able to optimally control the car in unknown circuits with the need of a little (if any) adaptation. The very feasibility to develop such a universal agent would open up opportunities to build a framework of adaptive racing games in which the human competes against the computerized racer in dynamic or a priori unknown (e.g., user-defined) tracks and dynamically changeable (moving cars, obstacles, etc.) environments. The proposed approach could also be applied for automated design of the control software for navigation of remotely operated vehicles capable to find an optimal solution to various tasks in uncertain and dynamic environment.

Achieving our objective implies that the following four tasks should be addressed: (i) formalizing the driving style by defining the set of its *key parameters*, (ii) defining the notion of *general driving rules* that, when applied in the context of the concrete features of the current surrounding of the car, would yield concrete values of the key parameters of the driving style, (iii) defining the set of relevant *features of the surrounding*, that would facilitate the creation of general driving rules, and (iv) developing an *algorithm paradigm* for automated determination of the fastest, yet general driving rules, which yield concrete, optimal values of the parameters of the driving style when applied for the currently negotiated section of the circuit.

The related work by Suzuki and Floreano [10] demonstrates the feasibility of applying an active vision for landmark navigation of a scale vehicle. Wloch and Bentley [15] applied genetic algorithms for automated optimization of the setup of the simulated racing car. Togelius and Lucas [14] used scale models of cars in their research to demonstrate the ability of the artificial evolution to develop optimal neurocontrollers with various architectures. Our previous work [13] shows that with addressing the video feedback latency through an anticipatory modeling a human competitive driving agent could be automatically evolved via genetic algorithms. However, in [13] we considered an agent obtained as a result of time-consuming simulated evolution on a single, a priori known circuit, and how well the agent would perform in unknown circuits was considered as beyond the scope of our work.

In this work we attempt to address the problem of generality of the optimized driving agent by proposing an evolutionary framework, able to develop optimal, yet general driving "rules", rather than concrete optimal values of the driving style parameters. We shall verify the generality of applying the evolved single rule both for all the sections of a single circuit, and for all the sections of several a priori known circuits with different characteristics.

The remaining of this document is organized as follows. Section 2 explains the configuration of the system. In Section 3 we formalize the driving style and introduce the notion of driving rules. Section 4 explains the main attributes of the algorithm paradigm employed for evolution of the fastest, yet general driving rules. Section 5 presents the experimental results, and Section 6 draws a conclusion.

2 Modeled Configuration of the Physical System

2.1 System Configuration

In our work we use a software model of the scaled (1/24) racing car. The main mechanical characteristics of the car are as elaborated in [12]. The perceptions of the agent are modeling the real physical system on which an image feed is obtained from a video camera mounted overhead. The camera features a CCD sensor and lenses with wide field of view (66 degrees), which allows to cover a sufficiently wide area of about 2800mm x 2100mm from an altitude of about 2200mm. The camera is modeled to operate at 640x480 pixels mode, scaling down the scene to about 4mm per pixel. The video sampling interval is 30ms, and the video feed latency is 90ms.

The car is operated by the agent via modeled ratio remote control with functionality including "forward", "reverse" and "neutral" throttle control commands and "left", "right" and "straight" steering controls.

2.2 Software Simulator

The developed software simulator of the car and the environment allows to (i) "compress" the runtime of the fitness evaluation in the eventual implementation of agent's evolution and (ii) to verify the very feasibility of certain circuit configurations without the need to be concerned about the risks of possible damage to the environment or the car [12]. Furthermore, the internal model of the car and the environment comprises the kernel of the developed simulator. This model is continuously applied by the driving agent in order to anticipate the intrinsic, yet unperceivable state of the car from the currently available (outdated) perceptions. The model, calibrated with the concrete values of car's parameters (e.g., mass, turning radius, max velocity, max acceleration, max deceleration, etc. [12]) takes into consideration (i) the Newtonian physics of the car including the handling attitude of car on cornering (neutral steering, oversteer and understeer), and (ii) the feedback latency of 90ms.

The complexity of the effects of handling attitudes of the cornering car (e.g., understeer, oversteer, and neutral steering) on the lap time renders the task of optimizing the driving style of the agent quite challenging [13], which additionally motivated us to consider an automated evolutionary approach to address it.

The realism of the model of the car is proved by the very limited need of adaptation to the physical system of the agent, initially evolved in the model [12]. Also, the realism is indirectly indicated by the consistence of both the obtained driving lines and the lap times of the physical car around predefined circuits [13], which is attributed to the negligibly small difference between the intrinsic, but unperceivable state of the car (i.e., position, orientation, and velocity) and the state, anticipated from the outdated perceptions.

3 Representation of Driving Rules

Achieving our objective to design a general driving agent, able to control the model of car in various circuits with diverse characteristics implies that we shall address the following four tasks:

(i)   How to formalize the driving style by defining the set of its key parameters,
(ii)   How to define the notion of general driving rule,
(iii) How to define the set of relevant features of the current surrounding of the car, that would facilitate the creation of general driving rules, and
(iv)   What algorithm paradigm could be used for automated evolution of fastest, yet general driving rules, which yield concrete, optimal values of the driving style parameters when applied for the currently negotiated section of the circuit?

In this Section we elaborate on the former three, while in the following Section 4 we discuss the latter of these tasks.

3.1 Parameterization of the Driving Style

We consider the driving style as the driving line, which the car follows before and around the corners in the circuits, combined with the breakdown of the speed, at which the car travels along this line. Our choice of parameters of the driving style is based on the view, shared among the high-performance drivers from various teams in different racing formulas, that (i) the track can be seen as a set of consequent turns they need to optimize divided by simple straights, and that (ii) the turns with the preceding straights should be treated as a single whole [1][3]. Based on these standpoints, we introduce the following set of four key parameters of the driving style, pertaining to each of the turns of the circuit:

(i)   Approach (homing) angle – the constant bearing of the apex of the turn. Higher values of the latter parameter yield wider driving lines featuring higher turning radiuses,
(ii)   Straight-line velocity - the velocity at which the car approaches the turn,
(iii) Turning velocity – the velocity inside the turn, and
(iv)   Throttle lift-off zone – the distance from the apex at which the car begins slowing down from the straight-line velocity to the turning velocity.

Notice that the first parameter solely defines the driving line, and the latter three parameters define the breakdown of the velocity along this line.

Viewing the desired values of these four parameters as values that the agent tries to maintain, the functionality of the agent can be algorithmically formalized in a way as shown in Figure 1. The usage of the values of the key driving style attributes are underlined in the figure and indicated as "desired". As Figure 1 illustrates, both the orientation (lines 7-10) and the speed (lines 12-15) of the car are continuously adjusted in order to match the desired values of the corresponding attributes. The open-loop adjustment of the car's velocity (lines 13 and 14) is implemented by macro-commands ShiftGear(Gear), implemented via pulse-width modulation of the sequence of "forward" and "neutral" throttle commands with duty cycle of 120ms (4 sampling intervals). The possible values of the input parameter Gear (and, consequently, the possible values of the driving style parameters "Straight-line velocity" and "Turning velocity") are 1, 2, 3 or 4, which correspond to the duty ratios of PWM of 0.25, 0.5, 0.75, and 1 respectively.

```
1. At each time step do begin
2. //--- Perceptions:
3. Obtain the agent's perceptions of car's state: position (P), orientation and speed (V);
4. Obtain the agent's perceptions of the environment: approach angle (Aₐ), and distance (A_D) to the current apex
5. //--- Reaction of the agent to the current perceptions
6. //--- A) Steering control:
7. if (Aₐ> Desired Aₐ)   and (abs(Aₐ - Desired Aₐ)> Desired Threshold Aₐ)
8.     then SetSteering(Left)
9.     else   if (Aₐ< Desired Aₐ) and (abs(Aₐ - Desired Aₐ)> Desired Threshold Aₐ)
10.                     then SetSteering(Right)      else SetSteering(Straight);
11.//--- B) Throttle control:
12. if A_D > Desired Throttle Lift-off Zone
13.     then ShiftGear (Desired Straight Line Gear)
14.     else ShiftGear (Desired Turning Gear);
15. end
```

Fig. 1. Functionality of driving agent

Compared to the typical approaches of polynomial (spline) interpolation of curves, the use of the single parameter "Approach angle" to solely define the driving line significantly simplifies the representation of the driving style which, in turn reduces the search space of the simulated evolution which is intend to employed for automated optimization of driving rules. Fixing the homing angle of the car implies that the driving line is interpolated as an equiangular spiral. The proposed approach is inspired by the smooth decrease of the turning radius (and the corresponding optimal cornering velocity) along the spiral, which gives the agent an opportunity to smoothly, and progressively apply both the steering and the brakes as the car negotiates the turn. Moreover, "the smooth is fast" is among the most important rules in high-performing driving as the smoothness of both the radial (due to steering) and lateral (due to acceleration or braking) forces allows for a better control of the car near the traction limits of the tires [1]. The layouts of the roads in some highway junctions which can be approximated by equiangular spirals [6] could be viewed as a verification of the technical plausibility of the proposed an approach.

3.2   Representation of the Driving Rules

The driving rule in the proposed approach defines how to compute the four key parameters of the driving style for the currently perceived surrounding of the car. We implement the driving rule as a set of four evolvable algebraic functions. Each one of the four parameters of the driving style is associated with exactly one algebraic function and the value of the corresponding parameter is set to the value of the function, evaluated for the concrete features of the currently negotiated section of the circuit. Therefore, the same driving rule might potentially yield different values of the four driving style parameters when evaluated in different sections of the circuit. The relevant features of the currently negotiated section of the circuit are obtained from the features of the waypoint (or turn) at which the car is currently homing (Figure 2): (i) the length of the vector from the previous to the current waypoints $dist\_p\_c$, (ii) the length of the vector from the current to the next waypoints $dist\_c\_n$, and (iii) the angle between these two vectors $angle\_p\_c\_n$.

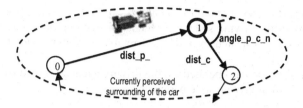

Fig. 2. Relevant features of the currently negotiated section of the circuit comprising three waypoints #0, #1 and #2. The car is currently homing at waypoint #1.

4 Evolution of Driving Rules

We assume that the four key parameters of the optimal driving style around different turns of a circuit might feature different values, and that these values should be evaluated from a single driving rule comprising four algebraic functions (expressions). Therefore, the objective of automatic design of optimal driving rules can be rephrased as an automatic discovery of such a set of four expressions that yield optimal values of the four driving style parameters for each of the turns in given circuit. In this section we elaborate on the main attributes of the genetic programming (GP) framework [8, 11], employed for automated discovery of general driving rules.

Table 1. Main Attributes of GP

Category	Value
Function set	$\{ +, -, *, / \ \}$
Terminal set	$\{$ dist_p_c $*$ dist_c_n, angle_p_c_n, random constant $[0..10] \}$
Population size	100 individuals
Selection	Binary tournament, selection ratio 0.1, reproduction ratio 0.9
Elitism	Best 4 individuals
Mutation	Random sub-tree mutation, ratio 0.01
Trial interval	Two laps around a predefined circuit
Fitness	Average lap time of two consecutive laps (ms) penalized with 0.6 ms for colliding with the "guardrails"
Termination criteria	Number of generations = 40

The genotype in the proposed GP encodes for the evolving set of four algebraic expressions corresponding to the four parameters of driving style. The function set of GP comprises the addition, subtraction, multiplication and protected division operation. The terminal set consist of the three variables which correspond the three relevant features of the currently negotiated section of the circuit (as shown in Figure 2) and a random constant within the range between 0 and 10. The main attributes of GP are summarized in Table 1.

Considering the evolving rule as a genotype and the evaluated concrete values of the four parameters of driving style as a phenotype, the fitness evaluation of GP is accomplished in the following two phases:

(i)   Developing the genotype into phenotype by setting the values of the *four* parameters of the driving style associated with each of the turns of circuit to the concrete values evaluated from the *four* expressions comprising the driving rule (Figure 3), and

(ii) A time trial of the car operated by the driving agent governed by the evaluated values of the four driving style parameters.

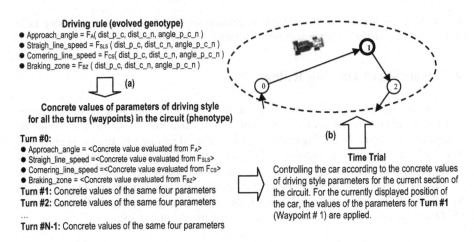

Driving rule (evolved genotype)
- Approach_angle = F_A(dist_p_c, dist_c_n, angle_p_c_n)
- Straigh_line_speed = F_{SLS} (dist_p_c, dist_c_n, angle_p_c_n)
- Cornering_line_speed = F_{CS}(dist_p_c, dist_c_n, angle_p_c_n)
- Braking_zone = F_{BZ} (dist_p_c, dist_c_n, angle_p_c_n)

(a)

Concrete values of parameters of driving style for all the turns (waypoints) in the circuit (phenotype)

Turn #0:
- Approach_angle = <Concrete value evaluated from F_A>
- Straigh_line_speed =<Concrete value evaluated from F_{SLS}>
- Cornering_line_speed =<Concrete value evaluated from F_{CS}>
- Braking_zone = <Concrete value evaluated from F_{BZ}>

Turn #1: Concrete values of the same four parameters
Turn #2: Concrete values of the same four parameters
...
Turn #N-1: Concrete values of the same four parameters

(b)

Time Trial
Controlling the car according to the concrete values of driving style parameters for the current section of the circuit. For the currently displayed position of the car, the values of the parameters for **Turn #1** (Waypoint # 1) are applied.

Fig. 3. The fitness evaluation accomplished in two phases: (a) developing the genotype (driving rule) into the phenotype (values of the driving style parameters), and (b) time trial with the driving agent being governed by the evaluated values of the four driving style parameters

5   Experimental Results

We conducted an experiment on evolution of the driving rules in order to verify the ability of the proposed evolutionary approach to automatically discover (if existent) a single driving rule that is general enough to be able to adequately control the car around all the turns in a predefined circuit. We considered the following four circuits featuring different characteristics: an O-shaped (two right, single-apex turns), 8-shaped (a right and a left, double-apex turns), S-shaped (a series of right and left turns) and 2S-shaped (concatenated two S-shaped) circuits. The results of fitness convergence aggregated over 20 independent runs of GP and the driving line of the car controlled by a sample best-of-run driving rule are illustrated in Figure 4. As Figure 4 shows, the average of the best lap time over all runs improves from 3850ms to 3600ms for O-shaped (Figure 4a), from 5410ms to 5000ms for 8-shaped (Figure 4b), from 4800ms to 4120ms for S-shaped (Figure 4c), and from 7810ms to 7060ms for 2S-shaped (Figure 4d). The result of the evolution of driving rules on all four circuits, intended to verify the generality of the evolved driving rules across several circuits with diverse characteristics is shown in Figure 4e). As Figure 4e) illustrates, the aggregated lap time improves in average from 25430ms to 21520ms within 40 generations.

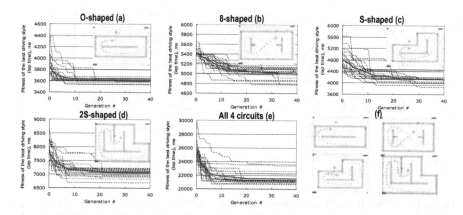

Fig. 4. The fitness convergence and the driving line of a sample best-of-run driving rules, evolved in 20 independent runs on O- (a), 8- (b), S- (c) and 2S-shaped circuits (d), respectively. The results of evolution on all four circuits are shown in (e), and the sample best-of-run driving lines are depicted in (f).

Table 2. The values of the driving style parameters, evaluated from sample evolved best-of-run driving rule controlling the car as illustrated in Figure 5

# CS	Straight-line Gear	Turning Gear	Approach angle, degrees	Throttle lift-off zone, mm
0	4	1	17	380
1	4	1	48	460
2	4	1	44	120
3	4	1	37	140
4	4	1	11	370
5	4	1	17	170

The emergent driving line with the breakdown of the velocity of the car governed by sample evolved best of run driving rule on 2S circuit is shown in Figure 5. The corresponding values of the driving style parameters, computed for each of the six waypoints of the circuit are shown in Table 2.

In order to estimate the degree of optimality of the evolved general driving rules, we conducted a comparative analysis of the lap times of the agent governed by these general rules with the lap times of the agent controlled by the values of driving style parameters, directly evolved for each of the turns of the considered circuits. In order to directly evolve these values, we employed genetic algorithms (GA) with the main attributes which are identical to those of the GP used for evolution of the driving rules. AS the results, summarized in Table 3 indicate, the driving rules evolved via GP on particular circuits performs equally well (and, occasionally even better) than the fixed, well-tailored (via GA) values of the driving style parameters. The maximal degradation of about 7% of the lap time associated with the driving rules in S-shaped circuit indicates the favorable generality of the evolved driving rules.

The slight superiority of the evolved driving rules over the optimized values of driving style parameters in O-shaped and 8-shaped circuits (3.1% and 0.2% faster,

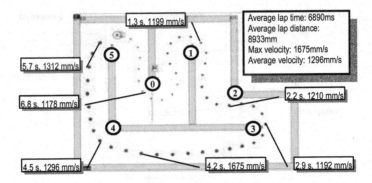

Fig. 5. The driving line and the breakdown of velocity of the car controlled by evolved sample best-of-run driving rule. The start-finish line of the illustrated 2S circuit is located next to the waypoint #0. The running direction is clockwise. The principal points of lap are marked with the timestamp information (the time into the lap) and the velocity of the car.

Table 3. Comparison of the lap times obtained through a direct optimization of the driving style parameters employing GA and the driving rules, evolved via GP

Circuit / Evolutionary approach	O-shaped Lap time, ms	O-shaped Degradation, %	8-shaped Lap time, ms	8-shaped Degradation, %	S-shaped Lap time, ms	S-shaped Degradation, %	2S-shaped Lap time, ms	2S-shaped Degradation, %
GA for a particular circuit	3721	---	5011	---	3848	---	6684	---
GP for a particular circuit	3604	-3.1	5002	-0.2	4120	7.0	7060	5.6
GP for all four circuits	4095	13.6	5430	8.4	4725	22.8	7219	8.0

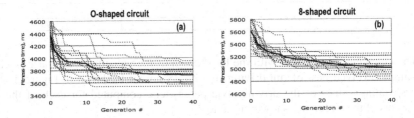

Fig. 6. Fitness convergence of 20 independent runs of GA evolving the values of the four driving style parameters in O-shaped (a) and 8-shaped (b) circuits respectively

respectively) can be explained by the reduction size of the evolutionary search of GP compared to that of GA. Indeed, as both the O-shaped and 8-shaped circuits feature identical turns, way the car negotiates these turns can be naturally expressed by a single driving rule. Conversely, GA needs to evolve the values of the driving style parameters for each of the turns separately, and the equality of the values of these parameters should be explicitly discovered. The presumed implications of the size of search space in both GP and GA for O- and 8-shaped circuits are illustrated by the corresponding fitness convergence characteristics. The fitness of GP converges faster (Figure 4a and 4b) than that of GA (Figure 6a and 6b), which suggests that a delaying the termination of the evolution (e.g., by setting the termination criteria of GA to more than 40 generations) would facilitate the additional improvement of lap times. Consequently, the superiority of GP over GA on O- and 8-shaped circuits should be acknowledged only in the condition of the considered budget of the evolutionary runs of no more than 40 generations.

6 Conclusion

We presented an automated design of driving agent, able to operate a software model of fast running car and verified the ability of the employed evolutionary paradigm to automatically discover a single driving rule that is general enough to adequately control the car around all the turns in a predefined circuit. The optimality of the evolved general driving rules is shown by the favorable comparison against the corresponding lap times obtained by direct evolutionary optimization of the values of the main driving style attributes.

In our future work we are planning to investigate how well the rules, evolved in predefined circuits would perform in a priori unknown circuits. Also, we contemplate a possible need of adaptation of the driving agent evolved in predefined circuits to the characteristics of unknown ones.

References

1. Bentley, R.: Speed Secrets: Professional Race Driving Techniques. Motorbooks International (1998)
2. Fogel, D.B.: Blondie24: Playing at the Edge of AI. Morgan Kaufmann, San Francisco (2001)
3. Frere, P.: Sports Cars and Competition Driving. Bentley Publishing (1992)
4. Funge, J.D.: Artificial Intelligence for Computer Games. Peters Corp. (2004)
5. Gillespie, T.: Fundamentals of Vehicle Dynamics. Society of Automotive Engineers International (1992)
6. Google Maps, Image of the junction near Matsubara city in Osaka Prefecture, Japan, http://maps.google.com/maps?ll=34.59,135.575&spn=0.00354,0.0042
7. IBM Corporation, Deep Blue (1997), http://www.research.ibm.com/deepblue/
8. Koza, J.R.: Genetic Programming: On the Programming of Computers by Means of Natural Selection. MIT Press, Cambridge (1992)
9. Robocup (2005), http://www.robocup.org/02.html
10. Suzuki, M., Floreano, D.: Active Vision for Neural Development and Landmark Navigation. In: 50th Anniversary Summit of Artificial Intelligence, pp. 247–248 (2006)
11. Tanev, I., Shimohara, K.: XGP: XML-based Genetic Programming Framework. In: Proceedings of the 34th Symposium of the Society of Instrument and Control Engineers (SICE) on Intelligent Systems, pp. 183–188 (2007)
12. Tanev, I., Joachimczak, M., Shimohara, K.: Evolution and Adaptation of an Agent Driving a Scale Model of a Car with Obstacle Avoidance Capabilities. In: Nolfi, S., Baldassarre, G., Calabretta, R., Hallam, J.C.T., Marocco, D., Meyer, J.-A., Miglino, O., Parisi, D. (eds.) SAB 2006. LNCS (LNAI), vol. 4095, pp. 619–630. Springer, Heidelberg (2006)

13. Tanev, I., Shimohara, K.: On Human Competitiveness of the Evolved Agent Operating a Scale Model of a Car. In: Proceedings of the IEEE Congress on Evolutionary Computation (CEC 2007), Singapore, September 25-28, 2007, pp. 3646–3653 (2007)
14. Togelius, J., Lucas, S.M.: Evolving Controllers for Simulated Car Racing. In: Proceedings of IEEE Congress on Evolutionary Computations (CEC 2005), Edinburgh, UK, September 2-5, pp. 1906–1913 (2005)
15. Wloch, K., Bentley, P.: Optimizing the Performance of a Formula One Car Using a Genetic Algorithm. In: Proceedings of the 8th International Conference on Parallel Problem Solving from Nature, Birmingham, UK, September 18-22, pp. 702–711 (2004)

BehaviorSim: A Learning Environment for Behavior-Based Agent

Fasheng Qiu and Xiaolin Hu

Computer Science Department, Georgia State University, Atlanta, GA 30303

Abstract. Behavior-based control is one of the fundamental control paradigms for autonomous agents to achieve adaptive behavior in a dynamical environment. Existing work has mainly focused on the research aspect rather than on the learning and educational aspect. This paper presents an effort to develop a learning environment for behavior-based agents. It allows educators and students to develop and exercise behavior-based control by setting up entities, behaviors, and behavior networks without involving significant programming effort. Specification of behavior-based agent system is presented and demonstrative examples are provided.

Keywords: Learning environment, BehaviorSim, behavior-based control, behavior network, action selection mechanisms.

1 Introduction

Behavior-based control is one of the fundamental control paradigms for autonomous agents to achieve adaptive behavior in a dynamical environment [1], [2]. It finds applications in many different fields such as robotics, AI, computer game design, and social crowd simulations. The behavior-based control features a set of behaviors working in parallel, each of which corresponds to a specific behavior of the agent. This results in a distributed control paradigm as oppose to the centralized deliberative control. Despite the tremendous work in behavior-based control, existing work has mainly focused on the research aspect rather than on the learning and educational aspect. It requires significant experience as well as programming skills in order to implement a behavior-based application, thus makes it difficult for beginners to learn and exercise the behavior-based control paradigm. This paper presents an effort to develop the BehaviorSim environment that supports learning of behavior-based control by defining simulated agents in an intuitive manner corresponding to the behavior-based paradigm. It allows educators and students to develop and exercise their own behavior-based applications by setting up entities, behaviors and other primitives without involving significant programming effort. The term "behavior-based agent" used in this paper is adapted from behavior-based robotics [3]. Here an agent is a general term that can be a robot, an animat, an artificial agent in AI, a character in game design, or a simulated individual in social crowd simulation.

One of the major challenges of developing a learning environment for behavior-based control is to adopt a general and well-defined architecture to capture the

M. Asada et al. (Eds.): SAB 2008, LNAI 5040, pp. 499–508, 2008.

essence of behavior-based control. Such an architecture should also be intuitive enough to allow beginners to easily understand and grasp it. Unfortunately, behavior-based control has taken many different formats and currently there is no such a "standard" behavior-based architecture (see a discussion in [1]). Several basic architectures exist, such as the Subsumption architecture [4], Activation spreading network [5], and Motor schema architecture [3], each of which has a different working mechanism. Some also include concepts, such as "goal" in the Activation spreading network [5], that are not explicitly supported by others. Furthermore, there is no common definition of what is a behavior and different works may define behaviors at different levels of details. For example, the work of [6] allows high level behavior modules (called options) to be broken down into low levels modules and basic behaviors. The approach that we adopted in this paper is mainly inspired from three sources of work: 1) the neurobiological study, in particular Edwards' work [7], of the mutual inhibition mechanism to account for animals' adaptive behavior in a dynamical environment, 2) the work of behavior-based robotics [3], and 3) the work of "Boids" [8] that demonstrates a variety of steering behaviors in a simulated environment. Inspired from these works, we define a general framework that captures the key components of behavior-based control (see details in Section 3). More advanced features, such as multi-layer control, hierarchical behavior decomposition, and concepts such as motivations and emotions are not explicitly considered for the purpose of easy to learn for new beginners. However, these features could be made explicit in future extensions of this work.

We embody these concepts in a mobile agent simulation environment where users can define and exercise their own behavior-based applications in an interactive way and immediately see how the agents work. The simulation environment is carefully designed so the defining of agents and their behaviors match closely to the major components of behavior-based control. Meanwhile, the environment is general enough to allow different users to define their own different applications (three demonstrative examples are given in section 6). It promotes an engaged learning environment where users can create, save, load, and modify their own behavior-based control applications.

2   Related Work

As a background of this work, we review several basic behavior-based architectures that are closely related to this work. Our review focuses on the aspect of action selection, i.e., how different behaviors are coordinated. According to [9], two primary coordination mechanisms are: competitive method where the output of the action selection is a single behavior; and cooperative method where the output is a combination of all the active behaviors. The Subsumption architecture [4], the activation spreading network [5], and some voting-based methods belong to the category of competitive coordination mechanism. The neurobiological work [7] studied an asymmetry mutual inhibition behavior network, where different behaviors mutually inhibit each other to compete for controlling a simulated crayfish model. At any time only one behavior takes control. Thus it also belongs to the competitive coordination mechanism. The Motor schema architecture developed in [3] is based on a cooperative action selection mechanism, where the output of the system is computed by vector summation of all the schema output vectors. Other approaches such as the ones using principles of potential fields and behavioral blending also belong to the cooperation coordination mechanism.

Compared to the research work in this field, less work has devoted to the learning and educational aspect of behavior-based control. In the limited work that we are aware of, the work of Reynolds' Boids [8] has inspired many to learn the behavior-based paradigm. The Boids website hosts a description of the model as well as an animation. It also shows a set of steering behaviors of autonomous agents such as pursue and evade, collision avoidance, and path follow. However, no environment is provided to allow users to define and experiment with their own simulated agents with behavior-based control. Several education-related works exist in the field of robotics. Blank et al. [10] introduced Pyro, a Python-based programming framework which provides a set of abstractions that allow students to write platform-independent robot programs. The work of [11] develops an agent-oriented behavior-based interface framework for educational robotics. Both of these works support behavior-based control. However the main focus of their work is on learning and programming educational robots. Agentsheets [12] is an agent-based simulation environment that provides user-friendly and intuitive interfaces for users to configure their own simulations by specifying scenes and the action rules for different agents. Another agent-based environment used for educational purpose is Netlogo [13], which provides a multi-agent programmable modeling environment where a scripting language is provided to code the agent behaviors as different procedures. These environments do not explicitly focus on agents with behavior-based control. This paper is extended from our previous work [14] that describes an early design and implementation of the BehaviorSim environment.

3   Behavior and Behavior Network

A major work in developing the BehaviorSim learning environment is to adopt a behavior-based architecture that captures the essence of behavior-based control. To do that we need to have an unambiguous view of what is a behavior and how multiple behaviors work together. As mentioned before, we adopted an approach inspired from the neurobiological work of mutual inhibition, behavior-based robotics, and the work of Boids. Specifically, we view a behavior as an independent computation module that can fulfill some particular task for an agent. Each agent can have multiple behaviors that run in parallel. These behaviors form a behavior network that defines how different behaviors compete or cooperate with each other for controlling the agent. This behavior network and its working mechanism represent the behavior-based control architecture and act as the decision making component of the agent. Below we describe the structures of behavior and behavior network used in BehaviorSim.

In general, a behavior is excited by some (external) sensory stimulus and/or (internal) states of the agent. It defines some actions to fulfill the task associated with this behavior. Based on this view, in BehaviorSim a behavior is characterized by three elements: an *activation* value, an *excitation* module, and an *action* module. The *activation* is a real number and represents the level of strength of this behavior. Its value is computed from two sources: 1) stimulus of the sensory inputs and/or internal states, and 2) inhibitions/excitations from other behaviors. This value is computed in every time step and used in the action selection of the behavior network. The *excitation* module defines how this behavior is excited by the sensory inputs and/or internal states. It returns a value (a real number) called *excitation*. In BehaviorSim, this module is specified by the user as a piece of code and is an essential part of a behavior. Two things are worthy to mention here. First, the concepts of sensing and internal states are not explicitly modeled by BehaviorSim. They are represented by methods and variables that the user can define and use based on their specific needs. Second, although BehaviorSim currently does not explicitly model concepts such as emotion and motivation, these concepts can be implicitly defined as properties of the agent and used by the excitation module to change the excitation of the behavior.

The *action* module of a behavior specifies the actions that will be carried out by the behavior if it is selected (a selected behavior is also referred to as *active*). Similarly, this module is defined by the user as a piece of code and is an essential part of a behavior. Typically, a behavior's action defines a "one-step" task, for example, returning a speed vector that drives the agent to move for one step (if it is active). However, there are situations where a behavior needs to supports a sequence of tasks in order. For example, considering a "dance" behavior of a mobile robot, if this behavior remains to be active, the robot should dance (move) in a particular order, e.g., moving left first, then right, then.... To support this kind of sequential task, a *task queue* can be set up in the action module. A task queue defines an array of tasks that will be sequentially executed as long as the behavior continues to be active. When all the tasks in the queue have been executed, the task queue is reset and then executed from the beginning again. A behavior's task queue can be *resumeable* or *non-resumeable*. A resumeable task queue allows a re-activated (from non-active) behavior to resume from the task that was previously interrupted. This is useful to keep track of the progress of a task sequence and to continue the task from where it stops. A non-resumeable task queue is always reset if it is interrupted. Similar to the works [15], BehaviorSim allows a task queue to be set up in a hierarchical manner and supports both sequential tasks and concurrent tasks. Pseudo code of how to set up a task queue is provided in section 4.

Multiple behaviors can form a behavior network that specifies how these behaviors compete or cooperate with each other. In BehaviorSim, a behavior network is characterized by three types of elements: a set of *behaviors*, an *action selection mechanism*, and a set of *coefficients (or weights)* that define the connections among the behaviors. The action selection mechanism defines how the set of behaviors work together. Currently, two action selection mechanisms, namely the *mutual inhibition/excitation* mechanism and the *cooperation* mechanism, are supported. Figure 1 illustrates these two mechanisms, where the blue circles are the behaviors.

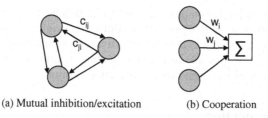

(a) Mutual inhibition/excitation   (b) Cooperation

Fig. 1. Two action selection mechanisms

$$Activation_j = Excitation_j(S_j, I_j) + \sum_{i}^{i \neq j} c_{ij} * Activation_i \qquad (1)$$

$$Sv = Sv_{winner} \quad \text{where} \quad winner = Select(Activation_1, ..., Activation_n) \quad (2)$$

$$Activation_j = Excitation_j(S_j, I_j) \qquad (3)$$

$$Sv = \sum_{i} k_i * w_i * Sv_i \quad \text{where} \quad k_i = \begin{cases} 1 & \text{if } Activition_i \geq T_{threshold} \\ 0 & \text{if } Activition_i < T_{threshold} \end{cases} \qquad (4)$$

The mutual inhibition/excitation mechanism is a competitive mechanism that allows only one behavior to be selected at any time step based on the winner-take-all principle. In this mechanism, different behaviors asymmetrically inhibit/excite each other through the inhibitory/excitatory *coefficients*, which define the level of inhibition/excitation. A coefficient is a real number ∈ [-1, 1] with a negative value meaning inhibition, a positive value meaning excitation, and 0 meaning no inhibition or excitation. The amount of inhibition/excitation from one behavior to another is computed according to the product of the first behavior's activation and the corresponding coefficient. Formula (1) shows how a behavior (*behavior_j*)'s activation is calculated. In this formula, c_{ij} is the coefficient from *behavior_i* to *behavior_j*; *Excitation_j(S_j, I_j)* represents *behavior_j*'s excitation, which is a function of the behavior's internal states S_j and sensory inputs I_j. After all behaviors' activations are calculated, the behavior network selects the behavior with the highest activation level as the winner behavior. This winner behavior controls the action of the agent for this time step. In the context of mobile agent, Formula (2) shows that the agent's speed vector *Sv* is defined by the action (speed vector) of the winner behavior. Note that the *Select()* method selects the winner behavior by comparing the activations of all behaviors. Different from the mutual inhibition/excitation mechanism that selects only one behavior, the cooperation mechanism combines the actions of multiple behaviors based on the vector sum principle. The amount of contribution of each behavior is defined by a weight $w \in [0, 1]$. Formula (3) shows that in this mechanism, the activation of a behavior only includes the excitation part (since no inhibition/excitation from other behaviors). Formula (4) shows the speed vector *Sv* of the agent is calculated as the vector sum of the actions of those behaviors whose activations are greater than or equal to a pre-defined threshold $T_{threshold}$.

The set of coefficients (weights) of the behavior network plays important roles in selecting the winner behavior or in vector summing the actions of multiple behaviors. In BehaviorSim, the values of these coefficients (weights) can be defined as constants, or be dynamically computed in every time step based on user-defined rules. For example, it makes sense to dynamically adjust the weights of different behaviors used in Formula (4) according to the current activations of those behaviors. The work of [16] shows another example where the mutual inhibition coefficients among behaviors are dynamically changed under different conditions.

4   System Specification

This section gives a formal description of a behavior-based agent system in BehaviorSim, followed by an algorithm for the action selection that describes how the simulation works.

Behavior-based Agent System Specification

In BehaviorSim, a behavior-based agent system is described by a tuple <*World, Entities*>, where *World* is the 2D environment and *Entities* is the set of entities.

- The *world* is described by a tuple <*Dimension, Type*>, where *Dimension* specify the dimension of the 2D map. *Type* = {*closed* | *open* | *rounded*}, where *closed* means there exists a "wall" surrounding the field that prohibits mobile agents from moving outside of the field; *open* means the field is open thus mobile agents can move outside of the field and disappear; *rounded* means that an agent moving outside an edge of the map will automatically appear on the other side of the map.
- The *Entities* includes a set of *stationary entities* and a set of *autonomous agents*.
  - A *stationary entity* represents a stationary environmental object such as food, obstacle, or a source for sensing or a destination point, meaningful for the autonomous agents. A stationary entity is described by a tuple <*EntityId, Position, Dimension, Category, Property*>, where *EntityId, Position, Dimension* are the identity, the position and the size of the entity; *Category* is the entity type, such as "food", "obstacle". Multiple entities can belong to the same category and thus share common information of the category. *Property* is a set of properties that describe domain specific information of the entity. For example, a *food* entity can have a property of *amount*.
    - Each *property* is described by a tuple < *Value, Dynamics*>, where *Value* is the value of the property and *Dynamics* is a function specifying how the value of that property may dynamically change in every time step. For example, considering the amount property of a food entity, the dynamics function *amount* = *amount* + *0.05* can be defined to indicate that the food amount will continuously grow (by 0.05) in every time step.
  - An *autonomous agent* represents a behavior-based agent such as a robot, an animat, or a game character. It is described by a tuple <*EntityId, Position, Dimension, Category, Property, Speed, Behavior network*>, where *EntityId, Position, Dimension, Category, Property* are the same as

those of stationary entities. *Speed* represents the current speed vector of the autonomous agent. *Behavior network* is responsible for the decision making of the agent.

> ➤ The *Behavior network* is described by a tuple *<Behaviors, Action selection mechanism, coefficients/weights>*, where *Behaviors* is a set of behaviors; *Action selection mechanism=* {*mutual inhibition/excitation | cooperation*}; and *coefficients/weights* are a list of coefficients/weights of the behavior network. Each *behavior* is described by a tuple *<Activation, Excitation, Action>*. See Section 3 for more details.

Both the *Excitation* and *Action* modules of a behavior can invoke method calls. BehaviorSim provides a set of primitive system APIs, such as *move(speed, direction)* to move an agent, that can be called by users when defining behaviors. Besides these system APIs, users can also define their own methods and use them in defining behaviors (each method essentially is to get or to set some properties of the corresponding stationary entities or mobile agents). For example, in the crayfish example shown in Section 6, a *eat()* method is defined and used by the action module of the eat behavior. This method decreases the *amount* property of the food and also increases the *energy* property of the crayfish. BehaviorSim also allows a user to set up a plan for a set of sequential tasks. This is achieved by a task queue. The pseudo-code bellow shows an example of defining a task queue. As can be seen, the task queue can be set up in a hierarchical way (see the **composite_task**, which has sub-tasks). Multiple tasks can also be added into the task queue as *concurrent tasks* (executed in one time step), or *sequential tasks* (executed in different time steps). In the code, **atomic_task1** and **atomic_task2** are concurrent tasks.

```
If (condition is true) {
    AddToTaskQueue(composite_task(...)) }
else {
    AddToTaskQueue(atomic_task1(...); atomic_task2(...))
    AddToTaskQueue(atomic_task3(...)) }
```

Action Selection
Action selection is conducted in a cyclic manner as the simulation proceeds. In every time step, each autonomous agent goes through the following major steps:

1 Compute the coefficients/weights if they are not constants.
2 Calculate excitation of each behavior: *Excitation_j(S_j, I_j)* in Formula (1) or (3).
3 For the *mutual inhibition/excitation* mechanism, calculate activation of each behavior using Formula (1). Skip this step for the *cooperation* mechanism.
4 For the *mutual inhibition/excitation* mechanism, select the one behavior (the winner behavior) with the highest activation. For the *cooperation* mechanism, select all the behaviors whose activations are greater than the threshold.
5 Execute the actions associated with the selected behaviors. For the *mutual inhibition/excitation* mechanism, only one behavior is selected, thus execute the action for that behavior. For the *cooperation* mechanism, vector sum all the

speed vectors returned from the selected behaviors' action modules. In both cases, if there is a task queue defined for a selected behavior, reset the task queue if the queue is empty. Otherwise, execute the next task in the queue.

6 The executed actions modify the properties of the corresponding entities.

5 The BehaviorSim Environment

The BehaviorSim software can be downloaded from http://www.cs.gsu.edu/~cscxlh/ BehaviorSim.htm. It allows users to define a simulation system including behavior-based agents and then run simulations to see the result. It has four major windows: a system editor window, a category definition window, a behavior network editor window, and a simulation window. The first three windows allow a user to setup the agent system, to define a category, and to define the behavior network for an agent, respectively. The simulation window allows a user to run simulations and see the results. Figure 2 shows the system editor window and the behavior network editor window. To set up an application in BehaviorSim, the following four steps are generally followed: 1). Define entity categories using the category definition window. Set up the common properties and methods for that category. 2). Set up the entities in the agent system using the system editor window. This can be done by dragging and dropping a corresponding category icon into the world (see Figure 2(a)). 3). Define the behavior networks for autonomous agents using the behavior network editor window as shown in Figure 2(b), where behaviors and the behavior network can be specified. 4). After all configurations are done, run simulations to see how agents behave. All the configurations can be saved as a XML file that can be loaded and modified in the future.

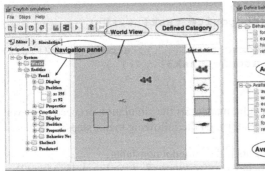

(a) System editor window

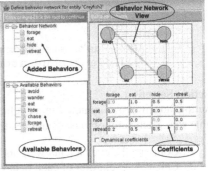

(b) Behavior network editor window

Fig. 2. User Interface of BehaviorSim

6 Demonstrative Examples

Three examples belonging to three different applications are developed to demonstrate the BehaviorSim environment. The first example is based on the work of

[7] to simulate a crayfish model in an environment with a food source, a shelter, and a predator. In this example, the crayfish has seven behaviors: *escape* (escape from predator), *retreat* (retreat to shelter), *defend* (defend against predator), *swim, hide* (hide in shelter), *forage* (forage for food), and *eat* (eat food to gain energy). These behaviors form a mutual inhibition behavior network and enable the crayfish to avoid the predator, to find and eat food, and to find and hide in a shelter. Details of these behaviors (including their excitation formulas) and the behavior network can be found in [7]. The predator is also controlled by a behavior network with three behaviors: *wander, avoid,* and *chase* (chase the crayfish if it is in its visibility range). The *food* and *shelter* are two static entities, with the food having a property of *food amount*. The second example is based on the work of [16]. It concerns a dynamical team formation including six robots, one of which is explicitly specified as the team leader through its identity. Each robot has four behaviors: *search* to search for neighboring robots before forming a team, *avoid* to avoid collision with other robots, *follow* to follow its front robot, and *wait* to wait for its back robot during convoy. These behaviors form a mutual inhibition behavior network and allow randomly deployed robots to dynamically form a team (based on identity) and then convoy in a line formation. The set of behavior network coefficients change before and after a robot finds its neighboring robots. Details can be found in [16]. The third example is based on the work of Boids [8] to show a flock of creatures, each of which has four behaviors: *random move, separation, alignment* and *cohesion*. Random move moves the creatures in a random direction. The other three behaviors are simplified version of the ones described in [8]. These behaviors form a cooperation behavior network with the weight of each behavior proportional to the excitation of the behavior. Space does not allow us to elaborate these examples. The major configurations of these examples are summarized in Table 1. For each example, a simulation movie as well as the system model (the XML file that describes the configurations of the agent system) can be found at: http://www.cs.gsu.edu/~cscxlh/BehaviorSim.htm.

Table 1. Three Demonstrative Examples

Examples	Stationary entities	Autonomous agents	Action Selection Mechanisms	Behaviors	Behavior network coefficient/weight
Crayfish behaviors simulation	1 food, 1 shelter	1 crayfish, 1 predator	Mutual inhibition	**Crayfish**: forage, hide, eat, retreat, escape, defend, swim **Predator**: chase, wander, avoid	Constant coefficients
Robot team formation	N/A	6 robots	Mutual inhibition	Avoid, follow, search, wait	Dynamic coefficients
Boids	N/A	15 boids	Cooperation	Random move, cohesion, alignment, separation	Dynamic weights

7 Conclusion

This paper presents an effort to develop BehaviorSim that is a learning environment for behavior-based agent. The environment is designed to allow users to define and play with the behavior-based control paradigm in a configurable manner. A specification of behavior and behavior network for the purpose of learning is developed. BehaviorSim is currently incorporated into a modeling and simulation class to be used by students. Future work includes improving the usability of the software and adding more advanced features.

References

1. Bryson, J.: Cross-Paradigm Analysis of Autonomous Agent Architecture. Journal of Experimental and Theoretical Artificial Intelligence 12(2), 165–189 (2000)
2. Mataric, M.J.: Behavior-Based Control: Main Properties and Implications. In: Proceedings, IEEE International Conference on Robotics and Automation, Workshop on Architectures for Intelligent Control Systems, May 1992, pp. 46–54 (1992)
3. Arkin, R.C.: Behavior-Based Robotics. The MIT Press, Cambridge (1998)
4. Brooks, R.A.: A Robust Layered Control System for a Mobile Robot. IEEE Journal of Robotics and Automation 2(1), 14–23 (1986)
5. Maes, P.: A Bottom-Up Mechanism for Behavior Selection in an Artificial Creature. In: From Animals to Animats. MIT Press, Cambridge (1991)
6. Lötzsch, M., Bach, J., Burkhard, H.-D., Jüngel, M.: Designing Agent Behavior with the Extensible Agent Behavior Specification Language XABSL. In: Polani, D., Browning, B., Bonarini, A., Yoshida, K. (eds.) RoboCup 2003. LNCS (LNAI), vol. 3020, pp. 114–124. Springer, Heidelberg (2004)
7. Edwards, D.: Mutual inhibition among neural command systems as a possible mechanism for behavioral choice in crayfish. J. Neurosci. 11, 1210–1223 (1991)
8. Boids, http://www.red3d.com/cwr/boids/
9. Petez, M.C.: A proposal of behavior-based control architecture with reinforcement learning for an autonomous underwater robot. PhD thesis, University of Girona (2003)
10. Blank, D. etc.: Pyro: A Python-based Versatile Programming Environment for Teaching Robotics. ACM Journal of Educational Resources in Computing 3(4) (2003)
11. Azhar, M.Q., Goldman, R., Sklar, E.: An agent-oriented behavior-based interface for educational robotics. In: Agent-Based Systems for Human Learning (ABSHL) Workshop at Autonomous Agents and Multiagent Systems (AAMAS-2006) (2006)
12. Repenning, A., Ioannidou, A., Zola, J.: AgentSheets: End-User Programmable Simulations. Journal of Artificial Societies and Social Simulation 3(3) (2000)
13. NetLogo 3.1.3 User Manual (2006), http://ccl.northwestern.edu/netlogo/
14. Lakhtanau, P., Hu, X., Qiu, F.: BehaviorSim: Towards an Educational Tool for Behavior-Based Agent. In: Proc. The 45th ACM Southeast Conference (ACMSE 2007) (2007)
15. Simmons, R., Apfelbaum, D.: A Task Description Language for Robot Control. In: Proceedings Conference on Intelligent Robotics and Systems (1998)
16. Hu, X., Edwards, D.: Context-Dependent Structure Control for Adaptive Behavior Selection. In: Proc. Workshop on Bio-inspired Cooperative and Adaptive Behaviours in Robots, in co-operation with The Ninth International Conference on the Simulation of Adaptive Behavior (SAB 2006) (2006)

Adaptive Behavioural Modulation and Hysteresis in an Analogue of a Kite Control Task

Allister Furey and Inman Harvey

Centre for Computational Neuroscience and Robotics
University of Sussex, Brighton, United Kingdom
a.d.j.furey@sussex.ac.uk, inmanh@sussex.ac.uk

Abstract. We define a simplified analogue of a kite control task that requires, in its simplest form, a situated artificial agent to switch between two mutually exclusive behaviours. In more complex versions of the task, the agent is required to adapt to changes within its environment that occur on different temporal scales. We describe the failure to evolve successful agents when a decision threshold is defined artificially and conversely the evolution of successful agents when they themselves are allowed to determine their own threshold through interaction with the environment. Agents are demonstrated capable of adapting both their switching behaviour and spatial domain according to environmental changes on three temporal scales, on the fastest of which, the agents behave in an opportunistic manner.

1 Introduction

There are novel wind energy systems in development using tethered airfoils – kites, as the wind intercepting element. The kites require active control that ideally modulates the flight behaviour according to environmental conditions, often to maintain the operation of the system within certain hard limits of viability. Additionally, the most efficient currently posited system configuration requires alternation between two discrete modes of operation. The point of switching between these two, mutually exclusive actions or 'behaviours' could be determined by a simple but rigid threshold or heuristic that is predetermined. However, it is likely to be beneficial to have an adaptive strategy that modulates behavioural hysteresis actively through some mechanism that takes into account the current state of the agent and its environment.

Here we present a simplified analogue or 'toy version' of the kite control task where an environmental resource with spatially heterogeneous distribution is exploited by an evolved agent. The task is designed so as to mirror key aspects of the kite control problem at a 'higher' level i.e. that of behavioural modulation and/or hysteresis without the complication and heavy computational burden of simulating the dynamics of kite flight, treated elsewhere [2,3]. The task is designed to be simple enough to allow evolution to proceed rapidly whilst still retaining a dynamical embodied interaction between the agent and its environment. Assessing the adaptive abilities of the evolved agents may indicate the potential of evolutionary robotics for generating kite control behaviours. Additionally, determining the best way in which to

M. Asada et al. (Eds.): SAB 2008, LNAI 5040, pp. 509–518, 2008.

evolve adaptive behaviour in this simple analogue task may inform future work in generating similar behaviours in real kite control.

2 Background and Motivation

The use of power kites for production of renewable energy from the wind is highly attractive due to the prospect of accessing the stronger and more consistent winds at altitudes above that of conventional wind turbines. This is especially so given that power generated increases with the cube of the wind velocity [4,6], meaning large gains in power output per unit area of the wind intercepting element can be realised. The lack of requirement for expensive and visually intrusive towers and much reduced territorial occupation [4,5] compounds the potential socio-economic advantages of such systems.

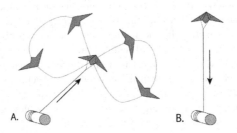

Fig. 1. Generation (reel out) phase A and retraction (reel in) phase B

Most current concepts (for review see [5]) for implementing kite energy operate with two functional phases as per Fig. 1. In the first (A), the kite is steered actively through the airspace in order to increase the apparent wind speed across the kite. The kite therefore pulls its lines out from a reel at ground level which is coupled to a dynamo, thus generating electricity. We have demonstrated previously in simulation [2,3] that an evolutionary robotics approach can be used to produce neural network controllers that both find an optimal flight trajectory for phase A and maintain it in the face of significant environmental perturbations. (see Fig. 2, right). In the second phase (B), the lines are retracted with the kite controlled so as to reduce resistance, the process is therefore cyclical and sustainable indefinitely whilst the wind allows. As long as A produces more energy than consumed in B then there is a net gain of energy. The kite control problem has some further key characteristics, firstly, the environment is highly dynamic, both the wind speed and direction can change rapidly. Secondly there are some hard constraints within which the agent must operate in order to keep the kite aloft and not overpower the hardware, as discussed below. Given these characteristics a high quality controller will have to perform a) in-behaviour adaptive modification in order to maximise productivity whilst b) meeting the viability limits imposed by the operational constraints.

Our exercise here of building a task analogous to the kite control problem is intended primarily to be informative as to the validity of an evolutionary robotics (ER) approach to tackling its behavioural aspects and to gain insight into how it might best be implemented. With this analogous task, without necessarily expecting direct cross-applicability, we aim to address two components of the kite control problem that may benefit from an adaptive controller. The first issue is one of switching between the two separate behaviours, that of generation and retraction. There is some time or performance cost to the switching from generation to retraction, given that the kite should be ideally be steered to zenith or the side of the wind window in order to minimise both the force and the time required to retract the lines. A heuristic of switching at predetermined points at some sensible intermediate values has been employed previously in the literature [4] constituting one potential strategy. Another possibility is much more frequent alternation, whereby retraction occurs within a given portion of every figure eight trajectory as posited by Houska [6]. The latter strategy avoids some of the problems of the former one, namely a reduction in yield at lower altitudes, and that of potentially reaching a hard line length limit and resulting wasted generation time or hardware damage. However, by not placing the kite at zenith or windows edge, greater energy needs to be expended to retract the kite and the momentum of the kite is reduced. We intend to approach this trade-off with our toy problem using standard ER techniques, and the as yet unaddressed question of whether to modify the switching points according to environmental conditions, a question closely coupled to the second issue.

This second issue is specifically related to the spatial distribution of the wind resource in relation to the kite tether point. As mentioned above, in normal wind conditions, the kite is steered in a looping pattern (see Fig. 2, right) directly downwind of the tether point (the ++ and + area in Fig. 2, left), in order to maximise the aerodynamic forces away from the reel and therefore the energy that is generated. However, in strong wind conditions, this strategy will generate forces that may damage the generating hardware, the kite, or its lines.

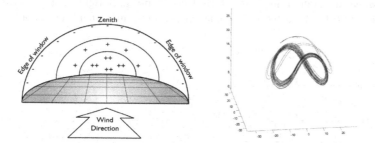

Fig. 2. (Left) The wind window concept. The most powerful portion of the window is signified by ++, less powerful by + and the least powerful by -. (Right) A plot of the trajectory of the kite wingtips in a 42 second trial as controlled an evolved neural network, starting from zenith.

This overpowered situation can be prevented by simply avoiding the most powerful region of flight space. As the wind speed increases, the extent of the avoided region should increase. Conversely with a weak wind resource the agent must confine itself to the most powerful region of flight space. These changes constitute short term, reactive modifications to the kite's spatial domain of activity and are the second level of adaptation that we shall address in the model. As we detail below, we intend to explore how changes in the intensity of the light source on different temporal scales affect the agent and its behaviour.

3 Methodology

3.1 Agent and Task Design

We have designed a task that captures some of the key properties of the kite generation strategy problem without the computational burden of modelling the aerodynamics of the kite. Our task is based around a simple 2 wheeled, 2 sensor agent (see Fig. 3, left) situated within an unbounded 2d area. The analogy for the flying lines is a battery, charged by an orientation independent 'solar panel' on the agent. As with the lines, there are upper and lower hard limits to the battery level. As success is judged by quantity of energy passed to the battery, the agent can score more highly by repetitively switching between charging and discharging the battery, corresponding to the reeling out and in of the lines respectively.

The agent's sensors are mounted 90 degrees apart on its perimeter, each has a 180 degree range and provides a signal describing its local light intensity. Within the simulated environment there is a point source light which constitutes an energy resource which the agent must exploit, the equivalent of the wind in the kite control problem. As the intensity of the light drops away exponentially with distance from the source, there is light intensity gradient which mirrors the wind window of the kite problem in that the energy resource is distributed spatially. This radial distribution regionalises the arena into 3 key areas as per Fig. 3 (right).

As would be expected in a kite hardware implementation, and in common with conventional wind turbines, we implement both a plateau of the generation/charge rate at some moderate to high light level (line B in Fig. 3, right), and a 'cut out' at a higher level still, where the environmental power source, if it were coupled to the

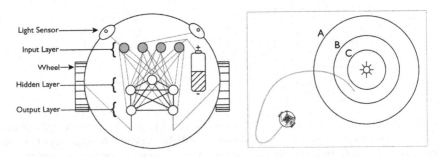

Fig. 3. (Left) A diagram of the agent and its neurocontroller architecture, light grey neural connections are feed forward, bold black neural connections are reciprocal with asymmetric weights. The battery connects to two input neurons, one connection carrying battery level data and the other, rate of change of battery level. (Right) Diagram of a portion of the experimental area, light source marked by the sun icon. Between line C and the sun marks the overpowered region, between B and C, the region where the charging rate plateaus. A marks the line at which the switching threshold is met.

generator, would overload and damage it. To mimic the reduction of line tension by spooling out the lines with the generator uncoupled, when a threshold point is reached corresponding to entering the area between line C in Fig. 3 (right), and the light, the battery continues to increase but the agent does not accumulate fitness until it leaves the region. If the threshold is exceeded when the battery is full, the agent is considered to be broken and the trial is ended prematurely, implications for the agents fitness as detailed below. Similarly, if the agents battery reaches zero, the agent has failed to collect sufficient energy from the environment and the trial is also ended.

In order to restrict the scope of the problem for ease of analysis and potentially increased evolution speed, we first simplify the problem of making some trade off between time cost of switching and energy required to switch by defining a lower force threshold at which switching automatically occurs. Once the agent crosses line A, at which point the battery charge rate input to the neural network will become negative, it commits to switch behaviour although the 'decision' to do so may have been made at some earlier point. As the boundary is determined by an environmental threshold independent of the agents activity, we term this regime 'threshold switching'. We subsequently assess a variation in which the rules are unchanged, except that the agent has some control over its own switching point. This is implemented implicitly by introducing a battery cost to movement, if the agent stays still, it will charge until full at any point in the environment at a rate determined by distance to the light, it is movement at a speed which consumes more energy than is recouped from the environment that will drive down the battery. The agent in this scenario is forced to make a more complex, but more realistic trade off in that moving far from the high-quality resource will speed up discharging, but mean that for portions of the return journey, time is spent near energy neutral and at suboptimal charging speeds. We term this regime 'enactive switching'.

3.2 Neural Network and Agent Dynamics

The neural network that drives the agents behaviour is a small continuous time recurrent neural network (CTRNN) of 4 input neurons, 3 hidden layer neurons and 2 output layer neurons, with both inhibitory and excitatory connections permissible. Each of the four input layer neurons receives input constituting one of the following four data values; left light sensor value, right light sensor value, proportion of battery capacity remaining, battery charge rate. Each input neuron connects to every hidden and output layer neuron, but receives no other input (see Fig. 3, left). The hidden and output layer neurons are fully interconnected, connection weights are asymmetric. A single neuron derives its dynamics from Eq.1:

$$\tau_j \dot{a}_j = -a_j + \sum w_{ij} \sigma(o_i - \theta_i) \tag{1}$$

θ is the bias term, w the weight and σ is the sigmoid function, τ_j is the neurons time constant, and o_i the old activation values from the previous timestep. In the simulation neuron activation values are integrated using Euler integration with a timestep of 0.1. The sigmoided values of the two output neurons form the motor activation values for the left and right wheels respectively. Average wheel speed determines the agent's velocity and difference in wheel speed determines the agents angle, dependent on the agents diameter. The agents position is also determined by Euler integration with a step size of 0.1. As the motivation of this work is to answer the questions whether ER can as tool generate the kind of switching behaviour required for kite control devices and how best to implement the ER strategy, we do not intend to analyse the network dynamics that result from the evolved weights.

3.3 Genetic Algorithm and Trial Configuration

We use a simple microbial genetic algorithm [1], with a population of genotypes or artificial 'DNA', randomly initialised, specifying possible parameter values. Pairs of neurocontroller parameter sets taken from the population of 20 have their performance evaluated. The winner overwrites the loser with small additive random mutations ±0.05 added to a random 50% subset of its parameter values. This process continues for several hundred iterations, each consisting of 10 competitive trials. For all tasks, fitness is judged by the amount that the battery is charged, over the time course of the trial. As the battery cannot be further charged once a hard upper limit is reached, and given that the trial length is set between 30 and 300 seconds such that the battery could be charged and discharged fully at least twice, the agent is effectively rewarded for switching between charging and discharging behaviour.

We perform three sets of experiments, the common elements to which are as follows: for each of the five trials that constitute one evaluation, the agent is reinitialised at a new random orientation, in a random location within a square of side 30 units. Its battery level is randomised between a minimum of 20 units and its maximum value of 60 units. Competing pairs share the same random initialisations for fair comparison.

In the first set of experiments, we evolve the agent (1000 generations) to maximise charging when the light intensity value of 5 remains virtually constant between experiments (+- noise of 0.2). In the second we use a successful population from the first experiment as a seed for evolution (500 generations) during which we vary the intensity of the light from evaluation to evaluation within a range of 1 to 10. In the third, (500 generations) the intensity of the light is varied within each experiment, with either continual, linear slow increases/decreases that run the whole length of the trial, constituting a gradual trend, or immediate discontinuous jumps between the original and higher or lower intensity values, constituting 'gusts' and 'lulls'. In this final experiment, we only judge the agent on its fitness in the second half of the experiment, the agent is therefore given an opportunity to have adapted to the changes in the environment after a period of unassessed interaction.

4 Results

The first and most salient result was the poor performance of the agents evolved under the threshold switching regime. Although some dozens of evolutionary runs were completed, most resulted in the agent approaching the light and either stopping in that position or moving in a small stereotypic trajectory near the light. This was also true when the population was seeded with successful phototactic agents. In doing so they successfully completed one behaviour but were unable to switch to another.

This incomplete strategy consisted of approaching the light and then successfully stopping in the plateau zone of the resource, but never moving from that position. Due to the failure on this simple task, threshold switching was not carried on to either of the other two tasks.

In contrast, the controllers evolved with 'enactive' switching evolved a variety of switching strategies all of which involved repetitively moving closer and further from the light. Interestingly, even though these agents were not evolved under the rules of threshold selection, they performed equally as well and with the same apparent strategy when operating under those rules. They seemed to be capable of generalizing from the smooth gradient of switching that they were evolved under to a sharp threshold, and are therefore surprisingly independent of the need to control their rate of discharge, once the behaviour is fully evolved. All the subsequent plots are of the 'enactive switching' evolved agents, but operating under threshold selection rules for clarity of interpretation. With the exception of one of several alternative suboptimal solutions shown in Fig. 4 (left), all plots are of the best performing agent from the best evolutionary run. After evolution, populations tended to be highly converged, with each agent from a given population pursuing near-identical strategies that were highly robust to changes in agent starting location or orientation and light dynamics, at least within ranges experienced in evolution.

As the centre and right plots of Fig. 4 demonstrate, the agents skim the edge or just enter the middle ring that marks the region where charging rate plateaus. They then continue the same trajectory until they make an abrupt turn once in the discharging portion of the arena.

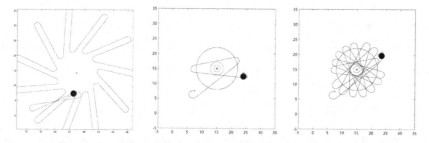

Fig. 4. An interesting but sub-optimal strategy is shown (left). The first few passes of the trajectory of the best performing agent with light source intensity of 5 are shown (centre) as well as the trajectory at the end of the experiment (right).

Although it is not immediately apparent from Fig. 4, Fig. 5 (right) shows how this cycle initially is skewed in favour of charging, resulting in a gradual accumulation of battery which then settles out and oscillates round some intermediate capacity value (c. 75% of capacity). After a further period of evolution with the light intensity varying between experiments by up to a factor of 10, the controllers were able to adapt well to changes in the light intensity. This seemed to be achieved by modulation of the turning angle on exit from the charging area, however as Fig. 6 (left) shows, the agent enters the charging zone at the same angle. Changing the discharge rate had a similar effect with wide turns keeping the agent in the discharging area longer when

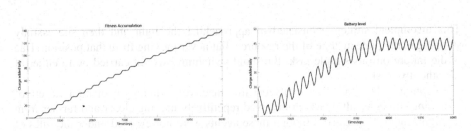

Fig. 5. Plots demonstrating the accumulation of fitness over the experiment shown in Fig. 4 (left) and the cycling of battery level in the same experiment (right)

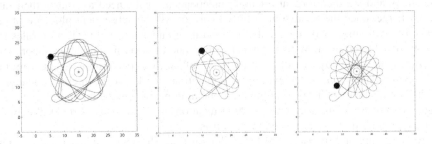

Fig. 6. Adaptation to an increased light intensity of 10 (left), lowered (centre) and elevated (right) discharge rates

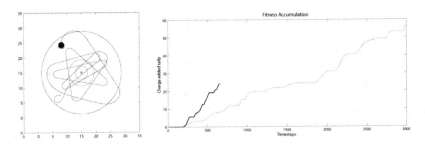

Fig. 7. (left) The modulation of trajectories with continuous and gradual increase in light intensity (1-20). The right plot compares the fitness of the agents with (dotted line) and without (solid line) additional evolution with dynamic light strength when exposed to rapid perturbations in light intensity.

the discharge rate was lowered (Fig. 6, centre) and tight turns occurring when the discharge rate is elevated (Fig. 6, right).

Finally, we compared the performance of the intensity-adaptive agents from the prior experiment with that of agents subject to further evolution in which the light intensity changed within the experiment on one of two temporal scales. When slow linear changes of intensity occurred, the performance of both sets of agents was essentially equal, all agents continuously modulating their trajectories as per Fig. 7 (left).

However when the intensity changes were rapid, gust-like perturbations, there was a marked improvement in the population that was further evolved (see Fig. 7 right),

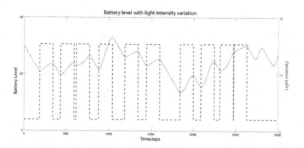

Fig. 8. A dual plot showing the change in battery level over time (solid line) concurrently with the sudden changes in the light intensity (dotted line)

which accumulates fitness more slowly than the agents naïve to intraexperimental changes in light intensity, but this trade-off allows it to survive indefinitely in the highly dynamic environment. Fig. 8 illustrates that the strategy of the non-naïve agents in response to gust-like perturbations in light intensity is an opportunistic one, in which during a gust the agent is always charging and immediately after a gust, always discharging. Additionally Fig. 8 seems to imply that the agent takes the strategy of charging at an appropriate rate for its battery level, presumably by regulating its turns to lead it closer or further from the source according to its battery state as it approaches the turn.

At the start, when its battery is relatively high, it takes relatively small quantities of the available increased resource within gusts, then after much charge is lost during the lull that begins at c.1600 timesteps, much more energy is extracted from the subsequent three gusts. When the variation stops at the end of the trial, the agent reverts to it's default behaviour. The fact that neither the charge speed nor the reversion to oscillatory behaviour seems consistent is likely due to the agents natural in/out oscillations and the delay inherent in turning where the agent is slowed due to the forward operation of only one wheel.

5 Discussion

Perhaps the most pertinent lesson of this exercise for future applications of evolutionary robotics for actual adaptive kite control was the relative success of evolution using the threshold and enactive rules sets for switching between charging and discharging behaviours. Imposing a sharp, externally-defined boundary between two behaviours prevented evolution of successful agents, contrary to their enactive peers which, through their own interactions with the environment determined their own behavioural boundary or lack of it. Given this experience, it would certainly seem prudent to assess an similar strategy for evolving reel-in and reel-out behaviours for kite control, especially given that a sharp boundary could be subsequently imposed on the enactive switching agents in the toy case with little impact on performance, which may be a necessity due to hardware constraints in kite energy systems.

The strategy of the most successful neurocontrollers as presented in Figures 4 through 8 was an elegant one that defied our prior expectations of how the problem would be solved. Instead of heading directly for the energy source and then doubling back for discharging, the supposedly discrete behaviours were subsumed into one action constituting a single pass. The 'decision' is then reduced to when, and at what angle to turn in order to start the next charge/discharge pair whose duration and outcome, in a static environment at least, is essentially predetermined. The ability of the evolved neurocontrollers to adapt to changes in light intensity and indeed discharging rates (Fig. 6) both between and within (Figs. 7,8) experiments dispelled our early concerns as to the fragility of this strategy. Given that here, supposed problems such as a trade off between persistence and dithering almost appear to be resolved almost by the environment itself, this type of behaviour seems best described as an ongoing interaction between the agent and its environment and less well described as one of discrete decisions at the top of a hierarchy being passed down to subordinate effector systems. This may be a product of the relative simplicity of the task and the lack of constraints on the network connectivity, indeed the imposition of symmetry or other neuroanatomically inspired constraints would be an interesting comparison, and could potentially generate more transparent internal dynamics.

The final experiment in which neurocontrollers were further evolved with exposure to dynamical environments produced the unexpected result of two co-existing strategies. The standard light passing behaviour switched seamlessly into one in which the environment was permitted to almost completely dominate whether the agent charged or discharged. This apparently submissive behaviour was much more robust than that of the naïve agents, with the agent exerting a more subtle control by increasing or decreasing its exposure to resource variations in order to regulate its state to a level that was notably more conservative than in the prior experiments (50% vs 75%). A thorough analysis of this final behaviour in response to a range of gust intensities and durations and indeed more realistic variations in intensity would be enlightening and is the subject of ongoing work.

In summary, this work has succeeded in its brief of informing future work in the control of kite energy systems at a behavioural level, specifically by suggesting that highly adaptive spatially embedded, behaviour-switching agents can be evolved if decision boundaries are determined by the agents themselves. The opportunistic behaviour seen in the most complex task bodes well for the application of ER to the development of kite control behaviours.

References

1. Harvey, I.: Artificial Evolution: a Continuing SAGA. In: Gomi, T. (ed.) Evolutionary Robotics: From Intelligent Robots to Artificial Life. Springer, Heidelberg (2001)
2. Furey, A.D.J., Harvey, I.: Evolution of Neural Networks for Active Control of Tethered Airfoils. In: Proc. 9th European Conference on Artificial Life, pp. 746–756 (2007)
3. Furey, A.D.J., Harvey, I.: Robust adaptive control for kite wind energy using evolutionary robotics. In: Proc. Biological Approaches for Engineering (in press, 2008)
4. Canale, M., Fagiano, L., Ippolito, M., Milanese, M.: Control of tethered airfoils for a new class of wind energy generator. In: Proc. 45th IEEE Conference on Decision and Control, 2046–4020 (2006)
5. Lansdorp, B., Ockels, W.J.: Comparison of concepts for high-altitude wind energy generation with ground based generator. In: Proc. 2nd China IREETEC, pp. 409–417 (2005)
6. Houska, B.: Robustness and stability optimisation of open-loop controlled power generating kites. PhD Thesis, University of Heidelberg (2007)

Self-adaptive Agent-Based Dynamic Scheduling for a Semiconductor Manufacturing Factory

Horng-Ren Tsai[1] and Toly Chen[2]

[1] Department of Information Technology, Lingtung University,
[2] Department of Industrial Engineering and Systems Management, Feng Chia University,
100, Wenhwa Road, Seatwen, Taichung City, Taiwan
tolychen@ms37.hinet.net
http://www.geocities.com/tinchihchen/

Abstract. A semiconductor manufacturing factory is a very complicated production system. Typical characteristics of a semiconductor manufacturing factory include: fluctuating demand, jobs with various product types and priorities, un-balanced capacity, jobs' reentrance to the bottleneck machines, hundreds of processing steps, alternative machines with unequal capacity, etc. Scheduling in a semiconductor manufacturing factory becomes a very difficult task owing to these characteristics. To enhance the performance of dynamic scheduling in a semiconductor manufacturing factory, a self-adaptive agent-based approach is proposed in this study. Firstly, a self-adaptive agent-based scheduling model, which integrates release control, dispatching and machine maintenance scheduling, is presented. Secondly, the negotiation protocol between agents is given. Thirdly, scheduling algorithms for decision making of agents are offered. Unlike in the past studies a single pre-determined scheduling algorithm is used for all agents, in this study every agent develops and modifies its own scheduling algorithm to adapt it to the outside conditions. Finally, production simulation is also applied in this study to generate some test data to evaluate the effectiveness of the proposed methodology.

1 Introduction

A semiconductor manufacturing factory is a very complicated production system. Typical characteristics of a semiconductor manufacturing factory include: fluctuating demand, jobs with various product types and priorities, un-balanced capacity, jobs' reentrance to the bottleneck machines, hundreds of processing steps, alternative machines with unequal capacity, etc. Scheduling in a semiconductor manufacturing factory becomes a very difficult task owing to these characteristics. Many studies have shown that directly applying common scheduling rules (such as first-in first out (FIFO), earliest due date (EDD), least slack (LS), shortest processing time (SPT), shortest remaining processing time (SRPT), FIFO+, SRPT+, and SRPT++) to a semiconductor manufacturing factory does not lead to very good results. Lu et al. [1] proposed two scheduling rules (fluctuation smoothing policy for variance of cycle time (FSVCT) and fluctuation smoothing policy for mean cycle time (FSMCT)) which

M. Asada et al. (Eds.): SAB 2008, LNAI 5040, pp. 519–528, 2008.

were tailored to a semiconductor manufacturing environment and achieved good performances in reducing the cycle time averages and standard deviations. However, every resource in a semiconductor manufacturing factory fabricating 12 or 18 inch wafers is very expensive, and therefore to construct an intelligent agent for each resource in order to fully use the resource is a direction that can be studied. In this way it is also possible to further elevate the performance of dynamic scheduling (such as reducing the cycle time averages and standard deviations, minimizing the maximal or average tardiness, minimizing the number of tardy jobs, the speed of scheduling or rescheduling, etc.) in the semiconductor manufacturing factory. Recently, agent-based planning or scheduling ideas have been concentrated on by academies and industries [2]. Agent technology has been widely used in all kinds of manufacturing systems successfully. To further enhance the performance of dynamic scheduling in a semiconductor manufacturing factory, a self-adaptive agent-based dynamic scheduling approach is proposed in this study. Firstly, a self-adaptive agent-based scheduling model, which integrates release control, dispatching and machine maintenance scheduling, is presented. Secondly, the negotiation protocol between agents (applying the extended contract negotiation protocol (ECNP) [3]) is given. Thirdly, scheduling algorithms for decision making of agents are offered. Unlike in the past studies (e.g. [3]) a single pre-determined scheduling algorithm is used for all agents, in this study every agent develops and modifies its own scheduling algorithm to adapt it to the outside conditions. In this respect, Monfared and Steiner [4] developed a framework for the design and development of adaptive scheduling and control systems, by incorporating concepts and techniques from the fuzzy set theory, the control theory, and the optimization theory. Through mathematical modelling and simulation, a real-time fuzzy adaptive scheduling and control system was constructed for a fully automated manufacturing system. Finally, production simulation is also applied in this study to generate some test data to evaluate the effectiveness of the proposed methodology. Some existing scheduling rules are also applied to the test data to make a comparison with the proposed methodology.

The remainder of this paper is organized as follows. Section 2 presents the self-adaptive agent-based scheduling model, which integrates release control, dispatching and machine maintenance scheduling. Subsequently in Section 3, the negotiation protocol between agents, ECNP, is introduced. Scheduling algorithms for decision making of agents are offered in Section 4. Every agent can develop and modify its own scheduling algorithm to adapt it to the outside conditions. To evaluate the effectiveness of the proposed methodology, production simulation is applied in Section 4 to simulate a semiconductor manufacturing environment and to generate some test data. Then the proposed methodology and several existing scheduling rules are all applied to the test data in Section 5. Based on the analysis results, some points are made. Finally, the concluding remarks and some directions for future research are given in Section 6.

2   Self-adaptive Agent-Based Dynamic Scheduling Model

The self-adaptive agent-based dynamic scheduling model is shown in Fig. 1, which integrates release control, dispatching and machine maintenance scheduling. There

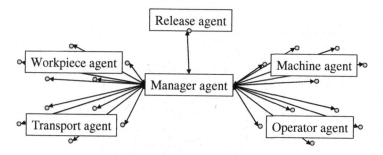

Fig. 1. The self-adaptive agent-based dynamic scheduling model

are six kinds of agents in the model, i.e. manager agent, release agent, workpiece agent, machine agent, transportation agent and operator agent:

1. Manager agent: The manager agent manages all kinds of agents in the self-adaptive agent-based dynamic scheduling mode. Each agent must register to the manager agent. The manager agent then keeps information about each agent and monitors the activities of all agents, but does not interfere with the autonomous decision of any agent. Besides, the manager agent keeps generating new workpiece agents according to the release plan made by the release agent, and canceling workpiece agents which controlled jobs have been finished.
2. Release agent: The release agent is used to control the release into the semiconductor manufacturing factory.
3. Workpiece agent: A workpiece agent controls a job in the semiconductor manufacturing factory, and is generated by the manager agent according to the release plan. The workpiece agent of a job owns all information about the job including product type, job size, priority, recipe, due date, etc. Workpiece agents have to negotiate with machine agents to realize the dynamic dispatching of jobs.
4. Machine agent: A machine agent controls one machine in the semiconductor manufacturing factory, and owns all information about the machine including availability, utilization, maintenance plan, queuing list, etc. A machine agent negotiates with incoming workpiece agents to determine whether accepting the processing of the workpiece or not. Besides, a machine agent has to arrange the sequence of processing accepted jobs on the machine, taking the outside conditions into consideration.
5. Transportation agent: A transportation agent controls one transportation tool in the semiconductor manufacturing factory to transport jobs between machines.
6. Operator agent: An operator agent controls one operator in the semiconductor manufacturing factory to load/unload jobs and to maintain machines.

Among them the self-adaptive behavior of a machine agent is focused in this study. The self-adaptive behavior of a machine agent means that the machine agent can develop and modify its own scheduling algorithm to adapt it to the outside conditions. There are two types of outside conditions that are considered by a self-adaptive machine agent (see Fig. 2):

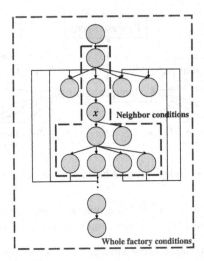

Fig. 2. The outside conditions of a machine agent x ($k = 2$)

1. The whole factory conditions: the most recent scheduling performance of the whole factory, e.g. the cycle time average and standard deviation of some most recently outputted jobs, the average tardiness of these jobs, the percent of tardy jobs, etc. If that is poor, then the machine agent might need to modify its scheduling algorithm to help improve the performance.
2. The k nearest neighbor conditions: the most recent scheduling performances of the k nearest neighbors, e.g. the stage cycle time averages and standard deviations of some most recently completed jobs by them, the deviations from the best and worst results, etc. If those are poor, then the machine agent might need to modify its scheduling algorithm to help improve the performances.

These outside conditions are of unequal importance, which is also decided by the machine agent itself.

3   The Negotiation Protocol

The negotiation protocol used in the proposed methodology is ECNP [3] that is described below:

1. A workpiece agent announces the task to relevant machine agents, including the content of the task, the original proposal (the latest start processing time of the task), token, and the deadline for coming to an agreement. The original proposal is valid only during its token.
2. The machine agents evaluate the proposal of the workpiece agent. Then they response to the workpiece agent with their counter-proposals and tokens in the workpiece agent's token. The machines agents' tokens should be less than the difference between the deadline for coming to an agreement and the workpiece agent's token. The machine agents keep their promises only during their own tokens.

3. The workpiece agent evaluates the responses from the machine agents, and then selects the machine agent with the best response. Then the workpiece agent sends acknowledgement to the selected machine agent in its token.

Compared with the traditional CNP, ECNP has advantages like higher efficiency negotiation, avoidance of valueless waiting, convergence of negotiation, avoidance of the deadlock of negotiation.

4 Self-adaptive Scheduling Algorithms for Decision Making of Machine Agents

The self-adaptive scheduling algorithm adopted by a machine agent is composed of a fuzzy back propagation network (FBPN) and a sorter. At first, the FBPN is used to generate a code for each job accepted by the machine agent. The configuration of the FBPN is established as follows:

1. Inputs: six parameters associated with each job including the product type, priority, job size, due date, processing time, and processing type (batch or single wafer). These parameters have to be normalized so that their values fall within [0, 1]. Then some production execution/control experts are requested to express their beliefs (in linguistic terms) about the importance of each input parameter in determining the sequence of processing on the machine. Linguistic assessments for an input parameter are converted into several pre-specified fuzzy numbers. The subjective importance of an input parameter is then obtained by averaging the corresponding fuzzy numbers of the linguistic replies for the input parameter by all experts. The subjective importance obtained for an input parameter is multiplied to the normalized value of the input parameter. After such a treatment, all inputs to the FBPN become triangular fuzzy numbers, and the fuzzy arithmetic for triangular fuzzy numbers is applied to deal with all calculations involved in training the FBPN.
2. Single hidden layer: Generally one or two hidden layers are more beneficial for the convergence property of the network.
3. Number of neurons in the hidden layer: the same as that in the input layer. Such a treatment has been adopted by many studies (e.g. [5, 6]).
4. Output: the code of the job.
5. Network learning rule: Delta rule.
6. Transformation function: Sigmoid function,

$$f(x) = 1/(1 + e^{-x}).\tag{1}$$

7. Learning rate (η): 0.01~1.0.
8. Batch learning.

At first, the FBPN is trained to fit a common scheduling rule, e.g. the FSMCT rule or the constant work-in-progress (CONWIP) rule. The training of a FBPN refers to Chen [6]. After that, the FBPN can be applied to generate the code of a job as follows.

At first, inputs are multiplied with weights, summated, and transferred to the hidden layer. Then activated signals are outputted from the hidden layer as:

$$\tilde{h}_j = (h_{j1}, \ h_{j2}, \ h_{j3}) = 1/(1 + e^{-\tilde{n}_j^h}), \tag{2}$$

where

$$\tilde{n}_j^h = (n_{j1}^h, \ n_{j2}^h, \ n_{j3}^h) = \tilde{I}_j^h(-)\tilde{\theta}_j^h, \tag{3}$$

$$\tilde{I}_j^h = (I_{j1}^h, \ I_{j2}^h, \ I_{j3}^h) = \sum_{all \ i} \tilde{w}_{ij}^h(\times)\tilde{x}_{(i)}, \tag{4}$$

and $(-)$ and $(\times)$ denote fuzzy subtraction and multiplication, respectively; $\tilde{h}_j$'s are also transferred to the output layer with the same procedure. Finally, the output of the FBPN is generated as:

$$\tilde{o} = (o_1, \ o_2, \ o_3) = 1/(1 + e^{-\tilde{n}^o}), \tag{5}$$

where

$$\tilde{n}^o = (n_1^o, \ n_2^o, \ n_3^o) = \tilde{I}^o(-)\tilde{\theta}^o, \tag{6}$$

$$\tilde{I}^o = (I_1^o, I_2^o, I_3^o) = \sum_{all \ j} \tilde{w}_j^o(\times)\tilde{h}_j. \tag{7}$$

The fuzzy-valued output $\tilde{o}$ is defuzzified according to the centroid-of-area (COA) formula:

$$o = COA(\tilde{o}) = (o_1 + 2o_2 + o_3)/4. \tag{8}$$

Then the defuzzified output o is the code of the job. The codes of all jobs are sorted by the sorter, and then the result determines the sequence of processing on the machine.

To self-adjust the scheduling algorithm, the whole factory conditions and the k nearest neighbor conditions of the machine are continuously monitored. If the recent performances of the whole factory or the k nearest neighbors are poor, then adjust the parameter values of the FBPN as follows:

$$\text{new } p = (1 - p_n^{w_n} p_g^{w_g})(p_{max} + p_{min} - \text{old } p) + p_n^{w_n} p_g^{w_g} \text{ old } p \tag{9}$$

where p_{max} and p_{min} are the maximum and minimum of parameter p, respectively; p_n is the normalized neighbor performance, while p_g measures the normalized global performance; $0 \leq p_n, p_g \leq 1$; w_n and w_g indicate the weights of p_n and p_g, respectively. It is a self-adjustment mechanism because every machine agent continuously considers the conditions of its neighborhood (see Fig. 3), and finally is able to construct a scheduling algorithm of its own.

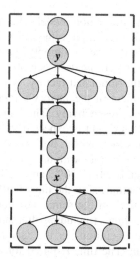

Fig. 3. The self-adjust mechanism considers only the neighbor of the machine agent

5 Some Test Data from a Simulated Semiconductor Manufacturing Factory

To evaluate the effectiveness of the proposed methodology, production simulation is applied in this study to simulate a semiconductor manufacturing environment and to generate some test data. A simulation model is often built to simulate the manufacturing process of a real semiconductor manufacturing factory [5-11]. To generate some test data, a simulation program coded using Microsoft Visual Basic .NET is constructed to simulate a semiconductor manufacturing factory with the following assumptions:

1. Jobs are uniformly released into the factory.
2. The distributions of the interarrival times of machine downs are exponential.
3. The distribution of the time required to repair a machine is uniform.
4. The percentages of jobs with different product types in the factory are predetermined. As a result, this study is only focused on fixed-product-mix cases. However, the product mix in the simulated factory does fluctuate and is only approximately fixed in the long term.
5. The percentages of jobs with different priorities released into the factory are controlled.
6. The priority of a job cannot be changed during fabrication.
7. A job has equal chances to be processed on each alternative machine/head available at a step.
8. A job cannot proceed to the next step until the fabrication on its every wafer has been finished. No preemption is allowed.

The basic configuration of the simulated semiconductor manufacturing factory is the same as a real-world semiconductor manufacturing factory which is located in the Science Park of Hsin-Chu, Taiwan, R.O.C. A trace report was generated every simulation run for verifying the simulation model. The simulated average cycle times have also been compared with the actual values to validate the simulation model. Assumptions (1)~(3), and (7)~(8) are commonly adopted in related studies (e.g. [5-8]), while assumptions (4)~(6) are made to simplify the situation. There are five products (labeled as A~E) in the simulated factory. A fixed product mix is assumed. The percentages of these products in the factory's product mix are assumed to be 35%, 24%, 17%, 15%, and 9%, respectively. The simulated factory has a monthly capacity of 20,000 pieces of wafers and is expected to be fully utilized (utilization = 100%). Three types of priorities (normal lot, hot lot, and super hot lot) are randomly assigned to jobs. The percentages of jobs with these priorities released into the factory are restricted to be approximately 60%, 30%, and 10%, respectively. Each product has 150~200 steps and 6~9 reentrances to the most bottleneck machine. The singular production characteristic "reentry" of the semiconductor industry is clearly reflected in the example. Totally 102 machines (including alternative machines) are provided to process single-wafer or batch operations in the factory. Thirty replications of the simulation are successively run. The time required for each simulation replication is about 12 minute on a PC with 512MB RAM and Athlon™ 64 Processor 3000+ CPU. A horizon of twenty-four months is simulated. The maximal cycle time is less than three months. Therefore, four months and an initial WIP status (obtained from a pilot simulation run) seemed to be sufficient to drive the simulation into a steady state. The statistical data were collected starting at the end of the fourth month. For each replicate, data of 30 jobs are collected and classified by their product types and priorities. Totally, data of 900 jobs can be collected.

6   Experimental Results and Discussion

To evaluate the effectiveness and efficiency of the proposed methodology and to make some comparisons with some existing scheduling rules – FIFO, EDD, SRPT, and FSVCT, all these methods were applied to five test cases containing the data of full-size (25 wafers per job) jobs with different product types and priorities. The results are summarized in Table 1. Since the information contained there may be too much to digest, we provide some bottom line comparisons.

According to experimental results, the following points are made:

1. As expected, FIFO performed well in reducing the cycle time average, but might show exceedingly bad performance in the standard deviation respect.

Table 1. Experimental results

	Cycle time average	Cycle time standard deviation
FIFO	249	84
SRPT	278	79
EDD	266	78
FSVCT	242	55
The proposed methodology	172	46

2. The proposed methodology achieved very good performance in both the cycle time average and standard deviation respects. The average advantages over the baseline approach, FIFO, are 31% and 45%, respectively. No other policy comes close to possessing such behavior.
3. Compared with the FSVCT policy aimed at reducing the cycle time variation, the proposed methodology still surpassed by 16% in reducing the cycle time variation.
4. The FSVCT policy performed the best among the four existing scheduling rules in both respects.
5. In the original ECNP-based approach, the scheduling approaches adopted by all agents were in fact the same, namely the CONWIP rule. Conversely, in the proposed methodology, every agent develops its own scheduling rule based on a FBPN. In other words, the values of the parameters in the FBPN of an agent are different from those of the other agents.

7 Conclusions and Directions for Future Research

A semiconductor manufacturing factory is a very complicated production system. Typical characteristics of a semiconductor manufacturing factory include: fluctuating demand, jobs with various product types and priorities, un-balanced capacity, jobs' reentrance to the bottleneck machines, hundreds of processing steps, alternative machines with unequal capacity, etc. Scheduling in a semiconductor manufacturing factory becomes a very difficult task owing to these characteristics. To further enhance the performance of dynamic scheduling in a semiconductor manufacturing factory, a self-adaptive agent-based dynamic scheduling approach is proposed in this study. The first part is a self-adaptive agent-based scheduling model, which integrates release control, dispatching and machine maintenance scheduling. Secondly, the negotiation protocol between agents applies ECNP. Thirdly, unlike in the past studies a single pre-determined scheduling algorithm is used for all agents, in this study every agent develops and modifies its own scheduling algorithm to adapt it to the outside conditions. Finally, production simulation is also applied in this study to generate some test data to evaluate the effectiveness of the proposed methodology. Some existing scheduling rules are also applied to the test data to make a comparison with the proposed methodology. According to experimental results, the self-adaptive agent-based dynamic scheduling approach achieved very good performance in both the cycle time average and standard deviation respects. The average advantages over the baseline approach, FIFO, are 31% and 45%, respectively.

However, to further evaluate the advantages and disadvantages of the proposed methodology, it has to be applied to a full-scale actual semiconductor manufacturing factory in the future.

Acknowledgement

This study is financially supported by National Science Council of Taiwan.

References

1. Lu, S.C.H., Ramaswamy, D., Kumar, P.R.: Efficient scheduling policies to reduce mean and variation of cycle time in semiconductor manufacturing plant. IEEE Transactions on Semiconductor Manufacturing 7(3), 374–388 (1994)
2. Albert, D.B.: A survey of factory control algorithms which can be implemented in a multi-agent hierarchy: dispatching, scheduling and pull. Journal of Manufacturing System 6, 276–289 (1996)
3. Li, L., Qiao, F., Wu, Q.: Agent-based dynamic scheduling for semiconductor wafer fab. In: Proceedings of IEEE Semiconductor Symposium, pp. 163–168 (2005)
4. Monfared, M.A.S., Steiner, S.J.: Fuzzy adaptive scheduling and control systems. Fuzzy Sets and Systems 115, 231–246 (2000)
5. Chang, P.-C., Hsieh, J.-C.: A Neural Networks Approach for Due-date Assignment in a Wafer Fabrication Factory. International Journal of Industrial Engineering 10(1), 55–61 (2003)
6. Chen, T.: A Fuzzy Back Propagation Network for Output Time Prediction in a Wafer Fab. Journal of Applied Soft Computing 2/3F, 211–222 (2003)
7. Chang, P.-C., Hsieh, J.-C., Liao, T.W.: A Case-based Reasoning Approach for Due Date Assignment in a Wafer Fabrication Factory. In: Aha, D.W., Watson, I. (eds.) ICCBR 2001. LNCS (LNAI), vol. 2080. Springer, Heidelberg (2001)
8. Chang, P.-C., Hsieh, J.-C., Liao, T.W.: Evolving Fuzzy Rules for Due-date Assignment Problem in Semiconductor Manufacturing Factory. Journal of Intelligent Manufacturing 16, 549–557 (2005)
9. Chang, P.-C., Liao, T.W.: Combining SOM and fuzzy rule base for flow time prediction in semiconductor manufacturing factory. Applied Soft Computing 6, 198–206 (2006)
10. Chen, T.: A Look-ahead Back Propagation Network to Predict Wafer Lot Output Time. WSEAS Transactions on Computers 5(5), 910–915 (2006)
11. Chen, T., Tsai, H.R., Wu, H.C.: Wafer Lot Output Time Prediction with a Hybrid Artificial Neural Network. WSEAS Transactions on Computers 5(5), 817–823 (2006)

Author Index

Lecture Notes in Artificial Intelligence (LNAI)

Vol. 5110: W. Hodges, R. de Queiroz (Eds.), Logic, Language, Information and Computation. XI, 298 pages. 2008.

Vol. 5097: L. Rutkowski, R. Tadeusiewicz, L.A. Zadeh, J.M. Zurada (Eds.), Artificial Intelligence and Soft Computing – ICAISC 2008. XVI, 1269 pages. 2008.

Vol. 5078: E. André, L. Dybkjær, W. Minker, H. Neumann, R. Pieraccini, M. Weber (Eds.), Perception in Multimodal Dialogue Systems. X, 311 pages. 2008.

Vol. 5064: L. Prevost, S. Marinai, F. Schwenker (Eds.), Artificial Neural Networks in Pattern Recognition. IX, 318 pages. 2008.

Vol. 5040: M. Asada, J.C.T. Hallam, J.-A. Meyer, J. Tani (Eds.), From Animals to Animats 10. XIII, 530 pages. 2008.

Vol. 5032: S. Bergler (Ed.), Advances in Artificial Intelligence. XI, 382 pages. 2008.

Vol. 5027: N.T. Nguyen, L. Borzemski, A. Grzech, M. Ali (Eds.), New Frontiers in Applied Artificial Intelligence. XVIII, 879 pages. 2008.

Vol. 5012: T. Washio, E. Suzuki, K.M. Ting, A. Inokuchi (Eds.), Advances in Knowledge Discovery and Data Mining. XXIV, 1102 pages. 2008.

Vol. 5009: G. Wang, T. Li, J.W. Grzymala-Busse, D. Miao, A. Skowron, Y. Yao (Eds.), Rough Sets and Knowledge Technology. XVIII, 765 pages. 2008.

Vol. 4994: A. An, S. Matwin, Z.W. Raś, D. Ślęzak (Eds.), Foundations of Intelligent Systems. XVII, 653 pages. 2008.

Vol. 4953: N.T. Nguyen, G.S. Jo, R.J. Howlett, L.C. Jain (Eds.), Agent and Multi-Agent Systems: Technologies and Applications. XX, 909 pages. 2008.

Vol. 4946: I. Rahwan, S. Parsons, C. Reed (Eds.), Argumentation in Multi-Agent Systems. X, 235 pages. 2008.

Vol. 4944: Z.W. Raś, S. Tsumoto, D.A. Zighed (Eds.), Mining Complex Data. X, 265 pages. 2008.

Vol. 4938: T. Tokunaga, A. Ortega (Eds.), Large-Scale Knowledge Resources. IX, 367 pages. 2008.

Vol. 4933: R. Medina, S. Obiedkov (Eds.), Formal Concept Analysis. XII, 325 pages. 2008.

Vol. 4930: I. Wachsmuth, G. Knoblich (Eds.), Modeling Communication with Robots and Virtual Humans. X, 337 pages. 2008.

Vol. 4929: M. Helmert, Understanding Planning Tasks. XIV, 270 pages. 2008.

Vol. 4924: D. Riaño (Ed.), Knowledge Management for Health Care Procedures. X, 161 pages. 2008.

Vol. 4923: S.B. Yahia, E.M. Nguifo, R. Belohlavek (Eds.), Concept Lattices and Their Applications. XII, 283 pages. 2008.

Vol. 4914: K. Satoh, A. Inokuchi, K. Nagao, T. Kawamura (Eds.), New Frontiers in Artificial Intelligence. X, 404 pages. 2008.

Vol. 4911: L. De Raedt, P. Frasconi, K. Kersting, S. Muggleton (Eds.), Probabilistic Inductive Logic Programming. VIII, 341 pages. 2008.

Vol. 4908: M. Dastani, A. El Fallah Seghrouchni, A. Ricci, M. Winikoff (Eds.), Programming Multi-Agent Systems. XII, 267 pages. 2008.

Vol. 4898: M. Kolp, B. Henderson-Sellers, H. Mouratidis, A. Garcia, A.K. Ghose, P. Bresciani (Eds.), Agent-Oriented Information Systems IV. X, 292 pages. 2008.

Vol. 4897: M. Baldoni, T.C. Son, M.B. van Riemsdijk, M. Winikoff (Eds.), Declarative Agent Languages and Technologies V. X, 245 pages. 2008.

Vol. 4894: H. Blockeel, J. Ramon, J. Shavlik, P. Tadepalli (Eds.), Inductive Logic Programming. XI, 307 pages. 2008.

Vol. 4885: M. Chetouani, A. Hussain, B. Gas, M. Milgram, J.-L. Zarader (Eds.), Advances in Nonlinear Speech Processing. XI, 284 pages. 2007.

Vol. 4874: J. Neves, M.F. Santos, J.M. Machado (Eds.), Progress in Artificial Intelligence. XVIII, 704 pages. 2007.

Vol. 4870: J.S. Sichman, J. Padget, S. Ossowski, P. Noriega (Eds.), Coordination, Organizations, Institutions, and Norms in Agent Systems III. XII, 331 pages. 2008.

Vol. 4869: F. Botana, T. Recio (Eds.), Automated Deduction in Geometry. X, 213 pages. 2007.

Vol. 4865: K. Tuyls, A. Nowe, Z. Guessoum, D. Kudenko (Eds.), Adaptive Agents and Multi-Agent Systems III. VIII, 255 pages. 2008.

Vol. 4850: M. Lungarella, F. Iida, J.C. Bongard, R. Pfeifer (Eds.), 50 Years of Artificial Intelligence. X, 399 pages. 2007.

Vol. 4845: N. Zhong, J. Liu, Y. Yao, J. Wu, S. Lu, K. Li (Eds.), Web Intelligence Meets Brain Informatics. XI, 516 pages. 2007.

Vol. 4840: L. Paletta, E. Rome (Eds.), Attention in Cognitive Systems. XI, 497 pages. 2007.

Vol. 4830: M.A. Orgun, J. Thornton (Eds.), AI 2007: Advances in Artificial Intelligence. XIX, 841 pages. 2007.

Vol. 4828: M. Randall, H.A. Abbass, J. Wiles (Eds.), Progress in Artificial Life. XII, 402 pages. 2007.